G. V. Vinogradov, A. Ya. Malkin

Rheology of Polymers

Viscoelasticity and Flow of Polymers

With 220 Figures

Springer-Verlag Berlin Heidelberg GmbH
1980

Professor G. V. Vinogradov, D.Sc. (Chemistry)
Institute of Petrochemical Synthesis, Moscow, USSR

Professor A. Ya. Malkin, D.Sc. (Physics and Mathematics)
Institute of Plastics, Moscow, USSR

ISBN 978-3-642-52206-2   ISBN 978-3-642-52204-8 (eBook)
DOI 10.1007/978-3-642-52204-8

© Springer-Verlag Berlin Heidelberg 1980
Originally published by Springer-Verlag Berlin Heidelberg New York in 1980
Softcover reprint of the hardcover 1st edition 1980

Preface

If one dismisses the Prophetess Deborah who in her famous song after the victory over the Philistines sang "The mountains melted before the Lord" and her contemporary (on our time scale), the Egyptian Amenemhet, who designed the water clock, which was in fact the prototype of the capillary viscometer, the beginnings of modern rheology should be linked up with the works of the classics of natural sciences of the 19th century: James Clerk Maxwell, Lord Kelvin, and Ludwig Boltzmann, whose names are associated with the origination of the fundamental concepts of rheology. The foundations of experimental rheology were also laid in the nineteenth century in the works of J. M. L. Poiseuille, T. Schwedoff, and others.

The next step in the advancement of rheology dates back to the twenties of this century when E. C. Bingham, G. W. Scott-Blair, A. Nadai, and M. Reiner developed the fundamentals of the engineering approach to the technological properties of real materials, thereby outlining the numerous potential applications of rheology. The progress of polymer rheology was especially vigorous after World War II when polymeric materials found their way into industry and the home. Today, rheology is 60-70 per cent concerned with investigations of this kind of materials.

Polymer rheology has evolved as an independent science over the last 10-15 years and is in its various aspects intimately entwined with molecular physics, continuum mechanics, and the processing of polymeric materials. At present, rheological investigations of polymers are progressing at an enormously rapid rate, embracing a wide range of materials. The common features in methodology permit the application of the theoretical and experimental approaches worked out specifically for polymers to the study of the mechanical properties of various materials: biological objects, paste-like substances, such as lubricants, food products, clays, and cement mixes. The applied aspects of rheology are tied up with the development of new technological processes for the processing of plastics, elastomers and fibres, the evaluation and optimization of the existing industries, and with the forecasting and appraisal of the performance of articles in various fields of modern engineering.

The wide scope of investigations coupled with the specific manifestations of the properties of many types and classes of modern polymeric materials has compelled the authors to focus their attention on the key topics of polymer rheology—the general approach to the description of the mechanical properties of continuous media (Chapter 1); the results of investigations of the flow behaviour of polymers of various compositions and structures (Chapter 2); the description of the viscoelastic properties of long-chain macromolecules (Chapter 3); the mechanics of and the molecular approach to the Weissenberg effect (Chapter 4); the rubber elasticity of polymers in flow (Chapter 5); the rheological properties of filled polymers (Chapter 6); and a rigorous exposition of the deformation and ultimate properties of polymers in uniaxial extension (Chapter 7). The choice of topics and the extent to which they are dealt with have necessarily been arbitrary and are in part a reflection of the authors' interests, though the authors have taken care to avoid giving too much prominence to their favoured subjects.

The limited size of the book has not allowed us to discuss the experimental methods of rheology and boundary hydrodynamic problems. The problems of the rheology of solid polymers and cured (cross-linked) elastomers have also had to be omitted from the text. Thus, the present monograph is restricted to the narrower field delineated by the original meaning of the word rheology as the science of the flow of various materials.

The present book is an attempt to generalize the existing conceptions of the rheological properties of polymeric materials and to present the results of investigations that have been carried out by the authors and their co-workers since 1962, chiefly in the Laboratory of Polymer Rheology at the Institute of Petrochemical Synthesis of the USSR Academy of Sciences. The first Russian edition of this book was published in 1977. As with any rapidly growing field, the elapsed time between the publication of the Russian edition and the preparation of the English version of the book witnessed the publication of new important results. The book has now been augmented and thoroughly revised. Substantial revisions have been made in Chapters 2, 3, and 7. The major changes also include the addition of a new chapter on the rheological properties of filled polymers (Chapter 6).

The authors wish to express their gratitude to numerous colleagues, in particular Dr. B. P. Diatchenko and Professor Z. P. Schulman, who provided helpful comments on smaller or larger portions of the Russian edition and whose valuable suggestions and criticisms have made it possible to correct some misprints and remove some previously undetected obscurities and errors. We should also like to offer our most sincere thanks to Dr. V. S. Volkov for his help in writing some of the sections of Chapter 3. Finally, we wish to take this opportunity to pay tribute to the translator, Mr. A. A. Beknazarov, who has put much effort into the translation of the book into English and who has patiently and painstakingly gone through the difficult job of inserting the numerous corrections and additions made by the authors during the preparation of this edition.

Errors will doubtless appear, so we would appreciate having them called to our attention. We will be grateful for any constructive criticism of this book.

June 25, 1979
Moscow

G. V. Vinogradov
A. Ya. Malkin

Contents

CHAPTER 1

Basic Concepts of Rheology

1.1. Introduction

Rheology is concerned with the description of the mechanical properties of the various materials under the various deformation conditions, when they simultaneously exhibit the ability to flow and accumulate recoverable deformations. The objective of rheology is to work out general principles and assumptions that could be used to deduce quantitative relationships among the quantities being measured.

In theoretical treatments it is customary to abstract from the specific features of the structure of materials to an extent permissible by the problems to be solved. Thus, polymeric systems are often regarded as continuous media. This allows one to manipulate with models of real materials, which are constructed on the assumption that the quantities characterizing their properties and behaviour are continually changed throughout the material. Therefore, it becomes necessary to consider the events that take place at every point in the body. This leads to the concepts of stress and strain as the characteristics of the dynamic and kinematic state of the material in the small neighbourhood of the chosen point of the continuum.

The specificity of the mechanical properties of a material derives from the peculiarities of its response to external forces. Such a response of the body takes the form of strains on loading or when a state of stress results from deformation. Therefore, the fundamental objective of theoretical rheology is the consideration of possible relations between the kinematics and dynamics of the body at a point. This is, of course, based on the specific features of the molecular structure of materials, but rheology is in no way explicitly concerned with the consideration of molecular structure.

This chapter deals successively with the basic conceptions and quantitative characteristics of the properties of materials, which are introduced for the description of their dynamics and kinematics as continua, and also with the analysis of the simplest relations between the parameters governing the processes of deformation. This is all based on phenomenological rather than molecular-kinetic conceptions.

Propositions and inferences of theoretical rheology are tested experimentally, which enables one to select and improve the methods of simulating the properties of real materials.

Originally, the term rheology referred to fluid systems but gradually the scope of the term became wider. At present, rheology is defined as a science concerned with the laws of deformation of various materials, including those for which the process of flow is not determinative. This is all the more essential that in practice (and this is specific to polymeric materials) it is difficult to draw a boundary line between materials capable of flowing and those exhibiting only recoverable deformations. It becomes therefore important to define general concepts and ideas applicable to bodies having different properties. It is precisely this circumstance that has given a considerable impetus to the development of rheology as a science dealing with the study of materials exhibiting properties intermediate between the properties of ideal elastic bodies and those of viscous liquids, which constitute the subject of the theory of elasticity and hydromechanics of viscous liquids. These two types of materials with specific properties are regarded as limiting in rheology.

1.2. Stresses

1.2.1. The Stress Tensor

If we choose arbitrarily an infinitesimal volume in the neighbourhood of a certain point in a body, then the force exerted by the other part on this volume element may be regarded as a set of forces applied to its boundary surface. When one is speaking of an infinitesimal volume or point, one implies volumes with dimensions substantially larger than those within which the role of short-range molecular forces becomes important. It is for this reason that we may assume that the traction exerted by the other part of the body on the chosen volume element acts across the surface of the latter.

Let us choose on the surface of the volume element under consideration a certain area ΔS defined by the direction of its normal, which is determined by unit vector $\mathbf{n}$. This vector is considered positive if it is pointing outwards from the chosen volume element. The area ΔS is acted on by a force $\Delta \mathbf{F}$. Then we may consider the limit of the ratio $\lim (\Delta \mathbf{F}/\Delta S)$, when the area ΔS reduces to a point. This limit is also a vector σ and it characterizes the intensity of the surface force acting on the volume under consideration. The magnitude of the vector σ depends on the position of the given point in the body and on the choice of the orientation of the area ΔS, i.e., on the direction of the vector $\mathbf{n}$. Therefore, in order to describe the loading conditions at a point, it is necessary to know the dependence of the vector σ on the direction of the area, i.e., the vector $\mathbf{n}$, and on the position of that point in space.

Suppose that the volume element chosen is a tetrahedron. One of its faces of area dS is arbitrarily oriented in space, this being characterized by the vector $\mathbf{n}$. The other three faces are oriented along the coordinate axes as shown in Fig. 1.1. The dimensions of the edges of the tetrahedron are indicated in the figure. This volume element is in equilibrium, i.e., the vector sum of the forces acting on the tetrahedron is equal to zero; obviously, the sums of the projections of all the forces along the coordinate axes are also equal to zero. Aside from the surface forces, the tetrahedron is also acted on by body forces.

The surface forces are proportional to the areas of the faces, which are infinitesimal quantities of second order, and the body forces are proportional to the volume of the body, which is an infinitesimal quantity of a higher order of smallness. Therefore, neglecting the infinitesimal quantity of higher order, the equilibrium condition may be written down as follows:

$$\sigma \, dS = \sigma_1 \left(\frac{1}{2} \, dx_2 \, dx_3 \right) + \sigma_2 \left(\frac{1}{2} \, dx_1 \, dx_3 \right) + \sigma_3 \left(\frac{1}{2} \, dx_1 \, dx_2 \right)$$

Since the quantities $dx_2\, dx_3/2dS$, $dx_1\, dx_3/2dS$, and $dx_1\, dx_2/2dS$ are equal to the cosines of the angles made by the vector **n** with the coordinate axes, we obtain the following expression which relates the vector σ on an arbitrarily oriented area **n** to the vectors $σ_1$, $σ_2$, and $σ_3$ acting on the areas directed along the coordinate axes:

$$σ = σ_1 \cos (\widehat{\mathbf{n},\ x_1}) + σ_2 \cos (\widehat{\mathbf{n},\ x_2}) + σ_3 \cos (\widehat{\mathbf{n},\ x_3})$$

From this it follows that the vector σ may be expressed as a function of the vector **n** by means of three quantities: $σ_1$, $σ_2$, and $σ_3$. This determines the

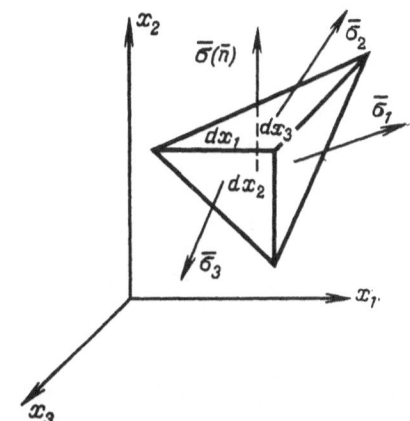

Fig. 1.1. A volume element in the form of a tetrahedron.

dependence of σ on the direction of the area characterized by the vector **n** or by the cosines of the angles $\widehat{\mathbf{n},\ x_1}$, $\widehat{\mathbf{n},\ x_2}$, and $\widehat{\mathbf{n},\ x_3}$, which is basically the same thing.

Each of the quantities, $σ_1$, $σ_2$, and $σ_3$, is defined in its turn by the projections onto the coordinate axes, which can be written in the form

$$σ_1\, (σ_{11},\, σ_{12},\, σ_{13});\ σ_2\, (σ_{21},\, σ_{22},\, σ_{23});\ σ_3\, (σ_{31},\, σ_{32},\, σ_{33})$$

where the quantities in parenthesis are sets of projections of the vectors $σ_1$, $σ_2$, and $σ_3$ onto the coordinate axes. In this notation, the first numerical subscript in the quantities $σ_{ij}$ specifies the projection of which vector is examined (i.e., specifies the orientation of the respective plane), and the second subscript specifies the axis onto which the vector $σ_{ij}$ is projected. Hence, the set of nine quantities $σ_{ij}$ depends on the position of the axes in space as well as on the orientation of the plane under consideration at a point. The set of the quantities $σ_{ij}$ completely define the operation of the forces in the neighbourhood of the given point. This set is called the stress, and its individual elements are termed the stress components.

The stress acting in the neighbourhood of a given point is of a rather complicated nature since a knowledge of nine quantities is required in order to describe it. It is a tensor (more exactly, a tensor of rank two). An exacting determination of quantities of this class is beyond the scope of the present book, but we shall consider some of the specific features and general regularities of operations that can be performed with tensors.

The whole set of the nine quantities that define the stress tensor $\{\sigma\}$ is usually written in the following form:

$$\{\sigma\} \equiv \begin{vmatrix} \sigma_{11} & \sigma_{12} & \sigma_{13} \\ \sigma_{21} & \sigma_{22} & \sigma_{23} \\ \sigma_{31} & \sigma_{32} & \sigma_{33} \end{vmatrix}$$

Quantities of tensorial nature will be enclosed in braces.

The diagonal terms in this matrix are called the normal components of the stress tensor since they correspond to the components of the vectors σ_1, σ_2, and σ_3 directed normal to the respective planes, and the remaining terms of the matrix are known as the tangential or shear stresses since they correspond to the vector components acting along the tangents to the respective planes, which are the boundary surfaces of the chosen elementary volume.

In a general case, all the components of the stress tensor $\{\sigma\}$ may depend on the coordinates of the chosen point in space and will therefore be functions of three variables, x_1, x_2, and x_3. The states in which the stress-tensor components are constant throughout the volume element considered are termed uniform stress states. If all the components in any one row in the matrix for $\{\sigma\}$ are equal to zero, i.e., one of the vectors σ_i is equal to zero, then it means that we have a state of biaxial or plane stress; if all the components in two rows are equal to zero, then we are dealing with a state of uniaxial stress. In the general case, when there is at least one non-zero component of the tensor $\{\sigma\}$ in each row, the dynamic state in the neighbourhood of the chosen point is a triaxial stress state.

1.2.2. The Complementary Shear Stress Condition. Rules of Transformation of Stress Components upon Rotation of Axes

The most important problem in stress theory is the establishment of the rules defining the change from one set of coordinate axes to another. These rules can easily be worked out on the basis of the general tensor theory, but for the sake of clarity this is done below in application to the plane stress state (Fig. 1.2).

Let us return once again to the analysis of the conditions of equilibrium. We shall expand the quantities $d\sigma_{ij}$ in a Taylor series, leaving only the first term since all the other terms are quantities of higher orders of smallness as compared with the first term; therefore

$$d\sigma_{12} = \frac{\partial \sigma_{12}}{\partial x_1}\, dx_1; \quad d\sigma_{21} = \frac{\partial \sigma_{21}}{\partial x_2}\, dx_2$$

We now write the equilibrium condition as the equality to zero of the sum of the moments of all the forces relative to the centre of the element shown in Fig. 1.2, taking into account that the force moment due to normal stresses is equal to zero since the lines of action of these forces pass through the centre:

$$\sigma_{21}\, dx_1 \frac{dx_2}{2} + \left(\sigma_{21} + \frac{\partial \sigma_{21}}{\partial x_2}\, dx_2\right) dx_1 \frac{dx_2}{2} - \sigma_{12}\, dx_2 \frac{dx_1}{2} -$$

$$- \left(\sigma_{12} + \frac{\partial \sigma_{12}}{\partial x_1}\, dx_1\right) dx_2 \frac{dx_1}{2} = 0$$

Here it is assumed that the dimension of the areas in the direction of the x_3 axis (perpendicular to the plane of the drawing) is the same and therefore all the terms of the above equality can be reduced to it.

Collecting like terms and neglecting terms of higher order of smallness, we obtain the following expression from the above equality:

$$| \sigma_{12} | = | \sigma_{21} |$$

which is known as the complementary shear stress condition or Cauchy's equality.

An analogous examination can be performed for other areas corresponding to other orientations of the planar element. Evidently the same result will

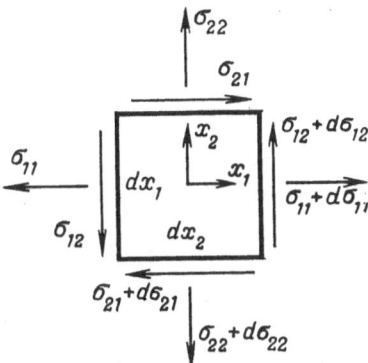

Fig. 1.2. Diagram illustrating the complementary shear stress condition.

obtain. This enables us to formulate a general proposition about the equality of pairs of shear stresses with identical indices, irrespective of their order, i.e., $\sigma_{ij} = \sigma_{ji}$. Thus, out of the nine stress tensor components only six are independent: three normal and three tangential components. The stress tensor is therefore symmetric.

The above-given equality, from which the complementary shear stress condition was deduced, is valid if there are no internal moments in an element, which may not always be the case; theories of the mechanical properties of polymers are known, which take account of the possibility of existence of internal moments. In this case, the stress tensor will not be symmetric and in order to describe it, use must be made of the complete array of 9 components.

Suppose now that the coordinate axes rotate relative to the original position defined by the x_1 and x_2 directions and will occupy a new position characterized by the x_α and $x_{\alpha+\frac{\pi}{2}}$ directions. We shall assume that the stresses defined with respect to the x_1 and x_2 axes are known and it is required to find the stresses acting on the planes oriented in the x_α and $x_{\alpha+\frac{\pi}{2}}$ directions

(Fig. 1.3). These stresses, which we shall denote by σ_α and τ_α for the plane x_α, are dependent on the angle α, and therefore the task is to find the dependences of σ_α and τ_α on the stresses referred to the x_1 and x_2 axes and the angle α. To solve this problem, we project, onto the axes x_α and $x_{\alpha+\frac{\pi}{2}}$, the vectors that determine the equilibrium condition for the element shown in Fig. 1.3, considering that the sum of the projections of all the forces acting on the sides of the triangle are equal to zero. We denote the side BC of the triangle as dx, then $AB = \cot \alpha \, dx$ and $AC = dx/\sin \alpha$. The equilibrium equation in the projections onto the axes x_α and $x_{\alpha+\frac{\pi}{2}}$ is written in the form

$$\begin{cases} -\sigma_{11} \sin \alpha \, dx + \sigma_{12} \cos \alpha \, dx - \sigma_{22} \cos \alpha \cot \alpha \, dx + \\ \quad + \sigma_{21} \sin \alpha \cot \alpha \, dx + \sigma_\alpha \dfrac{dx}{\sin \alpha} = 0 \\ \sigma_{11} \cos \alpha \, dx + \sigma_{12} \sin \alpha \, dx - \sigma_{22} \sin \alpha \cot \alpha \, dx - \\ \quad - \sigma_{21} \cos \alpha \cot \alpha \, dx + \tau_\alpha \dfrac{dx}{\sin \alpha} = 0 \end{cases}$$

After simple rearrangements we obtain the following expressions for the stresses σ_α and τ_α:

$$\sigma_\alpha = -\sigma_{12} \sin 2\alpha + \sigma_{11} \sin^2 \alpha + \sigma_{22} \cos^2 \alpha \tag{1.1}$$

$$\tau_\alpha = -\frac{1}{2} \sin 2\alpha \, (\sigma_{11} - \sigma_{22}) + \sigma_{12} \cos 2\alpha \tag{1.2}$$

The expressions obtained, however, allow us to establish some other important facts. Thus, let us find the value of the angle α_m at which $\tau_\alpha = 0$, i.e.,

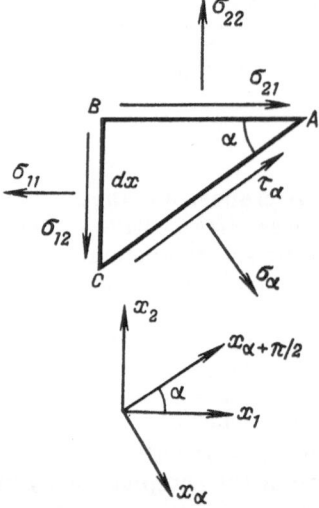

Fig. 1.3. Diagram illustrating the conditions for defining the principal stresses.

let us find the area whose orientation is determined by the value of the angle α_m and on which there are no shear stresses. Then, from the condition $\tau_\alpha = 0$ and formula (1.2) it follows that

$$\tan 2\alpha_m = \frac{2\sigma_{12}}{\sigma_{11} - \sigma_{22}}$$

or

$$\alpha_m = \frac{1}{2} \arctan \frac{2\sigma_{12}}{\sigma_{11} - \sigma_{22}} = \frac{1}{2} \operatorname{arccot} \frac{\sigma_{11} - \sigma_{22}}{2\sigma_{12}} \tag{1.3}$$

We shall now determine the condition of the extremum of σ_α as a function of α. To do this, we differentiate σ_α with respect to α and set the derivative $d\sigma_\alpha/d\alpha$ to zero. Having performed these manipulations, we see that the condition of the extremum of the quantity σ_α is determined by the same condition as the absence of shear stresses, i.e., $d\sigma_\alpha/d\alpha = 0$ at $\alpha =$

$= 1/2 \arctan [2\sigma_{12}/(\sigma_{11} - \sigma_{22})]$. Thus, the normal stress is extremal in the direction in which there are no shear stresses.

If we now consider the direction of the $x_{\alpha+\frac{\pi}{2}}$ axis, we can show, in the same manner, that $\tau_{\alpha+\frac{\pi}{2}} = \tau_\alpha$. Incidentally, this is an obvious corollary of the complementary shear stress condition. Hence,

$$\sigma_{\alpha+\frac{\pi}{2}} = \sigma_{12} \sin 2\alpha + \sigma_{11} \cos^2 \alpha + \sigma_{22} \sin^2 \alpha \qquad (1.4)$$

Proceeding from the equality of the derivative $d\sigma_{\alpha+\frac{\pi}{2}}/d\alpha$ to zero, we can find the extremum condition for the quantity $\sigma_{\alpha+\frac{\pi}{2}}$. Calculations show that this condition is the equality (1.3), and, hence, the extremum condition for the stress σ_α. This means that in the plane stress state at each given point there exist at all times two mutually perpendicular directions, x_{α_m} and $x_{\alpha_m+\frac{\pi}{2}}$ [the angle α_m is defined by equality (1.3)], in which the normal stresses are extremal and no shear stresses are present. One of these stresses is maximal and the other minimal. These extremal values of stresses are called *principal stresses*. Further in the text we shall denote them by σ_1 and σ_2 (the principal stresses σ_1 and σ_2 should not be confused with the above-considered vectors σ_1 and σ_2). Thus, at a given point we can always specify such directions in which there occurs the tension or compression of the material and no shear stresses are present.

1.2.3. Invariants of the Stress Tensor

As seen from the preceding discussion, upon rotation of the axes the values of all the stress components are changed. There exist, however, such combinations of them, which remain constant upon transformation of coordinates. Let us consider, for example, the sum of stresses $\left(\sigma_\alpha + \sigma_{\alpha+\frac{\pi}{2}}\right)$ with the angle α being chosen arbitrarily. Addition of the right-hand sides of formulas (1.1) and (1.4) leads to the sum $(\sigma_{11} + \sigma_{22})$. Since in a special case the angle α_m may serve as the angle α, we may write the following equalities:

$$I_1 = \sigma_\alpha + \sigma_{\alpha+\frac{\pi}{2}} = \sigma_{11} + \sigma_{22} = \sigma_1 + \sigma_2 \qquad (1.5)$$

Analogously, by substituting the corresponding expressions for the stress components and performing certain trigonometric transformations, it can be shown that the following equalities are fulfilled:

$$I_2 = \sigma_1 \sigma_{22} - \sigma_{12}^2 = \sigma_\alpha \sigma_{\alpha+\frac{\pi}{2}} - \tau_\alpha^2 = \sigma_1 \sigma_2 \qquad (1.6)$$

The resulting expressions, which remain unchanged upon a change of the coordinate system, are called *invariants*. The concept of invariants is of the same general nature as the above-indicated methods of transformation of coordinates, which constitute a special case of the transformation of the components of any tensor. We can assert that formulas (1.5) and (1.6) define the first and second invariants of any tensor.

Expressions (1.5) and (1.6) have been derived for the plane stress state. But quite analogous arguments, though requiring more complicated transformations, are valid for the general case of the three-dimensional stress state. Here there appears a third principal stress σ_3 and, instead of the statement as to the existence of two mutually perpendicular directions, in which the normal stresses are extremal and there are no shear stresses, it is necessary to draw the conclusion as to the existence (in a three-dimensional space) of three such mutually perpendicular directions. Furthermore, when generalizing the concept of invariants to the three-dimensional (triaxial) stress state formulas (1.5) and (1.6) will be somewhat changed because of the appearance of new stress components. We shall only give the final result without delving into the detailed derivation of the respective formulas:

$$I_1 = \sigma_1 + \sigma_2 + \sigma_3 = \sigma_{11} + \sigma_{22} + \sigma_{33} \tag{1.7}$$

$$I_2 = \sigma_1\sigma_2 + \sigma_1\sigma_3 + \sigma_2\sigma_3 = \sigma_{11}\sigma_{22} + \sigma_{11}\sigma_{33} + \sigma_{22}\sigma_{33} - (\sigma_{12}^2 + \sigma_{13}^2 + \sigma_{23}^2) \tag{1.8}$$

Apart from the first (linear) invariant I_1 and the second (quadratic) invariant I_2, which exist only for the plane stress state, in the three-dimensional case there can be constructed a third (cubic) invariant of the stress tensor, I_3, which is expressed in terms of the stress components in the following way:

$$I_3 = \sigma_1\sigma_2\sigma_3 = \sigma_{11}\sigma_{22}\sigma_{33} + 2\sigma_{12}\sigma_{13}\sigma_{23} - (\sigma_{11}\sigma_{23}^2 + \sigma_{22}\sigma_{13}^2 + \sigma_{33}\sigma_{12}^2) \tag{1.9}$$

The tensor invariants (and, in particular, the stress tensor invariants for the case considered above) are the most important characteristics of the tensor because they are independent of the choice of the coordinate axes with respect to which the tensor components are defined. Physical laws do not depend on the choice of the coordinate system used. Therefore, the relationship between the tensors, which determine the characteristics of a process, must be written as a relation between their invariants. For instance, if any physical property of a body, say viscosity, depends on stresses that arise on loading, then in the case of the three-dimensional state of stress (when different stress components are acting) this dependence must be expressed in terms of the invariants of the stress tensor.

1.2.4. Mohr's Circle

The above-given dependences of σ_α and τ_α on α may be regarded as parametric equations relating the variables σ_α and τ_α, where the role of a parameter is played by the angle α. Then, eliminating the quantity α from formulas (1.1) and (1.2), we can obtain the dependence $f(\sigma_\alpha, \tau_\alpha) = 0$ which is interpreted graphically in the coordinates σ_α and τ_α. For this purpose we represent the values of σ_α and τ_α as functions of the principal stresses σ_1 and σ_2.

If the state of stress is referred to the coordinate axes x_{α_m} and $x_{\alpha_m + \frac{\pi}{2}}$, then $\sigma_{12} = 0$. In this case the dependences of σ_α and τ_α on α will be expressed in the following way:

$$\begin{cases} \sigma_\alpha = \sigma_1 \sin^2\alpha + \sigma_2 \cos^2\alpha \\ \tau_\alpha = -\frac{1}{2}(\sigma_1 - \sigma_2)\sin 2\alpha \end{cases} \tag{1.10}$$

In accordance with the meaning of the notation in Eq. (1.10) the angle α determines the orientation of arbitrarily chosen planes at a given point relative to the direction of the principal stresses.

From this, by way of simple transformations, expressing $\sin^2 \alpha$ and $\cos^2 \alpha$ through $\cos 2\alpha$ and eliminating the trigonometric functions of the double angle, we get:

$$\left[\sigma_\alpha - \left(\frac{\sigma_1 + \sigma_2}{2}\right)\right]^2 + \tau_\alpha^2 = \left(\frac{\sigma_1 - \sigma_2}{2}\right)^2 \tag{1.11}$$

Expression (1.11) is the equation for a circle in the coordinates σ_α and τ_α. The centre of this circle lies at a point with the coordinates $\sigma_\alpha = (\sigma_1 + \sigma_2)/2$ and $\tau_\alpha = 0$, i.e., on the abscissa. The radius of the circle is equal to $(\sigma_1 - \sigma_2)/2$.

This circle is called *Mohr's stress circle*; it is shown in Fig. 1.4. There are points on it which correspond to the principal directions σ_1 and σ_2, and also

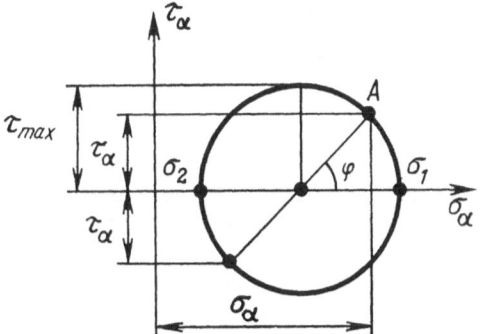

Fig. 1.4. Mohr's stress circle.

an arbitrary point A on the circle, the position of which relative to the abscissa is defined by the angle φ formed by the axis and the radius drawn from the centre of Mohr's circle. The angle φ is identified with the angle 2α since it is exactly in this case that the coordinates σ_α and τ_α of the arbitrary point are expressed in the manner stipulated by the system of equations (1.10).

The point which corresponds to the angle $(\varphi + \pi)$ corresponds to the orientation of the plane at the angle $(\alpha + \pi/2)$, i.e., the plane perpendicular to the one that has been considered. The corresponding construction is also shown in Fig. 1.4 in the region of negative values of τ_α. As should have been expected, on the mutually perpendicular planes the shear stresses are equal in magnitude and opposite in direction.

Thus, the set of points on Mohr's stress circle correspond to different orientations of the planes since $\varphi = 2\alpha$. The coordinates of these points, σ_α and τ_α, represent the values of normal and shear stresses in the direction at the angle α from the direction of the maximum stress σ_1. Therefore, Mohr's circle is a convenient graphical means of examining the dependences of the stress components on the choice of the directions at a given point in the body.

Through inspection of Mohr's stress circle one can easily find the maximum value of shear stress, $\tau_{\max}$. Evidently, it corresponds to the upper point of the circle and is equal to the radius of the circle, i.e.,

$$\tau_{\max} = \frac{\sigma_1 - \sigma_2}{2} \tag{1.12}$$

Examination of Mohr's circle allows one to clearly visualize the nature of the state of stress at a given point at any orientation of the axes, and also to elucidate the principal features of any particular loading conditions. Thus, for the special case shown in Fig. 1.4 as an illustration, one may immediately say that at the specified point there are **no** directions in which the compres-

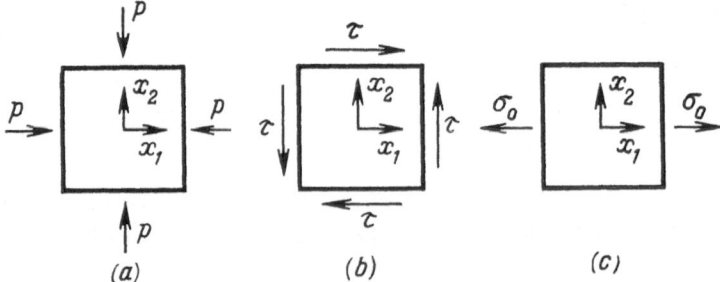

Fig. 1.5. Schematic representation of three principal types of loading:
a—hydrostatic compression; b—simple shear; c—uniaxial extension.

sion of the material could take place since Mohr's circle lies entirely to the right of the ordinate (in the region of positive values of σ_α).

Three principal cases of loading—hydrostatic compression, shear, and uniaxial extension—are schematically shown in Fig. 1.5.

1.2.5. Hydrostatic Compression

Figure 1.5a shows hydrostatic (or uniform) compression for the state of plane stress, which is readily generalized to the three-dimensional case. In the case of hydrostatic compression the principal stresses are equal in magnitude but opposite in sign with respect to the hydrostatic pressure p. Therefore, Mohr's stress circle for hydrostatic compression degenerates to a point with the coordinates $\sigma_\alpha = -p$ and $\tau_\alpha = 0$. Hence, no matter what the orientation of the planes is, there arise no shear stresses. In this case, all the directions are principal. Hydrostatic compression is represented by a stress tensor written* in the form

$$\{\sigma\} = \begin{vmatrix} \sigma_{11} & 0 & 0 \\ 0 & \sigma_{22} & 0 \\ 0 & 0 & \sigma_{33} \end{vmatrix} = \begin{vmatrix} -p & 0 & 0 \\ 0 & -p & 0 \\ 0 & 0 & -p \end{vmatrix} = -p \begin{vmatrix} 1 & 0 & 0 \\ 0 & 1 & 0 \\ 0 & 0 & 1 \end{vmatrix} = -p\delta_{ij}$$

The tensor δ_{ij}, called the Kronecker delta or unit tensor, is defined as the quantity equal to unity with the subscripts being equal ($i = j$) and to zero with the subscripts being different ($i \neq j$). The stress tensor describing the state of uniform compression is known as the *spherical stress tensor*.

1.2.6. Shear

This type of loading is presented in Fig. 1.5b for the plane stress state. The shear stresses τ act only on the planes parallel to the coordinate axes

---

* The operation of elimination of the constant quantity $-p$ from the matrix of the tensor components (see below) corresponds to the rule of multiplication of tensors by a constant scalar. In a general case, this can be formulated as follows: The tensor {A} with components a_{ij} is equal to the product of the tensor {B} with components b_{ij} by the scalar m if each of the components of the tensor {A} is equal to the component of the tensor {B} with the same subscripts, multiplied by m, i.e., {A} = m {B} if $a_{ij} = mb_{ij}$.

chosen. The stress tensor is then written in the form

$$\{\sigma\} = \begin{vmatrix} 0 & \tau & 0 \\ \tau & 0 & 0 \\ 0 & 0 & 0 \end{vmatrix}$$

According to formula (1.3) the magnitude of the angle α_m, which defines the direction of the principal stresses relative to the coordinate axes, is $45°$, and in accordance with Eqs. (1.1) and (1.4) the values of $\sigma_1 = \tau$ and $\sigma_2 = -\tau$, i.e., one of the principal stresses is the compressive stress and the other is the tensile stress. Naturally, at $\alpha_m = 45°$ the value of $\tau_\alpha = 0$. Thus, the application of only shear stresses inevitably results in the tension and compression of the elements of the body in the corresponding directions. The case under consideration is represented by Mohr's circle with the centre at the origin of coordinates and the radius equal to τ.

1.2.7. Uniaxial Tension

In the case of the uniaxial tension of an element of a body (Fig. 1.5c) the stress tensor is written in the form

$$\{\sigma\} = \begin{vmatrix} \sigma_0 & 0 & 0 \\ 0 & 0 & 0 \\ 0 & 0 & 0 \end{vmatrix}$$

It is obvious that the direction of the stress σ_0 and the directions normal to it correspond to the direction of the principal stresses. But, in contrast to the above-considered case of hydrostatic loading, here the principal stresses are not equal: one of them equals σ_0, and the other two are equal to zero. This is represented (in the two-dimensional case) by Mohr's circle with a diameter equal to σ_0, and this Mohr's circle lies so that it touches the ordinate axis. Therefore, shear stresses are acting in all directions, except the direction of the principal stresses. The maximum shear stress is equal to $\tau_{max} = \sigma_0/2$ in accordance with formula (1.12), and the respective orientation of the plane is determined by the angle $\alpha_m = 45°$. This means that in uniaxial tension in the directions at an angle of $45°$ to the direction of tension there arise maximum shear stresses tending to produce shear at this angle in the material.

The case of uniaxial tension may be examined by another method if we make use of the rule of addition of tensors.* This operation is necessary in order to single out the spherical component from the stress tensor corresponding to uniaxial loading. Indeed, on the basis of the above-given expression

---

* The rule of addition of tensors may be formulated thus: the tensor $\{A\}$ with components a_{ij} is equal to the sum of tensor $\{B\}$ with components b_{ij} and tensor $\{C\}$ with components c_{ij} if each of the components of tensor $\{A\}$ is equal to the sum of the components of tensors $\{B\}$ and $\{C\}$ with the same subscripts, i.e., $\{A\} = \{B\} + \{C\}$ if $a_{ij} = b_{ij} + c_{ij}$.

for $\{\sigma\}$ in the stress state we can perform the following identical transformations:

$$\{\sigma\} = \begin{vmatrix} \sigma_0 & 0 & 0 \\ 0 & 0 & 0 \\ 0 & 0 & 0 \end{vmatrix} = \begin{vmatrix} \dfrac{\sigma_0}{3} & 0 & 0 \\ 0 & \dfrac{\sigma_0}{3} & 0 \\ 0 & 0 & \dfrac{\sigma_0}{3} \end{vmatrix} + \begin{vmatrix} \dfrac{2\sigma_0}{3} & 0 & 0 \\ 0 & -\dfrac{\sigma_0}{3} & 0 \\ 0 & 0 & -\dfrac{\sigma_0}{3} \end{vmatrix} =$$

$$= \frac{\sigma_0}{3}\delta_{ij} + \frac{\sigma_0}{3}\begin{vmatrix} 2 & 0 & 0 \\ 0 & -1 & 0 \\ 0 & 0 & -1 \end{vmatrix} = \sigma_m\delta_{ij} + \frac{\sigma_0}{3}\begin{vmatrix} 2 & 0 & 0 \\ 0 & -1 & 0 \\ 0 & 0 & -1 \end{vmatrix}$$

The quantity σ_m is called the mean normal stress. If it is assumed that σ_m is equal in magnitude and opposite in sign to the hydrostatic pressure p, then it appears that uniaxial extension is by its physical significance a set of two states of stress: uniform dilatation under the action of the stress σ_m and the triaxial stress state which is defined by the second term in the formulas written above.

This interpretation of stretching is very important by its physical significance since it shows that the uniaxial stretching is, in fact, associated with the application of a negative uniform pressure σ_m and with the appearance of stress components in all three directions. The circumstance explains why on uniaxial stretching there occurs a change in the dimensions of the specimen not only in one but in all directions.

1.2.8. The Stress Deviator

The treatment of longitudinal elongation as a combination of two modes of deformation may be generalized to the case of any arbitrary state of stress. It may be represented as the superimposition of hydrostatic loading on such a state of stress in which all the components associated with uniform loading are absent. For this purpose, we make use of the definition of the mean normal stress σ_m as one-third of the sum of the diagonal components, i.e., according to Eq. (1.7) $\sigma_m = I_1/3$, and single out the spherical component from the complete stress tensor $\{\sigma\}$:

$$\{\sigma\} = \begin{vmatrix} \sigma_{11} & \sigma_{12} & \sigma_{13} \\ \sigma_{21} & \sigma_{22} & \sigma_{23} \\ \sigma_{31} & \sigma_{32} & \sigma_{33} \end{vmatrix} = \left(\frac{\sigma_{11}+\sigma_{22}+\sigma_{33}}{3}\right)\begin{vmatrix} 1 & 0 & 0 \\ 0 & 1 & 0 \\ 0 & 0 & 1 \end{vmatrix} +$$

$$= \begin{vmatrix} \dfrac{2\sigma_{11}-(\sigma_{22}+\sigma_{33})}{3} & \sigma_{12} & \sigma_{13} \\ \sigma_{21} & \dfrac{2\sigma_{22}-(\sigma_{11}+\sigma_{33})}{3} & \sigma_{23} \\ \sigma_{31} & \sigma_{32} & \dfrac{2\sigma_{33}-(\sigma_{11}+\sigma_{12})}{3} \end{vmatrix} =$$

$$= \sigma_m\delta_{ij} + \sigma'_{ij} = \frac{I_1}{3}\delta_{ij} + \sigma'_{ij}$$

The stress tensor $\{\sigma\}'$ with components σ'_{ij}, which is equal to the complete stress tensor minus the component representing uniform loading, is called the *stress deviator* (or *deviatoric stress*). Its components σ'_{ij} are expressed as written above. Evidently, the shear stress components in the complete stress tensor and its deviator are equal and the diagonal components of the stress deviator, σ'_{ij}, are defined as $(\sigma_{ij} - \sigma_m)$. The main feature of the stress deviator is that its first invariant is equal to zero, which is easily proved by a direct check-up—the addition of the components $(\sigma'_{11} + \sigma'_{22} + \sigma'_{33})$.

1.3. Deformations (Strains)

1.3.1. Geometrical Interpretation of Strain.
The Finite Strain Tensor

Let us consider in a certain volume element two infinitely closely spaced points A and B in the manner shown in Fig. 1.6. Suppose that the position of the points A and B in space is changed. Here we shall not discuss the

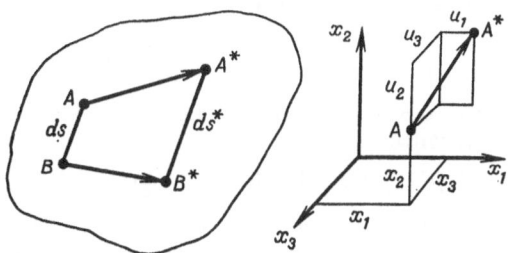

Fig. 1.6. Illustrating the strain concept.

possible causes of this change, the rate of the change, and the time period during which it occurred. We shall only be concerned with the geometrical picture of the phenomenon. In the most general case the displacements in space of the point B can be regarded as the movement of the point A, to which there is added the movement of B relative to A. The movement of the point B relative to A can be effected, first, by the rotation of the side BA relative to A as the centre of rotation and, second, by the change of the dimension of the segment BA.

In a general case, the movement of the points of a body in space is composed of the translational motion of the body as a whole (to this there corresponds the movement of the point A, which is regarded as the origin, relative to the fixed coordinate system), the rotation of the body relative to a certain centre (the point A may be provisionally assumed to be such a centre), and deformations—the changes of the distances between the points in the body. Thus, a distinction is made between the displacements of the body, which are not accompanied by a change of the distances between the points in it, and the deformations, which result in a change of the lengths of the segments connecting the points in the body. This proposition is strictly formulated in the form of the so-called Helmholtz theorem.

The motion of a body as a whole in space and its rotation relative to any point lies beyond the scope of rheology; it is studied in theoretical mechanics.

Of interest to the discussion that follows are only deformations—the relative displacements of points in the body (or, more exactly, in a continuum).

The changes of the distance between two points must be considered in application to an infinitesimal volume element. This means that if the coordinates of the point A are (x_1, x_2, x_3), then the coordinates of the point B, which is infinitely close to A, are $(x_1 + dx_1, x_2 + dx_2, x_3 + dx_3)$. According to the methods of analytical geometry, the square of the distance $(ds)^2$ between the points A and B is defined as the sum of the squares of the differences between their coordinates; therefore

$$(ds)^2 = (dx_1)^2 + (dx_2)^2 + (dx_3)^2$$

Suppose that, as a result of some motion in space, the points A and B are displaced to new positions A^* and B^*, which we shall define with respect to a space-fixed coordinate system. If the displacements of the point A along the three coordinate axes (see Fig. 1.6) are, respectively, u_1, u_2, and u_3, then the new coordinates of the point A will be $A^* (x_1 + u_1; x_2 + u_2; x_3 + u_3)$. By virtue of the continuity of the body, i.e., assuming that the movement in space has not brought about ruptures between the points A and B and these points remain to be infinitely close to each other, the coordinates of the new position of the point B may be represented in the following way:

$$B^* [(x_1 + u_1) + d (x_1 + u_1); \ (x_2 + u_2) + d (x_2 + u_2); \ (x_3 + u_3) + d (x_3 + u_3)]$$

Then the square of the distance between the new positions of the points A^* and B^* may be represented in the form

$$(ds^*)^2 = [d (x_1 + u_1)]^2 + [d (x_2 + u_2)]^2 + [d (x_3 + u_3)]^2$$

The change of the distance between the points A and B upon their displacement in space is characterized by the difference of the squares: $(ds^*)^2 - (ds)^2$. By expressing the quantities $(ds^*)^2$ and $(ds)^2$ in the manner indicated above, i.e., in terms of the differentials of the coordinates and displacements, we obtain:

$$(ds^*)^2 - (ds)^2 = (du_1)^2 + (du_2)^2 + (du_3)^2 + 2dx_1 \, du_1 + 2dx_2 \, du_2 + 2dx_3 \, du_3$$

The magnitude of the displacements u_i is a function of the coordinates of the point A prior to deformation since its movement in space depends on the original position. Therefore, we can write the following expressions for the exact differentials of the displacements:

$$du_i = \frac{\partial u_i}{\partial x_1} \, dx_1 + \frac{\partial u_i}{\partial x_2} \, dx_2 + \frac{\partial u_i}{\partial x_3} \, dx_3 \quad (i = 1, \, 2, \, 3)$$

Now we can perform all the necessary computations, squaring the corresponding sums. If the quantities in identical sets of pairs of the differentials dx_i and du_j are grouped together, then there will obtain the following expression for the difference of the squares of the distances between the points A and B before and after their movement:

$$(ds^*)^2 - (ds)^2 = \left[2 \frac{\partial u_1}{\partial x_1} + \left(\frac{\partial u_1}{\partial x_1} \right)^2 + \left(\frac{\partial u_2}{\partial x_1} \right)^2 + \left(\frac{\partial u_3}{\partial x_1} \right)^2 \right] (dx_1)^2 +$$

$$+ \left[2 \frac{\partial u_2}{\partial x_2} + \left(\frac{\partial u_1}{\partial x_2} \right)^2 + \left(\frac{\partial u_2}{\partial x_2} \right)^2 + \left(\frac{\partial u_3}{\partial x_2} \right)^2 \right] (dx_2)^2 +$$

$$+ \left[2 \frac{\partial u_3}{\partial x_3} + \left(\frac{\partial u_1}{\partial x_3} \right)^2 + \left(\frac{\partial u_2}{\partial x_3} \right)^2 + \left(\frac{\partial u_3}{\partial x_3} \right)^2 \right] (dx_3)^2 +$$

$$+\left[2\left(\frac{\partial u_1}{\partial x_2}+\frac{\partial u_2}{\partial x_1}\right)+2\left(\frac{\partial u_1}{\partial x_1}\cdot\frac{\partial u_1}{\partial x_2}+\frac{\partial u_2}{\partial x_1}\cdot\frac{\partial u_2}{\partial x_2}+\frac{\partial u_3}{\partial x_1}\cdot\frac{\partial u_3}{\partial x_2}\right)\right]dx_1\,dx_2+$$

$$+\left[2\left(\frac{\partial u_1}{\partial x_3}+\frac{\partial u_3}{\partial x_1}\right)+2\left(\frac{\partial u_1}{\partial x_1}\cdot\frac{\partial u_1}{\partial x_3}+\frac{\partial u_2}{\partial x_1}\cdot\frac{\partial u_2}{\partial x_3}+\frac{\partial u_3}{\partial x_1}\cdot\frac{\partial u_3}{\partial x_3}\right)\right]dx_1\,dx_3+$$

$$+\left[2\left(\frac{\partial u_2}{\partial x_3}+\frac{\partial u_3}{\partial x_2}\right)+2\left(\frac{\partial u_1}{\partial x_2}\cdot\frac{\partial u_1}{\partial x_3}+\frac{\partial u_2}{\partial x_2}\cdot\frac{\partial u_2}{\partial x_3}+\frac{\partial u_3}{\partial x_2}\cdot\frac{\partial u_2}{\partial x_3}\right)\right]dx_2\,dx_3$$

In abbreviated form the result obtained may be written as follows:

$$(ds^*)^2-(ds)^2=2\left[\varepsilon_{11}\,(dx_1)^2+\varepsilon_{22}\,(dx_2)^2+\varepsilon_{33}\,(dx_3)^2+\right.$$

$$\left.+\varepsilon_{12}\,dx_1\,dx_2+\varepsilon_{13}\,dx_1\,dx_3+\varepsilon_{23}\,dx_2\,dx_3\right] \tag{1.13}$$

where the quantities ε_{ij} are expressed in terms of the derivatives of the displacements and the squares of these derivatives in the same way as written above in the complete expression for the difference $(ds^*)^2 - (ds)^2$. As an example, we write out the expressions for two quantities, one with identical and the other with different subscripts:

$$\varepsilon_{11}=\frac{\partial u_1}{\partial x_1}+\frac{1}{2}\left[\left(\frac{\partial u_1}{\partial x_1}\right)^2+\left(\frac{\partial u_2}{\partial x_1}\right)^2+\left(\frac{\partial u_3}{\partial x_1}\right)^2\right]$$

$$\varepsilon_{12}=\left(\frac{\partial u_1}{\partial x_2}+\frac{\partial u_2}{\partial x_1}\right)+\left(\frac{\partial u_1}{\partial x_1}\cdot\frac{\partial u_1}{\partial x_2}+\frac{\partial u_2}{\partial x_1}\cdot\frac{\partial u_2}{\partial x_2}+\frac{\partial u_3}{\partial x_1}\cdot\frac{\partial u_3}{\partial x_2}\right) \tag{1.14}$$

The remaining quantities ε_{ij} are of analogous structure with the appropriate obvious replacement of the subscripts.

The quantities ε_{ij} are the characteristics of the deformation of the body in the neighbourhood of a given point. If at least one of the quantities ε_{ij} is different from zero it will be sufficient for the length of the linear element AB to be changed. The converse is also true: if all the quantities are equal to zero, then $ds^* = ds$, which is equivalent to the absence of deformation upon displacement of one point of the body relative to the other in space.

To describe the deformed state, use is made of the large-strain (or finite-strain) tensor, i.e., $\{\gamma\}$. Its components are defined through the quantities ε_{ij} in the following way:

$$\gamma_{ij}=\begin{cases}\varepsilon_{ij} & \text{at } i=j \\ \dfrac{1}{2}\varepsilon_{ij} & \text{at } i\neq j\end{cases} \tag{1.15}$$

The finite-strain tensor $\{\gamma\}$ is a geometrical characteristic of changes that have taken place in the vicinity of a given point in the body. The deformation is called unidimensional if the quantities γ_{ij} with only one pair of subscripts are not equal to zero (say, $\gamma_{12} = \gamma_{21}$), two-dimensional if the components with two different pairs of subscripts are not equal to zero (say, $\gamma_{12} = \gamma_{21}$ and $\gamma_{13} = \gamma_{31}$), and three-dimensional if all the non-diagonal components of the tensor $\{\gamma\}$ are different from zero. If $\gamma_{11} \neq 0$, then such a strain is termed uniaxial, although in this case one or both other diagonal components of the tensor $\{\gamma\}$ become automatically different from zero.

Some of the results of tensor theory, which have been considered above for the stress tensor, are also applicable to the finite-strain tensor. This refers, in particular, to the concept of the principal values of the large-strain tensor and to the corresponding three mutually perpendicular directions in a three-dimensional space. This also refers to the formulas defining the transformations of components upon rotation of the coordinate axes for the two-dimensional case (plane stress): the respective formulas remain quite valid upon replacement of σ_{ij} with γ_{ij} for the finite-strain tensor as well. Finally, in the same way as was done in formulas (1.7) through (1.9), we can construct the invariants of the finite-strain tensor, which will be denoted here by E_1, E_2, and E_3.

These quantities are defined in the following way:

$$E_1 = \gamma_1 + \gamma_2 + \gamma_3 = \gamma_{11} + \gamma_{22} + \gamma_{33} = \varepsilon_{11} + \varepsilon_{22} + \varepsilon_{33}$$

$$E_2 = \gamma_1\gamma_2 + \gamma_1\gamma_3 + \gamma_2\gamma_3 = \gamma_{11}\gamma_{22} + \gamma_{11}\gamma_{33} + \gamma_{22}\gamma_{33} - (\gamma_{12}^2 + \gamma_{13}^2 + \gamma_{23}^2) =$$

$$= \varepsilon_{11}\varepsilon_{22} + \varepsilon_{11}\varepsilon_{33} + \varepsilon_{22}\varepsilon_{33} - \frac{1}{4}(\varepsilon_{12}^2 + \varepsilon_{13}^2 + \varepsilon_{23}^2)$$

$$E_3 = \gamma_1\gamma_2\gamma_3 = \gamma_{11}\gamma_{22}\gamma_{33} + 2\gamma_{12}\gamma_{13}\gamma_{23} - (\gamma_{11}\gamma_{23}^2 + \gamma_{22}\gamma_{13}^2 + \gamma_{33}\gamma_{12}^2) =$$

$$= \varepsilon_{11}\varepsilon_{22}\varepsilon_{33} + \frac{1}{4}(\varepsilon_{12}\varepsilon_{13}\varepsilon_{23} - \varepsilon_{11}\varepsilon_{23}^2 - \varepsilon_{22}\varepsilon_{13}^2 - \varepsilon_{33}\varepsilon_{12}^2)$$

Since the quantities γ_{ij} are the characteristics of the state of strain, we shall use these quantities to express changes in the length of linear elements between the infinitely closely spaced points of the body. The relative change of the length of the segment AB discussed above (see Fig. 1.6) is defined as follows:

$$\varepsilon = \frac{ds^* - ds}{ds}$$

The extension ε is expressed in terms of the components of the large-strain tensor by the following equation:

$$\varepsilon\left(1 + \frac{\varepsilon}{2}\right) = \frac{(ds^*)^2 - (ds)^2}{2(ds)^2} =$$

$$= \frac{1}{(ds)^2}[\gamma_{11}(dx_1)^2 + \gamma_{22}(dx_2)^2 + \gamma_{33}(dx_3)^2 +$$

$$+ 2\gamma_{12}\,dx_1\,dx_2 + 2\gamma_{13}\,dx_1\,dx_3 + 2\gamma_{23}\,dx_2\,dx_3]$$

The quantities (dx_i/ds) represent direction cosines, whose values characterize the orientation of the linear element AB relative to the coordinate axes, before the element has been deformed. We shall use the following notation for these direction cosines:

$$\beta_1 = \frac{dx_1}{ds}; \quad \beta_2 = \frac{dx_2}{ds}; \quad \beta_3 = \frac{dx_3}{ds}$$

The formula for ε will thus become

$$\varepsilon\left(1 + \frac{\varepsilon}{2}\right) = \gamma_{11}\beta_1^2 + \gamma_{22}\beta_2^2 + \gamma_{33}\beta_3^2 + 2(\gamma_{12}\beta_1\beta_2 + \gamma_{13}\beta_1\beta_3 + \gamma_{23}\beta_2\beta_3)$$

Thus, the relative change of the length of any segment ds at a certain point is determined by the components of the strain tensor $\{\gamma\}$ and depends on the orientation of that side in space. Obviously, the quantities γ_{ij} do really determine the value of ε (but are not equal to it).

In a special case, when the linear element was directed parallel to some coordinate axis before deformation, two of the three direction cosines are equal to zero and the third to unity. Then the following expression obtains for the extension of such an element:

$$\varepsilon_i\left(1 + \frac{\varepsilon_i}{2}\right) = \gamma_{ii} \text{ or } \varepsilon_i = \sqrt{1 + 2\gamma_{ii}} - 1 \quad (i = 1, 2, 3) \tag{1.16}$$

We orient the coordinate axes parallel to the directions of the principal strains, i.e., along the principal elongations, which in this particular case will also be expressed by means of formula (1.15) since this formula holds for any system of coordinate axes. Using the general rules of finding the principal values of tensors, it can be shown that the principal extensions are expressed in terms of the components γ_{ij} in just the same way as the invariants of the finite-strain tensor are expressed in terms of γ_{ij}. From this it

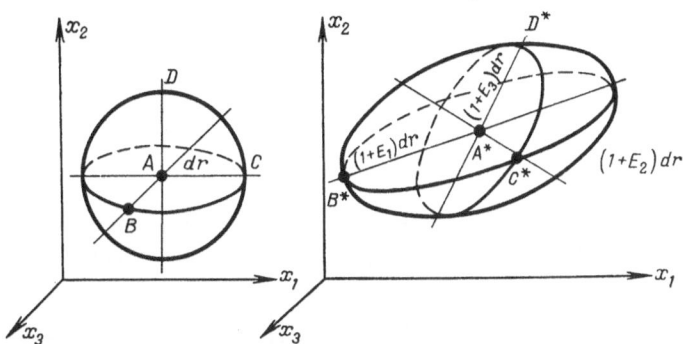

Fig. 1.7. Illustrating the concept of the principal strains.

follows that by their physical significance the quantities E_1, E_2, and E_3 represent the principal extensions at a given point in the body. Since the tensor invariants are independent of the direction of the coordinate axes, the principal extensions too are the strain characteristics which do not depend on the choice of coordinate system; this is in agreement with the physical conception of deformation since it is evident that the strain concept itself and the associated geometrical changes occurring at a point are not tied up with the choice of coordinate system.

Determination of the principal extensions allows one to obtain a very spectacular picture of deformation in the vicinity of a given point.

Let us choose in a solid continuum an infinitely small spherical element with a centre at a point A and radius dr (Fig. 1.7) defined in the coordinate system (x_1, x_2, x_3). Because of the displacements the following changes have taken place in the body: the point A has moved to a new position A^*, the directions of the radii AB, AC, and AD have changed to the directions A^*B^*, A^*C^*, and A^*D^*, respectively, and the sphere itself has turned into an ellipsoid with semi-axes whose lengths are $(1 + E_1)\, dr$, $(1 + E_2)\, dr$, and $(1 + E_3)\, dr$. The principal extensions characterize the change of the shape of the volume element. Besides, they determine the relative change of its volume γ_v, which is expressed as follows:

$$\gamma_v = \frac{\text{Volume of ellipsoid} - \text{Volume of sphere}^1}{\text{Volume of sphere}} = (1 + E_1)\,(1 + E_2)\,(1 + E_3) - 1$$

$$(1.17)$$

Thus, the relative change of the volume depends on the invariants of the strain tensor and, hence, is not associated with the choice of the coordinate axes, which is consistent with the physical meaning involved since, of course, the change of the volume must serve as a strain characteristic invariant with respect to the choice of coordinate axes.

1.3.2. Infinitesimal Strain

An important simplification of the above-considered general case of large strains is such a state of strain in which the derivatives of the displacements u_i are small and, for this reason, the quadratic terms in formula (1.13) and in the subsequent relations may be neglected as compared with the linear terms. Then the infinitesimal-strain (or small-strain) tensor $\{\gamma\}^0$ will be expressed in terms of the displacement derivatives in the following manner:

$$\{\gamma\}^0 = \begin{vmatrix} \dfrac{\partial u_1}{\partial x_1} & \dfrac{1}{2}\left(\dfrac{\partial u_1}{\partial x_2}+\dfrac{\partial u_2}{\partial x_1}\right) & \dfrac{1}{2}\left(\dfrac{\partial u_1}{\partial x_3}+\dfrac{\partial u_3}{\partial x_1}\right) \\[2ex] \dfrac{1}{2}\left(\dfrac{\partial u_2}{\partial x_1}+\dfrac{\partial u_1}{\partial x_2}\right) & \dfrac{\partial u_2}{\partial x_2} & \dfrac{1}{2}\left(\dfrac{\partial u_2}{\partial x_3}+\dfrac{\partial u_3}{\partial x_2}\right) \\[2ex] \dfrac{1}{2}\left(\dfrac{\partial u_3}{\partial x_1}+\dfrac{\partial u_1}{\partial x_3}\right) & \dfrac{1}{2}\left(\dfrac{\partial u_3}{\partial x_2}+\dfrac{\partial u_2}{\partial x_3}\right) & \dfrac{\partial u_3}{\partial x_3} \end{vmatrix}$$

The quantities contained in this matrix are the components γ_{ij}^0 of the tensor $\{\gamma\}^0$. The invariants of the tensor $\{\gamma\}^0$ are written in just the same way as

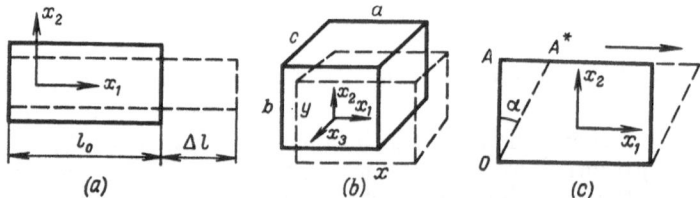

Fig. 1.8. Schematic representation of three principal types of deformation: a—uniaxial extension; b—hydrostatic compression; c—simple shear.

the invariants of the tensor $\{\gamma\}$ given above, the components γ_{ij} being accordingly replaced by the components γ_{ij}^0 (or ε_{ij} being replaced by ε_{ij}^0). We shall denote them as E_1^0, E_2^0, and E_3^0.

It is obvious that $\{\gamma\}^0$ is a simplified form for the case of small displacements (or small strains), i.e., when du_i/dx_j is much smaller than unity. This is what exactly occurs in most important cases of the deformation of solid materials such as metals and glasses. For polymers, however, the most essential specific feature is their ability to withstand large deformations. For this reason, in polymer rheology, with the exception of the limiting cases, it will not suffice to make use of the tensor $\{\gamma\}^0$ alone and one has to resort to the concept of finite strains and the corresponding tensor $\{\gamma\}$ in order to describe the state of strain.

We shall now take up a more detailed treatment of the simplest cases of deformation, which are of major interest for the determination of the rheological properties of polymeric systems—longitudinal elongation and volumetric and shear strain. For this purpose, reference is made to Fig. 1.8.

1.3.3. Longitudinal Elongation. Poisson's Ratio

Suppose that when being stretched a symmetrical body (say, with a square or circular cross-section) is elongated $\varkappa$ times (i.e., its extension ratio is equal to $\varkappa$). We assume that the deformation occurs uniformly along the entire length of the material. Then the extension (the elongation per unit length) (Fig. 1.8a) may be represented in the following way:

$$\varepsilon = \frac{\Delta l}{l_0} = \varkappa - 1 \quad \left(\varepsilon = \frac{\partial u_1}{\partial x_1} \right) \tag{1.18}$$

where l_0 is the original length; Δl is the increment in length, which has resulted from the stretching of the body.

The extension ε and the extension ratio, which is often called the draw ratio, $\varkappa$ are the natural characteristics of the strained state of the body, which may be employed to express the components of the tensor $\{\gamma\}$. According to formulas (1.14) and (1.15) the component γ_{11} is expressed in terms of $\varkappa$ or ε as follows:

$$\gamma_{11} = \varepsilon + \frac{1}{2} \varepsilon^2 = (\varkappa - 1) + \frac{1}{2} (\varkappa - 1)^2 = \frac{1}{2} (\varkappa^2 - 1) \tag{1.19}$$

since, by virtue of the uniformity of the deformation, the displacements u_2 and u_3 are independent of the coordinate x_1.

But the description of the strained state of the body is not limited to the introduction of the component γ_{11} since, apart from the stretching in the axial direction, it also undergoes compression in the lateral directions. The relation between the changes in the lateral and axial dimensions of the body cannot be established on the basis of a purely geometric picture of deformation since the lateral compression on uniaxial elongation is governed by the properties of the material. To characterize these properties, use is made of Poisson's ratio μ, which is defined as the ratio of the lateral contraction $\varepsilon^\perp$ to the longitudinal extension ε, i.e.,

$$\mu = \left| \frac{\varepsilon^\perp}{\varepsilon} \right|$$

In the case of small strains this definition coincides with the definition of μ^0 as the ratio of the components of the small-strain tensor:

$$\mu^0 = \left| \frac{\varepsilon^\perp}{\varepsilon} \right| \approx \left| \frac{\gamma_{22}^0}{\gamma_{11}^0} \right|$$

But with large strains $\varepsilon^\perp/\varepsilon \neq \gamma_{22}/\gamma_{11}$, and therefore the question of the definition of Poisson's ratio becomes ambiguous; this quantity may be defined, as before, as $| \varepsilon^\perp/\varepsilon |$ or Poisson's ratio may be taken as the ratio $| \gamma_{22}/\gamma_{11} |$, which we shall designate as μ^γ in order to differentiate it from μ.

We shall now consider the difference between the definitions of μ and μ^γ using, as an example, the important case of stretching without change of volume. Then the lateral strain is deduced from the volume constancy condition:

$$\varepsilon^\perp = 1 - \sqrt{\frac{1}{1+\varepsilon}} = 1 - (1+\varepsilon)^{-1/2} \approx \frac{1}{2} \varepsilon - \frac{3}{8} \varepsilon^2 + \ldots$$

Since

$$\frac{\partial u_2}{\partial x_2} = \frac{\partial u_3}{\partial x_3} = \sqrt{\frac{1}{\varkappa}} - 1$$

the finite-strain-tensor component γ_{22} and the component γ_{33}, which is equal to γ_{22}, are expressed with the aid of formulas (1.14) and (1.15) in the following way:

$$\gamma_{22} = \gamma_{33} = \frac{1}{2} \left(\frac{1}{\varkappa} - 1 \right) = -\frac{\varepsilon}{2(1+\varepsilon)}$$

From this we find the expressions for μ and μ^v:

$$\mu = \frac{1-(1+\varepsilon)^{-1/2}}{\varepsilon} \approx \frac{1}{2} - \frac{3}{8}\varepsilon + \ldots$$

and

$$\mu^v = \frac{1}{2(1+\varepsilon)\left(1+\dfrac{\varepsilon}{2}\right)} \approx \frac{1}{2} - \frac{3}{4}\varepsilon + \ldots$$

At small strains ($\varepsilon \ll 1$) both definitions of Poisson's ratio give the same result for the case of stretching without change in volume: $\mu = \mu^v = \mu^0 = = 0.5$. But, in accordance with what has been said above, at finite values of ε the quantities μ and μ^v have different values and must vary as functions of ε or $\varkappa$ in order to maintain the volume constant. This relationship between the extension ratio and Poisson's ratios, which are defined as $|\varepsilon^{\perp}/\varepsilon|$ or $|\gamma_{22}/\gamma_{11}|$, provided that the volume constancy condition is fulfilled, makes the relation between μ and μ^v and the volume change less clear. Below we shall describe a different method of determining Poisson's ratio, in which the requirement of volume constancy upon deformation will correspond to the condition of constancy of a certain parameter associated with the dimensional change in the body on stretching.

1.3.4. Measures of Strain

The parameters ε, $\varkappa$, γ^0_{ij} or γ_{ij} are various measures of strain. But they all do not satisfy the important requirement of additivity of two successive strains. Indeed, let a body of original length l_0 experience two successive elongations Δl_1 and Δl_2. We shall examine two cases: the body is stretched in a jumpwise manner or continuously. The results in both cases must be equivalent. We calculate the elongation per unit length:
(1) in the case of continuous straining:

$$\varepsilon_{1+2} = \frac{\Delta l_1 + \Delta l_2}{l_0} = \frac{(\Delta l_1)\, l_0 + (\Delta l_2)\, l_0 + (\Delta l_1)^2 + (\Delta l_1)\,(\Delta l_2)}{l_0\,(l_0 + \Delta l_1)}$$

(2) in the case of jumpwise straining

$$\varepsilon_1 + \varepsilon_2 = \frac{\Delta l_1}{l_0} + \frac{\Delta l_2}{l_0 + \Delta l_1} = \frac{(\Delta l_1)\, l_0 + (\Delta l_2)\, l_0 + (\Delta l_1)^2}{l_0\,(l_0 + \Delta l_1)}$$

A comparison of the resulting expressions shows that $\varepsilon_{1+2} \neq \varepsilon_1 + \varepsilon_2$, i.e. the total strain accomplished continuously is not equal to the sum of two successive strains, which is pointless. Note that the difference between ε_{1+2} and $\varepsilon_1 + \varepsilon_2$ disappears automatically if the elongations Δl_1 and Δl_2 are small as compared with l_0; then

$$\varepsilon_{1+2} = \varepsilon_1 + \varepsilon_2 = \frac{\Delta l_1 + \Delta l_2}{l_0}$$

An analogous conclusion can be made with respect to the tensor components γ_{ij} as well, which do not satisfy the requirement of the additivity of successive strains either.

In the light of what has been said above, there arises the problem of choosing a measure of strain that would be additive with respect to successive tensile strains. Such a measure can be obtained if one proceeds from the following reasoning. The value of extension ε was defined above as $\Delta l/l_0$, but it is not immaterial what the value of Δl is. If Δl is small, then it will be natural to refer it to l_0. But if Δl is not small, then at large extension ratios it will be more logical to refer an infinitesimal change of the length dl to the current value of the length l. This means that an infinitesimal extension $d\varepsilon^H$ determined in this way is given by

$$d\varepsilon^H = \frac{dl}{l}$$

Then, proceeding from the natural initial condition of the absence of deformation at $l = l_0$, we obtain the following expression for ε^H:

$$\varepsilon^H = \int_{l_0}^{l} \frac{dl}{l} = \ln \frac{l}{l_0} = \ln \varkappa = \ln(1+\varepsilon) \qquad (1.20)$$

The quantity ε^H as a measure of strain was first introduced by K. Roentgen; in the modern rheological literature it is called Hencky's measure of extension. If we now turn to the consideration of two successive tensile tests, then we shall have in the case of jumpwise straining

$$\varepsilon_1^H + \varepsilon_2^H = \ln \frac{l_0+\Delta l_1}{l_0} + \ln \frac{l_0+\Delta l_1+\Delta l_2}{l_0+\Delta l_1} = \ln \frac{l_0+\Delta l_1+\Delta l_2}{l_0}$$

and in the case of continuous extension

$$\varepsilon_{1+2}^H = \ln \frac{l_0+\Delta l_1+\Delta l_2}{l_0}$$

From this it follows that $\varepsilon_{1+2}^H = \varepsilon_1^H + \varepsilon_2^H$, i.e., the result is independent of the sequence of the tensile tests, and the Hencky measure of strain exhibits the property of additivity with respect to successive strains. Therefore, this definition is most conveniently employed in considering large longitudinal strains. This question will be discussed in greater detail in the analysis of the characteristics of the rheological properties of polymers on stretching (Chapter 7).

We now return to the concept of Poisson's ratio which we shall define as the ratio of lateral to axial strains. According to Hencky

$$\mu^H = \left| \frac{\varepsilon^{H,\perp}}{\varepsilon^H} \right| = \left| \frac{\ln(1+\varepsilon^\perp)}{\ln(1+\varepsilon)} \right|$$

When an incompressible body is extended, in which case no change in volume occurs,

$$\varepsilon^{H,\perp} = \ln(1+\varepsilon^\perp) = \ln\left[1+\left(\sqrt{\frac{1}{\varkappa}}-1\right)\right] = \ln \sqrt{\frac{1}{\varkappa}} = -\frac{1}{2}\ln \varkappa$$

Therefore, for an incompressible body $\mu^H = 0.5$ at any extension ratio. Thus, the volume remains constant on stretching when Poisson's ratio defined in terms of the Hencky extensions is equal to 0.5.

We have discussed above some measures of large strains. This discussion shows that, generally speaking, no single measure of strain exists and the strained state of the body can be described by various methods. Thus, in the case of stretching we may use ε, $\varkappa$ or any function of these as a characteristic of the strained state.

Reiner points out the following requirements which must be satisfied by any measure of strain: it must be dimensionless, which is associated with the requirement of the strain being independent of the dimensions of the body; at small strains any measure of strain reduces to the tensor $\{\gamma\}^0$ (to the quantity ε in a special case of longitudinal elongation). The last requirement is also satisfied by the Hencky strain measure, since at small ε the approximate equality $\ln(1 + \varepsilon) \approx \varepsilon$ is valid. To this there corresponds the identity of μ^H and μ^0 at small ε values. Later we shall also make use of some other measures of strain.

1.3.5. Volumetric Strains

Let us consider the case of volumetric strain (see Fig. 1.8b). For this purpose, we cut out a parallelepiped at a given point of a body, which is oriented along the principal axes. Let the sizes of its edges be a, b, c before deformation and x, y, z after deformation. To this there correspond the following values of the principal extension ratios:

$$\varkappa_1 = \frac{x}{a}; \quad \varkappa_2 = \frac{y}{b}; \quad \varkappa_3 = \frac{z}{c}$$

If the volume of the body does not change upon deformation, then $abc = xyz$, or

$$\varkappa_1 \varkappa_2 \varkappa_3 = 1 \tag{1.21}$$

We make use of the Hencky strain measure. Here the relative change of the volume V with an infinitesimal increment by dV can be expressed as dV/V.

Hence, the Hencky volumetric strain will be defined thus:

$$\gamma_V^H = \ln \frac{V}{V_0}$$

where V_0 is the volume of the unstrained body.

The principal Hencky strains are defined in the following way:

$$\gamma_1^H = \ln \varkappa_1 = \ln \frac{x}{a}; \quad \gamma_2^H = \ln \varkappa_2 = \ln \frac{y}{b}; \quad \gamma_3^H = \ln \varkappa_3 = \ln \frac{z}{c}$$

Therefore

$$\gamma_V^H = \ln \frac{xyz}{abc} = \ln \frac{x}{a} + \ln \frac{y}{b} + \ln \frac{z}{c} = \gamma_1^H + \gamma_2^H + \gamma_3^H$$

that is, the relative volume change is expressed as the first invariant E_1^H of the finite strain tensor defined according to Hencky.

It is evident that straining takes place without volume change if $E_1^H = 0$. This conclusion holds for any direction of the axes since the magnitude of the invariant does not depend on the selection of the coordinate axes.

The expression for the volume change, γ_V, has been obtained above [see formula (1.17)] in terms of the invariants of the finite strain tensor. If the strains are small, this expression can be simplified since the quadratic and cubic terms will in this case be substantially smaller than unity. Therefore

$$\gamma_V \approx E_1^0 = \gamma_1^0 + \gamma_2^0 + \gamma_3^0 = \varepsilon_1^0 + \varepsilon_2^0 + \varepsilon_3^0 \tag{1.22}$$

1.3.6. The Strain Deviator

Just as has been done for the stress tensor, we resolve the strain tensor into a spherical component and a deviator. To do this, we perform the following transformations:

$$\gamma_{ij}^{H} = \begin{vmatrix} \gamma_{11}^{H} & \gamma_{12}^{H} & \gamma_{13}^{H} \\ \gamma_{21}^{H} & \gamma_{22}^{H} & \gamma_{23}^{H} \\ \gamma_{31}^{H} & \gamma_{32}^{H} & \gamma_{33}^{H} \end{vmatrix} = \frac{\gamma_V^{H}}{3}\,\delta_{ij} + \gamma_{ij}^{H'}$$

where

$$\gamma_{ij}^{H'} = \begin{cases} \gamma_{ij}^{H} - \dfrac{\gamma_V^{H}}{3} & \text{at } i=j \\[2mm] \gamma_{ij}^{H} & \text{at } i \neq j \end{cases}$$

The quantity $\gamma_V^{H} = \gamma_1^{H} + \gamma_2^{H} + \gamma_3^{H}$ defines the relative volume change upon deformation. In the second term, the sum of the diagonal components, i.e., the first invariant, is equal to zero. Therefore the deformation described by the second term is pure shear which occurs without volume change, while the first term characterizes the volumetric component of the total deformation.

We shall now apply the above-described general procedure of resolving the tensor into the spherical component and the deviator for uniaxial elongation. This gives the following result:

$$\gamma_{ij}^{H} = \begin{vmatrix} \gamma_{11}^{H} & 0 & 0 \\ 0 & \gamma_{22}^{H} & 0 \\ 0 & 0 & \gamma_{33}^{H} \end{vmatrix} = \begin{vmatrix} \gamma_{11}^{H} & 0 & 0 \\ 0 & -\mu^{H}\gamma_{11}^{H} & 0 \\ 0 & 0 & -\mu^{H}\gamma_{11}^{H} \end{vmatrix} =$$

$$= \frac{1-2\mu^{H}}{3}\,\gamma_{11}^{H}\delta_{ij} + \frac{1+\mu^{H}}{3}\,\gamma_{11}^{H} \begin{vmatrix} 2 & 0 & 0 \\ 0 & -1 & 0 \\ 0 & 0 & -1 \end{vmatrix}$$

It follows from this that in a general case the longitudinal elongation results in the change of both the volume and shape of the body. If, however, $\mu^{H} = = 0.5$, then no volume changes occur and the total strain on elongational deformation is determined by the strain-tensor deviator alone and, hence, is only a set of shear deformations taking place in different directions.

1.3.7. Simple Shear

This type of deformation is schematically shown in Fig. 1.8c. In the direction of shear marked by an arrow there occurs a displacement u_1. Its gradient du_1/dx_2 is determined by the slope α which we denote by γ, i.e.,

$$\gamma = \tan \alpha = \frac{\partial u_1}{\partial x_2}$$

Since the lengths of linear elements which are oriented before deformation in the x_2 direction are changed on shear, there exists one more displacement component, u_2, which is different from zero. This is seen from the change of

3*

the length of the segment OA which becomes equal to OA^* after deformation, so that

$$\frac{OA^* - OA}{OA} = \sec \alpha - 1 = \sqrt{1 + \gamma^2} - 1$$

The resulting value of du_1/dx_2 for simple shear can be used to calculate the large-strain tensor $\{\gamma\}$. According to the above-given definition of the tensor $\{\gamma\}$ [see formulas (1.14) and (1.15)], for the case under consideration the following tensor components prove to be different from zero:

$$\gamma_{12} = \gamma_{21} = \frac{1}{2} \frac{\partial u_1}{\partial x_2} = \frac{1}{2} \gamma$$

$$\gamma_{22} = \varepsilon_{22} = \frac{1}{2} \left(\frac{\partial u_1}{\partial x_2} \right)^2 = \frac{1}{2} \gamma^2$$

Accordingly, the tensor $\{\gamma\}$ for simple shear is written thus:

$$\{\gamma\} = \frac{1}{2} \begin{vmatrix} 0 & \gamma & 0 \\ \gamma & \gamma^2 & 0 \\ 0 & 0 & 0 \end{vmatrix}$$

This tensor is graphically illustrated in Fig. 1.9 in which the components of the tensor $\{\gamma\}$ are marked by arrows (the $\frac{1}{2}$ factor is omitted in the shear deformations in constructing Fig. 1.9).

The appearance of a diagonal component in the strain tensor in simple shear is known as the Poynting effect which has been observed in the twisting of thin wires and which involves a change in the wire length. This effect is

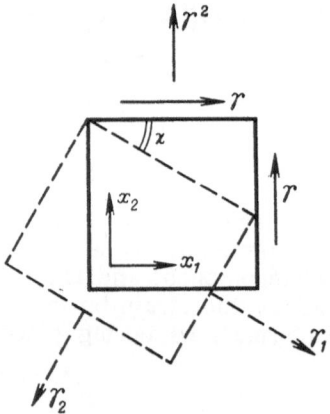

Fig. 1.9. Graphical illustration of the strain tensor concept for simple shear.

associated with the existence of non-zero components γ_{22} which are quadratic with respect to the magnitude of shear γ. This component of the tensor $\{\gamma\}$ is responsible for the observed change of the length in the direction of the axis of elastic bodies being twisted (twisting is an example of simple shear).

From Fig. 1.9 it is seen that to the directions of the x_1 and x_2 axes there correspond both tangential and diagonal components of the strain tensor. We shall therefore turn our attention to the principal axes, which we can do with the aid of the formulas derived for the state of plane stress. In order to calculate the principal values of the tensor $\{\gamma\}$ and find the directions of their orientation, we should utilize the previously obtained results for stresses [formula (1.3)], replacing the components σ_{ij} by γ_{ij}. This replacement enables one to calculate, by means of formula (1.3), the angle χ between the shear direction x_1 and the orientation of the principal axis, to which there corresponds the strain γ_1 (see Fig. 1.9). Substitution of the respective quantities into formula (1.3) yields the following expression:

$$\chi = \frac{1}{2} \arctan \left(\frac{2}{\gamma}\right) = \frac{1}{2} \operatorname{arccot} \left(\frac{\gamma}{2}\right)$$

Now, projecting all the components of the tensor $\{\gamma\}$ onto the directions χ and $(\chi + \pi/2)$ and setting up equations analogous to the equilibrium equations, just as was done in the case of the components of the stress tensor, we obtain the following formulas for the principal values of strain:

$$\gamma_1 = \frac{1}{2}(\cot^2 \chi - 1); \quad \gamma_2 = \frac{1}{2}(\tan^2 \chi - 1); \quad \gamma_3 = 0 \tag{1.23}$$

It is interesting to compare the resulting formulas (1.23) with expression (1.18) and analogous formulas relating the principal strains γ_i on extension to the corresponding extension ratios; these formulas may be written as follows:

$$\gamma_1 = \frac{1}{2}(\varkappa_1^2 - 1); \quad \gamma_2 = \frac{1}{2}(\varkappa_2^2 - 1); \quad \gamma_3 = \frac{1}{2}(\varkappa_3^2 - 1) \tag{1.24}$$

where $\varkappa_1$, $\varkappa_2$, and $\varkappa_3$ are the principal elongations (or the extension ratios in the direction of the principal axes).

A comparison of the corresponding pairs in expressions (1.23) and (1.24) allows one to find the values of the principal elongations:

$$\varkappa_1 = \cot \chi; \quad \varkappa_2 = \tan \chi; \quad \varkappa_3 = 1 \tag{1.25}$$

A special result obtained from the expressions derived for $\varkappa_1$, $\varkappa_2$, and $\varkappa_3$ is that in simple shear, as seen from formula (1.21), no volume change occurs because $\varkappa_1 \varkappa_2 \varkappa_3 = 1$.

1.3.8. Cauchy-Green and Finger Strain Tensors

Consideration of simple shear allows us to introduce two new tensors characterizing strain in the neighbourhood of a given point. They are frequently used in the literature in application to shear conditions and also for writing rheological equations of state (see below). Let us return to the analysis of formula (1.13) which is written now in the form

$$(ds^*)^2 - (ds)^2 = 2 \sum_{i=1}^{3} \sum_{j=1}^{3} \gamma_{ij} \, dx_i \, dx_j$$

in which case the change of the square of the length of a linear element may be represented in the following form:

$$(ds^*)^2 - (ds)^2 = \sum_{i=1}^{3} \sum_{j=1}^{3} (\gamma_{ij}^{G} - \delta_{ij}) \, dx_i \, dx_j \tag{1.26}$$

Here the quantities γ_{ij}^{G} are the components of the new tensor $\{\gamma\}^{G}$ called the Cauchy-Green strain tensor (its components will be designated by the superscript G). The quantities γ_{ij}^{G} are expressed in terms of the components of the finite-strain tensor as follows:

$$\gamma_{ij}^{G} = 2\gamma_{ij} + \delta_{ij}$$

That is, the shear (non-diagonal) components of the Cauchy-Green tensor are equal to the corresponding doubled components of the finite-strain tensor, and to the diagonal components there is added unity.

The difference between the previously considered large-strain tensor $\{\gamma\}$ and the Cauchy-Green strain tensor, except for the insignificant difference in the numerical coefficient, is the same as that between the relative dimensional change (i.e., the amount of dimensional change referred to the original size) and the degree of this change (i.e., the new size referred to the original one). In the absence of deformation all γ_{ij} are equal to zero and $\gamma_{ij}^{G} = \delta_{ij}$, i.e., are equal to unit tensor.

As regards simple shear, the components of the Cauchy-Green strain tensor for it will be written thus:

$$\{\gamma\}^{G} = \begin{vmatrix} 1 & \gamma & 0 \\ \gamma & 1 + \gamma^2 & 0 \\ 0 & 0 & 1 \end{vmatrix}$$

Of interest is also the "inverted Cauchy-Green tensor" or the Finger strain tensor $\{\gamma\}^{F}$, whose components will be denoted further in the text as γ_{ij}^{F}. They are defined from the following identity:

$$\sum_{i=1}^{3} \sum_{j=1}^{3} \gamma_{ij}^{G} \gamma_{ij}^{F} = \sum_{i=1}^{3} \sum_{j=1}^{3} \delta_{ij} = 3 \tag{1.27}$$

Without delving into the details of calculations, we write directly the components of the tensor $\{\gamma\}^{F}$ for simple shear:

$$\{\gamma\}^{F} = \begin{vmatrix} 1 + \gamma^2 & -\gamma & 0 \\ -\gamma & 1 & 0 \\ 0 & 0 & 1/\gamma \end{vmatrix}$$

Direct testing by substitution of the components γ_{ij}^{G} and γ_{ij}^{F} into formula (1.27) convinces us in the fulfillment of the written equality.

To complete the brief consideration of the Cauchy-Green and Finger strain tensors, we shall also give the principal components of these tensors expressed in terms of the principal elongations. For this purpose, we make use, for example, of formulas (1.24), from which, with account taken of the relation between γ_{ij}^{G} and γ_{ij}^{F}, it immediately follows that

$$\gamma_{1}^{G} = \varkappa_{1}^{2} \quad \gamma_{2}^{G} = \varkappa_{2}^{2} \quad \gamma_{3}^{G} = \varkappa_{3}^{2} \tag{1.28}$$

The principal components of the Finger strain tensor are defined as follows:

$$\gamma_{1}^{F} = \frac{1}{\varkappa_{1}^{2}} \quad \gamma_{2}^{F} = \frac{1}{\varkappa_{2}^{2}} \quad \gamma_{3}^{F} = \frac{1}{\varkappa_{3}^{2}} \tag{1.29}$$

Direct checking by substitution of these expressions into formula (1.27) confirms that it is really fulfilled, i.e., the Cauchy-Green and Finger strain tensors are "inverse" with respect to each other

In modern rheology theories, no use is practically made of any measures of large strains other than the tensors $\{\gamma\}^G$ and $\{\gamma\}^F$. But it should be pointed out that, in principle, other characteristics of the strained state may also be examined, particularly the various combinations of the tensors $\{\gamma\}^G$ and $\{\gamma\}^F$, and their functions. The treatment of the quantity $\{\gamma\}$ as the initial characteristic of the deformed state seems to be most pictorial since the components of the large-strain tensor directly express the change of the distance between the points upon their displacement in the body, i.e., the strain effect in the neighbourhood of a given point.

1.3.9. Simple Shear as a Deformation Accompanied by Rotation

In the case of simple shear considered above not only the lengths of linear elements are changed (which is really a deformation) but there also occurs the rotation of the principal axes, which is the rotation of an element of the body as a whole in the neighbourhood of a given point. This does not affect the above-given calculations and the treatment of the geometry of simple

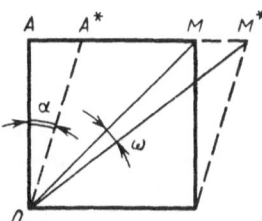

Fig. 1.10. Illustrating the effect of the rotation of a body in simple shear.

shear but it enables one to additionally elucidate certain specific features of the behaviour of the material under shear. For simplicity we shall confine ourselves to a discussion of small displacements, in which case use can be made of the tensor $\{\gamma\}^0$ in order to characterize the state of strain.

Simple shear is schematically shown in Fig. 1.10; the deformation consists of the displacement of the point A to the new position A^*, the magnitude of the angle α being assumed to be small. As seen from Fig. 1.10, in simple shear the diagonal OM is rotated and it occupies a new position OM^* after deformation. From this it follows that simple shear is accompanied by the rotation of the element of the body. The magnitude of rotation ω is determined by the angle α. It is not difficult to show that $\omega = \alpha/2$ if α is small.

Let us now determine the result of the superposition of two small shear deformations, one of which is defined as $\partial u_1/\partial x_2 = \alpha$ and the other as $\partial u_2/\partial x_1 = -\alpha$. The geometric picture of the displacements involved is depicted in Fig. 1.11, from which it is seen that the superposition of these two shear strains eventually leads only to the rotation of the element of the body as a whole. The magnitude of rotation ω is given by

$$\omega = \frac{1}{2}\left(\frac{\partial u_1}{\partial x_2} - \frac{\partial u_2}{\partial x_1}\right) \tag{1.30}$$

Therefore, in order to find the deformation proper for the case shown in Fig. 1.10. it is necessary to subtract the term due to the rotation from the displacement gradient $\partial u_1/\partial x_2$. In this connection the displacement gradient may be represented in the form of the following sum:

$$\frac{\partial u_1}{\partial x_2} = \frac{1}{2}\left(\frac{\partial u_1}{\partial x_2} + \frac{\partial u_2}{\partial x_1}\right) + \frac{1}{2}\left(\frac{\partial u_1}{\partial x_2} - \frac{\partial u_2}{\partial x_1}\right)$$

We have already considered the small-strain tensor $\{\gamma\}^0$, whose component γ^0_{ij} is defined as $\gamma^0_{ij} = 1/2\,(\partial u_1/\partial x_2 + \partial u_2/\partial x_1)$. This quantity is responsible for the deformation proper in the neighbourhood of a given point, regardless of the possible rotation. The quantity $\omega = 1/2\,(\partial u_1/\partial x_2 - \partial u_2/\partial x_1)$ represents the angle of rotation of the volume element in the neighbourhood of the given point without deformation. Therefore, the displacement gradient

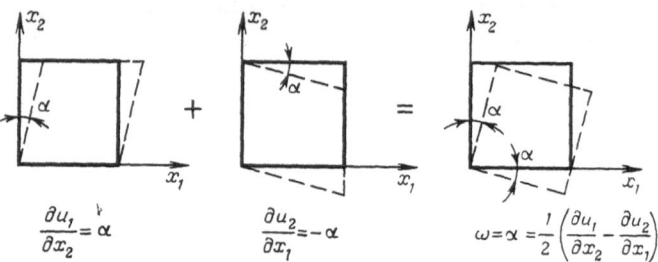

Fig. 1.11. Superposition of two modes of deformation in simple shear.

includes both the deformation and the rotation of the body. Hence, the presence of displacement gradients does not yet imply deformation since there is possible such a combination of the gradients when no deformation takes place and the body rotates as an entity. To this there corresponds the condition $du_1/\partial x_2 = -du_2/\partial x_1$.

1.3.10. Pure Shear

We may also visualize such shear conditions when no rotation occurs. In such a case $du_1/\partial x_2 = du_2/\partial x_1$. This type of deformation called *pure shear* is schematically shown in Fig. 1.12. As seen from this figure (and can be

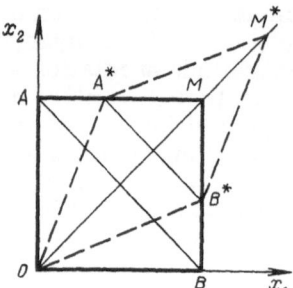

Fig. 1.12. Illustrating the concept of pure shear.

shown geometrically), in pure shear the diagonal of the square, AB, shifts parallel to itself and occupies a new position A^*B^*, and the diagonal OM does not change its position at all, being extended to OM^*. Therefore, in pure shear the elements of the body undergo no rotation.

From the viewpoint of continuum mechanics, the difference between simple and pure shear is not significant since the strain occurs in an infinitely small neighbourhood of the point and the rotation of the volume element as a whole has no effect on its deformation. But in reality, and this is important in particular in the case of polymeric systems, the elementary act of deformation involves not an infinitesimal volume but certain structural elements which may incorporate not too small a number of molecules or their constituent parts. The rotation of such volumes can result in a deformation inside them, which may be essential to the manifestation of the mechanical properties of polymer systems. Therefore, the correlations between the behaviour of polymers in simple shear and pure shear are nowadays under extensive investigation.

1.4. Kinematics of Deformation. Rate of Strain

In the preceding section we were concerned with the various states of strain, leaving untouched the problem of the rate of transition from one state to another. We shall now deal with the kinematics of deformation and its quantitative characteristics.

Figure 1.6 showed a point whose coordinates relative to a certain space-fixed coordinate system, or a coordinate basis, are determined by the quantities x_i ($i = 1, 2, 3$). This point is displaced in space and its coordinates x_i change at time t. Let this point have a coordinate system embedded in the body (convected coordinates), and let the coordinates of the point in this system be denoted as ξ_i ($i = 1, 2, 3$), i.e., the coordinates of this point in this coordinate system remain unchanged during the movement of the point. The coordinates x_i depend not only on time t but also on the choice of the point, i.e., on the values of ξ_i. The system of quantities x_i are the coordinates of the place occupied by the material point at the instant of observation t, and ξ_i are the Lagrangian (spatial) coordinates of the point. Thus, the coordinates x_i of the point are dependent on its three Lagrangian coordinates and time:

$$x_i = x_i (\xi_1, \xi_2, \xi_3, t)$$

We now consider a certain moment of time t_0 which is assumed as the initial time (zero time). At $t = t_0$ the coordinates of the place occupied by the point at the instant of observation were equal to X_i ($i = 1, 2, 3$), the magnitude of X_i being dependent on the choice of the point, i.e., on the coordinates ξ_i. The relation between x_i and X_i is expressed by the previously employed equation

$$x_i = X_i + u_i \tag{1.31}$$

where u_i are displacements which depend on the choice of the point and since we are now dealing with the process of deformation, u_i depends on the current time t as well.

The derivative of the coordinates X_i with respect to time t is equal to zero: $\partial X_i / \partial t = 0$ since X_i are the fixed values of the coordinates at a certain instant of time.

We shall now determine the velocity v_i of the point under consideration as the time derivative of its displacement: $v_i = du_i/dt$. This leads to an equivalent definition of the velocity v_i as the time derivative of the convected coordinates:

$$\frac{du_i}{dt} = \frac{d(x_i - X_i)}{dt} = \frac{dx_i}{dt} = v_i$$

We now establish the relation between the positions of the point at various instants of time. For this purpose, we consider the time derivative dX_i/dt, taking into account that X_i depends on ξ_i, and, by virtue of the existence of the dependence of x_i on ξ_i, the inverse dependence—the dependence of ξ_i on x_i and t—may also be considered. Then we obtain:

$$\frac{dX_i}{dt} = \frac{dX_i[x_1(t), x_2(t), x_3(t), t]}{dt} = \frac{\partial X_i}{\partial t} + \sum_{k=1}^{3} \frac{\partial X_i}{\partial x_k} \cdot \frac{\partial x_k}{\partial t} = 0$$

From this we get the following system of three equations:

$$\frac{\partial X_i}{\partial t} + \sum_{k=1}^{3} v_k \frac{\partial X_i}{\partial x_k} = 0 \quad (i = 1, 2, 3) \tag{1.32}$$

The resulting system of equations defines the interrelation of the original fixed and convected coordinates of the point (its kinematics) for various

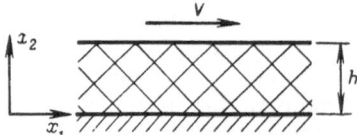

Fig. 1.13. Illustration of the kinematics of simple shear.

instants of time at certain specified initial conditions, i.e., at the known values of convected coordinates at time t_0.

To illustrate the foregoing, we examine the kinematics of simple shear for the case schematically shown in Fig. 1.13. Let the lower plane be fixed and the upper plane move with a velocity V; the distance between the planes is h. Then, $v_1 = Vx_2/h$; $v_2 = v_3 = 0$. It is apparent that the convected coordinates of the points, x_2 and x_3, do not change, i.e.,

$$X_2 = x_2, \quad X_3 = x_3$$

The system of equations (1.32) simplifies to give the following relation

$$\frac{\partial X_1}{\partial t} + v_1 \frac{\partial X_1}{\partial x_1} = 0$$

It is not difficult to see that the solution of Eq. (1.32) has the form

$$X_1 = x_1 - \frac{Vx_2(t - t_0)}{h} \tag{1.33}$$

which is the sought-for relation between X_1 and x_1. This solution satisfies the initial condition: the convected coordinates of the points at zero time $t = t_0$ are equal to X_i. Evidently, the term $Vx_2(t - t_0)/h$ is the displacement u_1.

The previously given definition of the strain tensor $\{\gamma\}$ (just as the definitions of other measures of strain considered above) related the change of the distance between the infinitely closely spaced points to some fixed position of these points in the body; here, as a result of the displacement of the body, the point at which the strain was considered changes its position and, hence, after deformation its position will be defined by new coordinates x_i. Furthermore, these coordinates x_i change continually in the process of deformation. Therefore, the strain tensor discussed above is defined by coordinates which are embedded at the given point and are moving and deforming solidally with it (the convected coordinates). But for a discussion of the kinematics of deforming bodies it is important to know how the deformation takes place at a given point in space, this point being characterized by some constant values of the convected coordinates.

We now turn to the determination of the components of the strain tensor and its time derivative, which is naturally called the rate of strain with respect to the space-fixed coordinate system. In essence, the task here is to convert the convected coordinate system characterized by the quantities ξ_i to the coordinate system x_i. We have already examined some special cases and procedures of transformation of tensor components from one coordinate system to another upon change of the direction of the axes. For the problem in order it is important to employ a general method of transforming tensor components from one coordinate system to another.

Let a certain strain tensor $\{\gamma\}$ be specified in the convected coordinate system by its components γ_{ij}. Then the relation between γ_{ij} and the components of this strain tensor in the coordinate system x_i is given by the following formula:

$$\gamma_{ij} = \sum_{\alpha=1}^{3} \sum_{\beta=1}^{3} \frac{\partial x_\alpha}{\partial \xi_i} \frac{\partial x_\beta}{\partial \xi_j} \Gamma_{\alpha\beta} \qquad (1.34)$$

where $\Gamma_{\alpha\beta}$ are the components of the strain tensor in the space-fixed coordinate system.

This formula is a special case of the general rule of transformation of the components of any strain tensor from some coordinates to others. Moreover, this relation may be looked upon as the definition of the tensor concept (more exactly, a tensor of rank two) because any physical object defined by its components $\Gamma_{\alpha\beta}$, which upon change of coordinate system is transformed so that in a new coordinate system its components can be calculated by formula (1.34), may be called a tensor. Therefore, if the components of a certain quantity are re-calculated from formula (1.34) upon change of coordinates, this quantity is a tensor by its nature.

Let us consider another formal rule which plays an important part in operations with tensors. We omit the summation signs in formula (1.34) and imply that when repeated subscripts are encountered in a formula [α and β are such subscripts in formula (1.34)], the summation should be carried out over them. For instance, the notation $a_n b_n$, in fact, represents the sum $a_1 b_1 + a_2 b_2 + a_3 b_3$. Formula (1.34) may then be rewritten in the following way:

$$\gamma_{ij} = \frac{\partial x_\alpha}{\partial \xi_i} \frac{\partial x_\beta}{\partial \xi_j} \Gamma_{\alpha\beta} \qquad (1.34a)$$

We shall now consider the rate of strain, i.e., a derivative of strain with respect to time in the convected and fixed coordinate systems. Differentiating, we obtain

$$\frac{d\gamma_{ij}}{dt} = \frac{\partial x_\alpha}{\partial \xi_i} \cdot \frac{\partial x_\beta}{\partial \xi_j} \frac{\partial \Gamma_{\alpha\beta}}{\partial t} + \left[\frac{\partial x_\alpha}{\partial \xi_i} \left(\frac{d}{dt} \cdot \frac{\partial x_\beta}{\partial \xi_j} \right) + \left(\frac{d}{dt} \cdot \frac{\partial x_\alpha}{\partial \xi_i} \right) \frac{\partial x_\beta}{\partial \xi_j} \right] \Gamma_{\alpha\beta}$$

The quantities appearing in this expression, namely, $d(\partial x_\beta/\partial \xi_j)/dt$ and $d(\partial x_\alpha/\partial \xi_i)/dt$, can be transformed in the manner shown, considering that $dx_i/dt = v_i$:

$$\frac{d\gamma_{ij}}{dt} = \frac{dx_\alpha}{\partial \xi_i} \cdot \frac{\partial x_\beta}{\partial \xi_j} \cdot \frac{d\Gamma_{\alpha\beta}}{dt} + \frac{\partial x_\alpha}{\partial \xi_i} \cdot \frac{\partial v_\beta}{\partial \xi_j} \Gamma_{\alpha\beta} + \frac{\partial x_\beta}{\partial \xi_j} \cdot \frac{\partial v_\alpha}{\partial \xi_i} \Gamma_{\alpha\beta}$$

Since the rate of strain depends on the convected coordinates, its derivatives may be rewritten thus:

$$\frac{\partial v_\beta}{\partial \xi_j} = \frac{\partial v_\beta}{\partial x_k} \cdot \frac{\partial x_k}{\partial \xi_j} \quad \text{and} \quad \frac{\partial v_\alpha}{\partial \xi_i} = \frac{\partial v_\alpha}{\partial x_k} \cdot \frac{\partial x_k}{\partial \xi_i}$$

because ξ_i and ξ_j are independent of t.

Changing in the second and third terms the notation of subscripts over which the summation is carried out, we finally obtain the following expression:

$$\frac{d\gamma_{ij}}{dt} = \frac{\partial x_\alpha}{\partial \xi_i} \cdot \frac{\partial x_\beta}{\partial \xi_j} \cdot \frac{d\Gamma_{\alpha\beta}}{dt} + \frac{\partial x_\alpha}{\partial \xi_i} \cdot \frac{\partial x_\beta}{\partial \xi_j} \cdot \frac{\partial v_k}{\partial x_\beta} \Gamma_{\alpha k} + \frac{\partial x_\alpha}{\partial \xi_i} \cdot \frac{\partial x_\beta}{\partial \xi_j} \cdot \frac{\partial v_k}{\partial x_\alpha} \Gamma_{k\beta}$$

or

$$\frac{d\gamma_{ij}}{dt} = \frac{\partial x_\alpha}{\partial \xi_i} \cdot \frac{\partial x_\beta}{\partial \xi_j} D_0\Gamma_{\alpha\beta} \tag{1.35}$$

Here the symbol D_0 denotes the following sequence of operations which should be performed over the components of the strain tensor $\Gamma_{\alpha\beta}$ written in the space-fixed coordinate system:

$$D_0\Gamma_{\alpha\beta} = \frac{d\Gamma_{\alpha\beta}}{dt} + \frac{\partial v_k}{\partial x_\beta} \Gamma_{\alpha k} + \frac{\partial v_k}{\partial x_\alpha} \Gamma_{k\beta} = \frac{\partial \Gamma_{\alpha\beta}}{\partial t} + v_k \frac{\partial \Gamma_{\alpha\beta}}{\partial x_k} + \frac{\partial v_k}{\partial x_\beta} \Gamma_{\alpha k} + \frac{\partial v_k}{\partial x_\alpha} \Gamma_{k\beta} \tag{1.36}$$

Formula (1.35) is completely analogous in structure to formula (1.34), which defined the rules of transformation of the tensor components in the space-fixed coordinate system to the components of that tensor in the convected coordinate system. Then we conclude that since $d\gamma_{ij}/dt$ represents the components of strain rate tensors in the convected coordinate system, the quantity $D_0\Gamma_{\alpha\beta}$ defines the components of this tensor in the x_i coordinate system. Thus, upon arbitrary deformation of the body the rate of strain can be calculated by means of formula (1.36), which has been derived by taking into account all possible transformations of the coordinates (i.e., their deformation and rotations). The resulting formulas define the rule of transformation from the convected coordinate system to the space-fixed coordinate system, relative to which the kinematics of the motion of the body are considered.

The previously performed transformations of the tensor $\{\gamma\}$ from the convected coordinate system to the space-fixed coordinate system are not restricted to the strain tensor. Therefore, all the results and formulas obtained apply equally well to any tensor specified in the convected coordinate system if it is to be transformed to the space-fixed coordinate system. But the application of the results obtained to the strain tensor allows one to derive very important simple formulas which define the components of the rate-of-strain tensor in the space-fixed coordinate system in terms of the velocity gradients, grad v. By the velocity gradient is meant a set of derivatives of the velocity components with respect to the coordinate axes:

$$\text{grad}\,v = \begin{vmatrix} \dfrac{\partial v_1}{\partial x_1} & \dfrac{\partial v_1}{\partial x_2} & \dfrac{\partial v_1}{\partial x_3} \\[2mm] \dfrac{\partial v_2}{\partial x_1} & \dfrac{\partial v_2}{\partial x_2} & \dfrac{\partial v_2}{\partial x_3} \\[2mm] \dfrac{\partial v_3}{\partial x_1} & \dfrac{\partial v_3}{\partial x_2} & \dfrac{\partial v_3}{\partial x_3} \end{vmatrix}$$

This definition of the gradient of the vector quantity (v) is a natural generalization of the ordinary procedure of calculating the gradient of a scalar, where the vector components are scalars.

In order to calculate the rate of strain in the fixed coordinate system, one should perform operations stipulated by formula (1.36) with respect to the components of the strain tensor $\Gamma_{\alpha\beta}$. Having carried out appropriate calculations, one can obtain the following expressions for the components of the strain-rate tensor $\{\gamma\}$ in the space-fixed coordinate system, which will be denoted here as $\dot{\gamma}_{ij}$:

$$\dot{\gamma}_{ij} = \frac{1}{2}\left(\frac{\partial v_i}{\partial x_j} + \frac{\partial v_j}{\partial x_i}\right) \tag{1.37}$$

It is easy to see that this result is analogous to the definition of the components (1.37) of the rate-of-strain tensor for the case of small strains described by the tensor $\{\gamma\}^0$, when with all the spatial effects being neglected the rate of strain is calculated as a partial derivative of displacements with respect to time.

The above reasoning concerning the difference in the determination of the quantities in the convected and space-fixed coordinate systems was put forward in a successive way by Oldroyd (1950) who introduced the respective concepts into the literature on rheology, though analogous ideas can be found in earlier works (Zaremba, 1903; Hencky, 1925). Oldroyd obtained a differential operator for differentiation with respect to time, of the type of formula (1.36), and applied it to the analysis of the rheological properties of liquids.

The physical meaning of the ideas used in the derivation of the above-considered expressions consists in assuming that upon deformation of the body it is necessary to take into account the movement of deforming elements in space. Then, if some quantity, say stress, is considered at a certain fixed point in space, and another is related to the behaviour of a given moving point of the body, it is necessary, in comparing these quantities, to reduce them to the same point: either of the material or, what is done more often, of space. Only in such a case is it physically justifiable to establish or assume the existence of any relations between the various quantities characterizing the properties of the material, the kinematics of its motion, and the forces arising during the deformation.

Although the idea itself of the necessity of reconciling the quantities determined in the convected and space-fixed coordinate systems seems to be self-evident and indisputable, concrete methods of transformation are far from obvious and unambiguous. The method proposed in the first work of Oldroyd has been considered above. As has already been pointed out, the calculation carried out takes into account all the changes of the coordinates due to their displacement and deformation in the neighbourhood of a given point and the rotation of the elements of the body. Incidentally, the fact that one has to take into account not only the movement of the coordinates in space but also their deformation when the quantities are transformed from the convected to the fixed coordinate system is not quite obvious and, what is more, is doubtful. The exclusion of the components associated with the deformation of the coordinates has resulted in deriving equations other than Eq. (1.36), equations that define the rules of transformation from the convected to the space-fixed coordinate system. The respective expressions have been introduced by De Witt who obtained a differential operator other than the operator D_O [see formula (1.36)], namely,

$$D_J \Gamma_{\alpha\beta} = \frac{d\Gamma_{\alpha\beta}}{dt} - \omega_{k\beta}\Gamma_{\alpha k} - \omega_{\alpha k}\Gamma_{k\beta} = \frac{\partial\Gamma_{\alpha\beta}}{\partial t} + v_k \frac{\partial\Gamma_{\alpha\beta}}{\partial x_k} - \omega_{k\beta}\Gamma_{\alpha k} - \omega_{\alpha k}\Gamma_{k\beta}$$

(1.38)

where

$$\omega_{ij} = \frac{1}{2}\left(\frac{\partial v_i}{\partial x_j} - \frac{\partial v_j}{\partial x_i}\right)$$

(1.39)

Such a rule of differentiation was first proposed as early as 1909 by Jaumann, and therefore the operator (1.38) is usually called the Jaumann operator or the Jaumann derivative with respect to time.

Later, the Jaumann derivative was used by Oldroyd who explained the physical meaning of the Jaumann derivative as follows. Suppose that there exists a certain object in the fixed coordinate system and the observer moves together with the absolutely rigid coordinate system which is not only displaced relative to the space-fixed coordinate system but rotates as well. Then, for the moving observer the rate of change of the object in question in time will be expressed by the Jaumann derivative.

The above-considered methods of transforming from the convected to the space-fixed coordinate system are not the only ones possible. Other methods have also been described in the literature, though they have not found wide application. However, the most significant fact that emerges from the analysis performed is thought to be not so much the form or method of transformation from one coordinate system to another as the necessity of such a change in the treatment of the properties of the body which is moving and is being deformed.

The various methods of transformation of the coordinate system lead, in the long run, to different predictions as to the properties of the material and its behaviour under particular loading conditions. Therefore, though from a theoretical viewpoint the choice of the "best" method of changing from one coordinate system to another is impossible, a comparison of theoretical predictions with experiment makes it possible to select those methods which do not contradict experiment. There may exist more than one such theories; therefore, for a more general theory to be chosen reliably, various experiments must probably be carried out because the predictions of different theories may prove to be the same for one series of experiments (or for the same types of motion) and quite different for others.

Concrete examples of the application of the operators considered above will be discussed further in the text (see Sections 1.9, 2.5.5 and 4.3) in connection with different strain and stress situations.

We shall indicate the difference between the concepts of the rate-of-strain tensor $\{\dot{\gamma}\}$ and the velocity gradient grad v, both defined in the space-fixed coordinate system. The components of the tensor $\{\dot{\gamma}\}$ are defined in terms of the derivatives of velocity components with respect to the coordinates as follows:

$$\{\dot{\gamma}\} = \begin{vmatrix} \dfrac{\partial v_1}{\partial x_1} & \dfrac{1}{2}\left(\dfrac{\partial v_1}{\partial x_2}+\dfrac{\partial v_2}{\partial x_1}\right) & \dfrac{1}{2}\left(\dfrac{\partial v_1}{\partial x_3}+\dfrac{\partial v_3}{\partial x_1}\right) \\[3mm] \dfrac{1}{2}\left(\dfrac{\partial v_2}{\partial x_1}+\dfrac{\partial v_1}{\partial x_2}\right) & \dfrac{\partial v_2}{\partial x_2} & \dfrac{1}{2}\left(\dfrac{\partial v_2}{\partial x_3}+\dfrac{\partial v_3}{\partial x_2}\right) \\[3mm] \dfrac{1}{2}\left(\dfrac{\partial v_3}{\partial x_1}+\dfrac{\partial v_1}{\partial x_3}\right) & \dfrac{1}{2}\left(\dfrac{\partial v_3}{\partial x_2}+\dfrac{\partial v_2}{\partial x_3}\right) & \dfrac{\partial v_3}{\partial x_3} \end{vmatrix}$$

This gives the relation between $\{\dot{\gamma}\}$ and grad v which may be represented in the form:

$$\text{grad } v = \{\dot{\gamma}\} + \begin{vmatrix} 0 & \dfrac{1}{2}\left(\dfrac{\partial v_1}{\partial x_2}-\dfrac{\partial v_2}{\partial x_1}\right) & \dfrac{1}{2}\left(\dfrac{\partial v_1}{\partial x_3}-\dfrac{\partial v_3}{\partial x_1}\right) \\[3mm] \dfrac{1}{2}\left(\dfrac{\partial v_2}{\partial x_1}-\dfrac{\partial v_1}{\partial x_2}\right) & 0 & \dfrac{1}{2}\left(\dfrac{\partial v_2}{\partial x_3}-\dfrac{\partial v_3}{\partial x_2}\right) \\[3mm] \dfrac{1}{2}\left(\dfrac{\partial v_3}{\partial x_1}-\dfrac{\partial v_1}{\partial x_3}\right) & \dfrac{1}{2}\left(\dfrac{\partial v_3}{\partial x_2}-\dfrac{\partial v_2}{\partial x_3}\right) & 0 \end{vmatrix}$$

The second term on the right-hand side of the equality corresponds to the rotational motions of the elements of the body without deformation since the equality given above is essentially the time-differentiated relation which has been written earlier for the components of the strain tensor in the analysis of simple shear [see formula (1.30)]. Here an analogous conclusion is made concerning the rate of strain. We can then say that the presence of velocity gradients in the material does not in itself imply its deformation since certain combinations of the components of the velocity gradient lead to the rotation of the body as a whole without its deformation, i.e., without change of the distance between the points of the material. To illustrate this, we can give an example of an absolutely rigid (non-deformable) body rotating relative to a certain centre. Since the velocities of the points situated at different distances from the centre of rotation are different, there exists a velocity gradient in the body (i.e., the velocity derivative in the direction of the centre of rotation is not equal to zero), but in this case the body is not deformed since it is supposed to be absolutely rigid. This example convincingly shows that there is a difference between the rate-of-strain and the velocity-gradient concepts.

At the same time there are cases where the values of the rate of strain and the velocity gradient are numerically equal. Thus, in uniaxial extension $\dot{\gamma}_{11} = $ grad v. In simple shear (see Fig. 1.13) the values of $\dot{\gamma}_{ij}$ and grad v differ by the factor 1/2, namely, grad $v = dv_1/\partial x_2$, and $\dot{\gamma}_{12} = \dot{\gamma}_{21} = (1/2) \times \partial v_1/\partial x_2$. The difference between $\{\dot{\gamma}\}$ and grad v remains valid for any rotational motion that occurs simultaneously with shear, say, in the case of annular flow of a liquid between two coaxial cylinders, one of which is rotating with angular speed ω and the second is fixed. Suppose that the gap between these cylinders, ΔR, is small. Then the rate of deformation corresponds to simple shear and is equal to $(\omega R/\Delta R)$, where R is the radius of the rotating cylinder. The component of the velocity gradient is calculated in the following manner:

$$\text{grad } \boldsymbol{v} = \text{grad } (\omega R) = \frac{d\,(\omega R)}{\langle dR} = \omega + R\,\frac{d\omega}{dR} \approx \omega + \frac{\omega R\vert}{\Delta R}$$

From this it is seen that in the case considered the rate of strain $\omega R/\Delta R$ and the velocity gradient differ by the value of the circular velocity ω.

Concluding this section, we shall give the expression for the invariants of the rate-of-strain tensor $\{\dot{\gamma}\}$, which will be denoted as T_1, T_2, and T_3. They are constructed in the same way as the previously considered invariants of the stress and strain tensors:

$$T_1 = \dot{\gamma}_{11} + \dot{\gamma}_{22} + \dot{\gamma}_{33}$$

$$T_2 = \dot{\gamma}_{11}\dot{\gamma}_{22} + \dot{\gamma}_{11}\dot{\gamma}_{33} + \dot{\gamma}_{22}\dot{\gamma}_{33} - \dot{\gamma}_{12}^2 - \dot{\gamma}_{13}^2 - \dot{\gamma}_{23}^2$$

$$T_3 = \dot{\gamma}_{11}\dot{\gamma}_{22}\dot{\gamma}_{33} + 2\dot{\gamma}_{12}\dot{\gamma}_{13}\dot{\gamma}_{23} - (\dot{\gamma}_{11}\dot{\gamma}_{23}^2 + \dot{\gamma}_{22}\dot{\gamma}_{13}^2 + \dot{\gamma}_{33}\dot{\gamma}_{12}^2)$$

It is obvious that any set of tensor invariants is also an invariant. Therefore, the following quantity is often used as the second invariant of the rate-of-strain tensor:

$$T_2' = 4T_1^2 - 8T_2 = 4\,(\dot{\gamma}_{11}^2 + \dot{\gamma}_{22}^2 + \dot{\gamma}_{33}^2) + 8\,(\dot{\gamma}_{12}^2 + \dot{\gamma}_{13}^2 + \dot{\gamma}_{23}^2)$$

It is usually employed in constructing rheological equations of state (constitutive equations) for a viscous liquid (see below).

1.5. Constitutive Equations

One of the main objectives of rheology is to establish the relations among the stress state of a body, the strains, and the rate of strain. Equations that give such a relation are called rheological equations of state or constitutive equations of a body. Determination of the dependence $f\,(\{\sigma\}, \{\gamma\}, \{\dot{\gamma}\}) = 0$ for a general case of loading conditions allows one to examine a set of special problems of the behaviour of the body under various kinematic and dynamic conditions.

Rheological equations of state are a mathematical reflection or mathematical models of the actual properties of the material. The general method of constructing constitutive equations consists in carrying out an experiment or a series of different experiments, which are described by appropriate relations. Then these relations are generalized with the aid of a constitutive equation, and on the basis of the resulting equation there are made predictions as to the behaviour of the material under conditions different from the experimental conditions studied. The next step is the testing of theoretical predictions. If the model does not provide a reasonable agreement with experiment, it must be reconsidered. Comparison of the behaviour of the model with experiment in essentially differing schemes of deformation enables one to judge about its generality the more reliably the wider is the range of experiments involved. But the tendency to describe the various experiments as exactly as possible often leads to the extremely increased complexity of the mathematical model. Therefore, the requirement of the generality of the model always contradicts the wish to construct a sufficiently simple model. The way out of this conflicting situation is the possibility of utilizing relatively simple rheological models for any small group of experiments for which they have been constructed and tested. But in such cases one must always be sure that the corresponding model is really used within its range of applicability and that no effects arise because of the rheological model having been used under such deformation conditions in which it has not been tested.

The possibility of generalizing the experimental facts obtained in particular experimental conditions is associated with the necessity to adhere to certain general rules. From this it follows that the relations between $\{\sigma\}$ and $\{\gamma\}$ may not be arbitrary. In the first place, the function $f(\{\sigma\}, \{\gamma\}, \{\dot{\gamma}\})$ is a physical law which reflects the actual properties of the material, these properties being independent of the manner in which this law is formally written. This gives rise to the requirement of the invariance of the physical law with respect to the change of coordinate system. As discussed above, the values of the tensor components are changed upon rotation of the axes but this does not entail the change of the properties of the material and of the physical relations that reflect these properties. Therefore, the physical features of strain must be expressed through the invariants of the respective tensors, these invariants being independent of the choice of coordinate system.

We shall now consider two physical concepts associated with the deformation of the body and expressing, in the most general form, the essence of what takes place during the process of deformation. In principle, one can think of two results of the application of a stress system to the material when the external forces perform a certain work. First, the work performed by the external forces can be stored in the material, so that an elastic energy expressed by the strain-energy function W is stored per unit volume in the material. Second, the work done by external forces can be dissipated irreversibly, which is determined by the rate of dissipation D in unit volume of the material per unit time. In order to incorporate the quantities W and D into the physical laws that define the behaviour of various materials upon deformation, they must be tied up with the invariants of the corresponding tensors. For this purpose the constitutive equation must make allowance for the transformation to the invariant form, in which the strain-energy function and the energy dissipation intensity are expressed in terms of the invariants of stress or kinematic tensors. The converse is also true: if the constitutive equation is written in invariant form, then it may also be represented as the relation between the components of the corresponding tensors. From this it follows that the scalar quantities entering into constitutive equations may not be arbitrary functions of any components of strain or stress tensors and must depend on the invariants of these tensors; otherwise, the requirement of the invariance of a physical law with respect to the choice of coordinate system would be violated. These scalar quantities are coefficients characterizing the particular properties of the material.

Introduction of the quantities W and D allows one to classify the various materials as follows. If on deformation $W \neq 0$ and $D = 0$, a material is called ideally elastic; when it is being deformed, the external work is not dissipated since all the work is stored in the material as the strain energy. If $W = 0$ and $D \neq 0$, the material is called viscous; when it is being deformed, all the external work is dissipated. Finally, if $W \neq 0$ and $D \neq 0$, the material is called viscoelastic and when it is being deformed some part of the external work is dissipated and the remainder is stored in the material.

After the load is removed, the elastic body undergoes changes in shape under the action of the elastic energy stored in it; these changes may be called elastic recovery. After the load has been removed the viscous material remains to be in the state in which it has been at the instant of load removal, because no energy sources exist, which would be able to cause further deformation of the material. Therefore, all the deformations in the viscous material are irrecoverable. After the external forces cease to act the viscoelastic material undergoes elastic recovery.

The principle of invariance of constitutive equations with respect to the transformation of coordinates, which has been discussed above, is not the only requirement that must be satisfied in their formulation. The second requirement is the principle of kinematic invariance. As a matter of fact, we have already considered it before in the analysis of the kinematics of a moving body where the necessity of changing from the convected to the space coordinate system was shown. This principle implies that when relations are established between various quantities, they must be referred to the same point in the body or in space. Therefore, if the time derivatives of any quantity are encountered in rheological constitutive equations, they should be calculated with account taken of the displacements of the body in space, i.e., account must be taken of the motion of the body as a whole and the rotation of the elements of the body in the neighbourhood of the points considered. Formally, this requires that not the partial time derivatives be included in constitutive equations but those derivatives which are calculated with account taken of the above-considered transformation of the coordinates of the points (relative to the fixed coordinate system) in time. Examples of such derivatives are the Oldroyd and Jaumann operators. Note that analogous (as to the sense) transformations must be performed not only with respect to time derivatives but also with respect to integrals that sum up the entire set of effects that have taken place at the preceding stages of deformation with respect to a given moment of time. Therefore, when such integrals are used in writing constitutive equations, one must take into account the rules of changing from the convected to the space-fixed coordinate system when the positions of the points of the body and the associated coordinates were changing during the movement at all the preceding moments of time (with respect to the current time). The relevant mathematical operations will be discussed below in the derivation of constitutive equations through the use of appropriate integrals.

Constitutive equations that describe well the behaviour of a given body under some deformation conditions may prove to be invalid for other conditions. The nature of the material may be of no less importance. Constitutive equations that describe satisfactorily the behaviour of various materials under various deformation conditions may be considerably complicated. Therefore, one has often to look for a compromise solution of this dilemma and pay special attention to the limits of applicability of this or that constitutive equation. The choice between a more complicated and exact constitutive equation and a simplified and less exact equation of state depends on the objectives set forth in each particular case.

The last question that should be considered in this section is as follows: How are the differences in physical nature among the various materials taken into account in constitutive equations? In accordance with all that has been said above this is done in two ways. First, the properties of various materials can be described by various constitutive equations. The primary classification based on this principle is the division of all materials into elastic, viscous, and viscoelastic. Within each of these groups of materials, in accordance with the form of the constitutive equation a differentiation can be made between the various types of bodies possessing specific properties and behaving differently under any particular deformation conditions. But if the materials belong to a single group, i.e., are described by similar constitutive equations, then the differences between them are reflected in the numerical values of the constants appearing in constitutive equations of similar form and individualizing the properties of the materials under study.

Such is the general approach to the rheological description of the properties of real materials. Below, in the subsequent sections, this approach will be discussed on concrete examples of materials possessing the various specific properties and, accordingly, exhibiting quite different effects on deformation. It must be remembered that any constitutive equation, regardless of its complexity, is nothing more than a mathematical model intended to reflect the specific features of the concrete properties of real materials. It is obvious that any model cannot be completely equivalent to a real material; it can only approximately describe its properties.

1.6. Elastic Bodies

1.6.1. An Ideal Elastic Body

We are concerned here with the possible stress-strain relations for the case where no dissipation of external work occurs when a stress system is applied to the body. It is important to emphasize that we shall only deal here with the equilibrium states of the materials being deformed and the time factor must not be taken into consideration.

Suppose that the application of stresses σ_{ij} in the neighbourhood of a given point results in deformation. The question is: What is the relation between the components σ_{ij} and γ_{ij}?

The simplest plausible supposition is that this relation is linear, i.e., $\sigma_{ij} = k\gamma_{ij}$, where k is a certain proportionality factor. We shall consider this case, using, as an example, uniaxial extension in the direction of the x_1 axis. There arise only stresses σ_{11}, while all the other components of the stress tensor, in particular σ_{22} and σ_{33}, are equal to zero. The body undergoes deformation not only in the x_1 direction but in perpendicular directions as well. Hence, the above relation $\sigma_{ij} = k\gamma_{ij}$ is justified for the component with the subscripts 11 but not for the components with the subscripts 22 and 33 because $\gamma_{22} \neq 0$ and $\gamma_{33} \neq 0$, while $\sigma_{22} = \sigma_{33} = 0$. Therefore, the assumption of the proportionality between σ_{ij} and γ_{ij} should be rejected since it does not hold for real bodies.

The physical grounds for correct relations between the components σ_{ij} and γ_{ij} are associated with the fact that uniaxial extension results, in general, in a volume change. Therefore, for the properties of the material to be characterized the use of the constant k alone is insufficient. An independent characteristic should be introduced, which would take account of the volumetric effects. We can then suppose that a constitutive equation must describe both the effects associated with the volume change in the body and the effect of the stress on the change of its shape. Accordingly, resolving the tensors $\{\sigma\}$ and $\{\gamma\}$ into the spherical and deviatoric components and comparing them, we can write the following linear constitutive equations as the simplest plausible supposition:

$$I_1 = kE_1 \tag{1.40}$$

$$\sigma'_{ij} = 2G\gamma'_{ij} \tag{1.41}$$

where k and G are certain coefficients.

If we assume that the hydrostatic pressure $p = -I_1/3$, then the first of these equations will be written in the form:

$$p = -KE_1 \tag{1.42}$$

where $K = -k/3$.

The meaning of the above relations becomes simple in the case of small deformations where $\{\gamma\} \approx \{\gamma\}^0$. Then the first invariant of the tensor $\{\gamma\}^0$ defines the relative change in volume, γ_v, and the coefficients K and G may be regarded, respectively, as the bulk modulus and the shear modulus.

The above-considered constitutive equations (rheological equations of state), (1.40) and (1.41), define the properties of the simplest elastic body known as the Hookean ideal elastic body. If it is assumed that these equations are valid under any loading conditions, then it will follow that the response of the material to a change of external conditions must be instantaneous since, according to the constitutive equations of the Hookean ideal elastic body, there must be a unique correspondence between the stress state and the strains, and the time factor is not included in the constitutive equations.

Still limiting ourselves to the consideration of small deformations, we apply the supposition made as to the linearity of the relations between the hydrostatic pressure and voluminal strains, on the one hand, and between the shear stresses and small shear deformations, on the other, to the analysis of uniaxial stretching. Suppose that to the stress σ_0 there corresponds an extension ε. The resolution of the tensors $\{\sigma\}$ and $\{\gamma\}^H$ into the volumetric component and the deviator for uniaxial extension has been performed earlier (see page 24 and 35). It now remains only to compare the corresponding components of the tensors $\{\sigma\}$ and $\{\gamma\}^0$ by means of relations (1.40) through (1.42). This yields the following result:

$$\frac{\sigma_1}{3} = K(1 - 2\mu)\,\varepsilon \tag{1.43}$$

$$\sigma_0 = 2G(1 + \mu)\,\varepsilon \tag{1.44}$$

We now introduce a further characteristic of uniaxial stretching—Young's modulus E, which, by definition, is equal to the ratio of the tensile stress σ_0 to the tensile strain ε produced by the stress applied:

$$E = \frac{\sigma_0}{\varepsilon} \tag{1.45}$$

Then, formulas (1.43) and (1.44) lead to the following relations between the constants of the material:

$$E = 3K(1 - 2\mu) = 2G(1 + \mu) = \frac{9KG}{3K + G} \tag{1.46}$$

$$G = \frac{E}{2(1 + \mu)} = \frac{3K(1 - 2\mu)}{2(1 + \mu)} = \frac{3EK}{9K - E} \tag{1.47}$$

$$\mu = \frac{1}{2} - \frac{E}{6K} = \frac{E}{2G} - 1 = \frac{3K - 2G}{6K + 2G} \tag{1.48}$$

$$K = \frac{E}{3(1 - 2\mu)} = \frac{2G(1 + \mu)}{3(1 - 2\mu)} = \frac{EG}{9G - 3K} \tag{1.49}$$

Thus, we have introduced four constants: Young's (tension) modulus E, Poisson's ratio μ, the bulk modulus K, and the shear modulus G. Of these only two constants are independent, the other two are calculated by means of formulas (1.46) through (1.49). The two independent constants are the only characteristics of the material, which determine, according to formulas (1.42) and (1.43), its volumetric deformations and the change in shape.

An important special case of an elastic material is an incompressible body for which Poisson's ratio is equal to 0.5. To this there corresponds an infinitely large value of the bulk modulus K, which follows from the first two equalities of (1.49). For an incompressible material there remains only one independent constant—Young's modulus or the shear modulus; the relation between E and G at $\mu = 0.5$ has the form

$$E = 3G \tag{1.50}$$

In the above reasoning it was assumed that one-third of the first invariant of the stress tensor, $I_1/3 = \sigma_m$, is equal to the hydrostatic pressure (with an opposite sign). This is however only a supposition. In the general case, the identity of the quantities σ_m and $-p$ is still subject to discussion. The point is that the concept of the mean stress derives from the consideration of the stress tensor, while the concept of the hydrostatic pressure is associated with the thermodynamic relations defining the state of the body rather than with the characteristics of the stress state in the neighbourhood of a given point in the continuum.

The assumption of the relation between σ_m and $-p$ must be the subject of a special consideration involving concrete materials.

The relation between the state of stress and the hydrostatic pressure is important in all cases where the pressure exerts an influence on the thermodynamic state of the body, say, on the phase equilibria. The hydrostatic pressure resulting from the action of stress may have an effect on phase transitions. One cannot however assume a priori that this pressure is equal to $-\sigma_m$.

1.6.2. The Constitutive Equation of an Elastic Body in Invariant Form

We shall now be concerned with the assumption, in invariant form, of the linearity between the deviatoric components of the stress tensor and small deformations, assuming, as before, that the material is incompressible.

The work of stresses referred to unit volume at infinitesimal strains is expressed in the following manner:

$$dW = \sigma'_{ij}d\gamma^0_{ij} = (\sigma'_{11}d\varepsilon^0_{11} + \sigma'_{22}d\varepsilon^0_{22} + \sigma'_{33}d\varepsilon^0_{33} + \sigma'_{12}d\varepsilon^0_{12} + \sigma'_{13}d\varepsilon^0_{13} + \sigma'_{23}d\varepsilon^0_{23})$$

where, as has been pointed out earlier, $\gamma^0_{ij} = \frac{1}{2}\varepsilon^0_{ij}(1 + \delta_{ij})$.

In the most general case, the strain energy W is a function of the three invariants E_1^0, E_2^0, and E_3^0. The deviatoric component of the stress tensor is thus expressed as follows:

$$\sigma'_{ij} = \frac{\partial W}{\partial \varepsilon_{ij}^0} = \frac{\partial W}{\partial E_1^0} \frac{\partial E_1^0}{\partial \varepsilon_{ij}^0} + \frac{\partial W}{\partial E_2^0} \frac{\partial E_2^0}{\partial \varepsilon_{ij}^0} + \frac{\partial W}{\partial E_3^0} \frac{\partial E_3^0}{\partial \varepsilon_{ij}^0} \tag{1.51}$$

These relations are of general importance and are valid for any elastic material with an arbitrary form of the function $W(E_1^0, E_2^0, E_3^0)$. Since we are concerned here with an incompressible material, $E_1^0 = 0$ and therefore the quantity W should be assumed to be a function only of E_2^0 and E_3^0. The simplest assumption boils down to the following: the strain-energy function W is linearly dependent on the second invariant, and the third invariant makes no contribution at all to the strain-energy function. This means that

$$W = -BE_2^0 \tag{1.52}$$

where B is a constant.

It is easy to see that $dE_2^0/\partial \varepsilon_{ii}^0 = -\varepsilon_{ii}^0$ and $dE_2^0/\partial \varepsilon_{ij}^0 = -\tfrac{1}{2}\varepsilon_{ij}^0 = -\gamma_{ij}^0 \; (i \neq j)$. Hence, we find the relation between the stress and strain components:

$$\sigma'_{ij} = B\gamma_{ij}^0$$

The resulting expression is quite equivalent to the above hypothesis of the linear relation between σ_{ij}' and γ_{ij} [Eq. (1.41)] if we assume that the constant B is equal to $2G$. Then, in uniaxial extension (when $\sigma_{11} = \frac{3}{2}\sigma_{11}'$) the elastic modulus $E = 3B/2 = 3G$.

The formulas and relations obtained are understandable and pictorial for the case of small strains. The generalization of the linear relations suggested [formula (1.41)] between σ_{ij} and the components of the strain tensor for large strains leads to nonlinear stress-strain relationships. Moreover, such a generalization is, in principle, ambiguous since, as has been pointed out above, there are various possible methods of describing finite strains. Instead of the infinitesimal strain tensor we now consider the finite-strain tensor $\{\gamma\}$.

In the most general form, the invariant representation of the dependence of the strain-energy function W on $\{\gamma\}$ has the form of the function $W(E_1, E_2, E_3)$. With small strains the first invariant E_1 is equal to zero since it has the meaning of volumetric strain which is absent in the case of the incompressible body considered. But with finite strains the invariant E_1 has no such simple physical meaning and the assumption of the absence of volume strains does not yet imply the constancy of the quantity E_1. The incompressibility condition, however, enables the exclusion from our consideration of one of the three invariants. This can be done with the aid of formula (1.17) since it follows from this formula that at $\gamma_v = 0$ the third invariant can be expressed in terms of the first and second invariants:

$$E_3 = \frac{1}{(1+E_1)(1+E_2)} - 1$$

Of course, the second invariant E_2 could also be omitted, assuming that W depends only on E_1 and E_3.

1.6.3. Finite Strains in Elastic Bodies

As the simplest assumption, we make use of the hypothesis of the linearity of the relation between the strain-energy function W and the first invariant of the finite-strain tensor:

$$W = AE_1 = A\,(\gamma_1 + \gamma_2 + \gamma_3) = A\left[\frac{\varkappa_1^2 - 1}{2} + \frac{\varkappa_2^2 - 1}{2} + \frac{\varkappa_3^2 - 1}{2}\right] =$$

$$= \frac{A}{2}\,(\varkappa_1^2 + \varkappa_2^2 + \varkappa_3^2 - 3) \tag{1.53}$$

This is equivalent to the corresponding linear relation written in terms of the components of the Cauchy-Green strain tensor:

$$W = \frac{A}{2}\,(\varkappa_1^2 + \varkappa_2^2 + \varkappa_3^2 - 3) = \frac{A}{2}\,(E_1^G - 3) \tag{1.53a}$$

where E_1^G is the first invariant of the Cauchy-Green strain tensor.

We shall now turn to the calculation of the stress-strain relation that follows from the hypothesis of the linear dependence of W on E_1, i.e., foɪmulɾ (1.53).

To do this, we express the strain-energy function in terms of stresses σ_i and extensions ε_i in the directions of the principal axes:

$$dW = \sigma_1\,d\varepsilon_1 + \sigma_2\,d\varepsilon_2 + \sigma_3\,d\varepsilon_3$$

It is not difficult to see that the infinitesimal increments in extension $d\varepsilon_i$ are defined in terms of the extension ratios $\varkappa_i$ in the following way:

$$d\varepsilon_i = \frac{dl_i}{l_i} = \frac{d\,(l_i/l_{i,\,0})}{l_i/l_{i,\,0}} = \frac{d\varkappa_i}{\varkappa_i}$$

Hence, it may be said that the strain-energy function is defined, in terms of the extension ratio, by the formula

$$dW = \sigma_1\,\frac{d\varkappa_1}{\varkappa_1} + \sigma_2\,\frac{d\varkappa_2}{\varkappa_2} + \sigma_3\,\frac{d\varkappa_3}{\varkappa_3} \tag{1.54}$$

From the incompressibility condition it follows that

$$d\,(\varkappa_1\varkappa_2\varkappa_3) = 0$$

or

$$\frac{d\varkappa_3}{\varkappa_3} = -\left(\frac{d\varkappa_1}{\varkappa_1} + \frac{d\varkappa_2}{\varkappa_2}\right)$$

The differential of the strain-energy function is then given by the following relation:

$$dW = (\sigma_1 - \sigma_3)\,\frac{d\varkappa_1}{\varkappa_1} + (\sigma_2 - \sigma_3)\,\frac{d\varkappa_2}{\varkappa_2} \tag{1.55}$$

The same quantity can be found on the basis of the hypothesis of the linear dependence of W on E_1 [formula (1.53)]. Here we shall assume that only two quantities, $\varkappa_1$ and $\varkappa_2$, are independent since $\varkappa_3 = (\varkappa_1\varkappa_2)^{-1}$. Then, using expression (1.53), we perform the following transformations:

$$dW = \left(\frac{\partial W}{\partial \varkappa_1}\right) d\varkappa_1 + \left(\frac{\partial W}{\partial \varkappa_2}\right) d\varkappa_2 = A\left[\left(\varkappa_1 - \frac{1}{\varkappa_1^3\varkappa_2^2}\right) d\varkappa_1 + \left(\varkappa_2 - \frac{1}{\varkappa_1^2\varkappa_2^3}\right) d\varkappa_2\right] =$$

$$= A\left[(\varkappa_1^2 - \varkappa_3^2)\,\frac{d\varkappa_1}{\varkappa_1} + (\varkappa_2^2 - \varkappa_3^2)\,\frac{d\varkappa_2}{\varkappa_2}\right]$$

Comparison of the last equality with the expression for the differential of the strain-energy function (1.55) gives the following system of two equations:

$$\begin{cases} \sigma_1 - \sigma_3 = A\,(\varkappa_1^2 - \varkappa_3^2) \\ \sigma_2 - \sigma_3 = A\,(\varkappa_2^2 - \varkappa_3^2) \end{cases} \qquad (1.56)$$

The result obtained is the solution of the problem of determination of stresses at specified elongations. However, the solution given is incomplete since there are two equations for the determination of three quantities: σ_1, σ_2, and σ_3. Such a result is natural since the stresses for an incompressible body at specified strains are only determined within an arbitrary constant. This is associated with the fact that the superimposition of the external pressure p_0 does not lead to the change of the strained state of the body but is added to the stresses. Therefore, the system of equations (1.56) determines the stresses only within a constant dependent on the external pressure. Then the solution of the problem of determination of stresses may be represented in the following form which satisfies Eqs. (1.56):

$$\begin{cases} \sigma_1 = A\varkappa_1^2 + C \\ \sigma_2 = A\varkappa_2^2 + C \\ \sigma_3 = A\varkappa_3^2 + C \end{cases} \qquad (1.57)$$

where C is a constant subject to determination in terms of the external pressure.

Let us now consider uniaxial stretching in the $\varkappa_1$ direction in the absence of external pressure; then, from the conditions of the problem it follows that $\sigma_2 = \sigma_3 = 0$. If the specified extension ratio $\varkappa_1$ is here equal to $\varkappa$, then $\varkappa_2 = \varkappa_3 = \varkappa^{-1/2}$ and the constant C is given by $C = -A\varkappa^{-1}$. The final solution of the problem of the stress in uniaxial stretching, depending on the quantity $\varkappa$, will then be expressed in the following manner:

$$\sigma_1 = A\left(\varkappa^2 - \frac{1}{\varkappa}\right); \quad \sigma_2 = \sigma_3 = 0$$

The thus obtained value of stress σ_1 is the true stress referred to the cross-sectional area of the specimen at the specified extension ratio $\varkappa$. But if the stress is referred to the initial cross-sectional area of the specimen, then such a conditional stress σ_c will be given by

$$\sigma_c = \frac{\sigma_1}{\varkappa} = A\left(\varkappa - \frac{1}{\varkappa^2}\right) \qquad (1.58)$$

Thus, the assumption of the linear dependence of the strain-energy function W on the first invariant of the finite-strain tensor has led us to believe that the dependence of the stress in stretching on the extension ratio is nonlinear, though in the limiting case of small elongations this linearity is naturally preserved.

The result obtained is well known in the theory of rubber viscoelasticity since in this theory the relation between the strain-energy function and the principal extension ratios can be obtained on the basis of molecular-statistical considerations exactly in the form suggested in the formulation of the hypothesis—Eq. (1.53).

We shall employ formula (1.53) for the consideration of a further important case of deformation, simple shear. The values of the principal extension ratios for simple shear effected by the amount γ have been computed in Sec. 1.3.6, where it was shown [see formula (1.25)] that $\varkappa_1 = \cot \chi$, $\varkappa_2 =$

$= \tan \chi$, and $\varkappa_3 = 1$. Here χ is the angle between the shear directions and the directions of the principal axes. Substitution of these expressions for $\varkappa_i$ into formula (1.53) leads to the following relation:

$$W = \frac{1}{2} A \gamma^2$$

Since the work involved in shear is expressed as $dW = \tau \, d\gamma$, where τ is the tangential stress acting on the planes across which the shear occurs, then

$$\tau = \frac{dW}{d\gamma} = A\gamma \qquad\qquad\qquad (1.59)$$

Thus, the dependence of the shear stress on the shear, as predicted by the hypothesis (1.53), is found to be linear and therefore the quantity A has the meaning of the shear modulus. The shear modulus on extension is however far from being equal to $3A$ and, in general, is devoid of the simple meaning imparted to it when the deformations are small. Nonetheless, irrespective of the nonlinearity of the behaviour of the material on extension, its properties are described only by one constant, the constant A, which characterizes the individual features of the material. It is important to note that the nonlinearity of the behaviour of the material in stretching is not associated with any "structural" effects and is a consequence of only the development of large elastic strains; such a nonlinearity may be called geometric.

The linearity of the relations between W and E_2^0 predicted by formula (1.52) leads to linear stress-strain relations for any type of deformation. This linearity is in agreement with Hooke's law, and the bodies whose properties are described by the strain-energy function (1.52) are called, as has already been pointed out, Hookean ideal elastic bodies or Hookean bodies. The strain-energy function forms the basis of the classical theory of elasticity concerned with small deformations of elastic bodies, particularly of metals. Its historical and practical importance is great since it is the strain-energy function that is the basis of most calculation methods in the theory of the resistance of materials.

The strain-energy function used in formula (1.53) appeared for the first time in the works of the authors of the kinetic theory of rubber elasticity and it may therefore be called the Kuhn-Mark function (the KM function).

The KM function is applicable, in the first place, to cross-linked (cured) elastomers (rubbers) which exhibit high elasticity, i.e., large recoverable deformations. This function is employed as a first approximation for describing the stress-strain relations for rubbers under equilibrium conditions. This limitation on the use of the strain-energy function, specified by expression (1.53), is very essential. The results of investigations of the properties of a material, which are described by the KM function in equilibrium conditions, cannot be used to predict its behaviour in other conditions, say, in transient deformation regimes.

1.6.4. The Reiner Elastic Body

There exist a number of materials whose specific mechanical properties are described well enough by the KM function. Facts are also known, which contradict the condition (1.53). This makes it necessary to complicate the dependence $W(E_i)$ by introducing a number of material constants. Some relatively simple expressions for the strain-energy functions are considered below.

To explain certain effects arising upon deformation of polymer systems, Reiner suggested supplementing the strain-energy function [formula (1.52)] with a second term linearly dependent on the third invariant E_3^0, namely, the infinitesimal-strain tensor $\{\gamma\}^0$. This function has the form

$$W = -BE_2^0 + AE_3^0 \tag{1.60}$$

where A and B are the constants of the material.

Using formula (1.51), we can obtain, from the Reiner function, the following expression for the components of the stress tensor:

$$\sigma_{ij} = -B\frac{\partial E_2^0}{\partial \varepsilon_{ij}^0} + A\frac{\partial E_3^0}{\partial \varepsilon_{ij}^0}$$

In calculating the derivatives of the invariants from the components of the infinitesimal-strain tensor we shall assume that the volume of the body being deformed remains unchanged. Then, since the normal components are determined within an arbitrary term, which is dependent on the external pressure, the following formulas are obtained for the components of the stress-tensor deviator:

$$\sigma'_{ij} = B\gamma_{ij}^0 + A\sum_{k=1}^{3} \gamma_{ik}^0\gamma_{kj}^0 = B\gamma_{ij}^0 + A\gamma_{ik}^0\gamma_{kj}^0 \tag{1.61}$$

The coefficient B, just as in the strain-energy function, has the meaning of the doubled shear modulus.

The coefficient A before the product of the components of the strain tensor (before the "quadratic term") is called the *modulus of cross elasticity*. Its physical meaning becomes clear when considering simple shear; then $\gamma_{12}^0 \neq 0$ and all the other components of the tensor $\{\gamma\}^0$ are equal to zero. The Reiner potential (1.60) predicts that in this case not only shear but also "cross" normal stresses will appear, which are directed normal to the direction of shear:

$$\sigma'_{11} = \sigma'_{22} = A\,(\gamma_{12}^0)^2; \quad \sigma'_{33} = 0$$

These normal stresses are really observed in the shear of polymers (the Weissenberg effect; for the relevant experimental facts, see Chapter 4). The Reiner potential describes correctly this effect as being quadratic with respect to deformations. Hence, it falls rapidly with diminishing deformation.

The introduction of the Reiner potential was the first attempt to describe the appearance of normal stresses in shear.

1.6.5. The Mooney-Rivlin Elastic Body

The application of finite strain theory to cured (cross-linked) elastomers has shown that the KM function must also be refined. This can be done by introducing the second invariant of the large-strain tensor into the expression for the strain-energy function. Indeed, suppose that the dependence of W on E_1 and E_2 is linear:

$$W = AE_1 + BE_2 \tag{1.62}$$

We shall perform the following algebraic manipulations, taking into account that for an incompressible body $\varkappa_1 \varkappa_2 \varkappa_3 = 1$:

$$E_1 = \gamma_1 + \gamma_2 + \gamma_3 = \frac{1}{2}(\varkappa_1^2 + \varkappa_2^2 + \varkappa_3^2 - 3)$$

$$E_2 = \gamma_1 \gamma_2 + \gamma_1 \gamma_3 + \gamma_2 \gamma_3 = \frac{1}{4}[\varkappa_1^2 \varkappa_2^2 + \varkappa_1^2 \varkappa_3^2 + \varkappa_2^2 \varkappa_3^2 - 2(\varkappa_1^2 + \varkappa_2^2 + \varkappa_3^2) + 3] =$$

$$= \frac{1}{4}\left[\left(\frac{1}{\varkappa_1^2} + \frac{1}{\varkappa_2^2} + \frac{1}{\varkappa_3^2} - 3\right) - 2(\varkappa_1^2 + \varkappa_2^2 + \varkappa_3^2 - 3)\right]$$

and, hence,

$$W = \frac{A}{2}(\varkappa_1^2 + \varkappa_2^2 + \varkappa_3^2 - 3) + \frac{B}{4}\left[\left(\frac{1}{\varkappa_1^2} + \frac{1}{\varkappa_2^2} + \frac{1}{\varkappa_3^2} - 3\right) - 2(\varkappa_1^2 + \varkappa_2^2 + \varkappa_3^2 - 3)\right]$$

or

$$W = C_1(\varkappa_1^2 + \varkappa_2^2 + \varkappa_3^2 - 3) + C_2\left(\frac{1}{\varkappa_1^2} + \frac{1}{\varkappa_2^2} + \frac{1}{\varkappa_3^2} - 3\right) \tag{1.62a}$$

in which expression the new constants C_1 and C_2 are expressed in terms of A and B in the following way:

$$C_1 = \frac{A-B}{2} \quad \text{and} \quad C_2 = \frac{B}{4}$$

Finally, considering that the sum $(\varkappa_1^{-2} + \varkappa_2^{-2} + \varkappa_3^{-2})$ is the first invariant of the inverted Cauchy-Green strain tensor, i.e., the Finger tensor, the function (1.62) may be represented as a linear combination of the first invariants of the tensors $\{\gamma\}^G$ and $\{\gamma\}^F$ in the following form:

$$W = C_1(E_1^G - 3) + C_2(E_1^F - 3) \tag{1.62b}$$

where E_1^F is the first invariant of the Finger tensor.

The function (1.62) was proposed independently by Mooney and Rivlin and is called the Mooney-Rivlin function (or the MR function).

To demonstrate the part played by the second term in the Mooney-Rivlin function, let us examine the case of uniaxial extension characterized, as before, by elongations in the x_1 direction, which is equal to $\varkappa$. Then, $\varkappa_2 = \varkappa_3 = \varkappa^{-1/2}$ and W will assume the form:

$$W = C_1\left(\varkappa^2 + \frac{2}{\varkappa} - 3\right) + C_2\left(\frac{1}{\varkappa^2} + 2\varkappa - 3\right)$$

Since in uniaxial extension $\sigma_1 \neq 0$ and $\sigma_2 = \sigma_3 = 0$, formula (1.55) simplifies to

$$dW = \sigma_1 \frac{d\varkappa_1}{\varkappa_1} = \sigma_1 \frac{d\varkappa}{\varkappa}$$

Differentiating the expression written for $W(\varkappa)$, we obtain:

$$dW = 2\left[C_1\left(\varkappa - \frac{1}{\varkappa^2}\right) + C_2\left(-\frac{1}{\varkappa^3} + 1\right)\right]d\varkappa$$

And comparing the last two formulas, we find that

$$\sigma_1 = 2\left[C_1\left(\varkappa^2 - \frac{1}{\varkappa}\right) + C_2\left(\varkappa - \frac{1}{\varkappa^2}\right)\right]$$

The conditional stress σ_c (referred to the initial cross-sectional area of the specimen and sometimes called the engineering stress) will then be represented in the form

$$\sigma_c = 2\left[C_1\left(\varkappa - \frac{1}{\varkappa^2}\right) + C_2\left(1 - \frac{1}{\varkappa^3}\right)\right] = 2\left(C_1 + \frac{C_2}{\varkappa}\right)\left(\varkappa - \frac{1}{\varkappa^2}\right) \tag{1.63}$$

The difference between formulas (1.58) and (1.63) consists in the appearance of the term $2C_2/\varkappa$. If the second invariant E_2 does not make a contribution to the strain-energy function, then $C_2 = 0$ and formula (1.63) turns into relation (1.58). The role of the term $2C_2/\varkappa$ can be estimated if one takes into account that usually $C_2 \approx 0.1\ C_1$ according to experimental data. It should, however, be noted that cases where other relations between the constants C_1 and C_2 hold true.

In the case of shear deformation, the application of the Mooney-Rivlin function yields a simpler result:

$$\tau = 2\,(C_1 + C_2)\,\gamma \tag{1.64}$$

Thus, if we put $2\,(C_1 + C_2) = A$, the Mooney-Rivlin function predicts the same linear stress-strain relation for shear as the Kuhn-Mark function [see formula (1.59)]; the role of the shear modulus is played by the sum $2\,(C_1 + C_2)$. Thus, the various functions predict the same form of the dependence of τ on γ for shear deformations. Therefore, the data obtained in the investigation of this type of deformation cannot serve as the criterion for the selection of the theory for a correct description of other types of deformation. This is a typical example of cases when for the validity of a theory to be judged, it is necessary to carry out experiments, using various deformation schemes, because different theories may provide identical predictions for some special cases, the results predicted for other experiments being substantially different.

Formulas (1.53) and (1.60) may be regarded as the first terms of the power series of a certain function $W\,(E_1, E_2)$. Addition of new terms of the series enables the agreement between theory and experiment to be improved. Thus, if the Mooney-Rivlin function describes experimental data not quite well, the next term of the series may be introduced, assuming that the dependence $W\,(E_1, E_2)$ must include, for example, the quadratic term with respect to the first invariant. Then the function W can be represented in the form:

$$W = C_1 E_1 + C_2 E_2 + C_3 E_1^2 = C_1^\bullet\,(E_1^G - 3) + C_2^\prime\,(E_1^F - 3) + C_3^\bullet\,(E_2^G - 3)^2 \tag{1.65}$$

In this case, the properties of the material are described with the aid of three constants: $C_1,\ C_2,\ C_3$ or $C_1^\prime,\ C_2^\prime,\ C_3^\prime$. In particular, for uniaxial stretching this function predicts the following dependence of the conditional stress σ_c on the extension ratio:

$$\sigma_c = \left(A + \frac{B}{\varkappa} + C\varkappa^2\right)\left(\varkappa - \frac{1}{\varkappa^2}\right) \tag{1.65a}$$

where the constants A, B, and C are expressed in terms of the coefficients C_1, C_2, and C_3. Here the quadratic term in the expansion of the function $W\,(E_1, E_2)$ leads to the appearance of the correction term $C\varkappa^2$. Using formula (1.66) and choosing three appropriate empirical constants, it is possible to describe well experimental data from the dependence $\sigma_c\,(\varkappa)$ for various materials.

The idea that the function $W\,(E_1, E_2)$ can be expanded into a power series and that any number of terms in this series can be used to reconcile theory with experiment allows the experimentally determined stress-strain relations to be described as accurately as possible for various loading conditions. The same idea can be formulated in a somewhat different manner if it is assumed that the KM function (1.53) does really hold true but the quantity A is not a constant and depends on the deformation conditions. Since relation (1.53) must be represented in invariant form, it follows from this that A should be regarded as a certain function of the invariants E_1 and E_2. As a matter of fact, it is quite equivalent to the representation of the function $W(E_1, E_2)$ in the form of the sum of the power series of the invariants E_1 and E_2 with constant coefficients.

Another variant of the generalization of the KM function has been proposed by Tschoegl*. It has the form

$$W = \frac{2A}{n} (\varkappa_1^2 + \varkappa_2^2 + \varkappa_3^2 - 3) \qquad (1.66)$$

The additional constant n makes expression (1.66) somewhat flexible, which allows one to describe well the mechanical behaviour of elastomers, including polymer melts. Expression (1.66) leads to the following expressions for the stress in uniaxial extension:

$$\sigma_1 = 2A (\varkappa^n - \varkappa^{-n/2}) \qquad (1.66a)$$

and in simple shear:

$$\tau = \frac{2A}{2^n \sqrt{4+\gamma^2}} [(\sqrt{4+\gamma^2} + \gamma)^n - (\sqrt{4+\gamma^2} - \gamma)^n] \qquad (1.66b)$$

According to the results of our experiments, these formulas have proved applicable to the description of the functions $\sigma_1(\varkappa)$ and $\tau(\gamma)$ at $n = 2.9$, i.e., n deviates substantially from the value required by the KM function.

The use of the various empirical formulas for the purpose of determining the strain-energy function has, however, an essential disadvantage—the lack of physical substantiation. The point is that the KM function has a clear statistical substantiation in the form of the entropy theory of rubber viscoelasticity. No such substantiation exists for a more complicated MR function. The utilization of other forms of the function is thought to be even more arbitrary. The task of theoretical substantiation of the various forms of the dependence of the strain-energy function on the invariants of the strain tensor is believed to be at present a rather topical problem in polymer physics, whose solution would enable one to choose, among a vast number of possible, empirically selected expressions, the one that would really be consistent with the nature of finite strains.

Attention should be focused on one circumstance which is substantially important for polymeric systems. The point is that it is precisely for such systems under the conditions of manifestation of rubber elasticity (large recoverable deformations) that the attainment of equilibrium states often requires a long period of time. This means that it is precisely for these systems transition processes and non-equilibrium states, which are not described in a general case by the above-considered theories of the strain-energy function, play an important part. They only correspond to limiting cases—the equilibrium states.

---

* N. Tschoegl, the lecture read at the Institute of Petrochemical Synthesis of the USSR Academy of Sciences, Moscow, November, 1973; Blatz P. J., S. C. Sharda, and N. W. Tschoegl, *Trans. Soc. Rheol.*, **18**, 145 (1974).

1.7. Viscous Liquids

If in the process of deformation the work done by external forces is entirely dissipated, so that $W = 0$, then this is a process of flow in pure form; when the action of external forces is discontinued, the deformation that has taken place is found to be irrecoverable and the state attained will be a state of equilibrium.

It has already been pointed out that the strain tensor can be decomposed into a spherical tensor and a deviator. The spherical tensor reflects the action of bulk homogeneous deformation, which is accompanied by the accumulation of potential energy. The flow process is therefore determined by the relation between the deviators of strain tensors and the rates of strain. The strain tensor cannot be used for this purpose because to each specified value of the strain tensor deviator there may correspond an unlimited set of states of strain in a liquid.

1.7.1. Newtonian Liquids

The simplest assumption concerning the possible relation between σ'_{ij} and $\dot{\gamma}_{ij}$ is Newton's hypothesis, according to which the following equality is fulfilled:

$$\sigma'_{ij} = 2\eta\dot{\gamma}_{ij} \tag{1.67}$$

where η is a coefficient called viscosity.

Liquids for which the relation between stresses and rates of strain is described by formula (1.67) are called *Newtonian liquids*. In the case of simple shear, where the shear stress τ and the velocity gradient (rate of shear) $\dot{\gamma} = dv_1/dx_2$ are operating, this formula gives the following simple relation:

$$\tau = \eta\dot{\gamma} \tag{1.68}$$

This relation is usually regarded as the formulation of Newton's hypothesis since formula (1.67) is, as a rule, tested under the conditions of simple shear when all the components of the tensor $\{\dot{\gamma}\}$, except $\dot{\gamma}_{12} = \dot{\gamma}_{21}$, are equal to zero.

Newton's hypothesis may be represented in invariant form. To do this, we should consider the dependence of the intensity of dissipation D on the invariants of the strain-rate tensor, T_1, T_2, and T_3. Then, instead of formula (1.67), the following relation may be taken as the basic relation:

$$D = -2\eta T_2 \tag{1.69}$$

from which there directly follows formula (1.67).

Equality (1.69) may also be regarded as the formulation of Newton's hypothesis and expression (1.69) may be called the Newton dissipative function.

The relation between the stress and the rate of strain is important in uniaxial extension. It can be obtained on the basis of the following considerations. The strain tensor in uniaxial stretching has the form

$$\{\sigma\} = \begin{vmatrix} \sigma_0 & 0 & 0 \\ 0 & 0 & 0 \\ 0 & 0 & 0 \end{vmatrix} = \frac{\sigma_0}{3}\delta_{ij} + \frac{\sigma_0}{3}\begin{vmatrix} 2 & 0 & 0 \\ 0 & -1 & 0 \\ 0 & 0 & -1 \end{vmatrix}; \quad \sigma'_{ij} = \frac{\sigma_0}{3}\begin{vmatrix} 2 & 0 & 0 \\ 0 & -1 & 0 \\ 0 & 0 & -1 \end{vmatrix}$$

The strain-rate tensor is represented in this case by its diagonal components only. Here, by virtue of the incompressibility condition $\dot{\gamma}_{11} + \dot{\gamma}_{22} + \dot{\gamma}_{33} = T_1 = 0$, and therefore $\dot{\gamma}_{22} = \dot{\gamma}_{33} = -\frac{1}{2}\dot{\gamma}_{11}$. So we have

$$\{\dot{\gamma}\} = \begin{vmatrix} \dot{\gamma}_{11} & 0 & 0 \\ 0 & \dot{\gamma}_{22} & 0 \\ 0 & 0 & \dot{\gamma}_{33} \end{vmatrix} = \begin{vmatrix} \dfrac{\partial v_1}{\partial x_1} & 0 & 0 \\ 0 & \dfrac{\partial v_2}{\partial x_2} & 0 \\ 0 & 0 & \dfrac{\partial v_3}{\partial x_3} \end{vmatrix} = \frac{1}{2}\frac{\partial v_1}{\partial x_1} \begin{vmatrix} 2 & 0 & 0 \\ 0 & -1 & 0 \\ 0 & 0 & -1 \end{vmatrix}$$

It is not difficult to see that the tensor written above is a deviator because its first invariant is equal to zero. Then, proceeding from the Newton hypothesis (1.67) and comparing σ'_{ij} with $\dot{\gamma}'_{ij}$, we obtain:

$$\frac{\sigma_0}{3} = 2\eta \left(\frac{1}{2}\frac{\partial v_1}{\partial x_1} \right) \quad \text{or} \quad \sigma_0 = 3\eta \frac{\partial v_1}{\partial x_1}$$

If now the proportionality factor connecting the tensile stress σ_0 and the longitudinal velocity gradient dv_1/dx_1 is designated by λ and called the extensional viscosity, then the last expression yields the following relation between shear viscosity η and extensional viscosity λ:

$$\lambda = 3\eta \tag{1.70}$$

The last equality is known as Trouton's formula and the quantity λ may be called Trouton viscosity.

1.7.2. Rivlin Fluids

The Newton hypothesis regarding the linear relationship between the dissipative function and the second invariant of the strain-rate tensor has been found to be a very convenient approximation, which describes the flow properties of the overwhelming majority of low-molecular-weight liquids. But investigations of the flow properties of polymer systems have revealed numerous effects which will be discussed in the subsequent chapters of this book and which do not correspond to the Newton hypothesis. Therefore, attempts have been made to generalize this hypothesis.

One of the first attempts of this kind was to construct a constitutive equation for a viscous liquid, similar to the expression for the Reiner strain-energy function, i.e., it was assumed that in a general case the dissipative function depends not only on the second but also on the third invariant of the strain-rate tensor, so that

$$D = -\eta T_2 + \eta_c T_3 \tag{1.71}$$

The dependence $D\,(T_2,\,T_3)$ as determined by formula (1.71) has been introduced by Rivlin and it may be called the Rivlin dissipative function. Then, by analogy with the above-mentioned cross elasticity coefficient, η_c is termed the cross viscosity coefficient.

From the Rivlin dissipative function there follows the appearance of normal stresses in shearing flow (the Weissenberg effect) in the same way as it is predicted by the Reiner function. In this case, the normal stresses must be proportional to the square of the rate of shear. However, the Rivlin dissipative function predicts the appearance of normal stresses in the shearing motion of a purely viscous (inelastic) liquid, which contradicts experimental data since the presence of normal stresses is ordinarily associated with the rubber elasticity of the liquid.

1.7.3. Non-Newtonian Fluids

The course of development of the concepts of nonlinear relationships describing the deformation of liquids was somewhat different from the path followed by the theory of large recoverable deformations. It was usually assumed that the dissipative function is completely defined by the second invariant of the strain-rate tensor; hence the magnitude of viscosity is dependent on it. So, we may write the following relationship:

$$\sigma'_{ij} = 2\eta \, (T_2) \, \gamma'_{ij} \tag{1.72}$$

which corresponds, in a general case, to the concept of the non-Newtonian (or anomalously viscous) liquid as a material whose viscosity depends on the deformation conditions. Since in simple shear $\dot{\gamma}_{21} = {}^1\!/_2 \partial v_1 / \partial x_2 = {}^1\!/_2 \dot{\gamma}$, we may write the following relation instead of formula (1.72):

$$\tau = \eta \, (T_2) \, \gamma \tag{1.73}$$

Relation (1.73) is a generalization of formula (1.68) for a non-Newtonian liquid. It is formula (1.73) that is often regarded as the basis for the definition of the concept of the non-Newtonian liquid.

A large variety of analytical conceptions have been proposed in the literature for the dependence $\eta \, (T_2)$; the most important of these will be discussed in the chapter devoted to the flow properties of polymers. At this point we shall only dwell on the problem of the correspondence between formula (1.73) and its invariant representation—formula (1.72).

The viscosity of liquids is usually measured under the conditions of unidimensional deformation, when only one component of $\dot{\gamma}_{ij}$, say, $\dot{\gamma}_{12} = {}^1\!/_2\dot{\gamma}$, undergoes change in simple shear. The concept of the constitutive equation of a liquid in invariant form, however, requires that experimental results be extended to the case of three-dimensional deformation. Accordingly, it is necessary that the validity of such a generalization be tested under deformation conditions different from those at which the shear or extensional viscosity has been measured. Therefore, if measurements are made for one type of unidimensional strain, then information is required about the specific behaviour of the material at least for one more type of deformation. What has been said can be illustrated by comparing the results of measurements of shear and extensional viscosity.

Let us write the results of measurements in simple shear in the following way:

$$\tau = k\dot{\gamma}^n; \quad \eta = \frac{\tau}{\dot{\gamma}} = k\dot{\gamma}^{n-1} \tag{1.72a}$$

These empirical formulas with the constants k and n are widely used in rheology and are called the power law. Its disadvantages and merits are considered in the chapter dealing with the viscosity of polymeric systems.

Here we shall be concerned with the representation of the power law in invariant form. For this purpose, we replace $\dot{\gamma}$ by $\sqrt{4\,|\,T_2\,|}$, which leads to the expression

$$\sigma'_{ij} = 2k\,(4\,|T_2|)^{\frac{n-1}{2}}\,\dot{\gamma}'_{ij} \qquad (1.72\text{b})$$

It has been pointed out above that instead of the invariant T_2 use is often made of the invariant T'_2. For an incompressible liquid, when $T_1 = 0$, the following relation between T_2 and T'_2 is fulfilled:

$$T'_2 = -8T_2$$

Therefore, the dependence of the shear stress τ on the shear rate $\dot{\gamma}$ for a liquid whose properties are described by the power law will be given by

$$\tau = k\left(\frac{1}{2}\,T'_2\right)^{\frac{n-1}{2}}\dot{\gamma}$$

In the generalized (three-dimensional) form the power law is formulated thus:

$$\sigma_{ij} = 2k\left(\frac{1}{2}\,T'_2\right)^{\frac{n-1}{2}}\dot{\gamma}_{ij} \qquad (1.72\text{c})$$

Any experimentally found dependences $\eta\,(\dot{\gamma})$ for the unidimensional case can be generalized in an analogous manner. Here the constitutive equation for a viscous liquid will be represented by formula (1.67), in which the coefficient η should be regarded as being dependent on T_2 (or on T'_2), the quantity $\dot{\gamma}$ being replaced by $4\,|\,T_2\,|^{1/2}$ or by $(1/2\,|\,T'_2\,|)^{1/2}$.

Let us compare the results of the application of the general formula (1.67) to shear and stretching. As pointed out above, for simple shear $T_2 = -\dot{\gamma}_{12}^2 = -\frac{1}{4}\dot{\gamma}^2$. Repeating the same reasoning that has been employed above in considering the longitudinal elongation of a Newtonian liquid, we obtain the following generalized expression for extensional viscosity:

$$\lambda = \frac{\sigma_0}{\dot{\varkappa}} = 3\eta\,(T_2)$$

where $\dot{\varkappa} = \partial v_1/\partial x_1$ is the longitudinal velocity gradient.

For the stretching of the axisymmetric stream of an incompressible liquid:

$$\dot{\gamma}_{11} = \dot{\varkappa}; \quad \dot{\gamma}_{22} = \dot{\gamma}_{33} = -\frac{1}{2}\dot{\varkappa}; \quad \dot{\gamma}_{ij} = 0 \ (i \neq j)$$

This means that $T_2 = -{}^3/_4\dot{\varkappa}^2$.

Suppose now that the experimentally measured function $\eta\,(\dot{\gamma})$ for shear is found to be decreasing, which corresponds to a large body of experimental data available for polymer systems. It means that when use is made of the generalization of the non-Newtonian law for the three-dimensional case the invariant function $\eta\,(T_2)$ must be the one that decreases. Then, as the longitudinal velocity-gradient increases the extensional viscosity must decrease too, and under the conditions in which the comparison is made. $\dot{\varkappa} = \dot{\gamma}/\sqrt{3}$, the Trouton law must always be fulfilled.

It is well known, however (see Chapter 7 which is devoted to the discussion of the actual behaviour of polymers in extension), that when polymers undergo shear their viscosity falls off, whereas on stretching the extensional viscosity may increase. This shows that the generalization of constitutive equations based only on measurements in shearing flow may lead to the contradiction to the results of experiments carried out by a different deformation scheme. It is obvious that the above-considered model of the non-Newtonian liquid with viscosity decreasing on shear cannot correctly describe the behaviour of polymer systems on extension. This stresses the necessity of a thorough testing of the generalization of rheological models. In the literature, use is rather frequently made of a generalization of the function $\eta\ (\dot{\gamma})$, determined in shear, by replacing $\dot{\gamma}$ by $(4\mid T_2\mid)^{1/2}$ with the subsequent use of the thus obtained "invariant" constitutive equations for the analysis of complex flows.

1.7.4. Viscoplastic Bodies

In conclusion, we shall briefly consider the concept of a viscoplastic body which may be regarded, within the conceptual framework under discussion, as a special case of a viscous liquid. When we introduced the function $\eta\ (T_2)$ for non-Newtonian viscous liquids or the function $\eta\ (\dot{\gamma})$ for simple shear, no special requirements were put forth to the form of this function. However, a special case of this function is possible, which is of independent interest. We shall examine it for simple shear. It may be assumed that there exist such materials which do not flow at all until a certain critical stress τ_0 is attained, and at $\tau > \tau_0$ viscous flow develops; the constitutive equation of such a material, which is known as a viscoplastic body, is written in the following way:

$$\dot{\gamma} = \begin{cases} 0 & \tau < \tau_0 \\ f(\tau) & \tau \geqslant \tau_0 \end{cases}$$

Here $f(\tau)$ is a certain function which is not preliminarily defined and whose boundary value at $\tau = \tau_0$ is equal to zero. In a special case the function $f(\tau)$ may be approximated [by means of a linear relationship.

Then, the constitutive equation of such a "linear" material called the Bingham viscoplastic body assumes the form

$$|\tau| = \tau_0 + |\dot{\gamma}|\,\eta_B \quad (|\tau| > \tau_0) \tag{1.74}$$

The quantity τ_0 is termed the yield value of the stress and the coefficient η_B is known as the Bingham (or plastic) viscosity. Like all the constitutive equations considered above, formula (1.73) and its three-dimensional analog are a mathematical approximation of the properties of real materials.

1.8. Linear Viscoelastic Bodies

All the results considered in the previous sections refer to the equilibrium or quasi-equilibrium deformation conditions, in which the time factor has no effect at all on the results of observations of the behaviour of an elastic body or liquid. At the same time, valuable information on the properties

of real materials, polymers in particular, can be gained from the observations
of the transient deformation conditions involving the transition from one
equilibrium state to another or periodic departures from equilibrium.

Four simplest transient regimes of deformation are known: constant
strains or rates of strain with observation of the development of the stress;
the action of constant stress with observation of the progress of deformation;
and, finally, periodic or dynamic experiments, so called in accordance
with the specified law. In the last case, the most convenient law is the harmonic law since the theory of harmonic oscillations is the highly developed
branch of applied mechanics.

1.8.1. The Concept of Linear Viscoelasticity

Here we shall be concerned with methods of describing transient deformation conditions without differentiating between shear and longitudinal
elongation. Therefore, strains and stresses will be designated respectively
by γ and σ, irrespective of whether these quantities refer to shear of elongation. Besides, we shall deal here only with small deformations, when the
effects of geometrical nonlinearity are insignificant.

Suppose that at time $t = 0$ a certain amount of deformation has occurred
in a body in a jumpwise manner, which is maintained constant afterwards,
so that at $t < 0$ the value of $\gamma = 0$, and at $t \geq 0$ the strain $\gamma = \gamma_0$. The
stresses that have developed in a jumpwise way do not change in an ideal
elastic solid body, whereas in real bodies there takes place a transient process: the stresses change more or less rapidly with time. This process is described by the relation

$$\frac{\sigma(t)}{\gamma_0} = \varphi(t) + G_\infty$$

In this expression, $\varphi(t)$ is the relaxation function. In a number of works,
the relation $\sigma(t)/\gamma_0$ is called the relaxation modulus. The general requirement to the function $\varphi(t)$ is the condition of its decrease or, more exactly,
of its non-increase. The coefficient G_∞ is chosen so as to satisfy the condition $\varphi(\infty) = 0$. This means that G_∞ characterizes the stresses that are
retained in the material after the relaxation is complete. Therefore, G_∞
is called the equilibrium (or residual) modulus of elasticity. Let the sum
$\varphi(0) + G_\infty$ be denoted as G_0 and called the instantaneous modulus of elas
ticity because this quantity characterizes the magnitude of stress which arises
at the instant ($t = 0$) specification of the strain γ_0. This stress, which develops in the body when the strain γ_0 has taken place, is made up of the decreasing (relaxing) components and a component which persists in the body for
an indefinitely long period of time after the completion of the transient
relaxation process.

With liquids $G_\infty = 0$ since the applied stress is not retained in them;
if $G_\infty \neq 0$, the body may be regarded as solid: the residual stresses are
retained for an indefinitely long time. Obviously, $G_\infty \leq G_0$ and the values
of the relaxation function lie between (G_∞ to G_0) and zero. Materials in
which the stress relaxation is observed in experimentally measured time
intervals are viscoelastic bodies.

The most important concept in the theory of viscoelastic materials is
the linearity concept. Linear viscoelastic bodies are those in which the func-

tion $\varphi(t)$ and the coefficient G_∞ are independent of the value of specified strain γ_0. Analogous definitions of linearity will be given later on for functions that characterize other transient regimes.

We shall now consider another transient regime. Let a stress σ_0 be applied to the body at $t = 0$. Then, in a general case, an instantaneous strain is developed in the body, after which both recoverable and irrecoverable deformations begin to appear. This can be written as follows:

$$\frac{\gamma(t)}{\sigma_0} = I_0 + \psi(t) + \frac{t}{\eta}$$

where I_0 is the instantaneous compliance; η is the coefficient of viscosity, and $\psi(t)$ is the creep or memory function.

By way of generalizing what has been said above about the relaxation function, a viscoelastic body will be called here linear if the creep function $\psi(t)$ and the coefficients η and I_0 are independent of the specified stress σ_0. The value of instantaneous compliance determines the deformation at the initial moment of time, at $t = 0$, and therefore $\psi(0) = 0$. To characterize the other extreme case, $t \to \infty$, we may introduce the concept of the equilibrium compliance I_∞, which is defined by the formula

$$I_\infty = I_0 + \psi(\infty)$$

This quantity corresponds to the maximum recoverable deformation $\gamma(\infty)$, which develops in the body at the specified stress.

The only general requirement to the function $\psi(t)$ is the non-decrease condition; in the general case $I_\infty \geqslant I_0$ and $\psi(t)$ increases from 0 to $(I_\infty - I_0)$ or remains to be equal to zero.

1.8.2. Sinusoidal Loading

Such straining conditions have come to be known in rheology as dynamic loading. Let the deformations change in time by the harmonic law:

$$\gamma(t) = \gamma_0 e^{i\omega}$$

Here γ_0 is the amplitude of deformation; i is an imaginary unit; ω is the circular frequency equal to $2\pi f$ (where f is the number of cycles per unit time).

The representation of $\gamma(t)$ in the form of $\gamma_0 e^{i\omega t}$ implies that the change of deformation may be regarded as taking place by the sine or cosine law, depending on whether the real or imaginary strain component is considered. As a matter of fact, this is immaterial, and the use of strain and stress values in a complex form facilitates mathematical operations.

The response of the body to the periodic change of deformation consists in the appearance of variable stresses $\sigma(t)$ and, in a general case, the stress is made up of two components: the first component is in phase with the strain and the second is out of phase with the strain.

In a general case, the change in stress with time may be represented as follows:

$$\sigma(t) = G_\infty \gamma(t) + \sigma_0' e^{i(\omega t + \delta)}$$

where δ is a certain magnitude of the *phase angle* (the phase difference between stress and strain) which is called the loss angle (the meaning of this term will be clarified at a later time); σ_0' is the amplitude value of the stress, which lags in phase with respect to the change in strain by the angle δ.

Let us recall that for liquids, i.e., for materials capable of flowing under the action of an infinitely small stress and incapable of retaining residual stresses, $G_\infty = 0$ and the first term in the sum written for $\sigma(t)$ disappears.

Let us define the basic characteristic of the harmonic regime of deformation as the ratio σ/γ, which will be termed here the complex modulus G^*, so that

$$G^* = \frac{\sigma(t)}{\gamma(t)} = G_\infty + \frac{\sigma_0'}{\gamma_0} e^{i\delta} = \left(G_\infty + \frac{\sigma_0'}{\gamma_0} \cos\delta\right) + i\left(\frac{\sigma_0'}{\gamma_0} \sin\delta\right) = (G_\infty + G') + iG''$$

where

$$G' = \frac{\sigma_0'}{\gamma_0} \cos\delta \quad \text{and} \quad G'' = \frac{\sigma_0'}{\gamma_0} \sin\delta$$

It is evident that the value of the complex modulus is determined by a constant coefficient—the *amplitude ratio*, σ_0'/γ_0, which is the ratio of the amplitude values of stress and strain, and the phase angle δ. The components of the complex modulus, G' and G'', are called, respectively, the *storage modulus* and the *loss modulus* (or the real part and the imaginary part of the complex modulus). From the definition of the quantity G^* it follows that the response of the body to the harmonic strain must be a change of stress by the harmonic law; the angle δ remains constant in each cycle. Otherwise, the definition of the complex modulus in terms of the amplitude ratio (σ_0'/γ_0) and the angle δ would become ambiguous.

We shall continue to develop the concepts of linearity as the proportionality between stresses and strains: a viscoelastic body is called linear if in harmonic oscillations the values of G' and G'' or (G_0'/γ_0) and δ are independent of the amplitudes of the specified strain. The ratio (σ_0'/γ_0) is the absolute value of the complex quantity $G' + iG''$:

$$\sqrt{(G')^2 + (G'')^2} = \frac{\sigma_0'}{\gamma_0}$$

and the phase angle δ is expressed in terms of the components of the complex modulus:

$$\delta = \arctan\left(\frac{G''}{G'}\right); \quad \tan\delta = \frac{G''}{G'}$$

In the case under consideration the components of the complex modulus, G' and G'', just as the phase angle, are independent of the amplitudes of stress and strain and, in this sense, characterize the properties of a linear viscoelastic body; they are determined by the frequency ω, in just the same way as the value of the relaxation and creep functions depends on time, but not on the values of stress and strain.

Let us now specify not the strain but the stress which varies by the harmonic law:

$$\sigma = \sigma_0 e^{i\omega t}$$

The body instantly responds to the change in stress by undergoing a strain equal to $I_0\sigma$, which occurs in phase with $\sigma(t)$. Besides, a viscous flow develops by the law

$$\frac{d\gamma}{dt} \eta = \sigma(t)$$

or

$$\gamma(t) = \int_0^t \frac{\sigma(t)}{\eta}\, dt = \frac{\sigma_0}{\eta} \int_0^t e^{i\omega t}\, dt = \frac{\sigma_0}{i\omega\eta} e^{i\omega t} = -i\, \frac{1}{\omega\eta}\, \sigma(t)$$

Finally, there may exist a third strain component, which is out of phase with the specified change in stress. Then, we may write the following expression for the time dependence of deformation:

$$\gamma_{\parallel}''(t) = I_0 \sigma(t) + \gamma_0' e^{i(\omega t - \delta)} - i\, \frac{1}{\omega\eta}\, \sigma(t)$$

in which expression γ_0' is the amplitude of strain, which is out of phase with $\sigma(t)$ by the phase angle δ.

Let us now define a new characteristic of the properties of the material, which is termed the *complex compliance* I^*, as the inverse of the complex modulus:

$$G^* I^* = 1$$

The assignment of stresses by the harmonic law leads to the following relations:

$$I^* = \frac{\gamma(t)}{\sigma(t)} = \left(I_0 - i\, \frac{1}{\omega\eta}\right) + \frac{\gamma_0'}{\sigma_0}\, e^{-i\delta} = \left(I_0 + \frac{\gamma_0'}{\sigma_0}\, \cos\delta\right)$$

$$-i\left(\frac{1}{\omega\eta} + \frac{\gamma_0'}{\sigma_0}\, \sin\delta\right) = (I_0 + I') - i\left(\frac{1}{\omega\eta} + I''\right)$$

where

$$I' = \frac{\gamma_0'}{\sigma_0}\, \cos\delta; \qquad I'' = \frac{\gamma_0'}{\sigma_0}\, \sin\delta$$

In the above expression for I^* the quantity I_0 corresponds to the instantaneous elastic deformations of the material, and $1/\omega\eta$ corresponds to viscous flow. Therefore, the viscoelastic behaviour of the material is governed by the values of I' and I''.

Assuming that $G_\infty \ll G'$, $I_0 \ll I'$ and $(\omega\eta)^{-1} \ll I''$, we obtain, from the equality $G^* I^* = 1$, the following relations between the components of the complex modulus and the complex compliance:

$$G' = \frac{I'}{(I')^2 + (I'')^2}; \quad G'' = \frac{I''}{(I')^2 + (I'')^2}; \quad I' = \frac{G'}{(G')^2 + (G'')^2}; \quad I'' = \frac{G''}{(G')^2 + (G'')^2}$$

It is evident that $I' \neq (G')^{-1}$ and $I'' \neq (G'')^{-1}$. The tangent of the angle δ (the loss factor) is expressed in terms of the ratio of the components of the complex compliance as well as in terms of the ratio G''/G', i.e.

$$\tan\delta = \frac{I''}{I'}$$

When the sinusoidally varying stress is specified, the change in the rate of strain can be followed, the strain rate $\dot{\gamma}$ being connected with γ in the following manner:

$$\dot{\gamma} = \frac{d\gamma}{dt} = \gamma_0 i\omega e^{i(\omega t - \delta)} = i\omega\gamma$$

Then, we may introduce, as a characteristic of the material, the ratio $(\sigma/\dot{\gamma})$ called the complex viscosity η^*. This quantity may be represented, the constant term η being omitted, in the form of the real and imaginary parts:

$$\eta^* = \frac{\sigma}{\dot{\gamma}} = \frac{\sigma_0 e^{i\delta}}{\gamma_0 i \omega} = \frac{\sigma_0}{\gamma_0 \omega} (\sin \delta - i \cos \delta) = \eta' - i\eta''$$

where

$$\eta' = \frac{\sigma_0}{\gamma_0 \omega} \sin \delta; \qquad \eta'' = \frac{\sigma_0}{\gamma_0 \omega} \cos \delta$$

Here $\gamma_0 \omega$ is the amplitude of the strain rate. From the formulas written above there emerge the following relations between the components of the complex viscosity and those of the complex modulus:

$$\eta' = \frac{G''}{\omega}; \qquad \eta'' = \frac{G'}{\omega}$$

The quantity η' is often for simplicity called just the *dynamic viscosity*.

Thus, of all the above characteristics of the viscoelastic properties of the material, which are measured under harmonic oscillations, any two may be

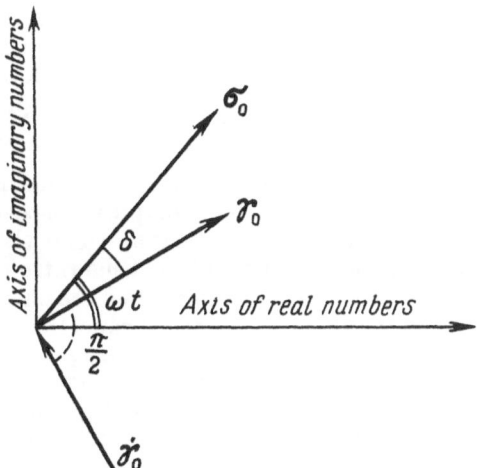

Fig. 1.14. Diagram illustrating the basic concepts referring to the deformation of viscoelastic bodies under conditions of harmonic oscillations.

independent, say, G' and G'', or I' and I'', or η' and η''; the remaining ones are expressed in terms of the two independent quantities chosen as the initial quantities, with the aid of simple algebraic relationships.

All the above-considered characteristics of a linear viscoelastic body depend on frequency and these relations may have a rather complicated form in the general case.

The conceptions that have been expounded above lend themselves to the simple graphical interpretation commonly employed in considering complex quantities. Indeed, let the stress and strain be represented by means of vectors of length $|\gamma_0|$ and $|\sigma_0|$ in coordinates in which the abscissa represents the axis of real values and the ordinate that of imaginary values (Fig. 1.14). The vectors rotate counterclockwise with an angular velocity ω rad/sec, forming angles equal to ωt and $(\omega t - \delta)$ with the axis of real values. Then, the projection of the vector σ_0 onto the abscissa, equal to $\sigma_0 \cos \omega t$, represents the real values of stress at a given moment of time. The vector γ_0 rotates, following the vector σ_0, with the same angular velocity, lagging by the angle δ. The above-given definition of linearity in harmonic oscillations requires that when the vector σ_0 is elongated a certain number of times the vector γ_0 be elongated by the same amount and the angle δ between the vectors remain unchanged during the rotation. Then, the relative position of the vectors σ_0 and γ_0 may be considered regardless of their orientation relative to the coordinate axes. Figure 1.14 also shows the vector $\dot{\gamma}_0$ directed at an angle of $90°$ to the vector γ_0. The length of the vector $\dot{\gamma}$ is equal to $\omega\gamma_0$, and the indicated orientation of $\dot{\gamma}$ follows from the condition that when $\gamma = \gamma_0 \cos(\omega t - \delta)$, then $\dot{\gamma} = \dot{\gamma}_0 [-\sin(\omega t - \delta)]$. These relationships are readily obtained from Fig. 1.14.

Now, on the basis of the graphical representation of the quantities σ, γ, and $\dot{\gamma}$, it is easy to define the earlier introduced concepts of the components of the complex modulus, complex compliance, and complex viscosity. The quantity $\sigma_0 \cos \delta$ is the projection of the vector σ_0 onto γ_0, and $\sigma_0 \sin \delta$ is the projection of the same vector onto the direction perpendicular to γ_0. Then, G' and G'' are taken to mean the ratios of the value of the vector γ_0 to the projections of the vector, respectively, onto the direction of the vector γ_0 and onto the direction perpendicular to it. If the vector γ_0 is projected onto σ_0, then completely analogous definitions can be given for the components I' and I''. Finally, if we consider the projections of σ_0 onto γ_0 and the direction perpendicular to $\dot{\gamma}$ and divide the values of the projections by the length of the vector $\dot{\gamma}_0$, then the corresponding ratios will be found to be equal to the components of the complex viscosity.

It is important to note that when the frequency is changed, not only the length of the vector γ but also the ratio of the lengths of the vectors σ_0 and γ_0 and the magnitude of the angle δ are changed. Therefore, the complete characteristics of the viscoelastic properties of the material as determined under harmonic oscillations are the frequency dependences of the components of the complex modulus or the frequency dependence of the angle δ.

The representation of the $\sigma(t)$ and $\gamma(t)$ curves in the form of harmonic functions allows a useful graphical interpretation of the function $f(\sigma, \gamma) = 0$ specified in parametric form. If $\sigma(t) = \sigma_0 \cos \omega t$ and $\gamma(t) = \gamma_0 \cos(\omega t - \delta)$, then, eliminating t from these equalities, we obtain the equation

$$\left(\frac{\sigma}{\sigma_0}\right)^2 + \left(\frac{\gamma}{\gamma_0}\right)^2 = \sin^2 \delta + 2 \left(\frac{\sigma}{\sigma_0}\right)\left(\frac{\gamma}{\gamma_0}\right) \cos \delta \qquad (1.75)$$

This expression is the equation for the ellipse. Designating the quantity (σ/σ_0) as x and (γ/γ_0) as y and normalizing the stresses and strains by their amplitudes, we transform Eq. (1.75) in the following manner:

$$x^2 + y^2 = 2xy \cos \delta + \sin^2 \delta \qquad (1.75a)$$

From the last equation it is easy to establish two special cases which are of particular interest: $\delta = \pi/2$, then Eq. (1.75a) assumes the form $x^2 + y^2 = \sin^2 \delta$, and the ellipse degenerates to a circle with a centre at the origin of coordinates; if $\delta = 0$, then Eq. (1.75a) simplifies to $x = y$ and the ellipse degenerates to a straight line passing through the coordinate origin. Further theoretical analysis shows that the area of the ellipse described by Eq. (1.75) is equal to $(\pi\gamma_0\sigma_0 \sin \delta)$.

Let us find out the physical significance of the data obtained for a viscoelastic body in connection with the concept of the strain-energy function and the dissipative function, i.e., we shall show that when a viscoelastic body executes harmonic oscillations, then really $W \neq 0$ and $D \neq 0$. We calculate the entire work A done per cycle of oscillations per unit volume:

$$A = \int_0^T \sigma(t)\, d\gamma$$

where the duration of the cycle $T = f^{-1}$ or $T = 2\pi/\omega$.

Let $\sigma = \sigma_0 \cos \omega t$ and $\gamma = \gamma_0 \cos(\omega t - \delta)$; then the above expression is calculated as follows:

$$A = - \int_0^{2\pi/\omega} \sigma_0 \cos \omega t\, [\gamma_0 \omega \sin(\omega t - \delta)]\, dt = - \frac{\sigma_0 \gamma_0}{2} \cos \delta \sin^2 \omega t \Big|_0^{\frac{2\pi}{\omega}} +$$

$$+ \frac{\sigma_0 \gamma_0}{2} \sin \delta \left(\omega t + \frac{1}{2} \sin 2\omega t \right) \Big|_0^{\frac{2\pi}{\omega}}$$

Upon substitution of the integration limits the first term is found to be equal to zero and the second term to $(\pi \gamma \sigma_0 \sin \delta)$. Hence,

$$A = \pi \gamma_0 \sigma_0 \sin \delta \tag{1.76}$$

From this it follows that the area of the ellipse constructed in accordance with Eq. (1.75) is numerically equal to the work done per cycle of harmonic oscillations and irreversibly lost (dissipated) upon deformation. Then the dissipative function D is calculated as the product of A times the number of cycles per unit time:

$$D = Af = A\frac{\omega}{2\pi} = \frac{\gamma_0 \sigma_0 \omega}{2} \sin \delta = \frac{\gamma_0^2 \omega}{2} G'' = \frac{\sigma_0^2 \omega}{2} I'' \tag{1.77}$$

In periodic deformations of a body no elastic energy can be accumulated per cycle, otherwise, with increasing number of cycles, it would increase indefinitely. In the first fourth of the cycle the elastic energy is stored in the material and in the second fourth it is consumed. This is seen from the character of the change of the first term in the expression for the entire work done per cycle. This term is maximal if the upper limit of integration is the time t equal to $\pi/2\omega$. In this case the second term is found to be equal to $^1/_4 \pi \gamma_0 \sigma_0 \sin \delta$, which corresponds to the energy dissipated in one-fourth of a cycle. Therefore, the first term gives the energy W_0 which is stored in unit volume of the body being deformed in harmonic oscillations. This quantity is given by

$$W_0 = \frac{\sigma_0 \gamma_0}{2} \cos \delta = \frac{\gamma^2}{2} G' = \frac{\sigma_0^2}{2} I' \tag{1.78}$$

The resulting formulas (1.76) and (1.78) enable us to reveal the physical significance of the material parameters G^*, I^*, and δ. The quantities G'' and I'' are the proportionality factors which define the mechanical dissipation at the given values of the parameters of the oscillation process when the amplitudes are equal to σ_0 and γ_0 at the frequency ω. Evidently, D increases with increase of the angle δ. Therefore, the quantities G'', I'', and δ determine the loss of work in harmonic oscillations, which justifies their frequently used names: G'' is known as the *loss modulus*, I'' as the *loss compliance*, and δ as the *loss factor*.

The quantity G' characterizes the elastic properties of the material since under the specified straining conditions an elastic energy proportional to

G' or I' is accumulated. Therefore, the coefficients G' and I' are more rea-
sonably called the storage modulus and the storage compliance, respectively,
meaning that the quantities G' and I' determine the amount of energy stored
in unit volume of the material in one-half of a deformation cycle.

1.8.3. Boltzmann's Superposition Principle. Boltzmann-Volterra Equations

The above-considered three principal deformation schemes—relaxation
at constant strain, creep at constant stress, and harmonic oscillations—are
the most important methods of investigating the viscoelastic properties of
polymeric materials. Each of these methods enables the determination of
a fundamental characteristic of the material: the relaxation function, the
creep function, or the complex modulus. Naturally, a question arises: Are
these three characteristics independent of one another or interrelated so that
we can find $\psi(t)$ or $G^*(t)$ on the basis, say, of the function $\varphi(t)$? The answer
to this question constitutes the essence of linear viscoelastic theory based
on the superposition principle formulated by Boltzmann and known as
Boltzmann's superposition principle.

Boltzmann's superposition principle states that all the effects of the mechanical history on the body may be considered independent and additive and
that the response of the body to these effects is linear.

Let any change in strain, $\Delta\gamma_i$, cause a change in stress, $\Delta\sigma_i$, the quantity
$\Delta\sigma_i$ being linearly related to $\Delta\gamma_i$. Then, for each instant of time t we may
write:

$$\sigma(t) = \sigma(t') + \Delta\sigma_i = \sigma(t') + \Delta\gamma_i[\varphi(t-t') + G_\infty]$$

where $\sigma(t)$ is the stress at time $t \geqslant t'$ and $\sigma(t')$ is the stress at time t'.

At different times the body may experience different effects. Then, from
the assumption of the additivity and independence of the separate effects,
we conclude that the total change in stress, $\Delta\sigma$, by time t is expressed through
the sum of the changes in stress $\Delta\sigma$, namely,

$$\Delta\sigma(t) = \sum_i \Delta\gamma_i[\varphi(t-t_i') + G_\infty]$$

Here $\Delta\sigma(t)$ is equal to the stress acting at time t if no stress was present in
the body at zero time.

If the stress is changed and we are dealing with the corresponding change
in strain, then, accordingly,

$$\gamma(t) = \gamma(t') + \Delta\gamma_k = \gamma(t') + \Delta\sigma_k\left[\psi(t-t') + \frac{t-t'}{\eta} + I_0\right]$$

$$\Delta\gamma(t) = \sum_k \Delta\sigma_k\left[\psi(t-t_k') + \frac{t-t_k'}{\eta} + I_0\right]$$

The mechanical effect may be altered continuously. In this case, the change
of strain (or stress) is found to be a continuous function of time, and the sum
must be replaced by an integral over the elapsed times at which the state of
the body could have undergone a change, from $t = -\infty$ to a given moment
of time t. This switchover from discontinuous changes of loading conditions
to continuous changes leads to integral equations, in the first of which the

argument is strains which change in time by an arbitrary law, and in the second, stresses:

$$\sigma(t) = \int_{-\infty}^{t} [\varphi(t-t') + G_\infty] \, d\gamma = \int_{-\infty}^{t} \frac{d\gamma(t')}{dt'} [\varphi(t-t') + G_\infty] \, dt' \qquad (1.79)$$

$$\gamma(t) = \int_{-\infty}^{t} \left[\psi(t-t') + \frac{t-t'}{\eta} + I_0\right] d\sigma = \int_{-\infty}^{t} \frac{d\sigma(t')}{dt'} \left[\psi(t-t') + \frac{t-t'}{\eta} + I_0\right] dt'$$

$$(1.80)$$

The above relations constitute the mathematical formulation of Boltzmann's superposition principle and are called the Boltzmann-Volterra integral equations because the theory of such equations was elaborated by Volterra. The first of these equations defines the stresses at time t as a function of the entire past history of strain changes, the second defines the strain as a function of the stress history. The interpretation of these equations may of course be reversed, assuming that at the specified function $\sigma(t)$ the first relation is an equation for determination of an unknown function $\gamma(t)$, and the second is an equation for determination of $\sigma(t)$ with the function $\gamma(t)$ being known. This enables the functions $\varphi(t)$ and $\psi(t)$ to be interrelated, as will be shown at a later time.

The above-written integral equations contain the relaxation and creep functions. These equations may be regarded as the generalizations of the simplest relaxation and creep experiments, which have been discussed above for the simplest loading conditions (constant strain or constant stress) and which are extended here to the arbitrary case of the change of strain or stress with time.

The function $\varphi(t - t')$ is a decreasing function, therefore the strain changes that have taken place at smaller values of t', when the argument $(t - t')$ is great, affect the value of $\sigma(t)$ less strongly than those changes which have occurred at later times when $(t - t')$ is lower. This means that the memory of the material for the earlier effects is weaker. An analogous reasoning applies to the second integral of (1.80) as well. Therefore, the relations written above are sometimes called the memory equations since they take account of the entire previous history of the material. The rate at which the past events are forgotten depends on the particular form of the functions $\varphi(t - t')$ and $\psi(t - t')$. If these are strong functions of their arguments, i.e., if φ decreases sharply with increasing $(t - t')$, and ψ increases sharply, the material then has a "poor memory" and responds rather strongly to external effects. But if these functions change slightly, the previous effects will tell for a long time on the state of the body in subsequent changes of its state.

Since each of relations (1.79) and (1.80) may be looked upon as an integral equation with respect to the function appearing under the integral, it is possible, as already mentioned above, to derive an equation relating the functions φ and ψ by substituting the expression for $\gamma(t)$ from (1.80) into (1.79). After all the mathematical manipulations have been performed, this equation assumes, according to Gross, the following form:

$$G_\infty I_0 - 1 + I_0 \varphi(t) + G_\infty \left[\frac{t}{\eta} + \psi(t)\right] + \int_0^t \varphi(t') \left[\frac{1}{\eta} + \frac{d\psi(t-t')}{d(t-t')}\right] dt' \qquad (1.81)$$

where, as before, G_∞ is the equilibrium elastic modulus. The resulting relation (1.81), which is known as the Volterra equation, permits us, at least in principle, to find one of the functions, $\varphi\,(t)$ or $\psi\,(t)$, from the other known one. This is an explicit statement that these functions are the independent characteristics of the material and are uniquely interrelated, so that the specification (or experimental determination) of one of them implies that the second function can be calculated from the other known function with the aid of Eq. (1.81). This result confirms the correlation between the behaviour of the material in relaxation and creep experiments.

We shall now deal with the relation between the complex modulus and the functions $\varphi\,(t)$ and $\psi\,(t)$, assuming that $\gamma = \gamma_0 e^{i\omega t}$. Employing expression (1.79), we get, according to the definition of the complex modulus:

$$G^* = \frac{\sigma\,(t)}{\gamma\,(t)} = (G_\infty + G') + iG'' = \frac{1}{\gamma_0 e^{i\omega t}}\left[i\omega\int_{-\infty}^{t}\gamma_0 e^{i\omega t}\varphi\,(t-t')\,dt'\right] + G_\infty =$$

$$= i\omega\int_{0}^{\infty}\varphi\,(x)\,e^{-i\omega x}\,dx + G_\infty = \left[\omega\int_{0}^{\infty}\varphi\,(x)\sin\omega x\,dx + G_\infty\right] +$$

$$+ i\left[\omega\int_{0}^{\infty}\varphi\,(x)\cos\omega x\,dx\right]$$

In the last expression the integration variable $(t - t')$ is replaced by x. We finally get:

$$\begin{cases} G'\,(\omega) = \omega\int_{0}^{\infty}\varphi\,(t)\sin\omega t\,dt \\[4mm] G''\,(\omega) = \omega\int_{0}^{\infty}\varphi\,(t)\cos\omega t\,dt \end{cases} \tag{1.82}$$

Thus, if the relaxation function $\varphi\,(t)$ is known, then the components of the dynamic modulus can be calculated from formulas (1.82). The result obtained allows an inversion to be made, i.e., the relations written may be regarded as equations for the function $\varphi\,(t)$ with the function $G'\,(\omega)$ or $G''\,(\omega)$ being known. The solution of the corresponding equations has the form*:

$$\begin{cases} \varphi\,(t) = \frac{2}{\pi}\int_{0}^{\infty}\frac{G'\,(\omega)}{\omega}\sin\omega t\,d\omega \\[4mm] \varphi\,(t) = \frac{2}{\pi}\int_{0}^{\infty}\frac{G''\,(\omega)}{\omega}\cos\omega t\,d\omega \end{cases} \tag{1.83}$$

---

* The solution of equations (1.82) for $\varphi\,(t)$ is a special case of the application of the theory of Fourier integrals. From this theory it follows that if the function $f\,(x)$ has been determined by the expression

$$f\,(x) = \sqrt{\frac{2}{\pi}}\int_{0}^{\infty}f_1\,(\alpha)\sin\alpha x\,d\alpha$$

then $f_1\,(\alpha)$ is expressed through $f\,(x)$ in the following manner:

$$f_1\,(\alpha) = \sqrt{\frac{2}{\pi}}\int_{0}^{\infty}f(x)\sin\alpha x\,dx$$

Thus, the Boltzmann superposition principle allows one to establish an unambiguous relationship between the behaviour of the material during relaxation and that under the conditions of harmonic oscillations.

Let us perform analogous transformations for the creep function, finding thereby its relationship to the components of the complex compliance. Let $\sigma = \sigma_0 e^{i\omega t}$; then from Eq. (1.80) we obtain the following expression for $\gamma(t)$:

$$\gamma(t) = i\omega\sigma_0 \left\{ \frac{1}{i\omega} I_0 e^{i\omega t} + \int_{-\infty}^{\cdot} \left[\frac{t-t'}{\eta} + \psi(t-t') \right] e^{i\omega t'}\, dt' \right\}$$

The calculations lead to the following expressions for the terms appearing in this expression: $I_0 \sigma_0 e^{i\omega t}$ for the first term, $1/i\omega\eta\sigma_0 e^{i\omega t}$ for the second term, and the third term (calculated by integration by parts) is

$$\sigma_0 e^{i\omega t} \int_0^\infty \frac{\partial\psi}{\partial t'} e^{-i\omega t'}\, dt'$$

From this, using the definition of the quantity I^* and resolving it into the components, we get

$$I^* = \frac{\gamma(t)}{\sigma(t)} = I_0 + \frac{1}{i\omega\eta} + \int_0^\infty \frac{\partial\psi}{\partial t'} e^{-i\omega t'}\, dt' =$$

$$= \left(I_0 + \int_0^\infty \frac{\partial\psi}{\partial t'} \cos\omega t'\, dt' \right) - i \left(\frac{1}{\omega\eta} + \int_0^\infty \frac{\partial\psi}{\partial t'} \sin\omega t'\, dt' \right)$$

Thus,

$$\begin{cases} I' = \int_0^\infty \frac{\partial\psi}{\partial t} \cos\omega t\, dt \\[4mm] I'' = \int_0^\infty \frac{\partial\psi}{\partial t} \sin\omega t\, dt \end{cases} \tag{1.84}$$

The last two formulas give the sought-for relation between the creep function and the components of the complex compliance, I' and I''. Naturally, these formulas permit the inverse transformation, as a result of which there is established the form of the function ψ, which is arbitrary with respect to time, with the known components of the complex compliance:

$$\begin{cases} \frac{\partial\psi}{\partial t} = \frac{2}{\pi} \int_0^\infty I'(\omega) \cos\omega t\, dt \\[4mm] \frac{\partial\psi}{\partial t} = \frac{2}{\pi} \int_0^\infty I''(\omega) \sin\omega t\, dt \end{cases} \tag{1.85}$$

The resulting relations between the characteristics of the harmonic deformation regime, G^* and I^*, and the relaxation and creep functions establish the correlation between the various characteristics of the behaviour of a linear viscoelastic body under the principal straining and stressing conditions. It has thus been shown that the functions G^*, I^*, $\varphi(t)$, and $\psi(t)$

introduced above are not independent characteristics of the body being deformed. Moreover, formulas (1.79) and (1.80) make it possible to determine the general character of the changes of stresses and strains under arbitrary loading conditions.

1.8.4. Relaxation and Retardation Spectra

Valuable inferences and relationships can be obtained by examining the mathematical properties of the relaxation and creep functions. This enables, in the first place, the introduction of the concept of the relaxation spectrum of a material.

The function $\varphi(t)$ decreases down to zero. Such a function may be represented in the form of the following integral:

$$\varphi(t) = \int_0^\infty F(\theta) e^{-t/\theta} d\theta \tag{1.86}$$

Here $e^{-t/\theta}$ is the kernel of the integrand, the function $F(\theta)$ is called the relaxation spectrum or the relaxation-time spectrum, the integration variable θ is known as the relaxation time. The physical significance of these terms will be disclosed further in the text.

Formula (1.86) is sometimes given in the form

$$\varphi(t) = \int_{-\infty}^\infty H e^{-t/\theta} d(\ln \theta) \tag{1.86a}$$

where $H = H(\ln \theta)$ is termed the logarithmic relaxation-time spectrum since the integration in expression (1.86) is carried out over the variable $\ln \theta$. It is evident that

$$H = \theta F(\theta) \tag{1.87}$$

Formulas (1.86) and (1.86b) constitute a formal definition of the relaxation spectrum of the material.

Instead of the variable θ, we can introduce the inverse quantity s into formula (1.86), which is called the relaxation frequency. Then

$$\varphi(t) = \int_0^\infty N_i(s) e^{-ts} ds \tag{1.86b}$$

The function $N(s)$ known as the relaxation frequency spectrum is related to $F(\theta)$ by the following algebraic relation:

$$F(\theta) = sN(s); \quad s = \frac{1}{\theta}$$

Formula (1.86b), which is another method of writing the basic formula (1.86), offers certain mathematical conveniences since, by its structure, it is found to be the Laplace integral transform of the function $N(s)$, i.e., $\varphi(t)$ is a Laplace representation of the relaxation frequency spectrum. The use of formula (1.86a) is convenient for the calculation of $\varphi(t)$ from $N(s)$ and vice versa. This is due to the existence of standard tables of Laplace integrals and inverse Laplace integrals of the various functions. Hence, if the analytical function $\varphi\,(t)$ has been specified or determined, then, using the results known from standard tables, we can readily find $N\,(s)$ and, hence, $F\,(\theta)$. The converse is also true.

The introduction of the logarithmic function $H\,(\ln\theta)$ in accordance with formula (1.86a) is expedient because relaxation measurements are often carried out in a very wide range of variation of the values of the argument, using, as a rule, logarithmic scales for representing the experimental data obtained.

An analogous reasoning is valid for the integral representation of the creep function. However, since $\psi\,(t)$ is an increasing rather than decreasing function, its integral formulation is the following expression:

$$\psi\,(t) = \int_0^\infty \Phi\,(\lambda)\,(1 - e^{-t/\lambda})\,d\lambda \tag{1.88}$$

where the function $\Phi\,(\lambda)$ is the retardation time spectrum. Sometimes, this formula is also employed in the logarithmic form:

$$\psi\,(t) = \int_0^\infty L\,(1 - e^{-t/\lambda})\,d\,(\ln\lambda) \tag{1.88a}$$

in which expression the logarithmic retardation time spectrum $L = L\,(\ln\lambda)$ is connected with $\Phi\,(\lambda)$ by the obvious relation:

$$L = \lambda\Phi\,(\lambda)$$

The frequency representation of the creep function is not practically used, though it is not difficult to construct by analogy with Eq. (1.86a).

The introduction of the functions $F\,(\theta)$ and $\Phi\,(\lambda)$ requires that their relationships to the earlier considered rheological characteristics of materials be established. Apparently, there must actually exist such a relationship since $F\,(\theta)$ and $\Phi\,(\lambda)$ are determined, respectively, through $\varphi\,(t)$ and $\psi\,(t)$, and the relaxation and creep functions are connected, as has been shown above, with the other characteristics of the material.

Let us now consider the correlation between the spectral functions $F\,(\theta)$ and $\Phi\,(\lambda)$ and the dynamic characteristics G^* and I^*. To do this, we express G in terms of $\varphi\,(t)$, replacing then $\varphi\,(t)$ by its integral transform (1.86):

$$G^*\,(\omega) = i\omega\int_0^\infty \varphi(t)\,e^{-i\omega t}\,dt = i\omega\int_0^\infty\int_0^\infty F\,(\theta)\,e^{-\left(i\omega t + \frac{t}{\theta}\right)}\,d\theta\,dt = i\omega\int_0^\infty \frac{F\,(\theta)\,d\theta}{i\omega + \frac{1}{\theta}}$$

Having performed transformations, we obtain the following final expressions for the frequency dependences of the components of the complex modulus:

$$G'\,(\omega) = \int_0^\infty F\,(\theta)\,\frac{(\omega\theta)^2}{1 + (\omega\theta)^2}\,d\theta = \int_{-\infty}^\infty H\,(\theta)\,\frac{(\omega\theta)^2}{1 + (\omega\theta)^2}\,d\ln\theta \tag{1.89}$$

$$G''\,(\omega) = \int_0^\infty F\,(\theta)\,\frac{(\omega\theta)}{1 + (\omega\theta)^2}\,d\theta = \int_{-\infty}^\infty H\,(\theta)\,\frac{(\omega\theta)}{1 + (\omega\theta)^2}\,d\ln\theta \tag{1.90}$$

These expressions enable us to find $G'\,(\omega)$ and $G''\,(\omega)$ with the function $F\,(\theta)$ being known. Structurally analogous expressions can be derived if, instead of $F(\theta)$, we make use of $N(s)$ as the initial function. Then we get

$$G'(\omega) = \int_0^\infty N(s) \frac{\omega^2}{\omega^2 + s^2}\, ds \tag{1.89a}$$

$$G''(\omega) = \int_0^\infty N(s) \frac{\omega s}{\omega^2 + s^2}\, ds \tag{1.90a}$$

Formulas (1.89) and (1.90) permit the inverse transformation, as a result of which it proves possible to calculate the relaxation spectrum from the various experimentally measured functions. Here we shall not consider the exact methods of calculating the spectrum because they all require the assignment of the characteristics of the material in analytical form, whereas experiments provide directly a set of discrete values of the characteristics. A further important feature of accurate methods of calculating the relaxation spectrum

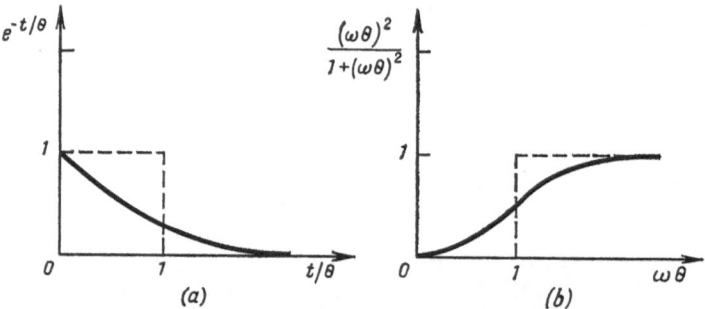

Fig. 1.15. Illustrating the calculation of the relaxation spectrum at a first approximation: a—stress relaxation; b—straining of a body under the conditions of harmonic oscillations.

is that the values of the function $\varphi(t)$ or $G'(\omega)$ must be known over the entire range of the arguments, i.e., from zero to infinity, respectively. Therefore, in practice, in order to calculate the relaxation spectrum, use is not made, as a rule, of accurate methods of the inversion of integral transforms, but instead, the various approximation methods of integral inversion are employed. The most important of these methods is the replacement of the kernels in the integral expressions (1.86), (1.89), and (1.90) by approximately equivalent functions which enable the spectrum to be calculated directly by differentiation of the integrals.

In order to show how such approximation methods of calculating the relaxation spectrum are realized, we shall consider in more detail the integrands in formulas (1.86) and (1.89). They are the product of the function that defines the relaxation spectrum and the function characterizing the relation between the argument of the spectral function and that of the function to be specified.

Figure 1.15 shows the kernels of the integrands (solid lines). There exists such a region of values of the arguments where the kernel is practically equal to unity and the spectrum practically determines the function to be specified. At the same time there is a region of values of the arguments where the kernel is close to zero and the spectrum makes a small contribution to the function characterizing the behaviour of the body. The fact that the transition from the argument values at which the kernel is close to unity to those at which the kernel is practically equal to zero occurs very rapidly is important. This forms the basis of the first approximation for the calculation of the function $F(\theta)$. The kernels in the integrands of (1.86) and (1.89) are replaced by the following approximation expressions:

$$e^{-t/\theta} \approx \begin{cases} 1 \text{ at } \theta > t \\ 0 \text{ at } \theta \leqslant t \end{cases}; \quad \frac{(\omega\theta)^2}{1+(\omega\theta)^2} \approx \begin{cases} 0 \text{ at } \theta \leqslant \frac{1}{\omega} \\ 0 \text{ at } \theta \geqslant \frac{1}{\omega} \end{cases}$$

In Fig. 1.15 this corresponds to the transition from a solid line to a broken line, i.e., to the replacement of the smoothly varying function to a jump function with the same area under it.

The integral expressions (1.86) and (1.89) are, respectively, transformed as follows:

$$\varphi(t) \cong \int_t^\infty F(\theta)\, d\theta$$

$$G'(\omega) \cong \int_{1/\omega}^\infty F(\theta)\, d\theta$$

Differentiating these relations with respect to the variable limit of integration, we obtain the following simple expressions which enable calculation of the spectral function $F(\theta)$ from the experimentally measured characteristics of the material:

$$F(\theta) \approx -\frac{d\varphi(t)}{dt}; \quad \theta = t \tag{1.91}$$

$$F(\theta) \approx \frac{\omega^2\, dG'(\omega)}{d\omega}; \quad \theta = \frac{1}{\omega} \tag{1.92}$$

What has been said above in regard to the relaxation spectrum and its relation to the relaxation function and the components of the complex modulus may be repeated for the retardation time spectrum and its relation to the creep function and the components of the complex compliance. This gives analogous analytical expressions. Thus, the dependence of the components of I^* on the spectral function $\Phi(\lambda)$ is expressed by the relations

$$I'(\omega) = \int_0^\infty \Phi(\lambda)\, \frac{1}{1+(\omega\lambda)^2}\, d\lambda + I_0 \tag{1.93}$$

$$I''(\omega) = \frac{1}{\omega\eta} + \int_0^\infty \Phi(\lambda)\, \frac{(\omega\lambda)}{1+(\omega\lambda)^2}\, d\lambda \tag{1.94}$$

The methods of inversion of the integral relations (1.88) and (1.93) for calculating the spectrum function $\Phi(\lambda)$ are quite analogous to those used for calculating $F(\theta)$, and therefore they may be omitted from our consideration; we shall only give the expression for $\Phi(\lambda)$ obtained to a first approximation from the function $I'(\omega)$:

$$\Phi(\lambda) = -\omega^2\, \frac{dI'(\omega)}{d\omega}; \quad \lambda = \frac{1}{\omega} \tag{1.95}$$

Finally, we shall mention the formula which relates the relaxation time spectrum to the retardation time spectrum. This formula was derived by Gross on the basis of the relation $G^*I^* = 1$, in which the complex quantities G^* and I^* were expressed in terms of the spectral functions. For the purposes of our discussion it is the very fact of existence of the Gross formula that is important since it completes the set of functions considered in linear viscoelasticity theory.

All the results obtained above may be summarized in the form of a chart (see Fig. 1.16), which reflects the structure of the theory of linear viscoelasticity.

This diagram shows how the quantities used in linear viscoelasticity theory are interrelated. Therefore, the experimental determination or specification of one of the functions means that it is possible to calculate the other characteristics of the material and enables one to predict, at least in principle, its behaviour under the various straining or stressing conditions; the latter is realized with the aid of the Boltzmann-Volterra equations (1.79) and (1.80).

Before proceeding to the discussion of concrete examples of the utilization of linear viscoelasticity theory and to the elucidation of the physical meaning of relaxation spectra let us dwell on certain limiting properties of the functions that have been introduced above.

At this point it is expedient to make use of the concept of the moment of the relaxation spectrum, which is defined by the exponent of the relaxation time, appearing as a cofactor in the integrand containing the function $F(\theta)$.

The first important special case of formula (1.86) is the limit of the relaxation function at $t = 0$. This allows the instantaneous modulus G_0 to be expressed in terms of the integral over the relaxation spectrum. Considering that at $t = 0$ the difference $G_0 - G_\infty = \varphi(0)$, we obtain:

$$G_0 = G_\infty + \varphi(0) = G_\infty + \int_0^\infty F(\theta)\, d\theta \tag{1.96}$$

This formula defines G_0 through the zero moment of the relaxation spectrum and the material constant G_∞.

We shall now be concerned with the deformation of a viscoelastic liquid, which occurs at constant rate of shear. Let the rate of shear $d\gamma(t')/dt'$ be expressed in the following way:

$$\frac{d\gamma(t')}{dt'} = \begin{cases} 0 & \text{at } t' < 0 \\ \dot{\gamma} = \text{const} & \text{at } t \geqslant 0 \end{cases}$$

Then, assuming $G_\infty = 0$, we can find, by means of formula (1.79), that the stresses are changing in time by the law:

$$\sigma(t) = \dot{\gamma} \int_0^t \varphi(t - t')\, dt' = \dot{\gamma} \int_0^t \varphi(x)\, dx$$

where the variable $(t - t')$ is replaced by x. Replacing $\varphi(x)$ by its expression in terms of the relaxation spectrum, we obtain:

$$\sigma(t) = \dot{\gamma} \int_0^t \int_0^\infty F(\theta)\, e^{-x/\theta}\, d\theta\, dx = \int_0^\infty F(\theta) \left[\int_0^t e^{-x/\theta}\, dx \right] d\theta$$

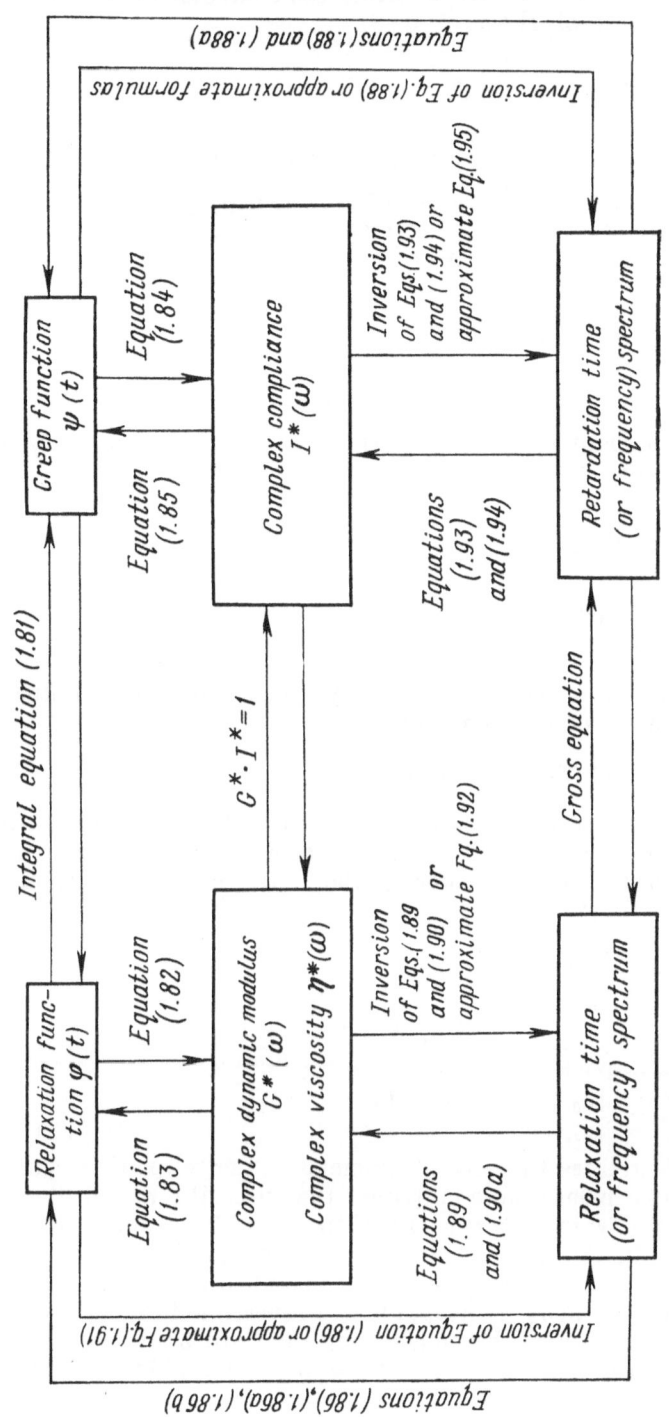

Fig. 1.16. Relations among various functions in the theory of linear viscoelasticity.

This yields the relation that describes the character of the change of stresses at constant **rate** of shear:

$$\sigma(t) = \dot{\gamma} \int_0^\infty \theta F(\theta)(1 - e^{-t/\theta})\, d\theta \tag{1.97}$$

This means that the stress is a linear function of the rate of shear, i.e , according to linear viscoelasticity theory the ratio $\sigma(t)/\dot{\gamma}$ for liquids is independent of the rate of shear at any moment of time.

The expression written for $\sigma(t)$ has a limit at $t \to \infty$, namely:

$$\sigma(\infty) = \lim_{t \to \infty} \sigma(t) = \dot{\gamma} \int_0^\infty \theta F(\theta)\, d\theta$$

This limit corresponds to the condition of steady-state flow Hence, the viscosity coefficient η, which by definition is equal to $\sigma(\infty)/\dot{\gamma}$, is expressed in terms of the relaxation spectrum as follows:

$$\eta = \frac{\sigma(\infty)}{\dot{\gamma}} = \int_0^\infty \theta F(\theta)\, d\theta \tag{1.98}$$

This formula imparts a clear physical meaning to the first moment of the function $F(\theta)$.

The meaning of the second moment of the relaxation spectrum for a viscoelastic body can be determined by considering the product $G^*I^* = 1$ written in terms of the components:

$$G_\infty + (G' + iG'')] \left[(I_0 + I') \quad i \left(\frac{1}{\omega\eta} + I'' \right) \right] = 1$$

The imaginary part of this expression is equal to zero. Substituting then the expressions for the components of the complex quantities in terms of the relaxation spectra and considering the limits of the corresponding expressions at $\omega \to 0$, we have:

$$I_0 + \psi(\infty) - \frac{1}{\eta^2} \int_0^\infty \theta^2 F(\theta)\, d\theta = 0$$

or

$$\int_0^\infty \theta^2 F(\theta)\, d\theta = \eta^2 [I_0 + \psi(\infty)] = \eta^2 I_\infty \tag{1.99}$$

This formula relates the second moment of the relaxation spectrum to the experimentally measured quantities: the viscosity η and the equilibrium compliance I_∞, which is a measure of the possible accumulation of reversible deformations in the body. Thus, the second moment of the function $F(\theta)$ defines the relation between the viscous and elastic properties of the material.

If a viscoelastic body is incapable of irreversible flow, i.e., is a solid, then $1/\eta = 0$. In such a case, by considering the limit of the real quantities contained in the formula $G^*I^* = 1$, at $t = 0$, we can obtain the following relation between the constants of the material:

$$\psi(\infty) = \frac{G_0 - G_\infty}{G_0 G_\infty} = \frac{\varphi(0)}{G_\infty [G_\infty + \varphi(0)]}$$

This relation establishes the connection between the constants of the material that characterize its behaviour in creep and relaxation experiments.

One more general relation can be deduced from formula (1.81) in the limiting case $t = 0$. We have:

$$G_\infty I_0 + \varphi(0) I_0 = G_0 I_0 = 1$$

or, for liquid viscoelastic liquids, for which $G_\infty = 0$:

$$\varphi(0) I_0 = 1$$

The result obtained allows us to perform the inversion of the relation between the values of the relaxation and creep functions at $t = 0$ and $t \to \infty$ and express the equilibrium residual value of the modulus for a viscoelastic solid in terms of the equilibrium compliance:

$$G_\infty = \frac{1}{\psi(\infty) + I_0} = \frac{1}{I_\infty}$$

The resulting relations between the constants characterizing the properties of viscoelastic materials and the integrals over the relaxation spectrum may be presented in the form of a table (see Table 1.1) with the aid of which one can establish, first, which of the constants for a viscoelastic liquid or a viscoelastic solid must be equal to zero and, second, which of the possible constants may be independent and which are determined through the other parameters of the material and, hence, cannot be regarded as the independent properties of the material.

We have just discussed the relations existing between the viscoelastic functions and the change of stress (or strain) in the simplest deformation regimes, for example, $\gamma = $ const or $\dot{\gamma} = $ const. Proceeding from the generalized formulation of the Boltzmann-Volterra principle, one can find such relationships for any arbitrary deformation conditions. Some of them may prove more useful for the determination of the viscoelastic functions than the simplest deformation regimes. Thus, certain methodological advantages (the exclusion of the initial transient conditions for discontinuous straining) are associated* with the use of such an experimentally specified dependence, $\gamma(t)$:

$$\gamma(t) = \begin{cases} \dot{\gamma}_0 t & \text{at } 0 \leqslant t < t_1 \\ \dot{\gamma} t_1 & \text{at } t \geqslant t_1 \end{cases}$$

where $\dot{\gamma}_0 = $ const and the characteristic time t_1 is chosen for a particular material so as to provide measuring convenience. This regime changes to stress relaxation at constant strain at $t_1 \to 0$ and to flow conditions with a constant rate of strain at $t_1 \to \infty$.

One can also use other loading conditions, the choice of which may be governed by methodological as well as theoretical reasons.

---

* Meissner J., *J. Polymer Sci.: Polymer Phys.* Ed., **16**, 915 (1978).

Table 1.1

	At the specified stress			At the specified strain		
	I_0	$\dfrac{1}{\eta}$	$I_\infty - I_0 = \psi(\infty)$	G_0	G_∞	$G_0 - G = \varphi(\theta)$
A viscoelastic solid	$\dfrac{1}{G_0} = \dfrac{1}{G_\infty} + \varphi(0)$	0	$\dfrac{1}{\eta^2}\displaystyle\int_0^\infty \theta^2 F(\theta)\, d\theta - I_0$	$\dfrac{1}{I_0}$	$\dfrac{1}{I_0 + \psi(\infty)} = \dfrac{1}{I_\infty}$	$\dfrac{\psi(\infty)}{I_0[I_0 + \psi(\infty)]} = \dfrac{I_\infty - I_0}{I_\infty I_0}$
A viscoelastic liquid	$\dfrac{1}{G_0} = \dfrac{1}{G_\infty + \varphi(0)} = \dfrac{1}{\displaystyle\int_0^\infty F(\theta)\, d\theta}$	$\dfrac{1}{\eta} = \displaystyle\int_0^\infty \varphi(t)\, dt = \dfrac{1}{\displaystyle\int_0^\infty \theta F(\theta)\, d\theta}$	$\dfrac{\varphi(0)}{G_\infty[G_\infty + \varphi(0)]} = \dfrac{G_0 - G_x}{G_0 G_\infty}$	$\dfrac{1}{I_0}$	0	$\dfrac{1}{I_0}$

1.8.5. The Simplest Models of Viscoelastic Materials and Their Generalizations

The theory of linear viscoelasticity has been expounded above as a phenomenological generalization of the qualitative concepts of a body capable of stress relaxation upon deformation or of exhibiting a delayed progress of strains after the stress has been applied. These concepts allow for a simple model interpretation based on the idea that any external force drives the system out of the equilibrium to which it tends to return at a rate proportional to the extent of departure from equilibrium. Suppose, for example, that the material is subjected to deformation at a certain rate $\dot{\gamma}$. Then the rate of change of the stress $\dot{\sigma}$ is made up of the component proportional to the rate of strain and the component proportional to the quantity that characterizes the extent of departure from equilibrium. In the case of mechanical effects the departure from equilibrium is determined by the stress. Therefore, the qualitative picture that has been depicted above, and which was first described by Maxwell, leads to the following equation:

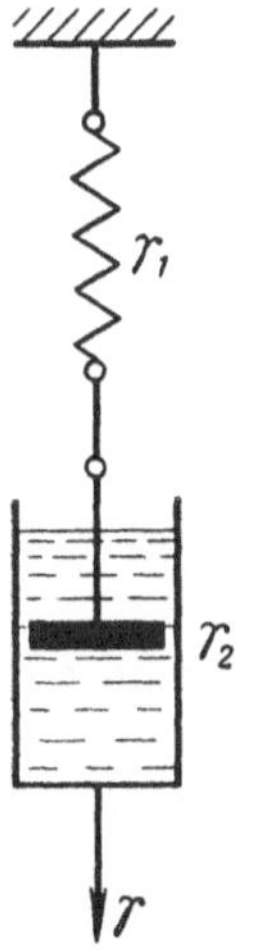

$$\dot{\sigma} = G\dot{\gamma} - \frac{1}{\theta}\sigma$$

Here G and θ are the proportionality factors governed by the properties of the material. This equation, which may be rewritten in the form

$$\frac{\dot{\sigma}}{G} + \frac{\sigma}{\eta} = \dot{\gamma} \qquad\qquad (1.100)$$

Fig. 1.17. The Maxwell model.

(where $\eta = \theta G$), may also be looked upon as a rheological equation of state of a certain material since it gives the relation between the stress and the rate of straining of the body.

Equation (1.100) can be represented by the simple mechanical model shown in Fig. 1.17 where it is assumed that the law of deformation of elastic elements (springs) γ_1 is given by the linear relation $\gamma_1 = \sigma/G$, and the law of deformation γ_2 of viscous elements (dashpots imagined as pistons moving in oil) is represented by the equation $\dot{\gamma}_2 = \sigma/\eta$. Since the total strain γ is the sum of the strain in the spring, γ_1, and the strain in the viscous element, γ_2, then $\gamma = \gamma_1 + \gamma_2$ or $\dot{\gamma} = \dot{\gamma}_1 + \dot{\gamma}_2$, and substitution of the values of $\dot{\gamma}_1$ and $\dot{\gamma}_2$, expressed in terms of stresses, leads to Eq. (1.100). The mechanical model depicted in Fig. 1.17 and consisting of a spring and a viscous element (a dashpot) connected in series is called the Maxwell model and the constitutive equation (1.100) is known as the Maxwell equation; accordingly, a viscoelastic material, whose properties are described by this constitutive equation, is called a Maxwell body.

Let us establish the properties of a Maxwell body. First, we find the relaxation function $\varphi(t)$ at constant strain. If $\gamma = \gamma_0 = \text{const}$, then the solution of Eq. (1.100) takes the form

$$\sigma = \sigma_0 e^{-t/\theta}$$

The constant of integration σ_0 equals $G\gamma_0$, where γ_0 is the given strain which is kept constant during relaxation. Then, it is seen that the Maxwell body relaxes to zero, i.e., this is a liquid and its relaxation function has the form

$$\varphi(t) = \frac{\sigma(t)}{\gamma_0} = Ge^{-t/\theta}$$

It is evident that the ratio $\sigma(t)/\gamma_0$ does not depend on the specified strain; therefore, according to the above-given definition, the Maxwell liquid is a linear viscoelastic body. Examination of the relaxation function yields the physical meaning of the constant θ: this quantity characterizes the rate of approach to equilibrium when the stresses disappear, and therefore it may be called the relaxation time. It is evident that the quantity θ is not equal to the time of transition to a state of equilibrium (which is theoretically equal to infinity for a Maxwell liquid) and only characterizes the rate of this process. Numerically, θ is equal to such a relaxation time during which the initial stress decreases e times.

From the expression for the viscosity determined through the relaxation function it is easily found that the viscosity of a Maxwell liquid is equal to

$$\frac{1}{\int\limits_0^\infty \psi(t)\,dt} = \frac{\theta}{I} = \eta$$

The same conclusion can be arrived at from the solution of Eq. (1.100) provided that $\dot{\gamma} = \text{const}$. The corresponding solution on the initial $(t = 0)$ condition of absence of stress has the form

$$\sigma(t) = \eta\dot{\gamma}(1 - e^{-t/\theta})$$

From this it follows that

$$\frac{\sigma(\infty)}{\dot{\gamma}} = \eta$$

Finally, the constant G has the physical meaning of the instantaneous modulus G_0 since $G_0 = \varphi(\theta) = G$ and $G_\infty = 0$.

On the basis of the formulas derived above, (1.82), we can find out how the Maxwell liquid behaves under harmonic oscillations. The corresponding formulas can also be obtained directly from the constitutive equation (1.100) if the harmonic law of the change of stress or strain is specified. The solutions of this equation for the indicated deformation regimes have the following form:

$$G'(\omega) = G\frac{I(\omega\theta)^2}{1+(\omega\theta)^2}; \quad G'' = G\frac{(\omega\theta)}{1+(\omega\theta)^2}; \quad \eta' = \frac{\eta}{1+(\omega\theta)^2}, \quad \tan\delta = (\omega\theta)^{-1}$$

The creep function of the Maxwell liquid is found from the following relation:

$$\frac{\gamma(t)}{\sigma_0} = \frac{1}{G_0} + \frac{1}{\eta}$$

It follows that with the assigned constant stress the Maxwell liquid exhibits an instantaneous jump in deformation, which is determined by the value of the instantaneous compliance $I_0 = 1/G_0$. Simultaneously, it begins to flow; its resistance to flow is given by the coefficient η. The Maxwell body does not displayl delayed strains, i.e., for it the viscoelastic component of the creep function is equal to zero.

In the case of arbitrary deformation conditions when $\dot{\gamma}\,(t)$ is a certain continuous function, the solution of Eq. (1.100) has the form:

$$\sigma\,(t)=G\int_0^t \dot{\gamma}\,(t')\,e^{-\frac{t-t'}{\theta}}\,dt'$$

Let us now consider the function $F\,(\theta)$ which defines the relaxation spectrum of the Maxwell liquid [see formula (1.86)]. This function must be determined from the equation

$$\varphi_i^r(t)=Ge^{-t/\theta}=\int_0^\infty F\,(t')\,e^{-t/t''}\,dt'$$

where the integration variable is denoted by t'. The solution of this equation for the function $F\,(\theta)$ has the form

$$F=G\delta\,(t'-\theta)$$

where $\delta\,(t'-\theta)$ is the Dirac delta function.*

This solution may also be interpreted as a "point" corresponding to the value of the argument $t'=0$. Thus, the relaxation spectrum of the Maxwell liquid is the value of the argument θ and the corresponding ordinate G. This gives the physical meaning of the above-given name for the function $F\,(\theta)$ as the relaxation spectrum: in a general case, this is a set of relaxation times θ and the corresponding values of the modulus G. Indeed, any arbitrary relaxation function can be represented, as accurately as desired, as the sum of exponential functions of the form

$$\varphi\,(t)=\sum_k G_k e^{-t/\theta_k}$$

Each of them has its own delta function in the relaxation spectrum, i.e., for the relaxation function expressed by the sum of exponents with values of relaxation times θ_k and of the corresponding moduli G_k the relaxation spectrum is represented by a set of points with arguments θ_k and ordinates G_k. This is what exactly leads to the concept of the relaxation spectrum as a function that characterizes the relaxational properties of the material.

Proceeding from the expressions written above for the components of the complex modulus of a liquid characterized not by a single relaxation

---

* The Dirac delta function $\delta\,(x-x_0)$ is a very peculiar function which is determined as follows: it is equal to zero everywhere, except for one point, $x=x_0$, and at this point it equals infinity; its basic property is the fulfilment of the following condition: $\int_{-\infty}^\infty \delta\,(x-x_0)\,dx=1$. This function may spectacularly be visualized as a "splash" in the neighbourhood of the point $x=x_0$; the width of this "splash" contracts to the point x_0 and its height increases indefinitely; in the limit this picture gives the delta function. Here and henceforth, use is made of a more general definition of the delta function which is taken to mean a functional of the form $\int_{-\infty}^\infty \delta\,(x-x_0)\,\varphi\,(x)\,dx=\varphi\,(x_0)$, which thereby associates to the function $\varphi\,(x)$ a certain number—the value of this function at the point of existence of the delta function, $\varphi\,(x_0)$. On the basis of this definition of the δ-function it is easy to see that $\int_0^\infty G\delta\,(x-\theta)\,e^{-t/x}\,dx=Ge^{-t/\theta}$, i.e., indeed the function $G\delta\,(t'-\theta)$ is the solution of the given equation.

time and a single modulus but by a set (spectrum) of relaxation times θ_k and of the respective moduli G_k, we obtain the following expressions for the functions $G'(\omega)$ and $G''(\omega)$:

$$G'(\omega) = \sum_k G_k \frac{(\omega\theta_k)^2}{1+(\omega\theta_k)^2}$$

$$G''(\omega) = \sum_k G_k \frac{(\omega\theta_k)}{1+(\omega\theta_k)^2}$$

where the number of terms is determined by the number of relaxation times of the material.

Since the viscosity and the instantaneous modulus of the liquid are determined by the relaxation function, these quantities can also be found for a body for which the relaxation function is described by the sum of exponents and the relaxation spectrum by a set of points G_k for the values of the argument θ_k. Calculations provide the following values for the viscosity and the instantaneous modulus:

$$\eta = \sum_k \frac{\theta_k}{G_k} = \sum \eta_k$$

$$G = \sum_k G_k$$

The concept of the phenomena that take place in the liquid as a set of relaxation processes, each of which is determined by its own values of the relaxation time θ_k and the modulus G_k, is interpreted by a mechanical model

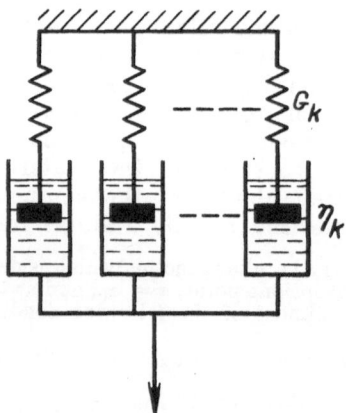

Fig. 1.18. Generalized Maxwell model.

which is a combination of Maxwell elements arranged in parallel; in each of them the modulus G_k and the viscosity η_k are different (see Fig. 1.18). We can imagine a case where in one of the Maxwell elements the viscosity η_1 is infinitely great; this corresponds to the appearance of an equilibrium value of the modulus G_∞, following which the model shown in Fig. 1.18 (the generalized Maxwell body) ceases to be a liquid and becomes a solid incapable of developing infinitely large deformations because its deformation at $\dot{\gamma} =$ const leads to an indefinite increase of stress in the elastic element of the model, G_∞, and, hence, to an indefinite increase of stress in the model as a whole.

Let us consider the properties of the simplest viscoelastic solid. For this purpose, we assume that in the generalized Maxwell model there are only two elements, the elastic modulus in the second being infinitely great. This degenerate model of the generalized Maxwell body is known as the Kelvin-

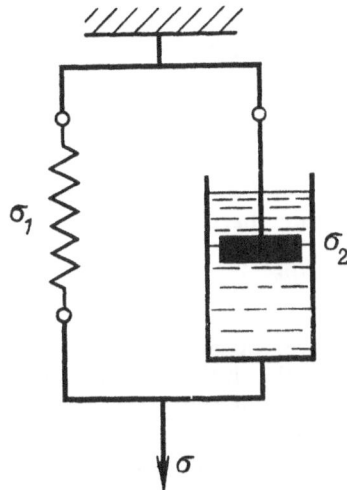

Fig. 1.19. The Kelvin-Voigt model.

Voigt model. It is shown in Fig. 1.19. Its physical significance consists in that the development of elastic deformations occurs with delay because it is retarded by the viscosity of the body. The constitutive equation for a visco-elastic solid described by the Kelvin-Voigt model is derived by inspection of Fig. 1.19. Obviously, the total stress σ applied to the model is composed of the stresses in its elements, i.e., $\sigma = \sigma_1 + \sigma_2$. Then, if $\sigma_1 = G_\gamma$ and $\sigma_2 = \eta\dot{\gamma}$, the constitutive equation will become

$$\sigma = G\gamma + \eta\dot{\gamma} \tag{1.101}$$

Here G and η are the constants of the model, which are far from being equivalent to the constants of the Maxwell model. Accordingly, these constants also differ in their physical significance, which is seen from the consideration of the basic features of the behaviour of the Kelvin-Voigt body under the simplest loading conditions.

Suppose, for example, that the Kelvin-Voigt body is subjected to a constant stress σ_0, beginning from the time $t = 0$. The boundary condition is the absence of strains at $t = 0$. Then, Eq. (1.101) takes the form

$$\gamma = \frac{\sigma_0}{G}(1 - e^{-t/\lambda})$$

where $\lambda = \eta/G$; this new parameter is called the retardation time since it characterizes the delayed response of the material to the applied stresses. Evidently, at $t \to \infty$ the solution of Eq. (1.101) turns into $\gamma(\infty) = \sigma_0/G$. From this we find the retardation function for the Kelvin-Voigt body:

$$\psi\,(t) = \frac{1}{G}\,(1 - e^{-t/\lambda})$$

the instantaneous compliance I_0 being equal to zero and the equilibrium compliance $I_\infty = \psi\,(\infty) = 1/G$. Thus, whereas for the Maxwell model G has the meaning of the instantaneous modulus and the equilibrium modulus is zero, for the Kelvin-Voigt model G has the meaning of the equilibrium modulus and the instantaneous modulus is infinitely large. Further, the viscosity of the Maxwell model is η and the viscosity of the Kelvin-Voigt model is infinitely large because it is a model of a solid body. The retardation time of the Maxwell body is equal to zero, i.e., it responds instantly to a change of load, and the relaxation time is equal to η/G; as regards the Kelvin-Voigt body, the relaxation time is equal to zero [which is directly seen from analysis of Eq. (1.101)], and the retardation time $\lambda = \eta/G$.

If in considering Eq. (1.101) under the $\sigma = 0$ conditions we assume that at $t = 0$ the deformation is equal to $\gamma\,(\infty) = \sigma_0/G$, the solution of the equation will then become

$$\gamma\,(t) = \frac{\sigma_0}{G}\,e^{-t/\lambda}$$

In other words, after the external load is removed there takes place an elastic recovery of the Kelvin-Voigt body, and in respect to time this process is a mirror image of the process of strain development at $\sigma_0 = $ const. It is also obvious that the deformation of the Kelvin-Voigt model is completely recoverable because $\lim \gamma$ at $t \to \infty$ under the conditions of elastic recovery is equal to zero.

Finally, let us examine the complete solution of Eq. (1.101) with an arbitrary law of the change of stress with time $\sigma\,(t)$. This solution has the form

$$\gamma\,(t) = \frac{1}{\eta}\int\limits_0^t \sigma\,(t')\,e^{-\frac{t-t'}{\lambda}}\,dt'$$

Completing the discussion of the Kelvin-Voigt model, we shall also give the expression for the frequency dependences of the components of the dynamic functions I' and I'':

$$I'\,(\omega) = \frac{1}{G}\,\frac{1}{1 + (\omega\lambda)^2}\,; \quad I''\,(\omega) = \frac{1}{G}\,\frac{(\omega\lambda)}{1 + (\omega\lambda)^2}\,; \quad \tan \delta = \omega\lambda = I''/I'$$

Passing over to the concept of the retardation time spectrum, we can show, in the same way as it was done for the Maxwell model, that the retardation spectrum of a body, whose properties are described by the Kelvin-Voigt model, is represented by a point with the argument λ. The further reasoning, which enables the generalization of the concept of a viscoelastic solid, is practically a repetition of the reasoning for a viscoelastic liquid, with the only difference that we are concerned here with the retardation time and not with the relaxation time. Accordingly, we can construct the generalized Kelvin-Voigt model (Fig. 1.20) possessing a retardation time spectrum, and the degeneracy of this model for a case, where the elastic modulus in one of the elements of the model is zero, turns this model of the viscoelastic solid into the model of the viscoelastic liquid. The simplest version of the last model corresponds to the Maxwell model. This completes the number of mechanical models and it turns out that the generalized Maxwell and Kelvin-Voigt models are equivalent to each other if in each of them one of the elements is degenerate: in one of the elements in the Maxwell model the viscosity becomes infinitely great, and in the Kelvin-Voigt model the modulus of one of the elements becomes equal to zero. It may thus be concluded that there exists a relationship between the relaxation and retardation time spectra. This has been pointed out earlier, in the discussion of the general theory of linear viscoelasticity.

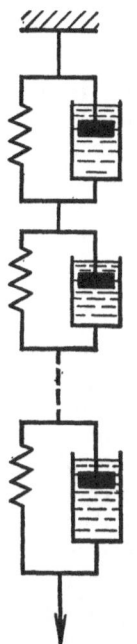

As has been shown in the discussion of the simplest viscoelastic liquid, the Maxwell body, to a single value of the relaxation time in the spectrum there corresponds the absence of retardation times. But if the spectrum contains several relaxation times, there appears a retardation time spectrum. In particular, retardation is detected in a liquid with two relaxation times. The following theorem can be proved: if the relaxation time spectrum of a viscoelastic body contains M discrete values (points), then the retardation time spectrum will consist of $(M-1)$ discrete values.

All the spectra considered above (of the Maxwell body, the Kelvin-Voigt body, and their generalized versions) were discrete or, as they are often called, linear. This was associated with the fact that the relaxation function was represented as the finite sum of the exponential functions, each with its particular values of the constants η_k and G_k.

This was represented by a set of discrete points in the relation spectrum with different values of arguments. In a general case, however, any relaxation function can be given as a continuous distribution of exponential functions,

Fig. 1.20. Generalized Kelvin-Voigt model.

so that to each of the values of relaxation time θ there will correspond a modulus $F(\theta)\,d\theta$ of its own, and the sum of exponential functions in this case will be replaced by an integral:

$$\varphi(t) = \int_0^\infty F(\theta)\, e^{-t/\theta} d\theta$$

It is evident that this expression for $\varphi(t)$ is in full agreement with the definition given above for the concept of the relaxation spectrum, the function $F(\theta)$ being continuous in this case. Thus, the concept of the relaxation function as the integral (1.86) is equivalent to the resolution of the relaxation process into a continuous series of processes with different relaxation times and the corresponding values of the moduli. Then, replacing the values of the moduli G_k in the expressions for $G'(\omega)$ and $G''(\omega)$ in a model with many relaxation times by $F(\theta)\,d\theta$, we arrive at the following formulas:

$$G'(\omega) = \int_0^\infty F(\theta)\, \frac{(\omega\theta)^2}{1+(\omega\theta)^2}\, d\theta \quad \text{and} \quad G''(\omega) = \int_0^\infty F(\theta)\, \frac{(\omega\theta)}{1+(\omega\theta)^2}\, d\theta$$

which are quite identical with the formulas derived in linear viscoelasticity theory for the general case of arbitrary spectra [see formulas (1.89) and (1.90)].

Examining the set of Maxwell elements with a continuous distribution of relaxation times under arbitrary deformation conditions, $\dot{\gamma}(t)$, we obtain the following integral:

$$\sigma(t) = \int\limits_0^\infty F(\theta) \int\limits_0^t \dot{\gamma}(t') e^{-\frac{t-t'}{\theta}} dt' \, d\theta \qquad (1.102)$$

which is nothing more than the expression of Boltzmann's superposition principle in terms of the relaxation spectrum, which can easily be seen if we substitute expression (1.86) for the relaxation function defined in terms of the relaxation time spectrum into formula (1.102) and take into account the possibility of existence of an equilibrium stress. Then we finally obtain:

$$\sigma(t) = \int\limits_0^t \dot{\gamma}(t') [G_0 + \varphi(t-t')] \, dt'$$

no deformations being present at $t < 0$.

Quite analogous arguments can be put forward as to the continuous distribution of retardation times, as a result of which it proves possible, on the basis of models, to derive all those formulas which have already been treated as the phenomenological concepts of linear viscoelasticity theory.

Thus, the assumption of the existence of a continuous distribution of relaxation or retardation times in the analysis of a combination of the Maxwell or Kelvin-Voigt models leads to the formulation of the Boltzmann-Volterra general principle, which yields all the above-considered relations of the linear theory of viscoelasticity. Of course, one should not think that the derivation of the Boltzmann-Volterra equations through analysis of the behaviour of a set of spring-dashpot models is based on conceptions other than the earlier used ideas of the mutual independency of successive effects on the material and of the linearity of its response to these effects. In implicit form the superposition principle is also employed in the modelling method since the assumption of the linearity of the properties of the spring and dashpot is in no way different from the assumption of the linearity of the response of the system to external effects. The idea that the stresses (or strains) in each of the elements of the generalized Maxwell or Kelvin-Voigt model develop independently of all the stresses (or strains) in the other elements, as a result of which the total stress or strain may be represented, respectively, as the sum of stresses or strains in the separate elements, signifies the application of the superposition principle in the modelling method.

Equations (1.79) and (1.80) may be regarded as integral constitutive equations for viscoelastic materials. But if the relaxation- or retardation-time spectra are discrete, the use of these integral constitutive equations is possible but inconvenient because one has to operate with delta functions in their practical utilization. Therefore if the relaxation-time spectrum is not continuous and consists of separate points ("lines"), the integral equations are usually replaced by differential equations, for which purpose use is made of the generalized Maxwell and Kelvin-Voigt models. We have already said above that such models correspond to the concept of a discrete relaxation- or retardation-time spectrum.

Let us consider, for example, a model with two relaxation times, which allows one to get an idea of the method of constructing differential constitutive equations for a material with a discrete relaxation-time distribution.

This model is shown in Fig. 1.21, which also gives the symbols of the constants. The constitutive equation for each element of the model has the form

$$\begin{cases} \dot{\gamma} = \dfrac{\dot{\sigma}_1}{G_1} + \dfrac{\sigma_1}{\eta_1}, \\[2ex] \dot{\gamma} = \dfrac{\dot{\sigma}_2}{G_2} + \dfrac{\sigma_2}{\eta_2} \end{cases}$$

where σ_1 and σ_2 are the stresses in the element of the model. In order to obtain the relation between the total stress $\sigma = \sigma_1 + \sigma_2$ and the strain γ,

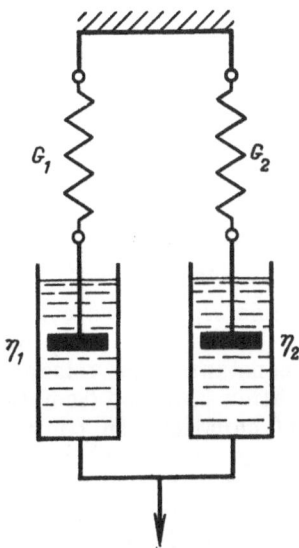

Fig. 1.21. A model of the type of the Maxwell model with two relaxation times.

the quantities σ_1 and σ_2 should be excluded. Appropriate manipulations lead to the following differential equation:

$$A_2 \ddot{\gamma} + A_1 \dot{\gamma} = B_2 \ddot{\sigma} + B_1 \dot{\sigma} + \sigma \tag{1.103}$$

Here the constants A_1, A_2, B_1, and B_2 are expressed in terms of the parameters of the model as follows:

$$A_1 = \eta_1 + \eta_2; \quad A_2 = \frac{G_1 + G_2}{G_1 G_2}\eta_1\eta_2; \quad B_1 = \frac{\eta_1}{G_1} + \frac{\eta_2}{G_2}; \quad B_2 = \frac{\eta_1\eta_2}{G_1 G_2}$$

It is obvious that the relaxation times of this model are equal to $\theta_1 = \eta_1/G_1$ and $\theta_2 = \eta_2/G_2$, and the retardation times are found by consideration of the $\sigma = $ const regime and analysis of the response of the model to constant stress. It is equal to

$$\lambda = \frac{A_2}{A_1} = \frac{\dfrac{1}{G_1} + \dfrac{1}{G_2}}{\dfrac{1}{\eta_1} + \dfrac{1}{\eta_2}}$$

The resulting differential equation (1.103) is a linear equation with constant coefficients. It may be written in the following way:

$$\sum_{n=1}^{2} A_n \frac{d^n}{dt^n} \dot{\gamma} = \sum_{n=0}^{2} B_n \frac{d^n}{dt^n} \sigma$$

The left- and right-hand sides of this equation contain the sums of differential operations or differential operators. It is essential that for a viscoelastic liquid the summation of the operations for the strain begins with $n = 1$, and for the stress, with $n = 0$, i.e., the minor term on the left-hand side of the equation is $\dot{\gamma}$, and that on the right-hand side is σ. The coefficient A_1 has the meaning of viscosity in steady-state flow (for a two-element model the viscosity $\eta = \eta_1 + \eta_2$).

If we now increase the number of Maxwell elements connected in parallel, to this there will then correspond the increase of the orders of the differential operators. For a viscoelastic liquid with an arbitrary number of discretely distributed relaxation times the constitutive equation may be represented in the form

$$\sum_{n=1}^{M} A_n \frac{d^n}{dt^n} \dot{\gamma} = \sum_{n=0}^{N} B_n \frac{d^n}{dt^n} \sigma \qquad (1.104)$$

where A_n and B_n are the constants associated with the parameters of the model; all the characteristic properties of the material are expressed in terms of these constants: the relaxation and retardation times, viscosity, etc.

The equation obtained (sometimes called the operator equation) is a linear differential equation. In consequence, the relaxation and creep functions prove to be independent of the specified stresses and strains, and the functions themselves have the form of the sum of the exponential functions. The operator equation (1.104) is a constitutive equation for a material with an arbitrary number of discretely distributed relaxation times. It may be looked upon as a special case of the integral equation (1.79) in which $\varphi(t) = \sum_k G_k e^{-t/\theta_k}$

and, hence, $F(\theta)$ in Eq. (1.86) may be regarded as the sum of delta functions. In practice, any relaxation function can always be represented in the form of the sum of a finite and usually small number of exponential functions. This leads to the possibility of using equations of the type (1.104) with a limited number of constants that characterize the properties of the material. Naturally, the use of a limited number of material constants is more convenient for calculations than the use of the continuous relaxation spectrum function in accordance with Eqs. (1.79) and (1.80), and this function must be determined over an infinitely wide range of variation of the argument, from 0 to ∞.

For theoretical calculations and estimations it is usually convenient to make use of an integral rheological equation of state since manipulations with continuous functions are facilitated by the application of many well-known results of mathematical analysis.

The integral constitutive equations (1.79) and (1.80) are the most general forms of linear stress-strain relations as in their derivation no assumptions have been made as to the nature of the relaxation and creep functions, and use has only been made of the superposition principle for linear responses of the material to external effects.

All the possible specific features of the properties of a linear viscoelastic material are defined more precisely within the framework of various types of relaxation spectra, i.e., the functions $F(\theta)$ and $\Phi(\theta)$ or the functions $\varphi(t)$ and $\psi(t)$. The same may be said about Eq. (1.104) since when a sufficiently high value of the upper summation limit is chosen it is possible to describe, as accurately as one wishes, any specific features of the properties of any particular material. This conclusion is not basically associated with the method of constructing the Maxwell and Kelvin-Voigt models and is based on the mathematical structure of the constitutive equations obtained.

One can conceive of the diverse combinations of the viscous (dashpots) and elastic (springs) elements and methods of connecting them into arbitrarily complicated chains and circuits, which could be used for a more or less pictorial model representation of the actual structure of a compound, but there remains valid the general statement that in any case the constitutive equation of an arbitrary circuit will be described by the constitutive equation (1.104) or by Eqs. (1.79) and (1.80) in a more general case.

1.8.6. Some Analytical Concepts of Viscoelasticity

The basic problem of the practical utilization of the general theory of linear viscoelasticity is the choice of the analytical representation of experimentally measured functions with the purpose of using the resulting expressions for calculating any other characteristics of the behaviour of a material under arbitrary deformation conditions. The main general difficulty in the solution of this basic problem is not so much the complex form of the functions measured as the fact that their values are always known in a more or less narrow but, in any event, always limited range of variation of the argument.

There exist two general ways to overcome this difficulty. One is to choose an analytical expression that will, to any desired accuracy, approximate experimental data, and to extrapolate that expression to the region of values of t from 0 to ∞ and to invert the integrals considered above. Here it is desirable to choose sufficiently simple or "convenient" (tabulated) analytical expressions. The second route consists of a graphical or any other numerical method of treatment of experimental data with the use of approximation methods of inversion of the integrals appearing in theoretical relations. The approximation methods of linear viscoelastic theory are widely used in practice and are considered in detail in a number of works.*

As regards the analytical methods, the use of the discrete relaxation spectrum described in the preceding section for the treatment of experimental data seems to be fruitful in the utilization of particular physical models, which sharply restricts the number of "free" constants determined by experiment. If formula (1.104) is considered to be purely empirical, then, of course, there can be chosen values of M and N large enough to describe experimental data with desired accuracy. However, this is beset with an essential difficulty due to the ambiguity of such a description. Indeed, the same data can be described by choosing the various values of A_n and B_n, depending on the

---

* Schwarzl, F. R. and L. C. E. Struik, in: *Advances in Molecular Relaxation Processes*, Elsevier Publishing Co., Amsterdam, 1, 201 (1967-68); Knoff, W. F. and I. L. Hopkins, *J. Appl. Polymer Sci.*, 16, 2963 (1972); Stanislav, J. and B. Hlaváček, *Trans. Soc. Rheol.*, 17, 331 (1973); Belov, M. A. and A. E. Bogdanovich, *Mekhanika Polimerov*, 5, 864 (1976).

choice of M and N. This leads to various predictions concerning other viscoelastic functions and to the loss of the physical significance of the constants found. But there exist no criteria for the selection of the values of M and N and, accordingly, of A_n and B_n, just as there are no estimates of errors that arise in using integral transformations because of the above-indicated ambiguity. Therefore, in spite of the seeming attractiveness of this approach, the scope of its practical utilization is limited, except for the case where its use is associated with the application of a physically substantiated mechanical model of a viscoelastic material containing a small number of arbitrary constants.

The use of continuous relaxation functions is resorted to mainly for analysis of the behaviour of solid polymeric materials at comparatively small deformations, when the relations of linear viscoelasticity theory are fulfilled. A very large number of such functions have been described in the literature. We shall therefore mention only some of them, those which are most frequently encountered in the description of the viscoelastic properties of various polymers. Thus, the analysis of relaxation phenomena is conveniently carried out with the frequency dependence of the complex compliance represented by the formula

$$I^* (\omega) = \frac{I_0}{(i\omega)^{\alpha+1} - \beta} \tag{1.105}$$

where I_0, α and β are empirical constants.

Expressions of the type of formula (1.105) are employed to represent experimental data on relaxation:

$$\varphi (t) = \frac{G_0 - G_\infty}{1 + (t/\tau_0)^a}$$

and creep:

$$\psi (t) = (I_\infty - I_0) \frac{(t/\tau_0)^b}{1 + (t/\tau_0)^b}$$

where τ_0, a, and b are empirical constants.

The convenience of all these formulas is due to the fact that their use enables the analytical calculation of the spectral function and the calculation of other characteristics of the viscoelastic behaviour of the polymer with the aid of that function.

Certain advantages are also exhibited by a similar, but a more general expression of the type

$$\frac{I^* (\omega) - I_0}{I_\infty - I_0} = [1 + (i\omega\tau_0)^{1-\alpha}]^{-\beta}$$

from which, at certain particular values of the constants α and β (for example, at $\beta = 1$ or at $\alpha = 0$ and simultaneously $\beta = 1$), there are obtained simple and extensively used representations of the viscoelastic properties of polymers. Without limiting the generality of this formula, one can, by selecting appropriate constants α and β, describe the sundry manifestations of the relaxation characterictics of polymeric materials with high accuracy.

The use of functions of the type of the Kohlrausch function may also be of practical value in describing relaxation:

$$\varphi (t) = (G_0 - G_\infty) e^{-(t/\tau_0)^n} \tag{1.106}$$

or creep:

$$\psi(t) = (I_\infty - I_0)[1 - e^{-(t/\tau_0)^m}] \tag{1.106a}$$

where τ_0, n, and m are empirical constants.

If we are dealing with a liquid ($G_\infty = 0$), whose relaxation properties are described by formula (1.106), we can express its viscosity in the following way:

$$\eta = \frac{G_0\tau_0}{n}\,\Gamma\left(\frac{1}{n}\right)$$

here $\Gamma\left(\frac{1}{n}\right)$ is the value of the gamma function at an argument equal to $1/n$, the value of n being, as a general rule, smaller than unity.

Calculations of spectral characteristics that are based on the use of functions of the Kohlrausch function type lead to rather complicated analytical expressions, and the corresponding formulas for $G^*(\omega)$ and $I^*(\omega)$ are unknown.

It should be noted in general that no sufficiently simple analytical expressions can be obtained for the entire set of viscoelastic characteristics of the material. Therefore, the following situation is typical here: the comparative simplicity and vividness of the analytical representation of any one of the viscoelastic functions chosen for the treatment of experimental data turn into an extremely unwieldy and complicated procedure and sometimes it becomes even impossible to obtain analytical expressions for the functions calculated in accordance with the scheme presented in Fig. 1.16. This once again stresses the expediency of using approximate calculation methods of transformation of some characteristics into others, which becomes especially attractive in view of the fantastic possibilities of modern computing technique.

An important exception that allows one to make successive use of analytical methods of calculation is the representation of the spectral function $H(\theta)$ in the form of a rectangle or wedge or their combination.* Indeed, if in a certain inverval $a \leqslant \theta \leqslant b$

$$H(\theta) = A = \text{const}$$

and beyond its limits $H(\theta) = 0$, then in the same interval**

$$L(\lambda) = \left\{A\left[\left(\ln\frac{b-\lambda}{\lambda-a}\right)^2 + \pi^2\right]\right\}^{-1}$$

And if in the interval $c \leqslant \theta \leqslant d$

$$H(\theta) = K\theta^{-1/2}$$

and beyond its limits $H(\theta) = 0$, then in the same interval**

$$L(\lambda) = \left\{K\lambda^{-1/2}\left[\left(\ln\frac{\sqrt{d}+\sqrt{\lambda}}{\sqrt{d}-\sqrt{\lambda}} \cdot \frac{\sqrt{\lambda}-\sqrt{c}}{\sqrt{\lambda}+\sqrt{c}}\right)^2 + \pi^2\right]\right\}^{-1}$$

Proceeding from these formulas, one can calculate the other viscoelastic functions of the polymer, and the summation of the expressions for the spectrum leads to the same summation of the functions being calculated.

---

* This approach to the description of dielectric and viscoelastic relaxations has been used by many authors, especially by A. Tobolsky and his school.
** Smith, T., *Trans. Soc. Rheol.*, 2, 131 (1958).

7*

The simplicity of approximation of the actual properties of the material with the aid of rectangular and wedge-like spectra (or of their combinations) is rather attractive. However, the corresponding functions describe the behaviour of real materials only semiquantitatively at best. Nonetheless, such simple functions are very useful since they can be used to test the existing approximate numerical methods of transformation of the various functions approximating the actual behaviour of the material with any desired accuracy.

1.8.7. Brief Conclusions

The phenomenological description of the properties of real materials—polymeric systems—is based on the concept of three idealized bodies: viscous bodies—on deformation the entire external work is dissipated; elastic bodies—all the work done by the external forces is stored; and linear viscoelastic bodies—the work of the external forces is partly stored and partly dissipated.

Linear viscoelasticity theory permits us to describe the behaviour of materials under various transient deformation conditions, i.e., when the time dependence of stresses or strains becomes crucially important. In the limiting case of long times the relations of this theory lead to the simplest dependences: the linear dependence of stresses on the rate of deformation for a linear viscoelastic liquid and the linear dependence of stresses on strains for a viscoelastic solid. Hence, under the conditions of the applicability of linear viscoelasticity theory the rheological properties of a liquid in steady-state flow obey Newton's law, and those of a solid under the conditions of equilibrium deformation obey Hooke's law.

Depending on the particular specific features of the relaxation properties of linear viscoelastic materials the ratios of the work dissipated to the work stored may be different but the relations between stresses and strains or rates of strain must remain linear.

The development of the quantitative concepts of the properties and behaviour of real materials is based on the idea of different combinations of the properties inherent in the above-considered idealized systems. This makes it possible to set up constitutive equations describing the behaviour of non-linear viscoelastic materials.

1.9. Nonlinear Theories of Viscoelasticity

The above-expounded ideas of elastic bodies, viscous liquids, and linear viscoelastic materials constitute the theoretical foundations of modern conceptions of the rheological properties of polymers. They rely on the model description of the behaviour of polymers as continua under the simplest deformation conditions. Thus, the model of an elastic body describes a set of the equilibrium states of the material, while the model of a viscous liquid simulates the behaviour of the material in steady-state shearing flow; the model of a viscoelastic body with a linear stress-strain relation imitates the various deformation regimes at small (tending-to-zero) stresses, strains, and rates of strain. All these cases are extreme cases among a large variety of possible deformation processes, but they are at the same time the most important since any sophisticated theory of the rheological properties of polymeric systems must satisfy the regularities of their behaviour under the simplest conditions indicated.

Fluid polymeric systems are characterized by the following three most important features of their properties, which must be predicted by constitutive equations: non-Newtonian behaviour in shearing flow, the possibility of large recoverable deformations, and the viscoelastic response to external forces—the retarded progress of deformation and relaxation processes. These properties have been discussed earlier (Sections 1.6-1.8). Below we shall expound the general principles and methods of joint description of these fundamental features of the rheological properties of polymeric systems which are regarded as non-linear viscoelastic materials.

The theoretical results to be discussed refer to liquids that do not change their volume of deformation, and therefore we shall deal here with the components of the stress-tensor deviator.

1.9.1. Stresses as a Functional of the Deformation History

The viscoelastic character of the response of the material to an external force signifies that the stresses acting in the immediate neighbourhood of a certain point of the material at a given moment of time depend on the strain history of the corresponding elementary volume. Therefore, the stresses π'_{kl} referred to the moving and deforming convected coordinate system with coordinates X_i, which is connected with the given point, are in a general case expressed as the functional f of the strain tensor $\{\gamma\}$ which depends on time t:

$$\pi'_{kl} = f \begin{bmatrix} t \\ \{\gamma\}(t) \\ -\infty \end{bmatrix} \qquad (1.107)$$

This constitutive equation describes, in a general form, the properties of a wide class of liquids, on deformation of which the stresses at a certain point are determined by the strain history only in its immediate neighbourhood. Liquids of this class were termed (rheologically) simple fluids by Noll.

The components of the stress tensor σ_{ij} referred to the fixed coordinate system x_i are expressed in terms of π'_{kl} with the aid of the rule of transformation of tensor components:

$$\sigma'_{ij} = \frac{\partial x_i}{\partial X_k} \frac{\partial x_j}{\partial X_l} \pi'_{kl}$$

This expression is the inverse of formula (1.34), according to which there is accomplished the change from the components $\Gamma_{\alpha\beta}$ of a certain tensor specified in the fixed coordinate system to the components of this tensor γ_{ij} in the convected coordinate system. Using the indicated change to formula (1.105), we get:

$$\sigma'_{ij}(t) = \frac{\partial x_i}{\partial X_k} \frac{\partial x_j}{\partial X_l} f \begin{bmatrix} t \\ \{\gamma\}(t) \\ -\infty \end{bmatrix}$$

The most important a priori feature that must be exhibited by the functional used to describe the rheological properties of polymeric systems is the principle of fading memory, which can be stated as follows: the influence of past deformations on the present stress is weaker for the distant past than for the recent past. The corresponding constitutive equations may be called the memory functionals.

The simplest functional possessing this property is the following linear functional:

$$\int_{-\infty}^{t} \varphi(t-\theta)\, d\gamma_{ij} = \int_{-\infty}^{t} \varphi(t-\theta)\, \dot{\gamma}_{ij}\, d\theta$$

where φ is the decreasing relaxation function.

This functional is the basis of the linear theory of viscoelasticity of small deformations.

The deformations in the written relation are referred to a point in the material (in the convected coordinate system). Therefore, the constitutive equation of a viscoelastic material with fading memory, which is based on this linear functional and referred to the space-fixed coordinate system, takes the form

$$\sigma'_{ij}(x_h, t) = 2 \int_{-\infty}^{t} \varphi(t-\theta) \frac{\partial X_m}{\partial x_i} \frac{\partial X_n}{\partial x_j} \dot{\gamma}_{mn}(X_k, \theta)\, d\theta \qquad (1.108)$$

where $\dot{\gamma}_{mn}$ are the components of the rate-of-strain tensor, which are determined at the instant of observation t at the point whose fixed coordinates at this time are (X_k).

This constitutive equation is an extension of the Boltzmann principle [see Eq. (1.79)] to the case of large deformations.

Further generalization of constitutive equations requires the introduction of non-linear functionals. In a general case, formula (1.107) may be written in the form of the expansion of the functional f into a series analogous to a Taylor's series in the expansion of functions. Then, the constitutive equation (1.107) written as the sum of integral functionals will assume the form

$$\pi'_{nm}(X_h, t) = 2 \int_{-\infty}^{t} \varphi_1(t-\theta)\, \dot{\gamma}_{mn}(\theta)\, d\theta +$$

$$+ \int_{-\infty}^{t} \int_{-\infty}^{t} \varphi_2(t-\theta_1; t-\theta_2)\, \dot{\gamma}_{an}(\theta_1)\, \dot{\gamma}_{ma}(\theta_2)\, d\theta_1\, d\theta_2 + \ldots \qquad (1.109)$$

Here the function $\varphi_1(t-\theta)$ is an ordinary relaxation function which is included in the equation of the linear theory of viscoelasticity, and $\varphi_2(t-\theta_1; t-\theta_2)$ is a new binary relaxation function which takes account of the superposition of the effect of various components of the tensor $\{\dot{\gamma}\}$ on the rate of stress relaxation. The subsequent terms include new relaxation functions of even higher orders. The series (1.109) is unlimited and contains, in a general case, integrals of orders as high as desired. It is the most general constitutive equation referred to the small neighbourhood of a point moving in space. Therefore, formula (1.109), in combination with the equation that defines the method of transformation from the components π'_{mn} to the quantities σ'_{ij}, is the most general constitutive equation of a non-linear viscoelastic material.

If this series is terminated after the sum of the first two terms, i.e., if we put

$$\pi'_{mn} = 2 \int\limits_{-\infty}^{t} \varphi_1 \dot{\gamma}_{mn}\, d\theta + \int\limits_{-\infty}^{t} \int\limits_{-\infty}^{t} \varphi_2 \dot{\gamma}_{an} \dot{\gamma}_{ma}\, d\theta_1\, d\theta_2 \qquad (1.110)$$

we shall obtain the integral constitutive equation of a memory viscoelastic material, whose properties are determined by two relaxation functions, φ_1 and φ_2.

The constitutive equation (1.110) is an analog of the equation for the Rivlin viscous fluid [see formula (1.71)] and the relation between the components of the stress and strain tensors for the Reiner elastic body [see formula (1.61)]. Thus, this constitutive equation is a generalization of the Reiner potentials and the Rivlin dissipative function to a viscoelastic material. Therefore, with short loading times the behaviour of the material whose rheological properties are defined by Eq. (1.110) is the same as that of the Reiner elastic body, and with long loading times, as that of the Rivlin viscous fluid. The character of stress variations with time is determined by the form of the relaxation functions—the linear function φ_1 and the binary function φ_2.

At small rates of deformation the main contribution to the stress in Eq. (1.109) is made by the first term; the second term and all the subsequent terms, which are responsible for the effects of second and higher orders, are negligibly small as compared with the first. Therefore, for the indicated restriction of deformation conditions formula (1.109) is reduced to the constitutive equation of a linear viscoelastic material.

Formula (1.110) with two terms does not provide a sufficient generality of the results for equilibrium and steady-state states of strain. Further generalization of this formula requires the inclusion of higher terms of the series, as follows from the general formula (1.109). However, this method of constructing general constitutive equations for viscoelastic materials is not used in practice since it is not clear how, based on experimental results, to find the binary and higher relaxation functions.

Another method of generalizing an integral constitutive equation of the memory type consists in using various measures of strain. This approach is based on the fact that in the limiting case corresponding to equilibrium deformation conditions the stresses depend on the strain tensors of various structures (see Sec. 1.6), and in transient, non-steady-state deformation conditions the time effects depend on the form of the relaxation function, which can be determined on the basis of measurements at small deformations.

In a general form, a constitutive equation of this type may be written thus:

$$\sigma_{ij} = \int\limits_{-\infty}^{t} \varphi\,(t-\theta)\, f\,[\gamma_{ij}\,(\theta)]\, d\theta \qquad (1.111)$$

where the form of the function $f\,(\gamma_{ij})$ is not determined in advance.

As special cases of the general non-linear relation (1.111) we give below two constitutive equations—one obtained by Spriggs and co-workers*:

---

* Spriggs, T. W., *Chem. Eng. Sci.*, **20**, 931 (1965); Spriggs, T. W., J. D. Huppler, and R. B. Bird, *Trans. Soc. Rheol.*, **10**, 191 (1966).

$$\sigma_{ij} = \int\limits_{-\infty}^{t} \varphi\,(t-\theta) \left[\left(1+\frac{\varepsilon}{2}\right) \gamma_{ij}^{F} + \frac{\varepsilon}{2}\,\gamma_{ij}^{G} \right] d\theta \qquad\qquad (1.112)$$

and the other derived by Bernstein, Kearsley, and Zapas (the BKZ equation)*:

$$\sigma_{ij} = \int\limits_{-\infty}^{t} [\varphi_{1}\,(t-\theta)\,\gamma_{ij}^{G} + \varphi_{2}\,(t-\theta)\,\gamma_{ij}^{F}]\,d\theta \qquad\qquad (1.113)$$

where φ, φ_{1}, and φ_{2} are relaxation functions of an arbitrary form; ε is the scalar parameter; γ_{ij}^{G} and γ_{ij}^{F} are the components of the Cauchy-Green and Finger strain tensors, respectively.

In the original BKZ theory the relaxation functions φ_{1} and φ_{2} were defined in terms of the strain-energy function of the material and its dependence on the invariants of the tensor $\{\gamma\}^{F}$, namely:

$$\varphi_{1} = \frac{\partial W}{\partial E_{2}^{F}}\,; \quad \varphi_{2} = \frac{\partial W}{\partial E_{1}^{F}}$$

Therefore the rheological properties of the material, as described by the BKZ constitutive equation, are determined by its strain-energy function, and here the strain should be referred to the current (given) moment of time corresponding to a certain variable state of the material.

These models may apply equally well to elastic bodies, when the tensors $\{\gamma\}^{G}$ and $\{\gamma\}^{F}$ are expressed in terms of the components of recoverable deformation, and to liquids.

In the latter case, considering that the deformation refers to the instantaneous state of the material, which is taken as the original state, the tensors $\{\gamma\}^{G}$ and $\{\gamma\}^{F}$ for simple shear occurring under the $\dot{\gamma} = $ const conditions are defined as follows:

$$\{\gamma\}^{G} = \begin{vmatrix} 1 & -\dot{\gamma}t & 0 \\ -\dot{\gamma}t & 1+(\dot{\gamma}t)^{2} & 0 \\ 0 & 0 & 1 \end{vmatrix}; \quad \{\gamma\}^{F} = \begin{vmatrix} 1+(\dot{\gamma}t)^{2} & \dot{\gamma}t & 0 \\ \dot{\gamma}t & 1 & 0 \\ 0 & 0 & 1 \end{vmatrix}$$

The exact form of the dependence of $\{\sigma\}$ on $\{\dot{\gamma}\}$ is determined by the choice of a constitutive equation and relaxation functions or an expression for the strain-energy function of the material.

1.9.2. Constitutive Equations of Systems with a Strain Dependent Spectrum

The use of various measures of strain and integral constitutive equations (1.111) may be regarded as being associated with the effect of an external force on the system, which results in a change of its relaxation properties.

---

* Bernstein, B., E. Kearsley, and L. Zapas, *Trans. Soc. Rheol.*, **7**, 391 (1963); **9**, 27 (1965); Zapas, L. and T. Craft, *J. Res. NBS*, **69A**, 541 (1965); Zapas, L., *J. Res. NBS*, **70A**, 525 (1966).

Indeed, formula (1.111) may be represented in the form

$$\sigma_{ij} = \int_{-\infty}^{t} \frac{\varphi_0 \, (t-\theta)}{F \, (\dot{\gamma}_{ij})} \, \dot{\gamma}_{ij} \, d\theta$$

where

$$F \, (\dot{\gamma}_{ij}) = \frac{\dot{\gamma}_{ij}}{f \, (\dot{\gamma}_{ij})}$$

The ratio $\varphi_0 \, (t - \theta)/F \, (\dot{\gamma}_{ij})$ may be regarded as a relaxation function which depends on the rate of deformation. This dependence must have a form invariant to the choice of coordinate axes, and therefore the effect of the stress intensity on the relaxation properties of the system must be expressed in terms of the invariants of the tensor $\{\dot{\gamma}\}$ or, as is usually assumed for viscous liquids, in terms of the second invariant T_2.

In a general case, the character of the effect of the stress intensity on the rheological properties of the system has not been determined, and the general constitutive equation for a material, whose relaxation properties are dependent on loading conditions, are represented as follows:

$$\sigma'_{ij} = \int_{-\infty}^{t} [\varphi \, (t-\theta); \, \dot{\gamma}_{ij}] \, \dot{\gamma}_{ij} \, (\theta) \, d\theta$$

A simple assumption that may serve as the basis of the theory of the rheological properties of a liquid with a relaxation spectrum dependent on straining conditions* is the assumption that with increasing intensity of loading on a polymeric system there occurs a partial breakdown of structural linkages. This leads to the suppression of slow relaxation processes, which is the stronger the higher the stress intensity. Since such a process is recoverable, the theory relates the non-linearity of the viscoelastic properties of polymeric systems to the thixotropic phenomena observed during the flow of polymers.

The relaxation-time distribution may be continuous, as in the above-considered integral constitutive equations, and discrete, as in models built up of Maxwell elements connected in parallel. For the sake of simplicity, let us consider the flow under the conditions of simple shear for a system with a continuous distribution of relaxation frequencies. In a certain differentially small part of the spectrum, whose relaxation frequency is in the range between s and $(s + ds)$, the effective modulus that characterizes this part of the spectrum is $N \, (s) \, ds$ and the viscosity is $N \, (s)/s \, ds$. The elastic energy $E \, (s) \, ds$, stored during shear flow in the structural elements responsible for relaxation with a frequency from s to $(s + ds)$, is equal to

$$E \, (s) \, ds = \frac{1}{2} \, N \, (s) \, ds \left[\int_{-\infty}^{t} \dot{\gamma} \, (\theta) \, e^{-(t-\theta)\,s} \, ds \right]^2$$

In the thixotropic theory of viscoelasticity, it is assumed that if this elastic energy reaches a certain critical value $E^* \, (s) \, ds$, the corresponding structural element will be "destroyed" and will no longer take part in the building-up of stresses. Then, the process of development of shear stresses and the changes

* Leonov, A. I. and G. V. Vinogradov, *Doklady Akad. Nauk SSSR*, 155, 406 (1964); Leonov, A. I., *Zhur. Prikl. Mekh. i Tekhn. Fiz.*, 4, 78 (1964); Leonov, A. I. and A. Ya. Malkin, *Zhur. Prikl. Mekh. i Tekhn. Fiz.*, 4, 107 (1965); *Izv. Akad. Nauk SSSR. Ser.: Mekh. zhidkosti i gaza*, 3, 184 (1968).

of the relaxational properties of the system will be written as follows:

$$\tau = \int\limits_{S(t)}^{\infty} N(s)\, ds \int\limits_{-\infty}^{t} \dot{\gamma}(\theta)\, e^{-(t-\theta)s}\, d\theta \qquad (1.114)$$

$$\frac{1}{2} N[S(t)] \left[\int\limits_{-\infty}^{t} \dot{\gamma}(\theta)\, e^{-(t-\theta)S(t)}\, d\theta \right]^2 = E^* \,|_{s=S(t)} \qquad (1.115)$$

Equation (1.114) is an extension of the Boltzmann-Volterra principle to the case of thixotropic viscoelastic materials with a continuous distribution of relaxation times. Equation (1.115) serves for the determination of the auxiliary function $S(t)$ which describes the change of the relaxation spectrum with time.

Thus, in the thixotropic theory of non-linear viscoelasticity, the properties of the system are described by two functions: the relaxation spectrum of the undeformed (undestroyed) system, $N(s)$, and the function that determines the distribution of critical energies, $E^*(s)$, among the structural units. Instead of the function $E^*(s)$, we may introduce the thixotropy function $\varphi(s)$, which is given by the following relation:

$$\varphi(s) = s \sqrt{\frac{2E^*(s)}{N(s)}}$$

In this case, equality (1.115) is written down as follows:

$$\varphi[S(t)] = S(t) \int\limits_{0}^{\infty} |\dot{\gamma}(\theta)|\, e^{-S(t)(t-\theta)} d\theta \qquad (1.116)$$

Equations (1.114) and (1.116) play an important part in this theory.

Proceeding from the theory of thixotropic viscoelasticity, let us examine the variation of shear stresses during the flow taking place with a constant velocity gradient.

For this case the basic relations of the theory—formulas (1.114) and (1.116)—assume the form

$$\tau(t) = \dot{\gamma} \int\limits_{S(t)}^{\infty} \frac{N(S)}{s} (1 - e^{-st})\, ds$$

$$\varphi[S(t)] = \dot{\gamma}[1 - e^{-S(t)t}]$$

Of special interest is the analysis of the relationships obtained at $t \to \infty$ i.e., upon transition to steady-state flow conditions. In this case

$$\eta(\dot{\gamma}) = \frac{\tau(\infty)}{\dot{\gamma}} = \int\limits_{S(t\to\infty)}^{\infty} \frac{N(s)}{s}\, ds \qquad (1.117)$$

$$\varphi[S(t \to \infty)] = \dot{\gamma} \qquad (1.118)$$

The function $S(t \to \infty)$ depends on $\dot{\gamma}$. Let us introduce the notation $\omega_0 = \lim\limits_{t\to\infty} S$. Formulas (1.117) and (1.118) may be combined:

$$\eta[\varphi(\omega_0)] = \int\limits_{\omega_0}^{\infty} \frac{N(s)}{s}\, ds$$

or, since $\varphi(\omega_0) = \dot{\gamma}$, it follows that

$$\eta(\dot{\gamma}) = \int\limits_{\omega_0(\dot{\gamma})}^{\infty} \frac{N(s)}{s}\, ds \qquad\qquad (1.119)$$

Here $\omega_0(\dot{\gamma})$ is a function showing the relaxation frequency at which the spectrum is truncated (i.e., up to what s-element the structural bonds responsible for the relaxation spectrum are destroyed) at a specified rate of deformation $\dot{\gamma}$.

The analytical form of the function $\omega_0(\dot{\gamma})$ is not determined in the theory; for a qualitative analysis of the various deformation conditions we may put

$$\omega_0 = a\dot{\gamma}$$

A comparison of theory with experiment, which has been made for a number of polymeric systems, has shown that the coefficient a depends but slightly on the velocity gradient; it increases from 1 up to about 4 with increasing $\dot{\gamma}$ within several decimal orders.

The use of the indicated approximation of the function $\omega_0(\dot{\gamma})$ means that under the conditions of transition from rest to steady-state shearing flow at $\dot{\gamma} = \text{const}$ there occur such changes in polymeric systems which lead to the disappearance of the structural elements responsible for relaxation processes taking place at frequencies smaller by order of magnitude than the specified velocity gradient.

The basic idea of the thixotropic theory of viscoelasticity regarding the effect of deformation conditions on the relaxation properties of viscoelastic materials may take various quantitative forms, which leads to different rheological relations. They should be regarded as constitutive equations for materials with a relaxation spectrum dependent on the deformation conditions. As compared with the original model of thixotropic viscoelasticity, further detailing refers to the nature of the effect of the rate of deformation on the relaxation spectrum of the system. Thus, the model, according to which at a frequency $s = \omega_0$ there occurs a stepwise truncation of the relaxation spectrum, is only the first approximation to the actual picture of the phenomena involved. The next approximation may take into account the smooth change of the spectrum in the region of frequencies of the order of $\dot{\gamma}$. This approach is realized, for example[*], in the following constitutive equation (written down in the convected coordinate system):

$$\sigma_{ij} = 2 \int\limits_{-\infty}^{t} \frac{\varphi_0(t-\theta)}{1+\left(\sqrt{\frac{1}{2}T_2'\theta}\right)^a}\, \dot{\gamma}_{ij}(\theta)\, d\theta \qquad\qquad (1.120)$$

Here a is an empirical parameter. It shows the effect on the relaxation spectrum of the intensity of loading, which is determined by the second invariant of the rate-of-strain tensor T_2'. The relaxation function of the system, which depends on the deformation conditions, $\varphi(t - \theta; T_2')$, is expressed in terms of the original relaxation function $\varphi_0(t - \theta)$ of the undeformed system as

---

[*] For a review of various integral models, see Bogue, D. C. and J. D. Doughty, *Ind. Eng. Chem. Fund.*, 5, 243 (1966).

follows:

$$\varphi\left(t-\theta; T_2'\right) = \frac{\varphi_0\left(t-\theta\right)}{1+\left(\sqrt{\frac{1}{2}\, T_2'\theta}\,\right)^a}$$

At very short times, $\theta \ll (T_2')^{-1/2}$, the value of $\varphi \approx \varphi_0$ and the system is characterized by the practically non-varying relaxation properties. At $\theta \gg (T_2')^{-1/2}$ the function φ is practically equal to zero, and this means that the corresponding (slow) relaxation processes are completely suppressed. In the region $\theta \approx (T_2')^{-1/2}$, according to Eq. (1.121) the relaxation processes are gradually suppressed ($\varphi < \varphi_0$), the extent of suppression being the stronger the greater is $(T_2')^{-1/2}$ as compared with θ.

On deformation there may take place not only the suppression of the relaxation spectrum, when $\varphi \ll \varphi_0$ at any value of θ, but the faster relaxation processes may become more intense, so that some of the regions of the relaxation spectrum are filled. This is actually observed on deformation of certain polymeric systems.

Models with a varying spectrum may be combined with other specific features of the system, which are responsible for the non-linearity of its properties. Thus, it should be taken into account that the above-considered constitutive equations refer to the elementary volume of the material, i.e., are written in the convected coordinate system. For the change to the space-fixed coordinate system, it is necessary to make use of the above-described formulas, just as has been done, for example, in the generalization of the integral equation of linear viscoelasticity theory for the case of large deformations [see formula (1.108)].

1.9.3. Differential Nonlinear Constitutive Equations

Just as the constitutive equation of a linear viscoelastic liquid may be represented in the form of the integral relation (1.79) or in an alternative form — as the differential (operator) equation (1.104), so for a non-linear model of a viscoelastic body the constitutive equation may be represented as integral operators — memory functionals or in the form of non-linear differential constitutive equations with a limited number of constants. The main condition that must be taken into consideration in generating differential constitutive equations is the necessity to use tensor quantities and their time derivatives and also to correlate the coordinate systems in which the rheological relations between the components of the stress and rate-of-strain tensors are defined.

For a viscous liquid it is assumed that the components of the stress tensor $\{\sigma\}$ are linearly dependent on the rate of strain $\{\dot{\gamma}\}$. Generalizing this model, according to Rivlin and Ericksen, we assume that $\{\sigma\}$ depends not only on $\{\dot{\gamma}\}$ but also on the higher derivatives of the strain tensor $\{\gamma\}$, so that this dependence can be represented as the sum of the series of the derivatives of the strain tensor. Instead of the tensor $\{\dot{\gamma}\}$, the Rivlin-Ericksen model makes use of the tensor $\{A\}^{(1)} = 2\,\{\dot{\gamma}\}$. Accordingly,

$$A_{ij}^{(1)} = 2\dot{\gamma}_{ij}$$

The higher derivatives of the tensor $\{A\}^{(n)}$ are determined with account taken of the displacement of the coordinates upon movement of a point and are calculated in accordance with the rules worked out by Oldroyd, so that

$$A_{ij}^{(n)} = \frac{\partial A_{ij}^{(n-1)}}{\partial t} + v_k \frac{\partial A_{ij}^{(n-1)}}{\partial x_k} + \frac{\partial v_k}{\partial x_j} A_{ik}^{(n-1)} + \frac{\partial v_k}{\partial x_i} A_{kj}^{(n-1)} \qquad (1.121)$$

The tensor $\{A\}^{(n)}$ is called the Rivlin-Ericksen kinematic n-order tensor.

The possible forms of the relation between the components of the tensor $\{\sigma\}$ and the derivatives of the tensor $\{A\}^{(1)}$ are represented as the sum of the following series:

$$\sigma'_{ij} = \eta_0 A_{ij}^{(1)} + \beta A_{ik}^{(1)} A_{kj}^{(1)} + \gamma A_{ij}^{(2)} + \dots \qquad (1.122)$$

This expression is a generalization of the operator equation (1.104) if the right-hand side of the equality retains only one term—the zero-order term, and the time derivatives are understood in the Oldroyd sense.

The constitutive equation (1.121) describes the properties of a viscoelastic liquid, whose specific behaviour is determined by the number of terms in the series or by its higher term. If the higher term is given by the Rivlin-Ericksen n-order tensor, then such a constitutive equation describes an n-order fluid. As a rule, second-order fluids are considered in the literature; they differ from a Newtonian fluid by the presence of quadratic terms. In a general case, the coefficients η_0, β, and γ may not be constant; they may depend in some way on the invariants of the tensors $\{A\}^{(1)}$ and $\{A\}^{(2)}$.

For shearing flow with the velocity gradient $\dot{\gamma}$ the derivatives of the tensor $\{A\}^{(1)}$ assume an especially simple form:

$$\{A\}^{(1)} = \begin{vmatrix} 0 & 1 & 0 \\ 1 & 0 & 0 \\ 0 & 0 & 0 \end{vmatrix} \dot{\gamma}; \quad \{A\}^{(2)} = \begin{vmatrix} 0 & 0 & 0 \\ 0 & 1 & 0 \\ 0 & 0 & 0 \end{vmatrix} 2\dot{\gamma}^2$$

and all the higher derivatives of $\{A\}^{(n)}$ at $n > 2$ are identically equal to zero. Therefore, in steady-state shearing flow the Rivlin-Ericksen fluid of any order exhibits exactly the same properties as a second-order fluid.

The integral constitutive equation for a memory fluid with a linear and a binary relaxation function, which is generalized according to Oldroyd, predicts for steady-state flow the same relationship between stresses and rate of strain as the differential constitutive equation for a second-order fluid. Here the coefficients η_0, β, and γ are found to be the integral characteristics of the relaxation properties of the system, namely:

$$\eta_0 = \int_0^\infty \varphi_1(\theta)\, d\theta$$

$$\gamma = -\int_0^\infty \theta \varphi_1(\theta)\, d\theta \qquad (1.123)$$

$$\beta = \int_0^\infty \int_0^\infty \theta_1 \theta_2 \varphi_1(\theta_1)\, \varphi_2(\theta_2)\, d\theta_1\, d\theta_2$$

The representation of $\{\sigma\}$ as the sum of the series of the derivatives of the tensor $\{A\}^{(1)}$ is not the only possible method of non-linear generalization of differential constitutive equations for a viscoelastic liquid. Using, as an example, the tensor $\{A\}^{(1)}$ as the original one, we can obtain the higher derivatives by any method capable of rendering the constitutive equation invariant to the change from the convected to the fixed coordinate system. Thus, instead of the tensors $\{A\}^{(n)}$, White makes use, in a number of works*, of the tensors $\{B\}^{(n)}$ obtained in the following way:

$$\{B\}^{(1)} = \{A\}^{(1)}$$

$$B_{ij}^{(n)} = \frac{\partial B_{ij}^{(n-1)}}{\partial t} + v_k \frac{\partial B_{ij}^{(n-1)}}{\partial x_k} - \frac{\partial v_k}{\partial x_j} B_{ik}^{(n-1)} - \frac{\partial v_k}{\partial x_i} B_{kj}^{(n-1)} \qquad (1.124)$$

that is, $\{B\}^{(n)}$ is also calculated with the aid of the Oldroyd operator but with a negative non-linear (cross) terms. The representation of the constitutive equation with the aid of the tensors $\{B\}^{(n)}$ is quite analogous to formula (1.121). For simple shear

$$\{B\}^{(1)} = \{A\}^{(1)} = \begin{vmatrix} 0 & 1 & 0 \\ 1 & 0 & 0 \\ 0 & 0 & 0 \end{vmatrix} \dot{\gamma}; \quad \{B\}^{(2)} = \begin{vmatrix} -1 & 0 & 0 \\ 0 & 0 & 0 \\ 0 & 0 & 0 \end{vmatrix} 2\dot{\gamma}^2$$

$$\{B\}^{(n)} = 0 \text{ at } n > 2$$

Further possibilities of generalization of constitutive equations of the differential type are associated, first, with the use of the complete operator constitutive equation (1.104) containing an arbitrarily large number of terms both on the left- and on the right-hand side and, second, with the use in this equation of differential operators of complex structure.

Apart from the above-considered Oldroyd operator with positive and negative non-linear terms and the Jaumann operator [see formula (1.37)], use is made of more complex operators—the non-linear Oldroyd operator $D_{O,n}$ and the Spriggs operator D_S, which generalizes a series of simpler operators.

The non-linear Oldroyd operator $D_{O,n}$ applied to the components a_{ij} of a certain tensor $\{a\}$ is written as follows:

$$D_{O,n}a_{ij} = D_O a_{ij} + \frac{1}{3} a_{mn}\dot{\gamma}_{mn}\delta_{ij} \qquad (1.125)$$

where D_O is the ordinary ("linear") Oldroyd operator.

Spriggs suggested** using a differential operator of the following structure:

$$D_S a_{ij} = D_J a_{ij} + (1+\varepsilon)\left(a_{ik}\dot{\gamma}_{kj} + a_{kj}\dot{\gamma}_{ik} + \frac{1}{3} a_{mn}\dot{\gamma}_{mn}\delta_{ik}\right) \qquad (1.126)$$

where D_J is the Jaumann derivative; ε is an arbitrary parameter.

    * White, J. L. and A. B. Metzner, *J. Appl. Polymer Sci.*, 7, 1867 (1963); White, J. L., *J. Appl. Polymer Sci.*, 8, 2339 (1964); White, J. L. and N. Tokita, *J. Appl. Polymer Sci.*, 11, 321 (1967); White, J. L., *Rubb. Chem. Techn.*, 42, 257 (1969).
    ** Spriggs, T. W., *Chem. Eng. Sci.*, 20, 931 (1965).

At $\varepsilon = -1$ the Spriggs operator degenerates into the Jaumann derivative; at $\varepsilon = 0$ it degenerates into the non-linear Oldroyd operator. Therefore,
with respect to the above-considered operators the operator D_J is a generalized
expression of transformations that are expected to take place upon a change
from the convected to the fixed coordinate system.

Some concrete results of the use of operators of different structure in differential models of viscoelastic materials will be obtained in subsequent chapters and used for a theoretical explanation of experimental results for the
stresses and their interrelations in simple shear and uniaxial extension. At
this point we shall limit ourselves to indicating the paths and methods of
constructing non-linear constitutive equations of the differential type, which
generalize the operator constitutive equation for a linear viscoelastic material.

The use of non-linear operators of various structure in the differential constitutive equation of a viscoelastic liquid implies the presence of a discrete
set of relaxation times. With an arbitrary choice of the constants A_n and
B_n [see Eq. (1.104)], the various relaxation times of the system are not interrelated and are the independent parameters of the material. The number of
such parameters may be, generally speaking, very large, which considerably
complicates the concrete utilization of equations of this type. Within the
framework of this approach a substantial simplification can be attained if
we assume that the different relaxation times are interrelated and the entire
relaxation spectrum is determined by the maximum value of the relaxation
time θ_m and the factor characterizing the density of lines in the discrete spectrum. As a concrete example of such a model, let us consider the Spriggs four-
constant model which is a set of viscoelastic elements connected in parallel.
The constitutive equation for each of these elements has the form

$$\sigma_{ij}^p + \theta_p \dot{D}_S \sigma_{ij}^p = 2\eta_p \dot{\gamma}_{ij} \tag{1.127}$$

where σ_{ij}^p is the stress in the pth element of the model; θ_p and η_p are its parameters.

The total stress

$$\sigma_{ij} = \sum_{p=1}^{\infty} \sigma_{ij}^p$$

It is obvious that the structure of Eq. (1.127) is quite analogous to the constitutive equation for the Maxwell body with the partial derivative d/dt
being replaced by the operator D_S.

The relaxation times θ_p are expressed in terms of the maximum relaxation
time θ_m by means of the formula

$$\theta_p = \frac{\theta_m}{p^\alpha} \tag{1.128}$$

where α is the parameter which characterizes the density of distribution of
relaxation times in the spectrum.

The values of η_p are defined in terms of the largest Newtonian viscosity η_0 as follows:

$$\eta_p = \eta_0 \frac{\theta_p}{\sum\limits_{p=1}^{\infty} \theta} = \eta_0 \frac{1}{p^{\alpha} \zeta (\alpha)}$$

where $\zeta (\alpha)$ is the Riemann zeta-function.*

At small rates of deformation $D_S = \partial/\partial t$, and the Spriggs model turns into an ordinary discrete model of a linear viscoelastic body.

With great loading intensities the substantial role is played by the non-linear terms of the operator D_S [see formula (1.126)] which includes an arbitrary parameter ε. Thus, the Spriggs model contains altogether four parameters: θ_m, α, η_0, and ε which are to be determined experimentally. As will be shown in subsequent chapters, this model allows one to describe correctly many of the essential specific features of the rheological properties of non-linear viscoelastic polymeric systems.

In nonlinear theories of the differential type, use is made, as a general rule, of the rate-of-strain tensor $\{\dot{\gamma}\}$ which is the time derivative of the large-strain tensor $\{\gamma\}$ (or of the tensor $\{\gamma\}^G$). But we may use, as a kinematic tensor, a derivative of another measure of strain, just as in the integral constitutive equations use was made of the tensors $\{\gamma\}^G$ and $\{\gamma\}^F$ [see Eqs. (1.112) and (1.113)].

Let a system be characterized by a discrete set of relaxation times θ_{α} and by the corresponding moduli G_{α}, and also let the stress be the sum of stresses associated with all the relaxation mechanisms. Then, with an arbitrary change, with time, of the strain (in the region of small deformations):

$$\sigma_{ij} = \sum_{\alpha} G_{\alpha} \int_0^t \dot{\gamma}_{ij} (t') \, e^{-\frac{t-t'}{\theta_{\alpha}}} \, dt'$$

which is a generalization of the dependence of σ_{ij} on $\dot{\gamma}_{ij}$ for a set of Maxwell elements arranged in parallel [see Eq. (1.100) and its solution]. For large deformations the form of the constitutive equation is retained but, according to Pao, instead of $\{\dot{\gamma}\}$ use may be made of the time derivative of the tensor $\{\gamma\}^F$, the strain being calculated with respect to the instantaneous state of the material. Then

$$\sigma'_{ij} (t) = \sum_{\alpha} G_{\alpha} \int_0^t \dot{\gamma}_{ij}^F e^{-\frac{t-t'}{\theta_{\alpha}}} \, dt' \qquad\qquad (1.129)$$

---

* The Riemann zeta-function $\zeta (\alpha)$ is given by the formula

$$\zeta (\alpha) = \sum_{n=1}^{\infty} n^{-\alpha}$$

Its important partial value which is used, for example, in dealing with the mechanical models of polymers is

$$\zeta (\alpha) = \sum_{n=1}^{\infty} n^{-2} = \pi^2/6$$

where, on simple shearing flow, to which there corresponds the recoverable shear strain γ_e,

$$\gamma_{ij}^{F} = \begin{vmatrix} 1+\gamma_e^2 & \gamma_e & 0 \\ \gamma_e & 1 & 0 \\ 0 & 0 & 1 \end{vmatrix}$$

and

$$\dot{\gamma}_{ij}^{F} = \begin{vmatrix} 2\gamma_e\dot{\gamma} & \dot{\gamma} & 0 \\ \dot{\gamma} & 0 & 0 \\ 0 & 0 & 0 \end{vmatrix}$$

This approach is, in many respects, analogous to that used in integral theories in which the stress is brought into agreement with the various strain measures. The constitutive equation (1.129), however, differs essentially from the integral equations (1.112) and (1.113) by the fact that not the strains themselves but their time derivative is used in the Pao theory. This enables the explicit inclusion of the recoverable deformation γ_e in the relation between the velocity gradient in steady-state flow and the components of the stress tensor.

Equation (1.129) allows for the natural generalization with replacement of summation by integration, which makes it possible to change over from the discrete to the continuous relaxation-time spectrum.

Among the latest theories, which take into account the ability of fluid polymeric systems to exhibit large recoverable deformations, the theory* based on the non-equilibrium thermodynamics deserves consideration. It describes the behaviour of incompressible polymeric materials by means of a limited number of parameters. In the simplest case of a single relaxation time, use is made of the following system of equations:

$$\left.\begin{aligned} \{\sigma\}+P\delta &= 2\,\{\gamma\}^{F}W_1 - 2\,[\{\gamma\}^{F}]^{-1}+2\eta s\exp\left(\frac{\beta}{G_e}W_s\right)\{\dot{\gamma}\} \\ W_J &= \partial W/\partial I_J;\ W_s = \frac{W\,(I_1,\,I_2)+W\,(I_2,\,I_1)}{2} \end{aligned}\right\} \tag{1.130}$$

$$D_J\,\{\gamma\}^{F} - \{\gamma\}^{F}\,\{\dot{\gamma}\} - \{\dot{\gamma}\}\,\{\gamma\}^{F} + 2\,\{\gamma\}^{F}\,\{\dot{\gamma}_J\} = 0 \tag{1.131}$$

$$\{\dot{\gamma}_J\} = \frac{2}{\theta}\exp\left(-\frac{\beta}{G_e}W_s\right)\left(\{\gamma\}^{F}-\frac{I_1}{3}\delta\right)W_{s,1} - \left([\{\gamma\}^{F}]^{-1}-\frac{I_2}{3}\delta\right)W_{s,2} \tag{1.132}$$

$$I_1 = Sp\,\{\gamma\}^{F};\ I_2 = Sp\,[\{\gamma\}^{F}]^{-1};\ Sp\,\{\dot{\gamma}\} = 0 \tag{1.133}$$
$$\det\{\gamma\}^{F} = 1$$

Here $\{\sigma\}$ is the stress tensor; $\{\gamma\}^F$, $\{\dot{\gamma}_J\}$, and $\{\dot{\gamma}\}$ are, respectively, the elastic strain tensor (the Finger measure of strain), the irrecoverable strain tensor, and the rate-of-strain tensor; δ is the unit tensor; I_1 and I_2 are the invariants of the tensor $\{\gamma\}^F$; $W\,(I_1,\,I_2)$ is the elastic-strain (strain-energy) function borrowed from the statistical theory of rubber elasticity; p is the isotropic pressure; G_e is the modulus of rubbery deformation; θ is the relaxation time; S is the ratio of the retardation to the relaxation time; η is the viscosity;

* Leonov, A. I., *Rheol. Acta*, 15, 85 (1976).

D_J is the Jaumann derivative with respect to time, a dimensionless, empirically chosen parameter which characterizes the effect of the orientation of macromolecular chains.

Formula (1.130) establishes the relation between the stress, recoverable deformation, and the rate of deformation. Formula (1.131) defines the relation between the recoverable deformation and the rate of deformation. The dependence of $\{\dot{\gamma}_f\}$ on $\{\gamma\}^F$ is given by formula (1.132).

In a number of works*, the formulas given above have been used to solve the various problems for shear and longitudinal deformations and to compare the results obtained with experimental data.

A comparison of theoretical and experimental data has demonstrated that they are in satisfactory agreement. The effect of the orientation factor at $\beta \neq 0$ on the relaxation time in simple shear and uniaxial extension has been qualitatively examined.**

We have considered above the principal methods of constructing nonlinear theories of viscoelasticity for polymers and have given examples of rheological equations of state (constitutive equations) which refer to various groups of theories; these examples do not cover the numerous theories proposed in the literature on rheology. We have not aimed at a detailed discussion of the theories and their comparison with experiment. The subsequent chapters of the book will deal with some of the theories, by way of comparison of experimental findings with theoretical predictions.

The principal task of the present section has been to outline the present-day trend in theoretical investigations devoted to the description of the rheological properties of polymeric systems.

Figure 1.22 presents the relationships between the rheological equations of state describing materials having different properties, and also the relations between the theories used for a quantitative description of the specific features of the viscoelastic properties of polymers.

A more detailed treatment of the foundations of rheology and also of the problems involved can be found in a number of monographs, particularly in:

1. Astarita, G. and G. Marrucci, *Principles of Non-Newtonian Fluid Mechanics*, McGraw-Hill Book Company, London, 1974.
2. Christensen, R. M., *Theory of Viscoelasticity*. Academic Press, New York, 1971.
3. Coleman, B. D., H. Markovitz, W. Noll, *Viscometric Flows of Non-Newtonian Liquids. Theory and Experiment*, Springer, Berlin, 1966.
4. Ferry, J. D., *Viscoelastic Properties of Polymers*, John Wiley and Sons, Inc., New York, 1970.
5. Green, A. E. and J. E. Adkins, *Large Elastic Deformations*, Oxford University Press, 1960.
6. Gross, B., *Mathematical Structure of the Theories of Viscoelasticity*, Paris, Hermann, 1953.
7. Lockett, F. J., *Nonlinear Viscoelastic Solids*, Academic Press, London, 1972.
8. Middleman, S., *The Flow of High Polymers*, Wiley Interscience, 1968.
9. Sedov, L. I., *An Introduction to Continuum Mechanics*, Fizmatgiz, Moscow, 1962 (in Russian); *Continuum Mechanics*, in two volumes, Nauka Publishers, Moscow, 1970 (in Russian).

---

* Leonov, A. I., E. N. Lipkina, E. D. Pasknin, and A. N. Prokunin, *Rheol. Acta*, **15**, 411 (1976); Prokunin, A. N., A. I. Isayev, and E. N. Lipkina, *Mekhanika Polimerov*, **4**, 699 (1977); Prokunin, A. N. and N. G. Proskurnina, *Inzhenerno-Fizicheskii Zhurnal*, **36**, 42, (1979).

** Leonov, A. I., E. N. Lipkina, and A. N. Prokunin, *Prikladnaya Mekhanika i Tekhnicheskaya Fizika*, **4**, 86 (1976).

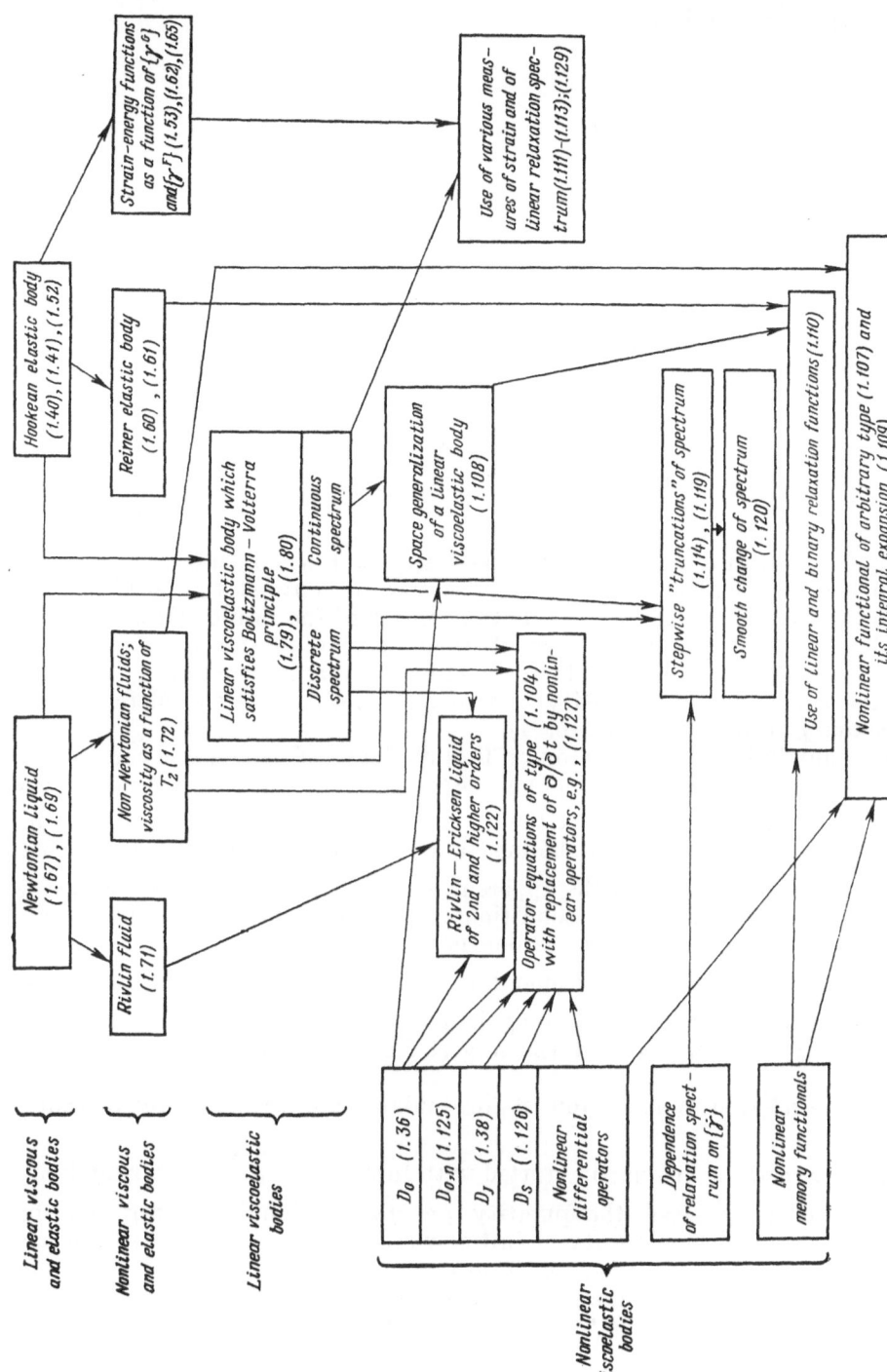

Fig. 1.22. Relationships among constitutive equations: for bodies having different properties (the numbers of the corresponding formulas are enclosed in parentheses).

CHAPTER 2

Shear Viscosity

2.1. Introduction

Among the various mechanical properties of polymer solutions, melts, and elastomers the most important property which readily lends itself to experimental investigation is viscosity in shearing flow. Usually, shear viscosity is defined as the proportionality factor relating the shear stress τ and the velocity gradient $\dot{\gamma}$ in simple shear:

$$\eta = \tau/\dot{\gamma}$$

In the literature devoted to the viscosity of low-molecular liquids, the coefficient η is called the dynamic viscosity in contrast to the kinematic viscosity ν equal to η/ρ, where ρ is the density of the liquid. However, taking into account the importance of dynamic methods of investigation of polymers, the proportionality factor η will be simply referred to as viscosity, and the term "dynamic viscosity" is reserved for η'; it was considered in viscoelasticity theory (see Chapter 1).

For many fluids called Newtonian liquids the coefficient η is a constant of a compound, which is dependent on temperature and pressure but not on the conditions of measurements. At the same time, the ratio $\tau/\dot{\gamma}$ may depend on the shear rate and stress and the mechanical history of the system. In such cases, the ratio $\tau/\dot{\gamma}$ is called the apparent viscosity. Materials whose viscosity depends on the deformation conditions are called non-Newtonian liquids.

Since the apparent viscosity is dependent on the rate of shear and shear stress, its limiting values corresponding to the conditions $\dot{\gamma} \to 0$ and $\dot{\gamma} \to \infty$ are important. The dependence $\eta(\dot{\gamma})$ is usually decreasing and therefore the quantity $\eta_0 = \lim_{\dot{\gamma} \to 0} \eta$ is called the largest Newtonian viscosity and $\eta_\infty = \lim_{\dot{\gamma} \to \infty} \eta$ is referred to as the lowest Newtonian viscosity. Since the measurement of viscosity is ordinarily started with low values of $\dot{\gamma}$ and τ, which are then gradually increased, the quantity η at $\dot{\gamma}$ and τ tending to zero may be termed the zero shear viscosity or initial viscosity ($\eta_{in} = \eta_0$). This term has the advantage that when it is used the other proportionality factors that relate the various variables, say the shear stress to the recoverable deformation and the normal stress difference to the square of the rate of shear, are uniquely termed.

The concept of viscosity has the unique meaning in steady-state flow when the entire increasing deformation is found to be irrecoverable. In nonsteady-state flow conditions, even when the inertia factor may be neglected, it is necessary to consider the fact that the rate of deformation $\dot{\gamma} = \dot{\gamma}_f + \dot{\gamma}_e$, where $\dot{\gamma}_f$ and $\dot{\gamma}_e$ are, respectively, the rates of irrecoverable and recoverable deformations. Regarding viscosity as the parameter that determines the ability of the material to dissipate the work in simple shear, it should be defined as the ratio $\tau/\dot{\gamma}_f$. This implies the division of the total deformation into the irrecoverable and the recoverable component. It should, however, be taken into consideration that in the literature the viscosity is sometimes defined as the ratio of the stress to the rate of total deformation.

The entire set of the dependences of the viscosity of polymer solutions and melts on temperature, pressure and deformation conditions determine their flow properties which are due to the nature and composition of the liquid concerned.

2.2. Temperature-Viscosity Relationships

The study of the temperature dependence of the viscosity of polymers is most important to the understanding of the mechanism of their flow process and to the elucidation of the relation between the structure of macromolecules and their behaviour on deformation. The temperature dependence of the viscosity of polymers has a substantial effect on their processibility since the sensitivity of viscosity towards a change in temperature governs not only the choice of the processing conditions but more often than not the quality of articles and the requirements set forth to the controlling and recording instruments.

At present, two approaches have been worked out in the consideration of the temperature dependence of viscosity. One of them is associated with the theory of absolute reaction rates and the other, with the free-volume theory. These two approaches consider the flow mechanism differently in many respects. Therefore, the final conclusions as to the form of the temperature dependence of viscosity are found to be different in these approaches.

According to the modern concepts, the elementary act of the flow process consists in that a molecular-kinetic unit overcomes the potential energy barrier upon transition from one equilibrium position to the next. For this to be done, it must possess sufficient energy and, besides, near the initial equilibrium position there must exist some free space—a "hole" to which there must correspond a new equilibrium position of the molecular-kinetic unit. The second requirement may be associated with the condition of the simultaneous alteration of the equilibrium positions of several structural-kinetic units. In such a case, the flow becomes a cooperative process.

If the probability of accumulation of the energy required to overcome the potential energy barrier (by analogy with chemical reactions it is called the energy of activation) is P_E, and the probability of formation of holes near the initial equilibrium position is P_v, then the total probability of a transfer— each elementary flow dispacement, P, will be given by

$$P = P_E P_v \tag{2.1}$$

The feature common to flow and chemical reactions is the transition of kinetic units (on a molecular basis) from one state of equilibrium to another, the potential energy barrier being overcome in both cases in elementary acts. Eyring, who has worked out the theory of absolute reaction rates and extended its basic propositions to diffusion processes and the flow of liquids, has made extensive use of the concepts of free volume and holes in liquids, but in fact the quantity P in Eq. (2.1) is identified by him with the quantity P'_E. Accordingly, in this theory the determination of the temperature dependence of viscosity boils down to the determination of the number of possible jumps of molecular-kinetic units over the potential barrier at various temperatures.

The general methods of the absolute rate theory lead to the following expression for the viscosity of liquids:

$$\eta = \left[\frac{hN}{V} \exp\left(-\Delta S^*/R\right) \right] \exp\left(\Delta H^*/RT\right) \tag{2.2}$$

where h is Planck's constant; N is the Avogadro number; V is the molar volume; ΔS^* and ΔH^* are, respectively, the entropy and heat of activation of the process of viscous flow; R is the gas constant; T is the absolute temperature.

Since the molar volume varies slightly with temperature, and the quantity ΔS^* is assumed to be independent of temperature, Eq. (2.2) may be rewritten as

$$\eta = B \exp\left(E/RT\right) \tag{2.3}$$

where E is the activation energy of the flow process; B is a constant.

An equation of the type (2.3) has been repeatedly discussed in the literature. It has been derived empirically by De Gusmàn (1913) and Arrhenius (1916); later it was derived theoretically by Frenkel in its kinetic theory of liquids (1925) and also by da Andrade (1934). Equation (2.3) will be called here the Arrhenius-Frenkel-Eyring formula (abbreviated to the AFE formula).

Another approach to the theory of the temperature dependence of viscosity is associated with the free volume concept. The idea that the fluidity of liquids is due to the presence of a free volume in them was first advanced by Batchinski (1913), who proposed a remarkably simple formula:

$$\eta^{-1} \sim (v - v_0) = v_f \tag{2.4}$$

that is, the fluidity equal to $1/\eta$ is directly proportional to the difference between the specific volume of the liquid, v, and the specific volume v_0 occupied by the molecules of the substance, or the viscosity is inversely proportional to the free volume v_f.

Formula (2.4) qualitatively correctly describes the temperature dependence of viscosity and has a profound physical significance. For many systems, however, especially for polymers in the fluid state, formula (2.4) proves to be nothing more than a crude approximation.

The free-volume theory as formulated at present is based on the observations by Doolittle, who has studied experimentally the viscosity of n-alkanes. He has found that the dependence of viscosity on free volume is expressed by the following equation:

$$\eta = A' \exp\left(B_0 v_0/v_f\right) \tag{2.5}$$

where A' and B_0 are constants.

An equation analogous in form is used by Fox and co-authors with the only difference that v_0 in Eq. (2.5) is replaced by v since $v_f \ll v$ (it will be shown quantitatively below).*

---

* Fox, T. G., S. Gratch, and S. Loshaek, in: *Rheology*, edited by F. R. Eirich, vol. I. Academic Press, New York, 1956.

For the temperature to be explicitly introduced into formula (2.5), the use of the temperature dependence of free volume is required. It will be natural to adopt the simplest form of this dependence. Thus, Williams, Landel, and Ferry* suggested that

$$v_f = v_{f,g} [1 + \alpha_0 (T - T_g)] \qquad (2.6)$$

where T_g is the glass-transition temperature; $v_{f,g}$ is the value of v_f at $T = T_g$; α_0 is the coefficient of thermal expansion of free volume.

This equation is valid at $T \gg T_g$. It is usually assumed that at $T < T_g$ the free volume remains unchanged and equal to $v_{f,g}$, which is due to the freezing of the conformational structure of polymeric chains in the region of temperatures lying below the glass-transition temperature. Therefore, if $T < T_g$, then the change of the volume occurs only as a result of the changes in the volume occupied by the molecules.

The specific volume of a liquid changes linearly with temperature:

$$v = v_g [1 + \alpha_l (T - T_g)] \qquad (2.7)$$

where α_l is the thermal-expansion coefficient of the liquid; v_g is the specific volume of the liquid at the glass-transition temperature.

Substituting Eqs. (2.6) and (2.7) into Eq. (2.5), we obtain:

$$\log \left(\frac{\eta}{\eta_g} \right) = \log a_T = \frac{B_0}{2.3} \left(\frac{\alpha_l}{\alpha_0} - 1 \right) \left(\frac{v_g}{v_{f,g}} \right) \frac{T - T_g}{\alpha_0^{-1} + (T - T_g)} \qquad (2.8)$$

where η_g is the viscosity of the polymer at the glass-transition temperature; $a_T = \eta/\eta_g$.

The constants appearing in formula (2.8) may be combined:

$$C_{1,g} = \frac{B_0}{2.3} \left(\frac{\alpha_l}{\alpha_0} - 1 \right) \left(\frac{v_g}{v_{f,g}} \right); \quad C_{2,g} = \alpha_0^{-1} \qquad (2.9)$$

Then Eq. (2.8) turns into the formula usually known as the Williams-Landel-Ferry equation (or the WLF equation):

$$\log \left(\frac{\eta}{\eta_g} \right) = \log a_T = \frac{C_{1,g} (T - T_g)}{C_{2,g} + (T - T_g)} \qquad (2.10)$$

It is ordinarily assumed that the WLF equation is valid for temperatures from the glass-transition temperature to $(T_g + 100°C)$.

In formula (2.10) the viscosity is referred to the quantity η_g, i.e., to the value of viscosity at the glass-transition temperature, and the temperature is compared with the glass temperature since the argument in Eq. (2.10) is the difference $(T - T_g)$. The constants $C_{1,g}$ and $C_{2,g}$ are expressed in terms of the thermal-expansion coefficients (α_l and α_0) and the values of the parameters at the glass temperature (v_g and $v_{f,g}$). Thus, the glass-transition temperature serves as the natural reference temperature. However, since it is difficult to measure viscosity at $T = T_g$, it is more convenient to choose a certain temperature $T_s > T_g$ as the reference temperature.

Let

$$T_s = T_g + \delta \qquad (2.11)$$

---

* Williams M. L. R. F. Landel, and J. D. Ferry, *J. Am. Chem. Soc.*, 77, 3701 (1955).

If now in Eq. (2.10) instead of T_g we introduce T_s and replace, accordingly, η_g by η_s, the WLF formula will become:

$$\log \left(\frac{\eta}{\eta_s} \right) = \frac{C_{1,s}(T-T_s)}{C_{2,s}+(T-T_s)} \tag{2.12}$$

The new constants $C_{1,s}$ and $C_{2,s}$ are expressed in terms of $C_{1,g}$, $C_{2,g}$, and δ as follows:

$$C_{1,s} = \frac{C_{1,g}C_{2,g}}{C_{2,g}+\delta}; \quad C_{2,s} = C_{2,g}+\delta$$

The temperature T_s may be chosen arbitrarily. It is often assumed that $T_s \approx T_g + 50°C$.

The condition of constancy of v_f at $T < T_g$ is not obligatory in the derivation of Eq. (2.10) since this derivation does not require the knowledge of $v_f(T)$ at $T < T_g$. Of interest is the assumption that the dependence $v_f(T)$ is linear in the region of temperatures $T > T_0$, where T_0 is a certain temperature lying below T_g, the value of v_f being equal to zero at $T = T_0$. If it is assumed that at $T > T_0$ the change of the specific volume is completely due to the change of free volume, then

$$\Delta v = v_f = \alpha_0' (T - T_0) \tag{2.13}$$

where α_0' is a constant.

In this case, from Eq. (2.5) there is obtained the following expression:

$$\log \eta = \log A' + (B_0 v_0/2.3\alpha_0') (T - T_0)^{-1}$$

or

$$\eta = A \exp [B/ (T-T_0)] \tag{2.14}$$

where

$$B = B_0 v_0/2.3\alpha_0'$$

Relation (2.14) is the well-known formula proposed by Fulcher for molten inorganic glasses and by Tammann and Hesse for supercooled organic liquids. Thus, this formula, which is often called the Fulcher-Tammann formula, may be looked upon as a corollary of one of the modifications of the free-volume theory. Indeed, it is not difficult to modify formula (2.10) into formula (2.14). This is done by means of the following transformations:

$$\log \eta = \frac{C_{1,g}(T-T_g)}{C_{2,g}+(T-T_g)} + \log \eta_g = (\log \eta_g + C_{1,g}) - \frac{C_{1,g}C_{2,g}}{T-(T_g-C_{2,g})} =$$
$$= \log A' + \frac{B}{2.3(T-T_0)}$$

From this it is immediately seen that the constants A' and B are expressed through the coefficients $C_{1,g}$ and $C_{2,g}$, and the quantity $C_{2,g}$ is related to the temperature difference:

$$C_{2,g} = T_g - T_0$$

The reverse conversion from formula (2.14) to a formula of the WLF equation type is also valid, which gives in a general case:

$$\log \eta = \log A + \frac{B (T-T_g)}{C+(T-T_g)} \tag{2.15}$$

The above-considered versions of the free-volume theory, while yielding identical expressions for the function $\eta (T)$, become different with respect to the temperature dependence of the occupied volume predicted by them. According to Fulcher and Tammann, at $T > T_0$ the value of $v_0 = $ const, whereas according to formulas (2.7) and (2.8)

$$v_0 = v - v_f = (v_g - v_{f,g}) + (v_g \alpha_l - v_{f,g} \alpha_0)(T - T_g)$$

The above-considered conceptions concerning the variation of the specific volume of the polymer and of its specific free volume with temperature is spectacularly illustrated by Fig. 2.1. Here the thick solid line shows the experimentally observed change of the specific volume with temperature. At $T < T_g$ the change of the volume occurs in proportion to the thermal-ex-

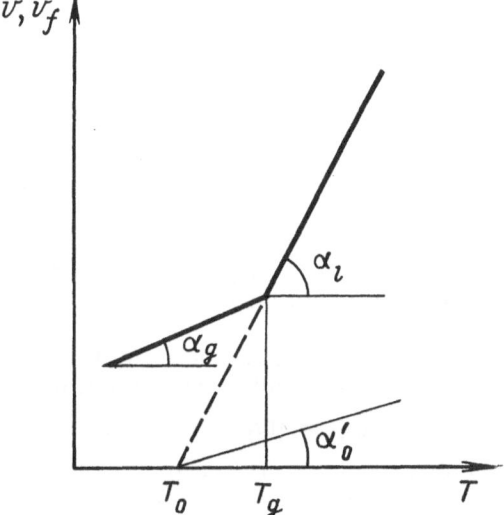

Fig. 2.1. Diagram illustrating the variation of the specific volume and specific free volume of polymers as a function of temperature.

pansion coefficient of the glass, α_g, and at $T > T_g$ the volume changes in proportion to the thermal-expansion coefficient of the liquid, α_l.

To the change of the specific volume along the broken line at temperatures below T_g there corresponds a hypothetical case of infinitely slow cooling, when at each temperature there is a particular equilibrium value of the volume. But in actuality, at a finite rate of cooling, because of the very high viscosity of the polymer in the region of temperatures of the order of T_g the volume has no chance to reach an equilibrium value and its change occurs as indicated by the solid line. Therefore, over the temperature range from T_g to T_0 the value of free volume may be found to be greater than the equilibrium value. Hence, the dependence of the glass-transition temperature on the rate of cooling is due to the difference between the free volume and its equilibrium value. Therefore, the point of inflection on the thick solid line in Fig. 2.1 may correspond to different values of the temperature which is experimentally adopted as the glass-transition temperature. An important consequence of Fig. 2.1 is the existence of the lower hypothetical limit of T_g since on equilibrium cooling down to T_0 the non-equilibrium portion of the free volume disappears completely. In Fig. 2.1 the dash line intersects the abscissa at a v value corresponding to the occupied volume at T_0.

As has been shown above, formula (2.14) is close in meaning to the WLF equation. If we assume that $T_0 = 0$, formula (2.14) turns into the simplest exponential AFE equation. Therefore, formula (2.14) is intermediate between the two extreme cases, to each of which there corresponds a limiting value of the temperature T_0, namely $T_0 = 0$ and $T_0 = T_g$. The introduction of the temperature T_0 enables one to describe satisfactorily the well-known numerous experimental data on the temperature dependence of the viscosity of various polymeric systems.

It is well known that the experimentally measured values of the glass-transition temperature are determined by the kinetics of cooling of the polymer. The more slowly its state is changed, the lower is the observed value of the glass-transition temperature. This fact forms the basis for the treatment of the glass-transition process as a kinetic transition.

The non-equilibrium nature of the glass-transition process, which is reflected in the dependence of the experimentally observed value of T_g on the kinetic factor, does not nevertheless deny the possibility of a thermodynamic treatment of this phenomenon. This possibility is due, first, to the existence of the equilibrium transition temperature T_0. Second, one may introduce* an additional structural parameter Z, which at $T > T_g$ is uniquely determined by thermodynamic variables—the temperature T and the pressure p, and at $T < T_g$ in a general case it plays an independent role, characterizing the degree of departure of the system from the state of equilibrium. The existence of a structure in "amorphous" polymers at $T < T_g$ is also confirmed by direct methods.** Then, on the comparison condition $Z = $ const there is possible a thermodynamic approach to the description of the glass transition and, in particular, there holds the formula that describes the shift of the transition temperature, depending on the hydrostatic pressure:

$$(dT_g/dp)_{Z=\text{const}} = \Delta\beta/\Delta\alpha$$

where $\Delta\beta$ and $\Delta\alpha$ are the jumps of the isothermal compressibility and heat capacity upon transition, respectively.

If the condition $Z = $ const is not fulfilled, then in a general case $dT_g/dp \neq \Delta\beta/\Delta\alpha$. The attainment of the limiting transition temperature T_0 is tied up, according to the standpoint expounded here, with the condition of the extremal nature of the additional (hidden) parameter Z since Z depends on the rate of cooling, thereby predetermining the well-known effect of the history of the glassy polymer on its physical and mechanical properties.

The values of thermodynamic parameters of the liquid, such as the entropy, exceed the corresponding values for the crystalline phase which is in equilibrium with the liquid. Here the difference between the indicated quantities diminishes with decreasing temperature. There is such a characteristic temperature, T_2 (this temperature being lower than T_g) at which this difference becomes equal to zero. Here the temperature T_2 has thermodynamic meaning of a phase transition of the second kind. Therefore, for T_2 there is fulfilled the Ehrenfest relation***

$$\frac{dT_2}{dP} = \frac{Tv\Delta\alpha}{\Delta C_p} = \frac{\Delta\beta}{\Delta\alpha}$$

where the left-hand side contains the derivative of T_2 with respect to the hydrostatic pressure; v is the specific volume; T is the temperature; and ΔC_p

* Gee, G., *Polymer* (England), **7**, 177 (1966).

** Zalwert, S., *Macromol. Chem.*, **177**, 3083 (1976).

*** DiMarzio, E. A., J. H. Gibbs, P. D. Fleming, and I C Sanchez, *Macromolecules*, 9, 763 (1976).

is the jump of heat capacity at the transition point. The concrete estimations carried out by Miller* on the basis of available experimental data for various glass-forming low-molecular-mass liquids and polymers have shown that the excess entropy ΔS disappears at a temperature T_2 close to T_0 contained in formula (2.14) for the temperature dependence of viscosity. An interesting exception is polystyrene for which $T_0 = 323$ K and $T_2 = 281$ (±15) K. The explanation of this fact is associated with the assumption that at T_0 there must disappear not the total excess entropy ΔS but its conformational component $\Delta S_c < \Delta S$, which is connected with the isomeric transitions upon rotation of the groups forming the polymeric chain relative to the bonds in this chain. For polyethylene and some other polymers, the difference between ΔS_c and ΔS is negligibly small, but in polystyrene and in other polymers containing phenyl rings the rotation of the bulky phenyl group relative to the main chain makes a significant contribution to the heat capacity and, hence, to ΔS but not to ΔS_c. Therefore, for such polymers $T_0 > T_2$.** The viscosity must become infinitely high when the conformational component of the excess entropy $\Delta S_c = 0$, i.e., at $T = T_0$, but from the foregoing it follows that the absence of segmental motions in the macromolecular chain may not be associated with the requirement $\Delta S = 0$, which is satisfied at a lower temperature T_2.

The general conclusion from the foregoing is that the characteristic temperature T_0 that determines the function η (T) can really be given a definite thermodynamic meaning associated with the concept of the excess conformational entropy.

An interesting interpretation may also be applied to the constant B in formula (2.14). Thus, for the homologous series of polyethylenes $(T_g - T_0)/2.3 \times \times B = 0.027$, and therefore $B = 640$ K. If we now consider the region of very high temperatures when $T \gg T_0$, the limiting value of activation energy will be defined as $E_\infty = 2.3\,RB = 12.1$ kJ/mole. At $T \gg T_0$ the free volume increases to such an extent that the change of the conformations of the macromolecular chain is found to be the limiting (rate-controlling) elementary act of viscous flow, which is attributed to the *trans-gauche* conversions. To these conversions there must correspond the activation energy obtained above for the region of very high temperatures. Indeed, the quantity E_∞ calculated for polyethylene is in good agreement with the value of the potential barrier of the indicated isomeric conversion for the methyl groups in ethane, which is known from independent measurements.

The above-considered thermodynamic and kinematic interpretation of the constants T_0 and B in the Fulcher-Tammann equation is of great practical and theoretical interest, though it needs to be confirmed more extensively by experimental data for various polymers.

Before concluding the discussion of the physical significance of the value of the temperature T_0, it should be pointed out that in amorphous polymers below the glass-transition temperature there has been detected, by means of various methods, in particular by the dynamic method, the existence of a low-temperature β-transition, which is interpreted by Boyer as the transition from one glassy state to another.*** Experimentally, this transition often

---

* Miller, A. A., *J. Polymer Sci.*, A-2, 4, 415 (1966); *J. Chem. Phys.*, 49, 1393 (1968); *Macromolecules*, 2, 355 (1969); 3, 674 (1970).

** Miller, A. A., *Amer. Chem. Soc. Polymer Preprints*, 18, 485 (1977).

*** Boyer, R. B., *J. Polymer Sci.*, C, 14, 3 (1966); *Polymer Eng. Sci.*, 8, 161 (1968).

manifests itself as the attainment of the brittle temperature. The possibility is not excluded that in its physical meaning this transition bears a relation to the characteristic temperature T_0, though this problem cannot yet be considered to have been clarified. According to a number of literature data $T_g/T_\beta \approx 1.15$-1.33, where T_β is the transition temperature determined from the position of the loss maximum on the temperature dependence of the dynamic properties of the polymer. For example, for polyethylene $T_g/T_0 = 1.25$; this value falls within the indicated range of values of T_g/T_β.

The WLF formula was originally thought to be empirical, but the obvious success of its utilization in a very large number of cases has led to the supposition that it can be substantiated on the basis of the physical conceptions of the structure of polymers. One of such theories has been proposed by Cohen and Turnbull.* They have derived formula (2.5) by proceeding from the molecular-kinetic concepts. In this theory it is assumed that the free volume per particle is v_f'. It is presumed that the distribution of free volumes, $P(v_f')$, satisfies the following requirement: the number of rearrangements of particles and holes must be the maximum possible, i.e., the entropy of the system must be maximal. From this quite natural thermodynamic condition there emerges the following form of the free-volume distribution:

$$P(v_f') = (c/\bar{v_f'}) \exp(-cv^*/\bar{v_f'})$$

where $\bar{v_f'}$ is the average value of the free volume per particle; c is a numerical coefficient ($1 \geqslant c \geqslant 1/2$) introduced in order to take into account the possible overlap of neighbouring free volumes.

For the elementary act of jumping it is necessary that a hole of size v^* be formed, the value of v^* being approximately equal to 10 v_f'. This relation is immaterial to the theory in its general form but it is of interest in comparing the scales of the quantities being considered. The probability of existence of holes for which $v_f' > v^*$ can be calculated as follows:

$$P(v_f' > v^*) = \int_{v^*}^{\infty} P(v_f') \, dv_f' \approx \exp(-cv^*/\bar{v_f'})$$

It is natural to presume that the fluidity is directly proportional to $P(v_f' > v^*)$ and the viscosity is inversely proportional to this quantity. Then

$$\eta = A \exp(cv^*/\bar{v_f'}) \tag{2.16}$$

where A is a constant. The identity of formulas (2.5) and (2.16) is quite obvious.

The attempt to establish a relation between the two basic theories, the activation theory and the free-volume theory, has been made by Macedo and Litovitz.** It is based on formula (2.1), according to which the probability of an elementary flow displacement, P, is determined by the probability P_v of formation of a hole and the probability P_E of accumulation of the energy required to cross the potential barrier of the transfer of particles from one equilibrium position to another. The quantity P_v is determined in the Cohen-Turnbull theory:

$$P_v = \exp(-cv^*/v_f') \tag{2.17}$$

---

* Cohen, M. H. and D. Turnbull, *J. Chem. Phys.*, **31**, 1164 (1959).
** Macedo, P. B. and T. A. Litovitz, *J. Chem. Phys.*, **42**, 245 (1965).

The quantity P_E is calculated in the activation theory. The main outcome of these calculations may be represented in a somewhat simplified form:

$$P_E \approx \exp\left(-E_v^*/RT\right) \tag{2.18}$$

where E_v^* is the activation energy at constant volume.

Since $\eta = A/(P_v \cdot P_E)$, it follows that

$$\eta = A \exp\left(\frac{cv^*}{v_f'} + \frac{E_v^*}{RT}\right) \tag{2.19}$$

Formula (2.19) is a natural generalization of the activation theory and the free-volume theory.

If, as has been done earlier, we assume that the expansion of the polymer occurs only due to the increase of free volume, while the volume occupied by molecules is not changed, then from Eq. (2.19) there is obtained a formula which is a natural generalization of the Fulcher-Tammann and Arrhenius-Frenkel-Eyring formulas:

$$\eta = A \exp\left(\frac{B}{T-T_0} + \frac{E_v^*}{RT}\right) \tag{2.20}$$

If $T \gg T_0$, then formula (2.20) turns into a simplest exponential relation, i.e., the liquid behaves as the Eyring liquid. At $E_v^*/RT \ll v^*/v_f'$, formula (2.20) changes to the Fulcher-Tammann formula. Obviously, if formula (2.19) and the condition $E_v^*/RT \ll v^*/v_f' \approx v_0/v_f'$ are taken into account, we can arrive at the WLF formula. In extreme cases the factor P_E or P_v plays the dominant role. In a general case, both factors must be considered.

Thus, in the region close to the glass-transition temperature, the free volume and its change with temperature play a determinative role. At sufficiently high temperatures the rate of activation processes becomes important. Therefore, in the region of temperatures close to T_g the flow properties of polymeric systems are well described by the universal WLF equation, and when the temperature rises the AFE equation is found to be more exact since it takes account of the activation character of the flow process.

Taking into account the complexity of a theoretical description of the dependence $\eta(T)$ for real polymers, empirical methods of calculation of viscosity at arbitrary temperatures may be of practical interest. One of such methods has been suggested by van Krevelen and Hoftyzer.* Although based on empirical correlations, it makes use, as the determining parameters, of the characteristic constants of the polymer, T_g (the temperature is estimated by the extent of remoteness from T_g) and the activation energy at $T \gg T_g$, where the flow is of the purely activation character. Therefore, basically both main factors that determine the temperature dependence of the viscosity of molten polymers are taken into account in this method in empirical form.

Interesting is a comparison of theoretical equations corresponding to the extreme situations for the temperature dependence of viscosity with experimental data. Typical examples of the temperature dependence of viscosity for a number of polymers of differing chemical nature are presented in Fig. 2.2 and those for a series of solutions of polystyrene in dibenzyl ether are given in Fig. 2.3. The data given are of illustrative value since the dependence on temperature is determined not only by the nature of the polymeric chain

---

* Krevelen, D. W. van and P. J. Hoftyzer, *Angew. Makromol. Chem.*, 52, 101 (1976); *Properties of Polymers*, Ch. 15, 2nd edition, Elsevier Publ. Co., Amsterdam, 1976.

but also by such factors as, for example, the presence or absence of side branches, etc. The data presented in Figs. 2.2 and 2.3, however, permit two points to be emphasized: first, the dependence $\eta(T)$ over a wide range of variation of temperature is not described by the AFE equation; second, if narrow ranges

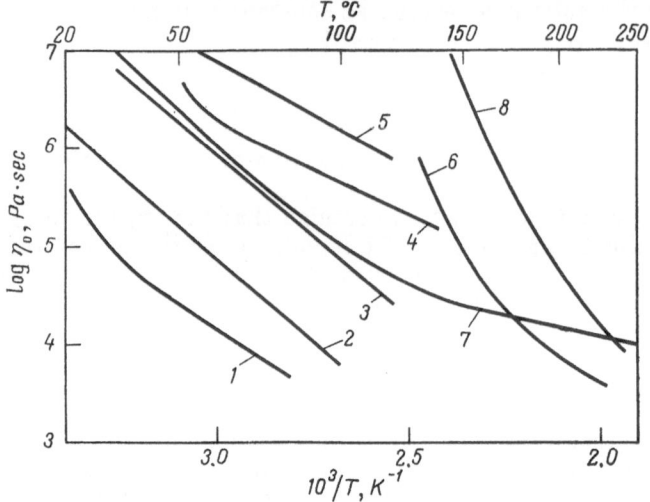

Fig. 2.2. Typical temperature dependences of the viscosity of some polymers:

1—polyisobutylene, molecular mass $\overline{M}_v \approx 8 \times 10^4$; *2*—ditto, molecular mass 1×10^5; *3*—butyl rubber (molecular mass 1×10^5) plasticized with 15 per cent of a mineral oil; *4*—natural rubber; *5*—styrene-butadiene rubber (87 : 23), molecular mass 3×10^5; *6*—high-pressure polyethylene; *7*—ethylene-propylene copolymer at 25°C; *8*—polystyrene, molecular mass 3.6×10^5.

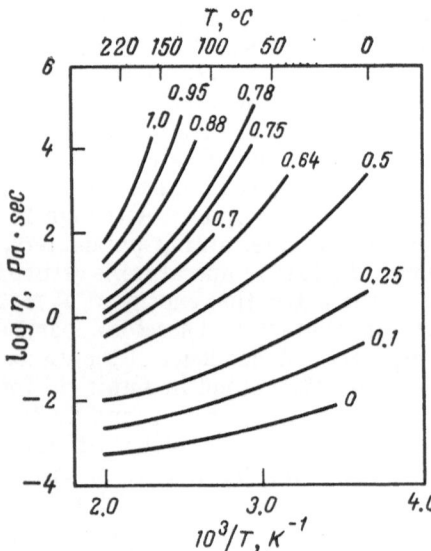

Fig. 2.3. Temperature dependence of the viscosity of solutions of polystyrene in dibenzyl ether. The numbers on the curves indicate the mass fractions of the polymer (after T. Fox and co-workers).

of temperature variation are considered, then any portion of the curve can be straightened up and the concept of the "apparent" activation energy calculated as $R \, [d \ln \eta/d \, (T^{-1})]$ may be used to characterize it. As a matter of

fact, in those cases where the activation energy does not depend on temperature, $R[d \ln \eta/d(T^{-1})]$ is not equal to the true activation energy. This can
easily be seen if we perform appropriate calculations, namely,

$$R\frac{d \ln \eta}{d(T^{-1})} = E(T) + \frac{1}{T}\frac{dE}{d(T^{-1})} \qquad (2.21)$$

Since the second term on the right-hand side of the equality is greater than
zero, the true activation energy at a given temperature is really smaller than
$R[d \ln \eta/d(T^{-1})]$.

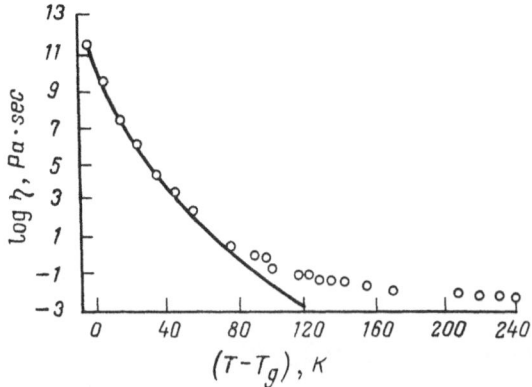

Fig. 2.4. Temperature dependence of the viscosity of 1,3,5-trinaphthylbenzene (after
T. Fox and coworkers). The solid line corresponds to Eq. (2.14) with the following values
of the parameters: $T_g = 343$ K; $T_0 = 193$ K; B—1.05×10^4.

As regards the comparison of experimental data on the temperature dependence of the viscosity of the polymer with the Fulcher-Tammann formula or

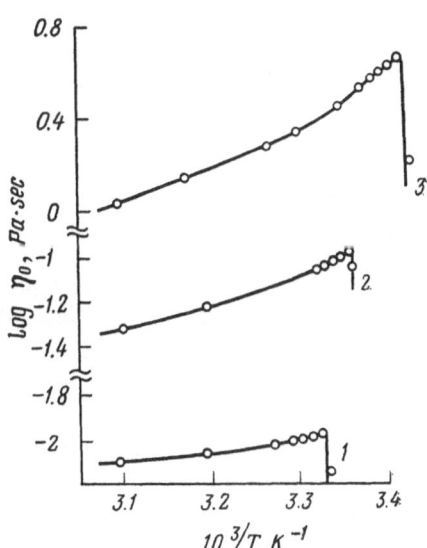

Fig. 2.5. Temperature dependence of
the viscosity of solutions of polystyrene in cyclohexane at the following
contents of the polymer (in volume
fractions): 0.052 (1); 0.101 (2); 0.198 (3).

with its analog, the WLF equation, one can always choose, over a temperature range of the order of 100°C, the constant T_0 in such a way as to describe

experimental data with the aid of these equations. But, even for low-molecular-mass liquids, whose viscosity can be measured over a wide range of temperatures, the Fulcher-Tammann formula proves unsatisfactory. This is seen from Fig. 2.4 where the solid line shows the dependence of $\log \eta$ on $(T - T_g)$ for 1,3,5-tri-α-naphthylbenzene; the constants of the indicated equation have been found for the region of temperatures $(T - T_g) < 50°C$.

Finally, it should be noted that there exist anomalous cases of the temperature dependence of viscosity, when the viscosity does not decrease but increases with rise of temperature. This effect is usually associated with specific interactions, say, on transition from a true to a colloidal solution near the phase-separation temperature. This has been demonstrated* for solutions of polystyrene in cyclohexane (Fig. 2.5) and other solutions in measurements of viscosity in the vicinity of the critical temperature.** The anomalous course of the temperature dependence of viscosity has also been observed*** for solutions of poly-2-hydroxyethyl methacrylate in aqueous solutions of urea (Fig. 2.6); it has been noted that this effect is connected with the existence of the upper critical mixing temperature, upon approach to which the interaction between two macromolecules is enhanced, and is characteristic of concentrated solutions with a negative enthalpy of mixing.

2.3. The Activation Energy of Flow

2.3.1. Newtonian Viscosity

The concept of the activation energy of flow is of fundamental importance to the activation theory of the flow of liquids. Of great interest is its experimental determination and the relationship between the activation energy and the structure and composition of polymers.

On the basis of Eq. (2.3) the activation energy is defined as the slope of the straight line that represents the dependence of viscosity on temperature in the coordinates $\ln \eta$ and T^{-1}. If this dependence is found to be nonlinear, it means that the activation energy (in the Eyring sense) is a function of temperature. In this case, as a general rule, as the temperature falls the potential energy barriers that are overcome in an elementary flow displacement mount. This can be accounted for either by the formation of new intermolecular bonds or by the increase of the number of bonds that must be overcome in the elementary flow act.

The WLF equation also shows that the activation energy of the flow process as defined by Eq. (2.3) must depend on temperature. Indeed, according to Eq. (2.12),

$$-R\frac{d \ln \eta}{d\,(T^{-1})}=2.3C_{1,s}C_{2,s}\,\frac{RT^2}{(C_{2,s}+T-T_s)^2} \qquad (2.22)$$

that is, the activation energy is not constant. This is a consequence of the fact that the use of the concept of the activation energy and the relations in

* Tager, A. A., V. E. Dreval, and K. G. Khabarova, *Vysokomol. Soedin.*, 6, 1593 (1964).

** Kobayashi, M., H. Ikeda, and Y. Masuda, *Nippon Kagaku Kaishi (J. Chem. Soc. Japan, Chem. and Ind. Chem.)*, 6, 866 (1977).

*** Quadrat, O. and P. Bradna, *Macromol. Chem.*, 178, 2953 (1977); Quadrat, O. and M. Bohdanecký, *Europ. Polymer J.*, 14, 335 (1978).

which it appears is justified only sufficiently far away from the glass-transition temperature.

In some cases there is observed a jumpwise change of E at a certain characteristic temperature. It means that at this temperature a structural (or relaxation) transition occurs, which leads to the appearance of a new mechanism of flow, as has been observed, for example, for polyvinyl chloride*. Phenomena of this kind may also arise in block copolymers and in the vicinity of the so-called $T_{l,l}$-transition observed in molten or amorphous polymers above their glass-transition temperature.**

According to the Eyring theory, the activation energy of flow must be of the same order of magnitude as the heat of vaporization E_{vap}. Indeed, it has

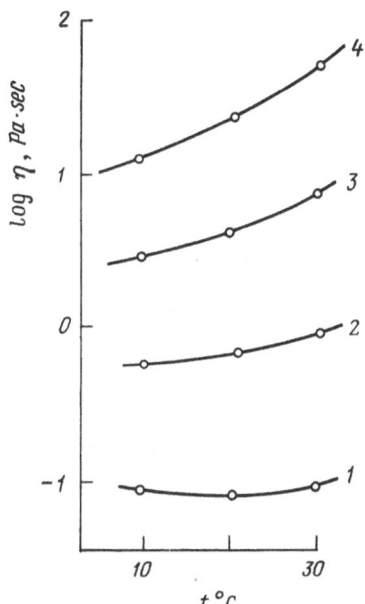

Fig. 2.6. Temperature dependence of the viscosity of solutions of poly-2-hydroxyethylmethacrylate in 8 M aqueous solution of urea. The polymer concentration, g/dl: 7.54 (1); 11.0 (2); 14.2 (3); 17.4 (4).

been found that for many low-molecular compounds the activation energy of viscous flow is about one-fourth of their heat of vaporization. A very important regularity has been established for the homologous series of n-alkanes. For the first members of this series $E = 1/4\ E_{vap}$. As the chain length increases, however, the value of E begins to gradually lag behind $1/4\ E_{vap}$, tending to a certain limit. This regularity is well seen in the left part of Fig. 2.7.

Kauzmann and Eyring (1940), who have established the indicated regularity, presume that as the length of the molecule increases the size of the molecular-kinetic unit becomes less than the total length of the molecule: the flow becomes segmental. If it is assumed that the work required for the formation of a hole big enough for a segment to occupy is equal, as before, to one-fourth of the heat of vaporization of a particle whose size is determined by the

---

* Münstedt, *J. Macromol. Sci.*, B14, 195 (1977); Collins, E. A., *Pure and Appl. Chem.*, **49**, 581 (1977).
** Gillham, J. K. and R. F. Boyer, *Amer. Chem. Soc. Polymer Preprints*, **17**, 171 (1976); **18**, 623 (1977); *J. Macromol. Sci.*, B13, 497 (1977).

length of the segment, then, using the data of Fig. 2.7a, one can find its length. The results of appropriate calculations are presented in Fig. 2.7b. The dotted line in Fig. 2.7b corresponds to the hypothetical case of the equality of the lengths of the segment and the molecule.

The values of activation energy of the flow of simple liquids are expressed in kJ/mole. The same unit of measurement is used for polymers. In the case

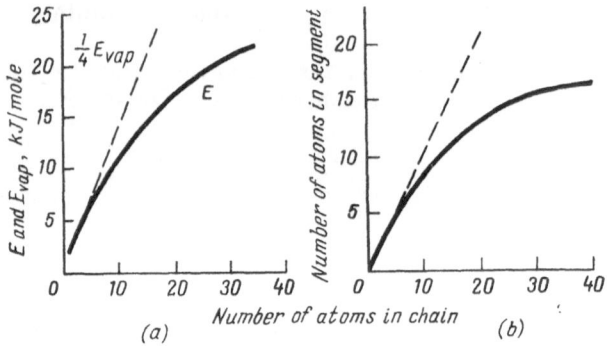

Fig. 2.7. Dependences of the activation energy of viscous flow, E, and one-fourth of the heat of vaporization E_{vap} (a) and also of the length of the segment,—a molecular-kinetic unit of flow, on the number of carbon atoms in molecules of n-alkanes (b). The dashed line in the graph on the right corresponds to the equality of the numbers of carbon atoms in the segment and in a molecule of n-alkanes.

of high-molecular-mass compounds the value of E_i must be referred to a mole of segments.

Figure 2.8 gives the results of an experimental determination of the activation energy of flow for the homologous series of polydimethyl siloxanes and polybutadienes containing 80 per cent of cis-groups.

From what has been said above and also from the data shown in Fig. 2.8 it follows that with increasing length of the macromolecule the segment be-

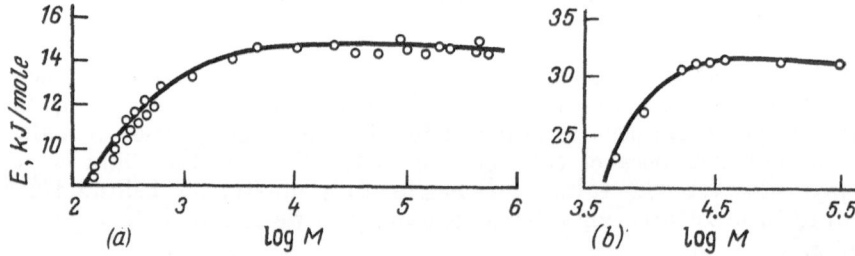

Fig. 2.8. Activation energy versus molecular mass for the viscous flow of polydimethyl siloxanes (a) [Kataoka, T. and S. Ueda, *J. Polymer Sci.*, B, 4, 317 (1966)] and cis-polybutadienes (b). [Vinogradov, G. V., A. Ya. Malkin, and V. G. Kulichikhin, *Vysokomol. Soedin.*, A10, 2522, (1968).]

comes an increasingly smaller fraction of its size. In very long molecules there is attained a limiting value of the length of the segment which consists of about 20-50 carbon atoms in the case of linear hydrocarbons. The activation energy of the viscous flow of a hydrocarbon of such a length is 25-29 kJ/mole. Indeed, for high-molecular-mass linear (low-pressure) polyethylenes the activation energy is of the order of 25-29 kJ/mole. Thus, we can speak of the activation energy of the flow of a polymer, without specifying its molecular mass since for polymers of sufficiently high molecular masses the activation energy is no longer dependent on molecular mass. It may thus be concluded that the activation energy of the flow of polymers does not depend on their molecular-mass distribution, at least if they do not contain noticeable amounts of low-molecular-mass fractions.

The activation energy of the flow of some polymers may be, in numerical values, commensurate with the chemical bond energy in macromolecules. This sometimes leads one to believe that the flow of polymers in such cases is mechano-chemical in nature, i.e., occurs as a result of the breakdown of macromolecules, the movement and recombination of radicals. Though mechano-chemical processes may really accompany the flow of polymers, a purely physical mechanism of flow is undoubtedly possible.

The lowest activation energy of flow is exhibited by high-molecular-mass compounds with a high chain flexibility and a weak intermolecular interaction. Such polymers are polydimethyl siloxanes (15 kJ/mole), cis-polybutadienes (about 20 kJ/mole), and linear polyethylenes (25-29 kJ/mole). Introduction of side branches into the polymeric chain increases the size of the segment and leads to an increase of the quantity E. In going from polybutadienes to polyisoprenes E increases to 55 kJ/mole. Almost the same result is obtained by the statistical copolymerization of polybutadiene with styrene. With polypropylene and polyoxymethylene E increases as compared with linear polyethylene by about 45 kJ/mole, and in the case of polyisobutylene it increases up to 54-59 kJ/mole. Replacement of the methyl radical in polypropylene by a phenyl radical (in polystyrene) doubles the activation energy of the flow of the polymer. The activation energy of flow is even higher in the case of polycarbonate (85 kJ/mole) and also of fluorinated derivatives of polyolefins and polyvinyl chloride. Thus, for the copolymer of tetrafluoroethylene with hexafluoropropylene and polyvinyl chloride the values of E are evaluated as 125 and 145 kJ/mole. Very high values of activation energy of flow are exhibited by derivatives of cellulose, which are characterized by especially high rigidity of macromolecular chains.

From what has been said above it follows that the value of E is affected by factors that determine the flexibility and interactions of macromolecules, primarily by the microstructure of the chain, the content of polar groups, the regularly or randomly distributed side branches and the quality of the solvent in concentrated solutions.

As regards the first of these factors, we may quote the considerable increase of the activation energy of the flow of polybutadienes with increasing content of cis-form in the macromolecular chain (Fig. 2.9).*

The quantity E is very sensitive to the presence of branches in the chain of linear polymers. Thus, in going from linear to high-pressure polyethylenes, in which there are 5-10 side branches per 100 atoms of the backbone chain, the value of E increases from 25-29 to 38-50 kJ/mole.

An interesting empirical correlation between E and the volume of side groups in macromolecular chains of linear polymers has been given by Porter and Johnson (Fig. 2.10).**

---

* Vinogradov, G. V., A. Ya. Malkin, and V. G. Kulichikhin, *Vysokomol. Soedin.*, A10, 2522 (1968); *J. Polymer Sci.*, A-2, 8, 333 (1970).
** Porter, R. S. and J. F. Johnson, *J. Polymer Sci.*, C, 15, 365, 373 (1966).

A more general approach to the correlation between E and the characteristic of linear polymers is based on the utilization of the parameter σ, which determines the thermodynamic flexibility of the macromolecular chain and is calculated from relation $\sigma^2 = \overline{L_0^2}/\overline{L_f^2}$ (where $\overline{L_0^2}$ is the mean-square distance between the ends of the unperturbed macromolecule in a theta-solvent; $\overline{L_f^2}$ is the same for a freely jointed chain at fixed valence angles). The indicated correlation is given in Fig. 2.11. It shows that the determining effect on the activation energy of the flow of polymers is exerted by the flexibility of the polymeric chain.

Since the lower polyolefins (polyethylenes and polypropylenes) are of great practical importance, attention has been paid to the comparison of their

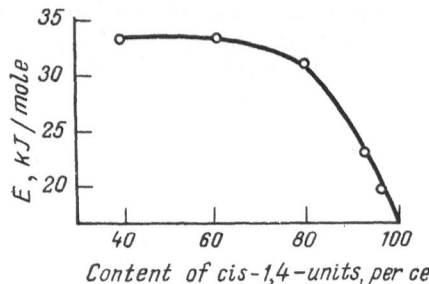

Fig. 2.9. Activation energy of viscous flow versus the content of *cis*-units in polybutadiene.

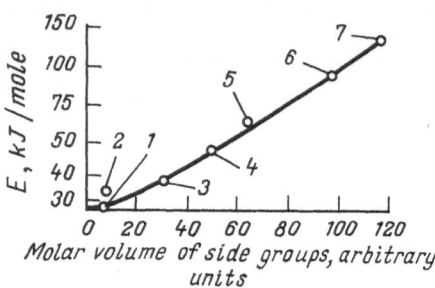

Fig. 2.10. Activation energy of viscous flow (on a logarithmic scale) of linear polymers against the volume of side groups:

1—polyethylene; 2—copolymer of ethylene with 1-butene; 3—polypropylene; 4—polyisobutylene; 5—polyvinyl acetate; 6—polystyrene; 7—poly-α-methylstyrene.

viscosity-temperature characteristics with the characteristics of the other members of the homologous series of poly-α-olefins. The relevant data obtained by a number of authors are presented in Fig. 2.12. Unfortunately, the values of the parameter σ are unknown for higher polyolefins, and it is therefore difficult to account quantitatively for the extremal character of the dependence of E on the length of the side alkyl radical. The same effect of side groups in comb-like polymers on the activation energy of flow is also typical for polyalkyl methacrylates and polyalkyl acrylates.* The values of E for polyolefins with large lengths of side radicals approach the value of E typical of linear polyethylene, and in the case of polyacrylates the activation energy approaches the value typical of branched polyethylene; in the latter case, this is probably due to the presence of a carboxyl group in polyacrylates.

An analogous extremal character is shown by the dependence of the activation energy on the length of side substituents in methacrylate-styrene copolymers.** But in general the existence of a maximum of E in comb-like polymers is associated with the fact that the volume of the monomeric unit increases with increasing length of the side chain. But with further increase of the length of the side radical the "loosening" of the polymer structure is

Platé, N. A., A. Ya. Malkin, V. P. Shibaev, T. Kh. Poolak, and G. V. Vinogradov, *Vysokomol. Soedin.*, A16, 437 (1974).
** Terrell, D. R. and H. Katiofsky, *J. Appl. Polymer Sci.*, 21, 1311 (1977).

found to be the prevailing factor, which is evidenced by the decreasing density of the polymer.

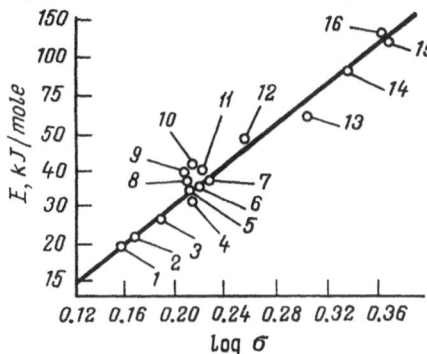

Fig. 2.11. Activation energy of viscous flow (on a logarithmic scale) of polymers versus the factor determining the thermodynamic flexibility of their macromolecules:

1—polydimethyl siloxane; 2—polybutadiene; 3—polyethylene glycol ether; 4—linear polyethylene; 5, 6—copolymers of ethylene with 1-butene; 7—polydecamethylene sebacate; 8—styrene-butadiene rubber; 9—polypropylene; 10—poly-ε-capramide; 11—natural rubber; 12—polyisobutylene; 13—polyvinyl acetate; 14—polystyrene; 15—poly-4-methyl-1-pentene; 16—poly-α-methylstyrene.

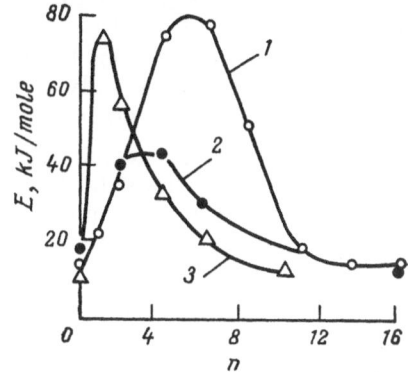

Fig. 2.12. Activation [energy of viscous flow of poly-α-olefins versus the number of carbon atoms n in side branches:

1—Porter, R. S., I. Wang, and I. R. Knox, J. Polymer Sci., B, 8, 10, 671 (1970); 2—Shirayama, K., T. Masuda, and S. Kita, Macromol. Chem., 147, 155 (1971); 3—Combs, R.L., D.F. Slonaker, H.W. Coover, J. Appl. Polymer Sci., 13, 519 (1969).

As regards the effect of the nature of the solvent on the activation energy of viscous flow, the value of E for concentrated solutions is found to be greater

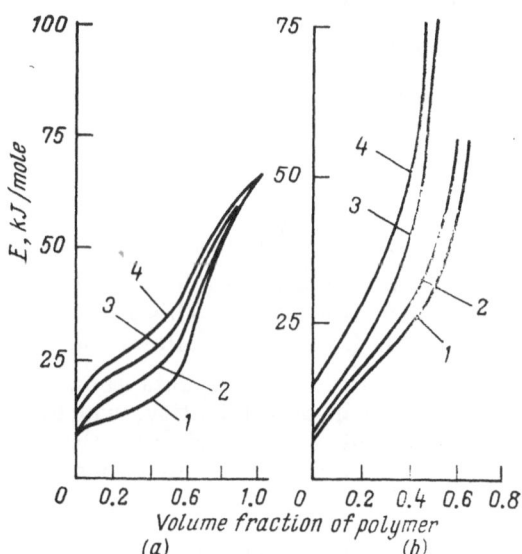

Fig. 2.13. Activation energy of viscous flow of polymer solutions:

a—polyisobutylene in toluene (1), isooctane (2), carbon tetrachloride (3), cyclohexane (4); b—polystyrene in ethyl acetate (1), ethylbenzene (2), carbon tetrachloride (3), decalin (4).

Volume fraction of polymer
(a)   (b)

in those (poor) solvents in which the polymeric chains interact strongly (Fig. 2.13).*

---

* Tager, A. A. and V. E. Dreval, J. Polymer Sci., C, 23, 181 (1968).

2.3.2. Non-Newtonian (Apparent) Viscosity

The above discussion of the temperature dependence of viscosity has been concerned with the initial (largest) viscosity, i.e., the existence of non-Newtonian flow conditions has not been taken into account at all. The apparent viscosity is a function not only of temperature but also of the shear stress and rate of shear. The temperature dependence of viscosity may be treated, assuming the condition $\tau = $ const or $\dot{\gamma} = $ const; generally speaking, the values of apparent viscosity may be compared under other conditions as well. The quantity E_τ may depend on the stress at which it is calculated, and the quantity $E_{\dot{\gamma}}$, accordingly, on the rate of shear. Both quantities certainly coincide in the region of Newtonian flow, where the activation energy is independent of the shear stress or rate of shear.

In the most general form, the relation between E_τ and $E_{\dot{\gamma}}$ has been found by Bestul and Belcher.* The total differential $d\eta$ of the function $\eta\ (\tau,\ T)$ is equal to

$$d\eta = \left(\frac{\partial \eta}{\partial \tau}\right)_T d\tau + \left(\frac{\partial \eta}{\partial T}\right)_\tau dT$$

If this equality is divided by dT and the resulting relation is considered at $\dot{\gamma} = $ const, then

$$\left(\frac{\partial \eta}{\partial T}\right)_{\dot{\gamma}} = \left(\frac{\partial \eta}{\partial \tau}\right)_T \left(\frac{\partial \tau}{\partial T}\right)_{\dot{\gamma}} + \left(\frac{\partial \eta}{\partial T}\right)_\tau$$

After rearrangements, taking into account that $(\partial \tau / \partial \eta)_{\dot{\gamma}} = \dot{\gamma}$, we obtain:

$$\left(\frac{\partial \eta}{\partial T}\right)_\tau \Big/ \left(\frac{\partial \eta}{\partial T}\right)_{\dot{\gamma}} = 1 - \dot{\gamma}\left(\frac{\partial \eta}{\partial \tau}\right)_T$$

This Bestul-Belcher formula is easily transformed into the expression

$$\frac{E_\tau}{E_{\dot{\gamma}}} = 1 - \dot{\gamma}\left(\frac{\partial \eta}{\partial \tau}\right)_T \tag{2.23}$$

where

$$E_\tau = \frac{\partial \ln \eta}{\partial (T^{-1})}\Big|_{\tau = \mathrm{const}} \quad \text{and} \quad E_{\dot{\gamma}} = \frac{\partial \ln \eta}{\partial (T^{-1})}\Big|_{\dot{\gamma} = \mathrm{const}}$$

Formula (2.23) gives the sought-for relation between E_τ and $E_{\dot{\gamma}}$.

At constant temperature the viscosity of polymeric systems usually decreases with increasing stress. This means that $(\partial \eta / \partial \tau)_T < 0$ and, accordingly, from formula (2.23) there follows the condition: $E_\tau > E_{\dot{\gamma}}$.

Numerous experimental data show that E_τ is, with an error of about 4 kJ/mole, a constant quantity, which is independent of τ. In many practically important cases, especially in the processing of polymers, it is, however, the quantity $E_{\dot{\gamma}}$ that becomes important since some technological processes are accomplished at constant rates of shear. In this case, $E_{\dot{\gamma}}$ expresses the temperature dependence of viscosity, but no particular physical meaning should be attached to it.

Formula (2.23) enables one to find the value of the ratio $E_\tau/E_{\dot{\gamma}}$ if the form

---

* Bestul, A. B. and H. V. Belcher, *J. Appl. Phys.*, 24, 696 (1953).

of the function $\eta(\tau)$ is known. In a relatively narrow range of τ values, sufficiently far away from the region of flow conditions, for which the viscosity may be assumed to be equal to its initial value, wide use is made of the power law:

$$\eta = K' \tau^{(n-1)/n}$$

where K' and n are constants.
Then

$$\left(\frac{\partial \eta}{\partial \tau}\right)_T = \left(\frac{n-1}{n}\right)\frac{\eta}{\tau} \; ; \; \frac{E_\tau}{E_{\dot{\gamma}}} = \frac{1}{n}$$

For melts of many polymers, at high rates of shear $n^{-1} \approx 3.5$. Hence, in these cases, the ratio $E_\tau/E_{\dot{\gamma}}$ increases from a value equal to unity in the region of Newtonian flow up to about 3.5. If polymer solutions have a region of the lowest Newtonian viscosity, then with increasing $\dot{\gamma}$ the ratio $E_\tau/E_{\dot{\gamma}}$ must pass through a maximum and then again become equal to unity.

2.4. Numerical Values of the Constants of the WLF Equation

The WLF equation and the constants appearing in it are of great value since they determine not only the temperature dependence of viscosity but also other characteristics of the viscoelastic properties of polymers, primarily the relaxation times. It should be kept in mind that the use of the WLF equation for the temperature dependence of relaxation times presupposes the possibility of describing the viscoelastic properties of polymeric systems by means of a single characteristic relaxation time.

Numerous experimental data on the temperature dependence of viscosity and relaxation times, which have been generalized by Ferry, show that at a first approximation the following very important conclusion is valid: the constants of Eq. (2.10) are universal with respect to a large group of polymers. Namely, when the glass-transition temperature is chosen as the reference temperature, these constants have the following values: $C_{1,g} = -17.44$ and $C_{2,g} = 51.6$. If the reference temperature is 50°C above T_g [i.e., δ in formula (2.11) is equal to 50°C], then $C_{1,s} = -8.86$ and $C_{2,s} = 101.6$.

The curve of $\log (\eta/\eta_g)$ against $(T - T_g)$ plotted in accordance with the universal values of the constants of Eq. (2.10) is shown in Fig. 2.14. It should however be noted that this graph can only be used for estimating, semiquantitative calculations of viscosity as a function of temperature. This is determined by the effect of the choice of the value of glass temperature on the results of calculations and also by the fact that the constants of Eq. (2.10) for concrete polymeric systems may differ from their universal (averaged) values.

The inaccuracy in the estimation of the reference temperature amounting to ±5°C leads to a three-fold change of the value of the ratio (η/η_g). The total range of variation of (η/η_g) shown in Fig. 2.14 corresponds to the region of applicability of the WLF equation, which extends from T_g to approximately $(T_g + 100°C)$. Over this temperature range the indicated ratio varies by about 10^{12} times. As compared with the range of variation of viscosity by 10^{12} times the error amounting to 3 times is insignificant, but for concrete calculations this error is of course very great.

For various polymers, the values of $C_{1,g}$ and $C_{2,g}$ may differ considerably from the universal values. Thus, Ferry quotes the following extreme values of $C_{1,g}$: −15.6 for polyvinyl acetate and polyurethane; −26.2 for a solution of cellulose nitrate; the order of the spread of the values of $C_{2,g}$ is analogous: 130 for poly-n-hexyl metacrylate; 20.2 for poly-1-hexene.

Thus, over a limited temperature range, an equation of the WLF type describes satisfactorily experimental data on the temperature dependence of the flow and relaxation properties of various polymer systems. However, if

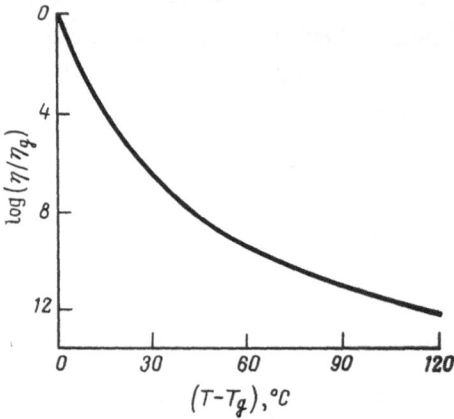

Fig. 2.14. Master viscosity-temperature curve for polymer systems according to the WLF equation.

a sufficiently precise concordance between experiment and the calculation formula is to be obtained, it is necessary that the values of the constants of the WLF equation be different from the averaged, universal values, i.e., the individual values for each system separately. Though quantitative calculations based on the "universal" WLF equation may sometimes present a risk, its use for a semiquantitative evaluation of certain characteristics of polymeric systems is thought to be important.

For such estimations to be carried out, use is made of formulas relating $C_{1,g}$ and $C_{2,g}$ to the specific volume of the polymer at various temperatures and to the values of the thermal expansion coefficient. In the first place, let $B_0 = 1$ in formula (2.5). Then, it is assumed that the expansion of the polymer above the glass temperature takes place due entirely to the increase of free volume, i.e., $v - v_g = v_f - v_{f,g}$. So, from formulas (2.7) and (2.8) it follows that

$$v_g \alpha_l = v_{f,g} \alpha_0$$

or

$$\alpha_l / \alpha_0 = v_{f,g} / v_g$$

The ratio $(v_{f,g}/v_g)$ denoted as f_g is a relative fraction of free volume in the polymer at the glass-transition temperature. From formula (2.9) and the expression for the coefficient $C_{1,g}$, which is deduced from this formula, it follows that

$$C_{1,g} = 0.435 \, (f_g - 1) \, f_g^{-1} \tag{2.24}$$

or

$$f_g \approx -0.435\, C_{1,g}^{-1}$$

Hence, assuming the universal value of the constant $C_{1,g}$, one can evaluate what fraction of the total volume at the glass temperature is the free volume. Calculations show that $f_g \approx 0.025$, i.e., to the transition of various polymers to the glassy state there corresponds the same value of f_g.

From the assumption that the expansion of a liquid is effected only by increasing the free volume it follows that $\alpha_0 = \alpha_l/f_g$. Employing the above-indicated universal value of f_g, one can easily find the universal value of the thermal expansion coefficients: $\alpha_l - \alpha_g \approx \alpha_l \approx 4.8 \times 10^{-4}$ K^{-1}.

Other approaches to the estimation of the value of f_g are also possible. Thus, the calculation of the free volume may be based on the determination of the difference between the specific volume of the system and the closest packing of its molecules, the closest packing corresponding to the specific volume of the substance in the crystalline state at 0 K. Haward and co-authors[*] suggest that the closest packing be regarded as a packing at which the internal pressure $p_i = (\partial U/\partial v) = 0$ (where U is the internal energy and v is the specific volume).

The condition $p_i = 0$ may be represented in the form

$$T\,(\alpha/\beta) - p = 0$$

where p is the external pressure; α and β are the thermal expansion coefficient and the isothermal compressibility coefficient, respectively.

The calculations performed for n-paraffins of the C_5-C_9 type have shown that the free volume thus calculated is in good agreement with the value of free volume obtained from the temperature dependence of their viscosity. An analogous calculation has been carried out for polystyrene, for which the results of a detailed investigation into the relationship among density, pressure, and temperature are known. This has made it possible to find p_i, to calculate the closest packing of molecules and, hence, to determine f_g. According to the data obtained by Haward and co-authors, the value of f_g for polystyrene is 0.13-0.17. This is substantially higher than the universal value of f_g (0.025), which is close to the value of f_g calculated on the basis of the temperature dependence of the relaxation properties of polystyrene. The considerable difference between the values of f_g found from measurements of viscosity and relaxation characteristics, on the one hand, and the thermal characteristics, on the other, raises the general question as to the agreement between these properties near the glass-transition temperature of polymers, though there might be another cause of the indicated discrepancy—the volume occupied by molecules at the glass temperature exceeds their volume corresponding to the closest packing.

Simha and Boyer[**] suggested that f_g be estimated in the following manner. They assumed that the occupied volume at 0 K can be found by extrapolation of the linear dependence of the specific volume on temperature, and that the expansion coefficient is equal to α_l. Then, $v_0^* = v_g - \alpha_l T_g$. Here v_0^* is the volume occupied by molecules at 0 K. The occupied volume $v_{0,g}$ at the glass temperature T_g is defined as $v_{0,g} = v_0^*\,(1 + \alpha_g T_g)$. Hence, the free volume

   [*] Haward, R. N., H. Breuer, and G. Rehage, *J. Polymer Sci.*, B4, 375 (1966).
   [**] Simha, R. and R. F. Boyer, *J. Chem. Phys.*, 37, 1003 (1962).

$v_{f,g}$ at the glass temperature is given by

$$v_{f,g} = v_g - v_{0,g} = v_g - v_0^* (1 + \alpha_g T) = v_g \frac{(\alpha_l - \alpha_g) T_g}{1 + \alpha_l T_g}$$

Thus, if it is assumed that $\alpha_l T_g \ll 1$, then

$$f_g = \frac{v_{f,g}}{v_g} \approx (\alpha_l - \alpha_g) T_g \tag{2.25}$$

This yields the Simha-Boyer rule based on the fact that at the glass temperature f_g = const. Estimation of the values of $\alpha_l - \alpha_g$ and T_g for a number of polymers gives, on an average, the numerical value of this constant:

$$(\alpha_l - \alpha_g) T_g \approx 0.113 \tag{2.26}$$

A comparison of numerous experimental data for a large number of polymers has shown* (Fig. 2.15) that the Simha-Boyer rule is an important ap-

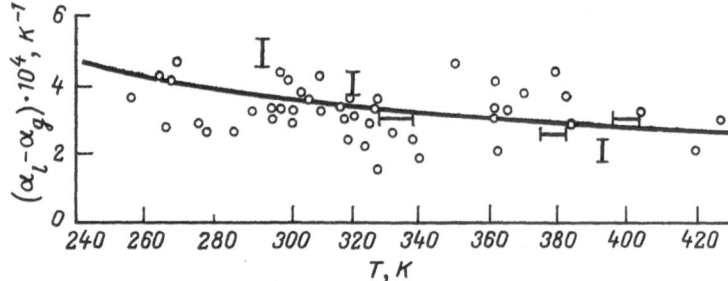

Fig. 2.15. Illustrating the Simha-Boyer rule (the solid line) and the spread of experimental data.

proximation. It reflects better the actual relations between the constants of a material than does the universal value of the difference $(\alpha_l - \alpha_g)$ calculated from the constants of the WLF equation. The universal value of $(\alpha_l - \alpha_g)$ is equal to 4.8×10^{-4} K^{-1}, while for various polymers the value of $(\alpha_l - \alpha_g)$ varies between 2.45×10^{-4} and 9.3×10^{-4} K^{-1}.

The Simha-Boyer rule is fulfilled in the best way for flexible-chain polymers containing no bulky pendant groups. Introduction of such groups leads to systematic deviations from the rule of constancy of the quantity $(\alpha_l - \alpha_g) T_g$. According to Simha and Weil**, this effect is associated with the nonlinear character of the change of the volume at $T < T_g$. Therefore, the value of the relative free volume can be determined in different ways, depending on the method of describing the dependences of α_l and α_g below T_g. In this case, the condition of glass transition is most exactly determined by the constancy of the ratio $(v - v_{0,l})/v$ lying within the limits 0.085 and 0.114. The limiting value of the volume $v_{0,l}$ is obtained by extrapolation of the dependence $v(T)$ from the region of $T > T_g$ to $T = 0$. However, the problem of the correctness of the method of extrapolation in the region of $0 < T < T_g$ remains open, though according to Simha and Weil the extrapolation must be based on the representation of the dependence $v(T)$ in the form $\ln v \propto T^{3/2}$ and $\alpha_l = \alpha_{l,g} (T/T_g)^{0.5}$, where $\alpha_{l,g}$ is the value of α_l at the glass temperature. Integration of the last relation from $T = 0$ to $T = T_g$

* Krause, S., et al., J. Polymer Sci., A3, 3573 (1965).
** Simha, R. and C. E. Weil, J. Macromol. Sci., B4, 215 (1970).

yields the value of the product $(\alpha_l T)$ at T_g, which is equal to the same value (0.18) for certain series of polymers of differing structure.

Thus, the free-volume concept permits a number of useful semiquantitative estimations which relate the glass-transition temperature to the material parameters being measured.

Apart from $C_{1,g}$ and $C_{2,g}$, the WLF equation contains one more constant— this is the viscosity at the glass temperature, η_g. It is often assumed that η_g is of the order of 10^{12} Pa·sec. Moreover, the glass-temperature concept itself was tied up with the conception that the viscosity of a liquid reaches the indicated value. This viewpoint cannot, however, be considered legitimate for polymers. Indeed, for sufficiently high molecular mass T_g is independent of molecular mass. The viscosity, however, is strongly dependent on molecular mass M and varies proportionately with $M^{3.5}$ (see Sec. 2.6). It follows that the region of glass-transition temperatures is not the region of the "isoviscous" states of polymers of different molecular masses.

The determination of the glass temperature must be based on the relaxation characteristic of the polymer (on a certain time-scale) but not on the viscosity values as the temperature-viscosity characteristic of the melt depends on the molecular mass of the sample, and T_g does not.

2.5. Non-Newtonian Viscosity in Steady-State Flow

2.5.1. Introduction

The phenomenon of non-Newtonian viscosity was discovered as early as in the first investigations of the flow properties of polymer solutions. It consists in that the ratio $\tau/\dot{\gamma} = \eta$, which is called the apparent viscosity, depends on the shear stress and rate of shear so that with increase of τ and $\dot{\gamma}$ the value of η usually decreases. This phenomenon is of great practical interest as the basic characteristic of the hydrodynamic behaviour of polymer systems in actual flow conditions. Non-Newtonian flow is of no less importance to theory since it is intimately connected with the structural specificity of polymer systems. Therefore, non-Newtonian flow behaviour has been thoroughly studied, both as a mechanical effect and as a phenomenon associated with the composition and structure of polymers.

The steady shearing flow conditions are described by the function $f(\tau, \dot{\gamma}) = 0$. Its graphical representation is called the *flow curve*. Relevant experimental data are rather frequently represented also in the form of the dependences $\eta(\dot{\gamma})$ or $\eta(\tau)$. Since in the case of non-Newtonian flow behaviour the quantity $\dot{\gamma}$ varies more strongly than τ, the dependence $\eta(\dot{\gamma})$ is found to be weaker than $\eta(\tau)$.

All these dependences constitute an important physical characteristic of polymeric systems. They are determined by their nature, temperature, and pressure. The use of various types of viscometers* enables one to easily cover

---

* Malkin, A. Ya. and A. E. Chalykh, *Diffusion and Viscosity of Polymers. Methods of Measurement*, Khimiya Publ., Moscow, 1979 (in Russian).

the range of variation of $\dot{\gamma}$ by 10^6 to 10^8 times, but measurements were also repeatedly conducted when the quantity $\dot{\gamma}$ varied by 10^{10} to 10^{12} times. In such cases the flow function must always be invariant to the method of measurement.

In a general case, the dependences $\dot{\gamma}(\tau)$, $\eta(\dot{\gamma})$, or $\eta(\tau)$ are S-shaped, covering the boundary values $\eta = \eta_0$ at $\tau \to 0$ and $\dot{\gamma} \to 0$ and $\eta = \eta_\infty$ at $\tau \to \infty$ and $\dot{\gamma} \to 0$, which corresponds to the linear Newtonian flow behaviour when the condition $\dot{\gamma} \propto \tau$ is fulfilled. It is necessary to emphasize that the proportionality between τ and $\dot{\gamma}$ and the fact that the apparent viscosity is

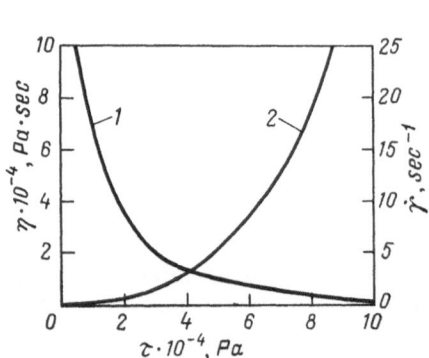

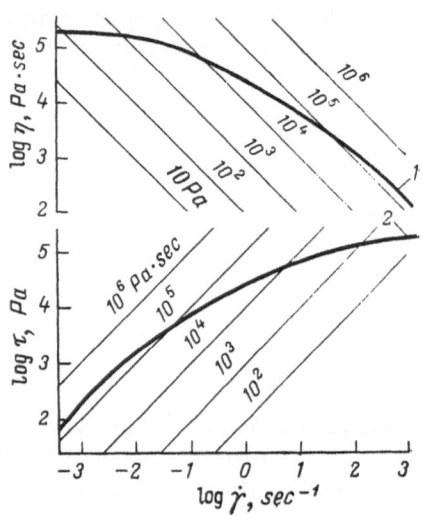

Fig. 2.16. Apparent viscosity (*1*) and rate of shear (*2*) versus shear stress for a representative sample of polypropylene (in linear coordinates).

Fig. 2.17. Apparent viscosity (*1*) and shear stress (*2*) versus rate of shear for a representative sample of polypropylene in log-log coordinates.

independent of $\dot{\gamma}$ and τ correspond to the limiting conditions for the relationships considered here. Hence, the proportionality between τ and $\dot{\gamma}$ and the constancy of η over certain ranges of $\dot{\gamma}$ and τ values are invariably approximate relationships. Their reliability is in many respects connected with the accuracy of the experimental methods employed.

The flow functions and, hence, the flow curves, which satisfy, to some degree of approximation, the condition of attainment of constant values of η_0 and η_∞, have come to be known as complete flow curves. Ordinarily, only a limited portion of a complete flow curve is obtained.

When use is made of linear scales on the coordinate axes, the $\eta(\dot{\gamma})$ and $\tau(\dot{\gamma})$ curves look like those shown in Fig. 2.16. Since measurements may cover a wide range of the values of the variables, the linear scales on the coordinate axes give steeply falling (or rising) curves. Therefore, preference is given to the use of a log-log scale. Figure 2.17 shows, in such coordinates, the data which have been used for constructing Fig. 2.16. The thin straight lines drawn at an angle of 45° to the coordinate axes are the lines of the constant values of τ or η.

Typical complete flow curves are shown in Fig. 2.18.

In a narrow range of variation of the variables the flow curves in a log-log scale may be approximated by a straight line with a slope n, which is called the flow index. Sometimes use is made of a more general concept of the flow behaviour index, meaning $n = d \log \tau / d \log \dot\gamma$, irrespective of whether it is constant or not.

The phenomenon of non-Newtonian flow behaviour of polymeric systems is of such a great theoretical and practical interest that it is not surprising that

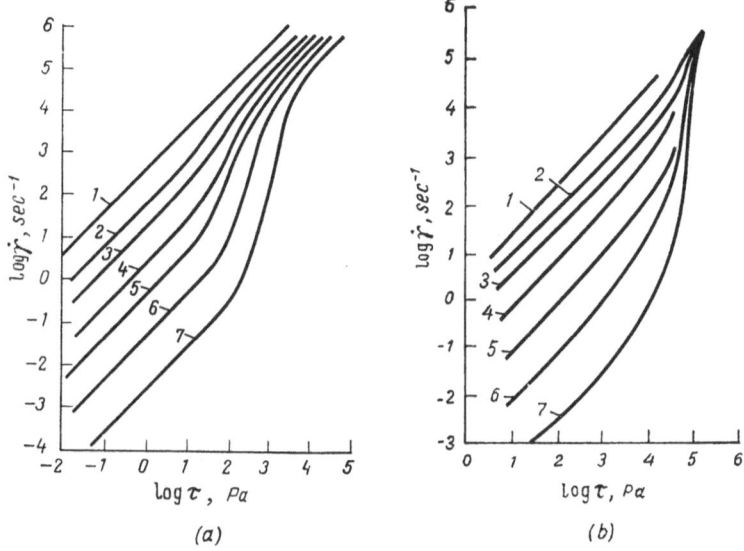

Fig. 2.18. Complete flow curves:

a—solutions of polyisobutylene in decalin with different concentrations of polymer in solutions, per cent: 1—0; 2—0.5; 3—11; 4—2; 5—3; 6—5; 7—9 [Brodnyan, J. G., F. H. Gaskins, and W. Philippoff, Trans. Soc. Rheol., 1, 109 (1957).]; b—polydimethyl siloxanes with molecular masses: 1—10,300; 2—22,000; 3—33,400; 4—57,000; 5—105,000; 6—206,000; 7—482,000. [Kataoka, T. and S. Ueda, J. Polymer Sci., A3, 2947 (1965).]

numerous theories have been developed, which rely on the various mechanical and physical hypotheses and conceptions, in which the non-Newtonian viscosity is tied up with the mechanism of viscous flow of polymers.

All the known theories of non-Newtonian viscosity in polymeric systems may be provisionally divided into the following large groups which differ in physical approach to the effect under discussion. These are the theories worked out by Eyring and his collaborators, which tie up non-Newtonian flow with the effect of the shear stress or rate of shear on the height of the potential energy barrier, which impedes the jump of molecular-kinetic units from one equilibrium position to another; the structural theories of non-Newtonian flow, which regard the dependence of apparent viscosity on flow conditions as a consequence of reversible (thixotropic) breakdown-recovery of the structure of the polymer; the hydrodynamic theories, in which the decrease of the apparent viscosity of a polymeric system with increasing rate of shear is accounted for by the change of the form and, hence, of the resistance to the movement of macromolecules; a number of phenomenological theories may be called formal since they are restricted to a mathematical description of non-Newtonian behaviour in connection with other mechanical properties of polymer melts and solutions, without referring them to the physical nature of the phenomenon.

While giving preference to this or that theory, one should not dispense with the ideas underlying the other theories.

2.5.2. The Eyring Theory and Its Generalization

The dependence of the apparent viscosity on the rate of deformation is treated by Eyring within the framework of the general theory of absolute reaction rates. According to the conceptions of this theory, the elementary

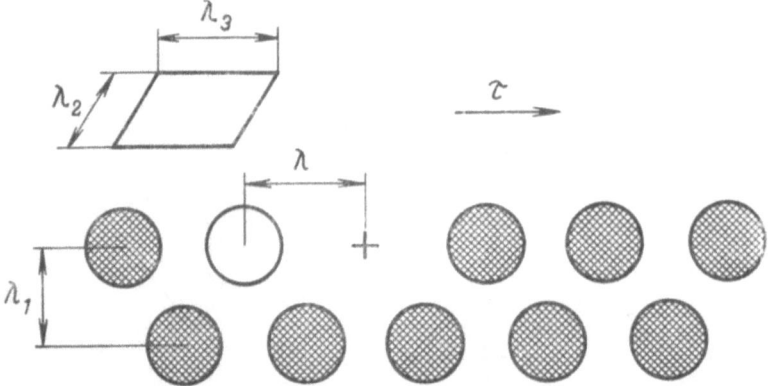

Fig. 2.19. Illustrating the jump of a molecular-kinetic unit over the potential energy barrier in a liquid (after H. Eyring).

flow process in a liquid, which is considered a quasicrystalline body, consists in crossing the energy barrier by a molecular-kinetic unit having sufficient energy to do it. These jumps occur constantly in all directions with equal probability, so that no definite direction of flow is present. The jumping frequency in the absence of an external force field depends on the height of the potential energy barrier and the size of molecular-kinetic units and is determined by the specific structure of the liquid and the temperature. According to the concepts of the absolute reaction-rate theory, the jumping frequency may be expressed as follows:

$$v_0 \propto \exp\left(-E/kT\right) \tag{2.27}$$

where E is the activation energy of flow, determined by the height of the potential barrier; T is the absolute temperature; k is Boltzman's constant.

The scheme of jumps over the potential barrier in a model system, which is a set of spherical particles, is shown in Fig. 2.19. The choice of a spherical particle as the model of the molecular-kinetic unit is unimportant since the shape of particles is not taken into account in calculations, and here it is only done for the sake of vividness.

The application of an external stress τ creates a force acting on each molecular-kinetic unit and equal to $\tau\,(\lambda_2\lambda_3)$, since $(\lambda_2\lambda_3)$ is the average surface area per each molecular-kinetic unit. Let the distance between two adjacent equilibrium positions be λ, then the force $\tau\,(\lambda_2\,\lambda_3)$ performs work at a distance of $\lambda/2$, i.e., upon movement to the top of the symmetrical potential barrier. This work $A = 0.5\,\tau\,(\lambda\lambda_2\lambda_3)$. Thus, the action of the external force is equivalent to the lowering of the height of the potential barrier in the direction of the force by the quantity A. It is not difficult to see that upon

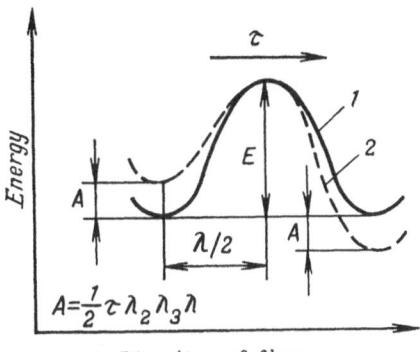

Fig. 2.20. Diagram showing the deformation of the potential barrier under the influence of shear in a liquid (after H. Eyring):
1—in the absence of an external force; *2*—under the influence of an external force.

movement of a molecular-kinetic unit in the direction opposite to the direction of the force an extra energy also equal to A must be consumed, i.e., the height of the potential barrier increases by $0.5\,\tau\,(\lambda\lambda_2\lambda_3)$. What has just been said is illustrated by Fig. 2.20.

Because of the change of the height of the potential energy barrier the jumping frequency is changed both in the direction of the force and in the opposite direction. This causes a shift of the dynamic equilibrium between the jumps in different directions and leads to the appearance of the flow direction.

The jumping frequency in the direction of the force and in the opposite direction (ν_+ and ν_-, respectively) is easy to calculate:

$$\nu_+ \propto \exp\left[-(E-A)\,/kT\right]$$

$$\nu_- \propto \exp\left[-(E+A)\,/kT\right]$$

or

$$\nu_+ \propto \nu_0\exp\left(A/kT\right)$$

$$\nu_- \propto \nu_0\exp\left(-A/kT\right)$$

With each jump a molecular-kinetic unit moves the distance λ, so that the total displacement per unit time is equal to

$$\Delta u = \lambda\nu_0\left[\exp\left(A/kT\right)-\exp\left(-A/kT\right)\right]=2\lambda\nu_0\sinh\left(\tau\lambda\lambda_2\lambda_3/2kT\right) \qquad (2.28)$$

where ν_0 is the number of jumps per unit time in the liquid when there is no external force acting.

The quantity Δu, which is given by formula (2.28), is basically the difference between the velocities of two adjacent layers since under the action of stresses

τ there takes place nothing more than a relative displacement of adjacent layers of the liquid. If now we determine the value of apparent viscosity by means of the following obvious equality:

$$\eta = \frac{\tau}{\Delta u/\lambda_1}$$

then it is easy to see that

$$\eta = \frac{\tau \lambda_1}{2\lambda v_0 \sinh (\tau \lambda \lambda_2 \lambda_3/2kT)} \tag{2.29}$$

This expression explicitly implies the dependence of the apparent viscosity on the deformation conditions, i.e., on the applied stress, since $\eta = \eta (\tau)$.

The quantity λ_1 is not strictly equal to $\lambda/2$, but in any case these are quantities of the same order of magnitude, and therefore at a first approximation it may be assumed that $\lambda_1 = \lambda/2$; the product $\omega = \lambda_1 \lambda_2 \lambda_3$ will then be the effective volume of the molecular-kinetic unit. Therefore, Eq. (2.29) may be written in the form

$$\eta = \eta_0 [z/\sinh z] \tag{2.30}$$

where $z = \tau \omega/kT$; $\eta_0 = kT/4v_0 \omega$ is a constant representing the zero viscosity.

Of special interest is the investigation of the behaviour of the functions $\eta (z)$ at $z \to 0$ or, what is the same thing, at $\tau \to 0$.

In this case, from formula (2.30) it follows that at sufficiently small τ values there is satisfied the condition $\eta = \eta_0 = \text{const}$, i.e., $\eta_0 = \lim_{\tau \to 0} \eta$ and η_0 is really the initial viscosity.

The values of z at which there begin noticeable departures from Newtonian flow conditions can be estimated by means of Eq. (2.30). According to the Eyring formula, $\eta \approx \eta_0$ if $z \ll \sqrt{6}$. In practice, the liquid behaves as a Newtonian fluid, in accordance with the Eyring theory, at $z < 0.3$.

Formula (2.30) may be regarded as a single-parameter dependence of (η/η_0) on z or τ. The parameter is the quantity (ω/kt), where ω is the volume of a kinetic unit, which is amenable to approximate evaluation at least.

If we assume that the length of a molecular-kinetic unit corresponds to the length of a kinetic flow segment, then for carbon-chain polymers this will correspond to not more than at least 30-40 carbon atoms of the main chain, or to 35-40 Å. The distance between the chains in the polymer is about 5 Å. It follows that $\omega \approx 10^{-21}$ cm^3. Hence, even at τ of the order of 10^6 Pa, which is close to the limits of experimental possibilities in viscosity measurements the quantity z remains to be sufficiently small for non-Newtonian behaviour to manifest itself. The non-Newtonian behaviour is however often observed at immeasurably smaller shear stresses. The explanation for this contradiction as suggested by Bartenev[*] consists in that the Eyring theory refers to systems of unchanging structure, while during the flow of polymeric systems there may occur substantial structural changes, i.e., he proposed using the structural (thixotropic) theory of non-Newtonian flow behaviour.

Independently of the theoretical substantiation, Eq. (2.30) can be used for the treatment of experimental data, in which case it is represented in the

---

[*] Bartenev, G. M., *Vysokomol. Soedin.*, 6, 2155 (1964).

form:

$$\eta/\eta_0 = B\tau/\sinh(B\tau) \tag{2.31}$$

Here B is an empirical constant which may not be connected with the structure of the system under investigation and with those physical ideas which formed the basis for the activation theory in the derivation of Eq. (2.30). Then formula (2.31) must be considered purely empirical.

Expressions of the type (2.31) are rather often used in the literature for a quantitative description of experimental data on the viscosity of polymeric systems. A liquid whose flow properties are described by a formula of the type (2.31) is characterized by the initial viscosity η_0. But at $\tau \to \infty$ the value of $\eta = 0$, which is unrealistic.

Equation (2.31) may be written in different ways. Thus, for calculations it is more convenient to make use of the dependence of viscosity not on the shear stress but on the rate of shear. Then

$$\eta = (B\dot{\gamma})^{-1} \ln[A\dot{\gamma} + \sqrt{(A\dot{\gamma})^2 + 1}] = (A/B)(A\dot{\gamma})^{-1}\operatorname{arsinh}(A\dot{\gamma}) =$$

$$= \eta_0 (A\dot{\gamma})^{-1} \operatorname{arsinh}(A\dot{\gamma}) \tag{2.32}$$

We can obtain the dependence of the shear stress on the rate of shear, i.e., the function $\tau(\dot{\gamma})$:

$$\tau = B^{-1} \ln[(A\dot{\gamma}) + \sqrt{(A\dot{\gamma})^2 + 1}] = B^{-1} \operatorname{arsinh}(A\dot{\gamma}) \tag{2.32a}$$

Or, on the contrary, the dependence of the rate of shear on the shear stress:

$$\dot{\gamma} = A^{-1} \sinh(B\tau) \tag{2.32b}$$

Formulas of the type (2.32) qualitatively correctly define the specific features of non-Newtonian flow behaviour of polymeric systems. Therefore, choosing the value of the parameter B, it was possible, not infrequently, to describe, by means of formulas (2.32), the experimental data obtained in not too wide a range of variation of deformation conditions.

But in many cases, when only one parameter is chosen, it is not possible to represent correctly the course of the flow curves. This becomes especially spectacular when the range of shear is extended. Therefore, attempts have been made to generalize the Eyring formula*. The idea underlying such a generalization consists in assuming the presence of different types of molecular-kinetic units, each with its own characteristic size x_i and characteristic relaxation time θ_i. There also exist molecular-kinetic units, a set of which behaves as a Newtonian liquid. In such a case, the following expression is obtained:

$$\eta = \sum_{i=1}^{N} \frac{x_i \theta_i}{B_i} \frac{\operatorname{arsinh}(\dot{\gamma}\theta_i)}{(\dot{\gamma}\theta_i)} \tag{2.33}$$

At $N = 1$, formula (2.33) reduces to the original Eyring formula.

In the generalized Eyring equation, which is known as the Powell-Eyring formula, it is assumed that a polymer consists partly of particles which, regardless of the stress, behave as a Newtonian liquid, and partly of particles

---

* Ree, T. and H. Eyring, *J. Appl. Phys.*, 26, 793 (1955).

which are responsible for the non-Newtonian behaviour of the system:

$$\eta = \frac{x_1\theta_1}{B_1} + \frac{x_2\theta_2}{B_2} \frac{\text{arsinh}(\dot{\gamma}\theta_2)}{(\dot{\gamma}\theta_2)} \tag{2.34}$$

or

$$\eta = \eta_\infty + \eta_1 \left[\text{arsinh}\left(\dot{\gamma}\theta_2\right)\right] / \left(\dot{\gamma}\theta_2\right)$$

where η_∞, η_1, and θ_2 should be regarded as the three experimentally determined constants

It is essential that at $\dot{\gamma} \to 0$ there is satisfied the condition $\eta = \eta_0 = = \eta_\infty + \eta_1 = \text{const}$ and since $\eta_\infty \ll \eta_1$, it follows that $\eta_0 = \eta_1$, i.e., the quantity η_1 has the meaning of the initial viscosity. At $\dot{\gamma} \to \infty$ the value of $\eta = \eta_\infty = \text{const}$, where η_∞ is the lowest Newtonian viscosity. From this it follows that the Powell-Eyring formula provides a qualitatively correct description of the behaviour of non-Newtonian liquids. The equation that has been proposed as the next generalization of formula (2.31) is the Ree-Eyring equation in which three types of "flowing groups" are considered, etc. Finally, the generalization of the ideas of Ree and Eyring has been brought up* to its logical end by assuming the existence of a continuous distribution of relaxation times among molecular-kinetic units and, hence, the sum in formula (2.33) must be replaced by an integral:

$$\eta = \dot{\gamma}^{-1} \int_0^\infty F(\theta) \, \text{arsinh}(\dot{\gamma}\theta) \, d\theta$$

Thus, the phenomenon of non-Newtonian viscosity is tied up with the existence of a continuous set (spectrum) of molecular-kinetic units having different characteristic relaxation times.

Many-parameter equations of the type of the Powell-Ree-Eyring equation, and especially the above formula which contains a continuous distribution of parameters, can describe practically any experimental data on the dependence of the apparent viscosity on the shear stress and rate of shear. This is attained by the appropriate choice of the numerical values of the parameters. In a number of experimental works one can find examples of the successful use of formulas (2.32) through (2.34) for describing the flow properties of various polymeric systems.

Equation (2.30) rests on serious theoretical considerations, but Eq. (2.31) is employed as purely empirical. This refers all the more to the Powell-Ree-Eyring formulas. Equations containing three and, all the more, six-seven parameters subject to experimental determination should not be regarded as having a molecular-kinetic substantiation. The main limitation on the applicability of the Eyring theory and its various extensions to polymeric systems is associated with the fact that they all describe the properties of purely viscous liquids. In this group of theories, the flow properties are not tied up with the other fundamental properties of polymers, with their viscoelasticity and rubber elasticity. Therefore, the equations discussed above are not brought into agreement and it seems even impossible to compare non-Newtonian behaviour with the other characteristics of polymer systems. As a result, the theoretical approach developed in the activation theory is

---

* Faucher, J. A., J. Appl. Phys., 32, 2336 (1964).

incapable of providing the desired physical generality of description of the entire set of mechanical properties of polymers, because of the specific features of their structure.

2.5.3. Structural Theories of the Non-Newtonian Flow of Polymeric Systems

The idea that the decrease of the apparent viscosity with increasing shear stress is due to the gradually progressing structural breakdown upon deformation of non-Newtonian systems has been developed since the first works in which this effect was described. Moreover, one of the first investigators of the process of non-Newtonian flow (Ostwald, 1925) used the term "structural branch" for the nonlinear portion of the flow curve and "structural viscosity" for the apparent viscosity which depends on the shear rate and stress. These terms are sometimes used in the current literature.

In earlier works, the dependence of viscosity on the shear rate and stress was usually studied on two-phase disperse systems. The idea that the solid phase forming a spatial structure in the liquid phase may be broken down under the influence of external forces (stresses) is quite natural. For such systems, the structural destruction should primarily mean the breakdown of the coagulating structures formed by the interacting particles of the dispersed phase.

The relation between non-Newtonian flow and thixotropy has been elaborated in a number of works*, chiefly in application to disperse systems.

In polymeric systems, the presence of a spatial structure is due, first of all, to the shape of macromolecules. Their enormous length coupled with the flexibility and the presence of intra- and intermolecular interactions—all this leads to the formation of the various structural units: entanglements (nodes, junctions) which result from intermolecular interactions, and formations of the type of nuclei of the crystalline phase.

The phenomenon of non-Newtonian viscosity may be regarded as the effect associated with the change of the structure of the system under the influence of straining, with account taken of the nature of the interaction of particles forming this structure. The only important point here is the presence of a certain number of specific contacts and linkages between the particles, as a result of which the system as a whole is a three-dimensional network. Deformation destroys some of these linkages, which makes possible a steady flow of the system.

The successive development of the conceptions of flow as a process analogous to a chemical reaction taking place in the forward (the rupture of bonds) and in the reverse (the re-formation of bonds) direction, leads to the possibility of formulation of the corresponding kinetic equation. It is assumed* that the deformation affects the dynamic equilibrium in structured thixotropic systems so that it increases the rate of bond rupture.

The Non-Newtonian Flow and Thixotropy.* The process of spontaneous thixotropic breakdown—the recovery of bonds in a structured system in the

---

* Denny, D. A. and R. S. Brodkey, *J. Appl. Phys.*, 33, 2269 (1962); Kim, H. T. and R. S. Brodkey, *A.I. Ch.E. Journal*, 14, 61 (1968); Lin O. C. C., *J. Appl. Polymer Sci.*, 19, 199 (1975).

absence of an external effect—may be described as follows:

$$-\frac{dN_t}{dt} = k_1 N_+^n - k_2 N_-^m \tag{2.35}$$

where N_+ and N_- are the concentrations of unbroken and broken bonds; k_1 and k_2 are the rate constants of the processes of breakdown and recovery of bonds; n and m are constants which, by analogy with chemical reactions, determine the reaction order for the breakdown and recovery of bonds; t is the time.

The viscosity of the system is determined by the number of bonds retained in the system at a given instant of time. It is natural to assume that

$$N_t = [\eta(t) - \eta_\infty] / (\eta_0 - \eta_\infty)$$

and, accordingly,

$$N_- = [\eta_0 - \eta(t)] / (\eta_0 - \eta_\infty)$$

where $\eta(t)$ is the viscosity at time t.

For flow at a rate of shear $\dot{\gamma}$ Eq. (2.35) assumes the form

$$\frac{1}{\eta_0 - \eta_\infty}\left(-\frac{d\eta}{dt}\right) = k_1 \left(\frac{\eta - \eta_\infty}{\eta_0 - \eta_\infty}\right)^n \dot{\gamma}^p - k_2 \left(\frac{\eta_0 - \eta}{\eta_0 - \eta_\infty}\right)^m \tag{2.36}$$

where p is a constant which determines the effect of the rate of shear on the process of breakdown of molecular bonds in the system.

When a steady flow is attained, there must exist an equilibrium between the processes of rupture and recovery of bonds in the system. To this there corresponds the condition: $(d\eta/dt) = 0$. Then, the viscosity under steady-state flow conditions is determined from the equation

$$k_1 \left(\frac{\eta - \eta_\infty}{\eta_0 - \eta_\infty}\right)^n \dot{\gamma}^p = k_2 \left(\frac{\eta_0 - \eta}{\eta_0 - \eta_\infty}\right)^m \tag{2.37}$$

Thus, when the constants k_1, k_2, n, m, p, η_0, and η_∞ are known, one can evaluate the dependence of the apparent viscosity on the rate of shear.

The presence of a large number of constants that are not amenable to theoretical estimation is an essential disadvantage of the theory under consideration. They must be determined from experimental results. Moreover, the same experimental data can be satisfactorily described if different values are assumed for the constants m and n. The value of the constant p may vary over a rather wide range and the possibility of its theoretical evaluation is generally extremely doubtful.

Finally, another shortcoming of the theory is the assumption that all the bonds in a structured system are identical, which contradicts the actual picture.

The Non-Newtonian Flow and Truncation of the Relaxation Spectrum.* The integral phenomenological characteristic of systems that exhibit viscoelasticity is their relaxation spectrum. Therefore, it seems natural to assume that the changes in viscosity upon deformation of polymers are due to reversible structural transformations.

In linear viscoelastic theory [see formula (1.98)], the viscosity is expressed by an integral of the type $\int_0^\infty \theta F(\theta)\, d\theta$, which is independent of deformation

* Leonov, A. I. and G. V. Vinogradov, Doklady Akad. Nauk SSSR, 155, 406 (1964); Leonov, A. I., Zhurn. Prikl. Mekh. i Tekhn. Fiz., 4, 78 (1964).

conditions. If the external effect alters the relaxation properties of the system, then, by way of generalizing formula (1.98), we can express the viscosity in the following manner:

$$\eta(\dot{\gamma}) = \int_0^\infty \theta F(\theta, \dot{\gamma})\, d\theta \qquad (2.38)$$

Here $F(\theta, \dot{\gamma})$ is the altered effective relaxation spectrum which depends on the rate of shear and is characterized by the fact that at $\gamma \to 0$ the spectrum $F(\theta, \dot{\gamma})$ changes into the initial spectrum $F_0(\theta) = F(\theta, 0)$ corresponding to the initial, undeformed state of the system. To this state there corresponds $\eta = \eta_0 = \int_0^\infty \theta F_0(\theta)\, d\theta$.

A formula analogous to formula (2.38) may also be written in terms of the relaxation frequency spectrum:

$$\eta(\dot{\gamma}) = \int_0^\infty s^{-1} N(s, \dot{\gamma})\, ds$$

In the simplest case, the spectrum $F(\theta, \dot{\gamma})$ has the form

$$F(\theta, \dot{\gamma}) = \begin{cases} F_0(\theta) & \text{at } \theta < \theta_m(\dot{\gamma}) \\ 0 & \text{at } \theta \geqslant \theta_m(\dot{\gamma}) \end{cases} \qquad (2.39)$$

Here $\theta_m(\dot{\gamma})$ is a function which defines the conditions of stepwise truncation of the initial relaxation spectrum under the action of deformation at a relaxation time θ_m.

Using the concept of the thixotropy function, we may write the following expression for viscosity:

$$\eta(\dot{\gamma}) = \int_0^{\theta_m(\dot{\gamma})} \theta F_0(\theta)\, d\theta \qquad (2.40)$$

To a first approximation it may be assumed that $\theta_m = \dot{\gamma}^{-1}$, i.e., that during the steady flow of viscoelastic polymeric systems there occurs the breakdown of those structural elements in the system which are responsible for relaxation processes taking place with frequencies numerically lower than the specified rate of shear $\dot{\gamma}$.

Theory is in better agreement with experimental data if it is assumed[*] that $\theta_m^{-1} = a(\dot{\gamma})/\dot{\gamma}$. Here $a(\dot{\gamma})$ is the rate-of-shear function which varies from 1 to about 5 upon variation of $\dot{\gamma}$ by several decimal orders. A more rigorous quantitative treatment shows[**] that the change of the initial relaxation spectrum under the influence of the deformation of polymeric systems

---

[*] Leonov, A. I. and A. Ya. Malkin, *Zhurn. Prikl. Mekh. i Tekhn. Fiz.*, 4, 107 (1965); *Izv. Akad. Nauk SSSR, Ser.: Mekh. zhidkosti i gaza*, 3, 184 (1968).
[**] Malkin, A. Ya., in: *Advances in Polymer Rheology*, edited by G. V. Vinogradov, Khimiya Publishers, Moscow, page 171, 1970 (in Russian).

takes place by way of parallel displacement of the long-time boundary of the spectrum, as schematically shown in Fig. 2.21.

The Non-Newtonian Flow and the Entanglement Network in Polymeric Systems. A concrete molecular-kinetic consideration of the structural conceptions of the nature of the non-Newtonian viscosity of polymers is associated with the concept of their network structure. Spatial network junctions arise as a result of the contact between long flexible macromolecules. In a state of rest they are formed and destroyed by thermal motion, so that at constant temperature there is established a dynamic equilibrium.

Graessley maintains* that the process of formation of network junctions upon contact of macromolecules is characterized by an induction period,

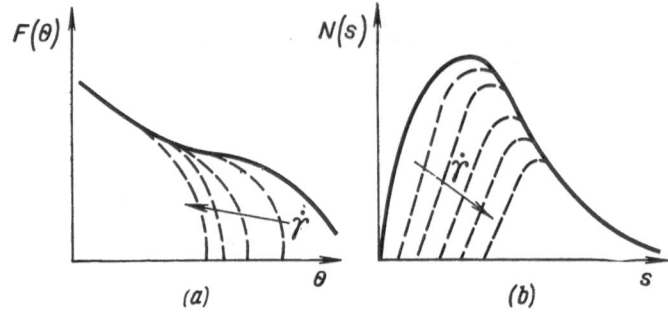

Fig. 2.21. Variation of relaxation-time (a) and frequency (b) spectra under the influence of shear. The solid lines refer to the initial spectra; the dashed lines refer to effective spectra changed under the influence of an increase in the rate of shear.

after the completion of which the number of nodes increases very rapidly, this being followed by the asymptotic approach to a state of equilibrium. Each macromolecular coil is characterized by a definite equivalent volume. When the coils move relative to one another, their volumes may overlap. Spatial network junctions are formed when the duration of this overlap process exceeds θ. The length of the contact between macromolecules of given size is determined by the rate of shear. Hence, $\dot{\gamma}$ is a parameter that defines the state of dynamic equilibrium of the system. Theory shows that it is uniquely characterized by the value of $x = \dot{\gamma}\theta/2$. Here, it is assumed that the viscosity is proportional to the density of nodes. Then, for a monodisperse polymer the dependence $\eta\,(\dot{\gamma})$ is expressed by the formula

$$\eta/\eta_0 = h\,(x)\,[g\,(x)]^{3/2} \tag{2.41}$$

where the functions $h\,(x)$ and $g\,(x)$ have the following form:

$$h\,(x) = \frac{2}{\pi}\left[\operatorname{arccot} x + \frac{x\,(1-x^2)}{(1+x^2)^2}\right]$$

$$g\,(x) = \frac{2}{\pi}\left(\operatorname{arccot} x + \frac{x}{1+x^2}\right)$$

Here $x = \dot{\gamma}\theta/2 = (\eta/\eta_0)\,(\dot{\gamma}\theta_0/2)$, where θ_0 is the "initial" characteristic value of the time constant, which corresponds to the condition $\dot{\gamma} \to 0$. Using the

* Graessley, W. W., *J. Chem. Phys.*, 43, 2696 (1965); 47, 1942 (1967); *J. Polymer Sci.*, A-2, 6, 1887 (1968); *Macromolecules*, 2, 49 (1969).

function (2.41) and the relations written, one can construct the master curve of (η/η_0) vs. $(\dot{\gamma}\theta_0/2)$. The main specific features of the function $\eta\,(\dot{\gamma})$, which has been found by Graessley, are the following: at low $\dot{\gamma}$ there exists a limit-ing value of initial viscosity η_0; at high values of $\dot{\gamma}$ the dependence $\eta\,(\dot{\gamma})$ is given by a power function with an exponent equal to $-9/11$ (i.e., $\eta \propto$ $\propto \dot{\gamma}^{-0.818}$).

These results have been extended by Graessley to polydisperse polymers, for which θ becomes dependent not only on $\dot{\gamma}$ but also on the specific features of the molecular-mass distribution (MMD) of the polymer. The outcome of this is that the shape of the flow curves is found to be dependent on the form and width of the MMD. The analytical form that expresses the effect of MMD on the function $\eta\,(\dot{\gamma})$ is very complicated, but the works of Graessley contain the corresponding formulas and tables for values of the function under discussion. Concrete calculations carried out for various real poly-mers* and also for a series of materials with arbitrary model MMD have shown** that the theory reflects correctly the observed form of the depen-dences $\eta\,(\dot{\gamma})$, and also the basic facts referring to the effect of MMD on the dependence $\eta\,(\dot{\gamma})$, such as the considerable change of the form of flow curves in going from mono- to polydisperse polymers even when the MMD is but slightly broadened, the sharp change of the beginning of the region of non-Newtonian flow behaviour as a function of the content of even small amounts of high-molecular-mass fractions, etc.

The successes attained in describing the dependences $\eta\,(\dot{\gamma})$ for mono- and polydisperse polymers with the aid of the Graessley theory of viscosity have given rise to the more general question of the relation between $\eta\,(\dot{\gamma})$ and the relaxation properties of the material. The answer here is the assump-tion*** according to which the following relation is fulfilled:

$$\eta\,(\dot{\gamma}) = \int_0^\infty \theta F\,(\theta)\,h\,(x)\,[g\,(x)]^{3/2}\,d\theta \tag{2.42}$$

At $\dot{\gamma} \to 0$, $h\,(x)\,[g\,(x)]^{3/2} = 1$ and therefore, as usual, $\eta\,(\dot{\gamma} \to 0) = \eta_0 =$ $= \int_0^\infty \theta F\,(\theta)\,d\theta$. The extension of the formula written above to the region of high $\dot{\gamma}$ is an assumption which is in satisfactory agreement with experi-mental data.

Interaction of Macromolecules and the Non-Newtonian Flow.

In another structural (molecular-kinetic) theory of non-Newtonian viscos-ity discussed in the literature****, on the basis of the presumed dependence of the molecular forces on the distance between the interacting particles, an attempt is made to elucidate the effect of the rate of shear and molecular parameters on the viscosity of polymeric systems.

Under shear the distribution function for the positions of the segments rela-

   * Saeda, S., *J. Polymer Sci., Polymer Phys. Ed.*, 11, 1465 (1973).
  ** Cote, J. A. and M. Shida, *J. Appl. Polymer Sci.*, 17, 1639 (1973).
 *** Cote, J. A. and M. Shida, *Trans. Soc. Rheol.*, 17, 401 (1973).
**** Williams, M. C., *A. I. Ch. E. Journal*, 12, 1064 (1966); 13, 534 (1967).

tive to the centre of mass of the macromolecule is distorted. The final
result is the expression for the dependence of viscosity on the rate of shear
in the form

$$(\eta - \eta_\infty) / (\eta_0 - \eta_\infty) = f(\dot{\gamma}\theta_W) \tag{2.43}$$

where θ_W is a constant which has the dimensions of time and which characterizes the individual properties of the substance.

The form of the function (2.43), according to Williams, is shown in Fig. 2.22.
A specific feature of this function is the transition to the proportionality
between η and $\dot{\gamma}^{-1}$ at high values of $\dot{\gamma}$.

Formula (2.43), which is the main result of the molecular-structural
theory of non-Newtonian viscosity is very close in its structure to the for-

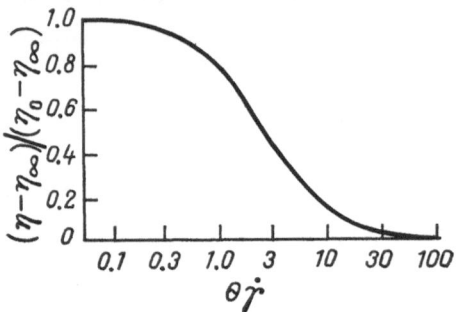

Fig. 2.22. The development of non-
Newtonian flow corresponding to the
Williams theory.

mulas obtained earlier, which also contain one constant having the dimensions of time. An important warning must be made in connection with this
dependence. At high shear stresses, in many polymers there is observed the
flow spurt effect (see Fig. 2.31 and the relevant discussion) caused by the
tear-off of the polymer from the duct walls as a result of its transition to the
forced rubbery state. To this phenomenon there also corresponds the condition of proportionality between η and $\dot{\gamma}^{-1}$. Therefore the fulfillment of the
indicated condition cannot be regarded as an unambiguous evidence of the
validity of the Williams theory.

It should be stressed that different approaches to the structural treatment
of the phenomenon of non-Newtonian viscosity of polymeric systems are
not conflicting and mutually exclusive. On the contrary, they rather supple-
ment one another and enable one to analyse the effect under discussion
on different levels. For example, the model of entanglements does not deal
with the question of the nature of the "entanglement", while the physical
model ignores the geometry of the phenomenon and defines the entanglement
as a set of molecular interactions. The thixotropic theory, while not claiming
the elucidation of the physical mechanism governing the observed change
of the viscoelastic properties of the system, establishes general relationships
between the effect of non-Newtonian viscosity and the changes of the
relaxation properties of the system.

2.5.4. Hydrodynamic Theories

When a suspension containing rigid anisodiametric particles is subjected
to deformation, these particles become oriented in flow with a deviation
of the distribution of the axes of the particles from the equilibrium distri-

bution. This changes the conditions of the streamlining of particles by the flow. Under the influence of Brownian motion they depart from the orientation imposed by the flow. As a result, there is attained a certain equilibrium orientation of particles, the departure from the maximum possible degree of orientation being governed by the ratio of the intensity of the flow effect (i.e., the velocity gradient) and Brownian motion. This equilibrium orientation at the specified rate of shear characterizes the intensity of energy dissipation and, hence, the dependence of the apparent viscosity on the rate of shear.

If the suspended particles are not absolutely rigid, the flow causes a change in the degree of asymmetry of the molecular coil and of the equilibrium orientation in the flow. This is accompanied by a change in the intensity of dissipation of the external work due to the deformation of the particles in the stream. These hydrodynamic effects associated with the deformation of the molecular coil lead to the change of the viscosity of the polymer.

The best-known description of the dependence η $(\dot\gamma)$ within the framework of hydrodynamic theory was offered by Bueche.* A detailed analysis of his theory will be given in Chapter 3. Here we shall confine ourselves to the indication that Bueche considers a suspension of macromolecular coils placed in a viscous liquid and rotating in the stream with a frequency $\omega =$ $= \dot\gamma/2$. In such a rotation the segments of macromolecules perform oscillations relative to the centre of mass, which leads to the dissipation of external work. The Bueche theory takes account of the viscoelastic behaviour of segments in the flow which is caused by the appearance of a relaxation spectrum in the polymeric system. From this follows the presence of a connection between non-Newtonian viscosity flow and the viscoelasticity and relaxation spectrum of polymeric systems. An important feature of the model treatment of polymeric systems in the Bueche theory is that the entire relaxation spectrum is uniquely determined by a single relaxation time θ_m^B since all the other relaxation times θ_p are expressed in terms of θ_m^B in a simple way: $\theta_p = \theta_m^B p^{-2}$, the quantity θ_m^B representing the longest relaxation time in the relaxation spectrum calculated by Bueche.

The final basic formula of the Bueche theory is written as

$$\frac{\eta - \eta_s}{\eta_0 - \eta_s} = 1 - \frac{6}{\pi^2} \sum_{p=1}^{N} \frac{(\dot\gamma\theta_p^B)^2}{p^2\,[p^4 + (\dot\gamma\theta_p^B)^2]} \left[2 - \frac{(\dot\gamma\theta_p^B)^2}{p^4 + (\dot\gamma\theta_p^B)^2}\right] \qquad (2.44)$$

or

$$\frac{\eta - \eta_s}{\eta_0 - \eta_s} = f_B\,(\dot\gamma\theta_p^B) \qquad (2.45)$$

where η_s is the viscosity of the solvent.

The summation in formula (2.44) takes account of all relaxation times defined in terms of θ_p^B. Note also that the function f_B in formula (2.45) is a decreasing function.

Formula (2.45) shows that the dependence of viscosity on the rate of shear is uniquely determined by the quantity θ_m^B which characterizes the properties of the polymeric system. Here the parameter θ_m^B has a simple physical meaning. It has been emphasized above that in many theories the parameter that determines the dependence of viscosity on the rate of shear is a certain

* Bueche, F., J. Chem. Phys., 22, 1570 (1954).

character stic relaxation time. In all these cases, just as in the treatment of the temperature dependence of viscosity in the WLF method, the effect of the change of external conditions on the flow properties of the polymeric system may be taken into account by changing a single relaxation time that determines the position of the relaxation spectrum.

For molten polymers, it is assumed that $\eta_s \to 0$. Then

$$\frac{\eta}{\eta_0} = 1 - \frac{6}{\pi^2} \sum_{p=1}^{N} \frac{(\dot{\gamma}\theta_p^B)^2}{p^2 [p^4 + (\dot{\gamma}\theta_p^B)^2]} \left[2 - \frac{(\dot{\gamma}\theta_p^B)^2}{p^4 + (\dot{\gamma}\theta_p^B)^2} \right] \qquad (2.44a)$$

The changeover to formula (2.44a) for polymer melts gives rise to certain doubts since this is associated with the replacement of the model with noninteracting macromolecular coils by systems of the network type, which must have an effect on the behaviour of macromolecules in the flow.

According to formula (2.44a), at $\dot{\gamma} \to 0$ the viscosity $\eta = \eta_0$; if $\dot{\gamma} \to \infty$, the following equality holds:

$$\frac{\eta}{\eta_0} = 1 - \frac{6}{\pi^2} \sum_{p=1}^{N} \frac{1}{p^2}$$

For large N values in the last formula $\sum_{p=1}^{N} p^{-2} \approx \pi^2/6$. This leads to the condition: $\eta_\infty \ll \eta_0$ and if $N \to \infty$, then $\eta_\infty = 0$.

2.5.5. Non-Newtonian Flow Behaviour and Nonlinear Viscoelasticity

The development of theories of nonlinear viscoelasticity leads, in a natural way, to the dependence of apparent viscosity on rate of shear. Here no assumptions are made as to the structure of the material being deformed and no changes taking place in it are considered. These theories are exclusively based on the mathematical properties of the operators and functions used, from the large number of which only those are selected, which describe correctly the properties of the material in the simplest deformation schemes.

A consideration of the mechanical properties of the simplest model of a viscoelastic liquid (the Maxwell model), or its generalizations written in the form of discrete linear differential operators, does not make it possible to describe the experimentally observed dependence of apparent viscosity on rate of shear or, what amounts to the same thing, the nonlinear dependence of shear stress on rate of shear. Indeed, according to the theory of linear viscoelasticity or the operator equation (1.104) the relation between shear stresses and rate of shear under steady-state flow conditions is always linear. But it has already been pointed out in Chapter 1 that in dealing with the kinematics of large deformations it is necessary to correlate the reference systems in which the stresses and strains are defined and the constitutive equation for a point in the material is written.

The equation of linear viscoelasticity theory is formulated for an element of a viscoelastic liquid. This element is moving in space, and therefore for the calculation of the parameters referring to the spacial coordinate system, it is necessary to make use of the appropriate coordinate transformations. In the case of the onstitutive equation given in the form of a linear differential operator, this leads to the necessity of replacement of an operation of partial differentiation by other, more structurally complicated differential operators involving various linear and nonlinear operations carried out over the components of stress and strain tensors.

Let us write the generalized operator equation (1.104) for a linear viscoelastic liquid in the form

$$\left(1 + \sum_{n=1}^{M} a_n \frac{\partial^n}{\partial t^n}\right) \sigma_{ij} = 2\eta_0 \left(1 + \sum_{n=1}^{N} b_n \frac{\partial^n}{\partial t^n}\right) \dot{\gamma}_{ij} \tag{2.46}$$

where, as usual, $\dot{\gamma}_{ij} = \frac{1}{2}(\partial v_i/\partial x_j + \partial v_j/\partial x_i)$, and a_n and b_n are constants, whose numerical values determine the viscoelastic properties of the liquid.

Equation (2.46) contains an operation of partial differentiation of the quantities σ_{ij} and $\dot{\gamma}_{ij}$ with respect to time. For a Maxwell liquid with one relaxation time Eq. (2.46) reduces to

$$\left(1 + \theta \frac{\partial}{\partial t}\right) \sigma_{ij} = 2\eta_0 \dot{\gamma}_{ij} \tag{2.47}$$

where θ is the relaxation time; η_0 is the Newtonian viscosity.

Hence, Eq. (2.47) is a special case of the operator equation (2.46) at $a_1 = \theta$, $a_n = 0$ ($n > 1$) and $b_n = 0$ (at any n value).

The constitutive equation (2.47) is quite analogous to Eq. (1.100) used in Chapter 1 for the analysis of the mechanical properties of the Maxwell liquid.

As has been pointed out in Chapter 1, the basic idea of the generalizations of viscoelasticity theory, which have been worked out for the purpose of taking account of nonlinear effects, consists formally in the replacement of the operator $(\partial/\partial t)$ by various differential operators associated with the change from the convected to the space-fixed coordinate system.

Let us consider in more detail the use of some differential operators. Suppose that there takes place the simple shear of a viscoelastic liquid in the direction of the x_1 axis, so that the velocity gradient in the direction of the x_2 axis is $\dot{\gamma}_0 = \partial v_1/\partial x_2$ (where v_1 is the velocity). A steady-state flow process is assumed.

In this case, it is obligatory that $\sigma_{13} = \sigma_{23} = \sigma_{31} = \sigma_{13} = 0$ (see Chapter 4). Then, in a general case, the rate-of-strain and stress tensors may be represented in the form:

$$\dot{\gamma}_{ij} = \frac{1}{2} \begin{vmatrix} 0 & \dot{\gamma}_0 & 0 \\ \dot{\gamma}_0 & 0 & 0 \\ 0 & 0 & 0 \end{vmatrix}; \qquad \sigma_{ij} = \begin{vmatrix} \sigma_{11} & \sigma_{12} & 0 \\ \sigma_{21} & \sigma_{22} & 0 \\ 0 & 0 & 0 \end{vmatrix}$$

where, by virtue of the complementary shear stress condition $\sigma_{12} = \sigma_{21}$.

Direct application of formula (1.36) to σ_{ij} and $\dot{\gamma}_{ij}$ yields the following results:

$$D_0 \sigma_{ij} = \begin{vmatrix} 0 & \sigma_{11} & 0 \\ \sigma_{11} & 2\sigma_{12} & 0 \\ 0 & 0 & 0 \end{vmatrix} \dot{\gamma}_0; \qquad D_0^2 \sigma_{ij} = \begin{vmatrix} 0 & 0 & 0 \\ 0 & \sigma_{11} & 0 \\ 0 & 0 & 0 \end{vmatrix}$$

$D_0^n \sigma_{ij_A}^{\cdot} = 0$ at $n > 2$;

$$D_0 \dot{\gamma}_{ij} = \frac{1}{2} \begin{vmatrix} 0 & 0 & 0 \\ 0 & \dot{\gamma}_0^2 & 0 \\ 0 & 0 & 0 \end{vmatrix} ; \qquad D_0^m \dot{\gamma}_{ij} = 0 \text{ at } m > 1$$

Therefore, the strain equation for the Maxwell model as generalized by Oldroyd will be written in the form:

$$\begin{cases} \sigma_{22} + 2\theta \dot{\gamma}_0 \sigma_{12} = 0 \\ \sigma_{11} = \sigma_{33} = 0 \\ \sigma_{12} + \theta \dot{\gamma}_0 \sigma_{11} = \eta_0 \dot{\gamma}_0 \end{cases}$$

The last system of equations leads to the following expressions for the components of the stress tensor:

$$\begin{cases} \sigma_{12} = \eta_0 \dot{\gamma}_0 \\ \sigma_{22} = -2\theta \eta_0 \dot{\gamma}_0^2 \\ \sigma_{11} = \sigma_{33} = 0 \end{cases} \tag{2.48}$$

Thus, the use of the operator D_0 in the Maxwell model does not provide a description of the phenomenon of non-Newtonian viscosity ($\sigma_{12}/\dot{\gamma}_0 = \eta_0 = \text{const}$) but predicts the appearance of normal stresses; here[*] the first normal stress difference $\sigma_1 \equiv \sigma_{11} - \sigma_{22} = 2\theta \eta_0 \dot{\gamma}_0^2$, and $\sigma_{11} - \sigma_{33} = 0$.

The use of higher derivatives, i.e., the consideration of a many-parameter model which obeys equation Eq. (1.100) with ($\partial^n/\partial t^n$) being replaced by D_0^n, leads to the following form of the first of the above-written equations:

$$\sigma_{22} + 2a_1 \dot{\gamma}_0 \sigma_{12} + 2a_2 \dot{\gamma}_0^2 \sigma_{11} = b_1 \eta_0 \dot{\gamma}_0^2$$

The other equations remain unchanged, and since $\sigma_{11} = 0$, then, as before $\sigma_{12} = \eta_0 \dot{\gamma}_0$. Therefore, taking into account the higher members of the many-parameter model of a viscoelastic body, as generalized according to Oldroyd, does not introduce anything new as compared with the Maxwell model with respect to shear stresses, but gives a somewhat modified expression for σ_{22}, namely,

$$\sigma_{22} = \eta_0 (b_1 - 2a_1) \dot{\gamma}_0^2$$

The application of formula (1.38) to the tensors σ_{ij} and $\dot{\gamma}_{ij}$ yields the following results:

$$D_J \sigma_{ij} = \begin{vmatrix} -\sigma_{12} & \frac{1}{2}(\sigma_{11} - \sigma_{22}) & 0 \\ -\frac{1}{2}(\sigma_{11} - \sigma_{22}) & \sigma_{22} & 0 \\ 0 & 0 & 0 \end{vmatrix}$$

_____

[*] The use of a somewhat different form of the Oldroyd operator leads to the same expression for the first normal stress difference, but instead of $\sigma_{11} = \sigma_{33}$ there is fulfilled the equality $\sigma_{22} = \sigma_{33} = 0$.

$$D_{\mathbf{J}}^2 \sigma_{ij} = \begin{vmatrix} -\dfrac{1}{2}(\sigma_{11}-\sigma_{22}) & -\sigma_{12} & 0 \\ -\sigma_{12} & \dfrac{1}{2}(\sigma_{11}-\sigma_{22}) & 0 \\ 0 & 0 & 0 \end{vmatrix}$$

and

$$D_{\mathbf{J}}^n \sigma_{ij} = -\dot\gamma_0^2 D_{\mathbf{J}}^{n-2}\sigma_{ij}$$

The last equality is also valid at $n = 2$ if $\sigma_{11} = -\sigma_{22}$, but this relation must yet be proved.

For $\dot\gamma_{ij}$

$$D_{\mathbf{J}}\dot\gamma_{ij} = \frac{1}{2}\begin{vmatrix} -1 & 0 & 0 \\ 0 & 1 & 0 \\ 0 & 0 & 0 \end{vmatrix}\dot\gamma_0^2$$

$$D_{\mathbf{J}}^2\dot\gamma_{ij} = -\frac{1}{2}\begin{vmatrix} 0 & 1 & 0 \\ 1 & 0 & 0 \\ 0 & 0 & 0 \end{vmatrix}\dot\gamma_0^3$$

and

$$D_{\mathbf{J}}^3\dot\gamma_{ij} = -\dot\gamma_0^2 D_{\mathbf{J}}^{n-2}\dot\gamma_{ij}$$

Then, the strain equation for the Maxwell model, as generalized according to De Witt with the aid of the Jaumann derivative, takes the following form:

$$\begin{cases} \sigma_{11} - \theta\dot\gamma_0\sigma_{12} = 0 \\ \sigma_{22} + \theta\dot\gamma_0\sigma_{12} = 0 \\ \sigma_{33} = 0 \\ \sigma_{12} + \dfrac{1}{2}\theta\dot\gamma_0(\sigma_{11}-\sigma_{22}) = \eta_0\dot\gamma_0 \end{cases}$$

From this we can obtain the following expression for the components of the stress tensor:

$$\begin{cases} \sigma_{11} = -\sigma_{22} = \dfrac{\theta\eta_0\dot\gamma_0^2}{1+(\dot\gamma_0\theta)^2} = \sigma_{12}\theta\dot\gamma_0 \\ \sigma_{33} = 0 \\ \sigma_{12} = \dfrac{\eta_0\dot\gamma_0}{1+(\dot\gamma_0\theta)^2} \end{cases} \qquad (2.49)$$

Thus, the utilization of the Jaumann operator in the Maxwell model of a viscoelastic body leads to the dependence of the apparent viscosity on the rate of shear:

$$\eta = \eta_0/[1+(\dot\gamma_0\theta)^2] \qquad (2.50)$$

The diagonal components of the stress tensor, which are responsible for the appearance of normal stresses, are found to be equal to:

$$\sigma_{11} = -\sigma_{22} = \eta_0\theta\dot\gamma_0^2; \quad \sigma_{33} = 0$$

or

$$\sigma_1 \equiv \sigma_{11} - \sigma_{22} = 2\eta_0 \theta \dot{\gamma}_0^2$$

The validity of the equality $\sigma_{11} = -\sigma_{22}$ is clearly seen in the case of the Maxwell model. In quite an analogous manner it is proved that when a many-parameter model of a viscoelastic body, as generalized according to Jaumann, is used, the relation $\sigma_{11} = -\sigma_{22}$ holds too. Therefore, for the sake of simplicity, the strain equation for a many-parameter model of a viscoelastic body is written below with direct account of this equality, in which case it assumes the form

$$\begin{vmatrix} \sigma_{11} & \sigma_{12} & 0 \\ \sigma_{21} & -\sigma_{11} & 0 \\ 0 & 0 & \sigma_{33} \end{vmatrix} A(\dot{\gamma}_0) + \begin{vmatrix} -\sigma_{12} & \sigma_{11} & 0 \\ \sigma_{11} & \sigma_{12} & 0 \\ 0 & 0 & 0 \end{vmatrix} B(\dot{\gamma}_0) =$$

$$= \eta_0 \dot{\gamma}_0 \left\{ \begin{vmatrix} 0 & 1 & 0 \\ 1 & 0 & 0 \\ 0 & 0 & 0 \end{vmatrix} C(\dot{\gamma}_0) + \begin{vmatrix} -1 & 0 & 0 \\ 0 & 1 & 0 \\ 0 & 0 & 0 \end{vmatrix} D(\dot{\gamma}_0) \right\}$$

where

$$A(\dot{\gamma}_0) = \sum_{n=0} (-1)^n a_{2n} \dot{\gamma}_0^{2n}$$

$$B(\dot{\gamma}_0) = \sum_{n=1} (-1)^{n-1} a_{2n-1} \dot{\gamma}_0^{2n-1}$$

$$C(\dot{\gamma}_0) = \sum_{n=0} (-1)^n b_{2n} \dot{\gamma}^{2n}$$

$$D(\dot{\gamma}_0) = \sum_{n=1} (-1)^{n-1} b_{2n-1} \dot{\gamma}_0^{2n-1}$$

For the Maxwell model, these functions are expressed as follows:

$$A = 1; \quad B = \theta \dot{\gamma}_0; \quad C = 1; \quad D = 0$$

The above equation which relates σ_{ij} to γ_{ij} takes the following form in the component notation:

$$\begin{cases} \sigma_{11} A(\dot{\gamma}_0) - \sigma_{12} B(\dot{\gamma}_0) = -\eta_0 \dot{\gamma}_0 D(\dot{\gamma}_0) \\ \sigma_{12} A(\dot{\gamma}_0) + \sigma_{11} B(\dot{\gamma}_0) = \eta_0 \dot{\gamma}_0 C(\dot{\gamma}_0) \end{cases}$$

This leads to the following expression for the dependence of the apparent viscosity on the rate of shear:

$$\frac{\eta}{\eta_0} = \frac{\sigma_{12}}{\dot{\gamma}_0 \eta_0} = \frac{A(\dot{\gamma}_0) C(\dot{\gamma}_0) + B(\dot{\gamma}_0) D(\dot{\gamma}_0)}{[A(\dot{\gamma}_0)]^2 + [B(\dot{\gamma}_0)]^2} \qquad (2.51)$$

The diagonal components of the stress tensor are equal to:

$$\sigma_{11} = -\sigma_{22} = \frac{B(\dot{\gamma}_0) C(\dot{\gamma}_0) - A(\dot{\gamma}_0) D(\dot{\gamma}_0)}{[A(\dot{\gamma}_0)]^2 + [B(\dot{\gamma}_0)]^2} \eta_0 \dot{\gamma}_0$$

$$\sigma_{33} = 0 \qquad (2.52)$$

Thus, the use of the Jaumann derivative in the operator equation of the theory of linear viscoelasticity, (1.104), leads to the prediction of a definite form of the dependence of the apparent viscosity and normal stresses on the rate of shear. Some other interesting aspects associated with the structure of the formulas obtained will be discussed later on, in Chapter 3 which is devoted to the analysis of the dynamic properties of polymers.

The use of the operators (1.36) and (1.38) in Eq. (1.104) is not the only possible way of generalizing the theory of linear viscoelasticity with the purpose of predicting the nonlinear properties of polymeric materials. Moreover, the use of the operator D_O does not yield the desired result at all since the viscosity $\eta = \eta_0$ proves to be constant.

Had the problem been restricted to the choice of the analytical expression for the dependence $\eta(\dot{\gamma})$, formula (2.51) would have invariably led to the solution of this problem since it contains a large (generally speaking, an infinitely large) number of arbitrary constants. It is however necessary that a constitutive equation containing these constants calculated, say, from the experimentally determined dependence $\eta(\dot{\gamma})$, predict correctly the behaviour of the material under other deformation conditions, in particular, under small-amplitude harmonic oscillations.

Without delving into the analysis of relevant experimental facts, we shall point out that the constitutive equation (1.104) containing Jaumann derivatives proves to be quantitatively unsatisfactory for the description of many important effects specific to polymeric materials. Therefore, in the literature there are proposed and discussed other forms of generalization of the constitutive equation (1.104), which are associated with the use of more complicated differential operators that satisfy the principle of invariance upon change from one coordinate system to another. The number of such operators is, generally speaking, unlimited. We shall give here the results of the application of two more differential operators of more complex structure for the purpose of illustrating the potentialities and results of such a theoretical approach.

As has already been mentioned, in his early works Oldroyd proposed the differential operator $D_{O,n}$ of the following structure:

$$D_{O,n} A_{ik} = D_O A_{mn} + \frac{2}{3} A_{mn} \dot{\gamma}_{mn} \delta_{ik} =$$

$$= \frac{\partial}{\partial t} A_{ik} + v_p \frac{\partial A_{ik}}{\partial x_p} - A_{ip} \frac{\partial v_k}{\partial x_p} - A_{kp} \frac{\partial v_i}{\partial x_p} + \frac{2}{3} A_{mn} \dot{\gamma}_{mn} \delta_{ik} \qquad (2.53)$$

where A_{ik} are the components of the tensor defined in the convected coordinate system.

The inclusion of this operator in the constitutive equation (1.104) gives the result obtainable by a method analogous to that employed above for the operators D_O and D_J:

$$\begin{cases} \dfrac{\eta(\dot{\gamma}_0)}{\eta_0} = \dfrac{\sigma_{12}}{\eta_0 \dot{\gamma}_0} = \dfrac{Q(\dot{\gamma}_0) S(\dot{\gamma}_0) + P(\dot{\gamma}_0) R(\dot{\gamma}_0)}{[R(\dot{\gamma}_0)]^2 + [S(\dot{\gamma}_0)]^2} \\[3mm] \sigma_{11} = -2\sigma_{22} = \dfrac{PS - RQ}{P^2 + S^2} 2\sqrt{\dfrac{2}{3}} \dot{\gamma}_0 \\[3mm] \sigma_{22} = \sigma_{33} \end{cases} \qquad (2.54)$$

In these expressions R, S, P, and Q are the functions of the constant velocity gradient $\dot{\gamma}_0$, which are expressed in the following way:

$$R(\dot{\gamma}_0) = \sum_{n=0} (-1)^n a_{2n} \left(\sqrt{\frac{2}{3}} \dot{\gamma}_0\right)^{2n}$$

$$S(\dot{\gamma}_0) = \sum_{n=1} (-1)^{n-1} a_{2n-1} \left(\sqrt{\frac{2}{3}} \dot{\gamma}_0\right)^{2n-1}$$

$$P(\dot{\gamma}_0) = \sum_{n=0} (-1)^n b_{2n} \left(\sqrt{\frac{2}{3}} \dot{\gamma}_0\right)^{2n}$$

$$Q(\dot{\gamma}_0) = \sum_{n=1} (-1)^{n-1} b_{2n-1} \left(\sqrt{\frac{2}{3}} \dot{\gamma}_0\right)^{2n-1}$$

It is obvious that these expressions are completely identical with the corresponding expressions for the functions $A(\dot{\gamma}_0)$, $B(\dot{\gamma}_0)$, $C(\dot{\gamma}_0)$, and $D(\dot{\gamma}_0)$, obtained by using the Jaumann operator if one takes account of the difference in the argument $\dot{\gamma}_0$ for the Jaumann operator and $(\sqrt{2/3}\dot{\gamma}_0)$ for the operator $D_{0,n}$. Therefore, with the same values of the coefficients a_n and b_n in the constitutive equation the above-compared methods of describing the function $\eta(\dot{\gamma}_0)$ provide identical qualitative results, but the curves of viscosity versus rate of shear are found to be shifted relative to one another along the $\dot{\gamma}_0$ axis by $\sqrt{2/3}$ times.

An operator of even more complicated structure has been proposed by Spriggs.* He makes use of constitutive equations corresponding to a set of Maxwell elements. The differential operator D_S is written thus:

$$D_S A_{ik} = D_J A_{ik} + (1+\varepsilon) \left(A_{ip}\dot{\gamma}_{pk} + A_{pk}\dot{\gamma}_{ip} + \frac{2}{3} A_{mn}\dot{\gamma}_{mn}\delta_{ik}\right)$$

Here D_J is the Jaumann operator, and ε is an arbitrary empirical constant included in characteristic functions that describe the relation among the components of normal stresses in simple shear, and also in other constitutive equations. An important assumption made by Spriggs is the one concerning the regular distribution of relaxation times θ_p. The point is that when the operator equation (1.100) is used, the relaxation times are expressed in terms of the constants of this equation. Since these constants are arbitrary, the choice of relaxation times is also arbitrary. As a large number of empirical constants are contained in the equation, this complicates the practical utilization of the equations if it does not exclude this possibility at all. If molecular models are employed (see, for example, Chapter 3), it turns out that relaxation times always vary in a regular manner. To characterize the relaxation-time distribution, the following relation has been proposed:

$$\theta_p = \theta_m p^{-\alpha}$$

where α is an empirical constant; θ_m is the longest relaxation time.

The constant α determines the rate of decrease of relaxation times in the spectrum of the polymeric system.

Apart from the three empirical constants introduced above (ε, θ_m, and α), the theory also considers the fourth constant—the initial viscosity η_0. Using

---

* Spriggs, T. W., *Chem. Eng. Sci.*, **20**, 931 (1965).

the above-considered scheme of calculation of derivatives with the aid of differential operators of complicated structure, the following expressions can be obtained for the apparent viscosity:

$$\frac{\eta\,(\dot{\gamma}_0)}{\eta_0} = \frac{1}{\zeta\,(\alpha)} \sum_{p=1}^{\infty} \frac{p^{\alpha}}{p^{2\alpha} + (c\dot{\gamma}_0\theta_m)^2} \tag{2.55}$$

and for the normal-stress components:

$$\sigma_1\,(\dot{\gamma}_0) \equiv \sigma_{11} - \sigma_{22} = \frac{2\theta_m\eta_0\dot{\gamma}_0^2}{\zeta\,(\alpha)} \sum_{p=1}^{\infty} \frac{1}{p^{2\alpha} + (c\dot{\gamma}_0\theta_m)^2} \tag{2.56}$$

$$\sigma_2\,(\dot{\gamma}_0) \equiv \sigma_{22} - \sigma_{33} = -\frac{\varepsilon}{2}\,\sigma_1\,(\dot{\gamma}_0) \tag{2.57}$$

In these formulas $\zeta\,(\alpha)$ is the Riemann zeta function*, c is a constant which is expressed in terms of ε as follows:

$$c^2 = \frac{2 - 2\varepsilon - \varepsilon^2}{3} \tag{2.58}$$

From this equation there are obtained, as a special case, the above-considered results which follow from the application of the operator $D_{0,n}$. For this to be done, we must put $\varepsilon = 0$; then, D_S changes to $D_{0,n}$ and $c^2 = 2/3$.

Thus, the resulting formulas predict the same form of the dependence $\eta\,(\dot{\gamma}_0)$ as the operators discussed above, but with an arbitrary shift of the curves relative to one another along the $\dot{\gamma}_0$ axis, which is determined by the choice of the constant c. This fact is a consequence of the uncertainty of the relation between the normal stresses, while in the equation given above this relation was specified in advance as soon as the form of the constitutive equation was chosen.

To establish the exact correlation between the forms of the expressions obtained above and relation (2.46), it is necessary to express relaxation times in terms of the constants a_n and b_n of the operator equation, the arbitrariness of the choice of these constants being also restricted by the requirement of a definite order in the distribution of relaxation times θ_p.

Some additional considerations associated with the results obtained will be presented in a discussion of the dynamic properties of polymeric systems.

Formula (2.55) is more preferable than the formulas given earlier because it contains only four arbitrary constants to be determined experimentally. These four constants must completely determine the concrete features of the behaviour of the material under various deformation conditions. A comparison with experiment has shown that the accuracy provided by this equation is not sufficiently high. However, of interest here is the general approach to the analysis of the dependence $\eta\,(\dot{\gamma}_0)$ and the elucidation of its relationship to the rheological equation of state of the material.

The main result that follows from the theories discussed above consists in that, using the concept of constitutive equations in the convected coordinate system and taking into account the necessity of correlating the reference systems in writing these equations, it is possible to predict, on the basis of "geometric" considerations, the existence of the effect of non-Newtonian viscosity. However, here no quantitative agreement can be obtained

---

* See footnote to page 112.

between theoretical formulas (in any case, the simplest ones) and experiment. From a formal standpoint, the detailing of the theory requires the introduction of new, more complicated methods of writing of constitutive equations. This means that the phenomenon of non-Newtonian flow behaviour does not boil down to purely geometric conceptions of the processes of rotation and transfer of the elements of the material in space. It may be said that the introduction of complex differential operators is a formal method of reflecting those physical (structural) changes which occur in the material simultaneously with the movement of its particles in space. These changes make their contribution to the observed effect of non-Newtonian viscosity.

In concluding the treatment of the problem of constitutive equations obtained on the basis of theories of nonlinear viscoelasticity, it should be pointed out that it is important that the number of constants in them be minimal. This considerably facilitates their experimental determination and, hence, their practical utilization. The most important method of determining the constants in constitutive equations is the analysis of small-amplitude harmonic loading conditions. Then, all generalizations of the operator equation (1.104) reduce to an analogous equation with partial time derivatives σ_{ij} and $\dot{\gamma}_{ij}$. The determination of the constants a_n and b_n and, through them, of a set of relaxation times becomes a simple task of the harmonic analysis of data obtained in measurements of the dynamic properties of the material. Many important cases of such an analysis will be dealt with in a chapter dealing with the description of the dynamic properties of polymeric systems.

2.5.6. Empirical Formulas for Apparent Viscosity

A large body of experimental data has been accumulated on the dependence of apparent viscosity on deformation conditions. The lack of a rigorous theory of non-Newtonian viscosity that would be in good quantitative agreement with experiment has given rise to numerous empirical formulas. Many of these, though being rather simple, correctly describe the specificity of the observed non-Newtonian viscosity of polymeric systems. Such empirical formulas are extensively employed in solving the various hydrodynamic problems associated with the flow of polymeric systems and, in particular, they are of value for the design calculation of the processing equipment.

In using empirical formulas one must always be careful: the use of these formulas beyond the limits of the region of deformation conditions for which they have been preliminarily tested is risky and largely intolerable. For example, even under the conditions of simple shear one must not employ empirical formulas intended for steady-state flow for the description of prestationary deformation conditions.

Most of the frequently used empirical formulas for the dependence of viscosity on shear stress or rate of shear are various versions of the power law. By the power law is meant the following formula:

$$\tau = k\dot{\gamma}^n \tag{2.59}$$

where k and n are constants. The quantity n is called the flow index. For non-Newtonian fluids $n < 1$, and for Newtonian liquids $n = 1$. From formula (2.59) it follows that $\eta = k\dot{\gamma}^{n-1}$ and $\eta = k^{1/n}\tau^{(n-1)/k}$. Formula (2.59) is widely used by many authors for the solution of various kinds of hydro-dynamic problems associated with the flow of polymer solutions and melts. It should however be noted that this formula fails to describe satisfactorily the behaviour of polymers at τ and $\dot{\gamma}$ tending to zero, which, among other things, does not make possible its use for finding the initial viscosity. But in the range of values of shear stress and rate of shear, which are most vital in technology, formula (2.59) describes the behaviour of many polymeric liquids sufficiently well. Besides, it is convenient owing to the simplicity of the graphical determination of the flow index in the representation of experimental data in doubly logarithmic coordinates.

It is not difficult to construct an analytical expression for the dependence $\eta(\dot{\gamma})$, such that it will satisfy the following requirements: there must exist finite values of viscosity at $\dot{\gamma} \to 0$ and at $\dot{\gamma} \to \infty$, and in the intermediate range of values of the argument the dependence $\eta(\dot{\gamma})$ must be defined by the power law. Such a formula has the form

$$\eta = \eta_\infty + (\eta_0 - \eta_\infty)/(1 + k\dot{\gamma}^n)$$

and since in the overwhelming majority of cases $\eta_\infty \ll \eta_0$ this formula takes the form

$$\eta = \eta_0/(1 + k\dot{\gamma}^n)$$

A convenient method of detailing the power law is to represent the dependence $\tau(\dot{\gamma})$ in the form of the sum of several terms of a single type. Increasing the number of such terms can provide any desired accuracy of the agreement between the analytical description of the flow curve and the concrete experimental data.

2.5.7. Concerning the Determination of the Newtonian Viscosity

An important point in the treatment of the phenomenon of non-Newtonian viscosity of polymers is the assumption of the existence of the limiting value of viscosity when the shear stress and rate tend to zero. In real measurements this must correspond to a certain region of sufficiently small values of τ and $\dot{\gamma}$ at which the viscosity is constant and equal to its limiting value—the initial viscosity. There is no contradiction at all in that, on the one hand, the initial viscosity corresponds to the limited value of the function and, on the other, a certain range of deformation conditions for polymeric systems, in which the viscosity is practically constant, is recorded experimentally. It will be useful to illustrate this by a simple example.

Suppose, for example, that $\eta(\tau)$ is specified in the form of a rather sharply decreasing function

$$\eta = \eta_0 \exp(-a\tau)$$

where a is an empirical constant.

This function at $\tau \to 0$ has a limit equal to the viscosity η_0. At small τ values the function in question may be represented by the first two terms of the series:

$$\eta = \eta_0 \exp\left(-a\tau\right) \approx \eta_0 \left(1 - a\tau\right)$$

Suppose that viscosity measurements are made in the range of τ values that vary by 100 times, say from $(10^4 a)^{-1}$ to $(10^2 a)^{-1}$. Here the apparent viscosity changes as compared with η_0 not more than by 1 per cent. This lies beyond the resolution limits of practically all methods of measuring the viscosity of polymers and, hence, over a rather wide range of variation of stresses (by 100 times) the viscosity remains constant, i.e., there is observed a region

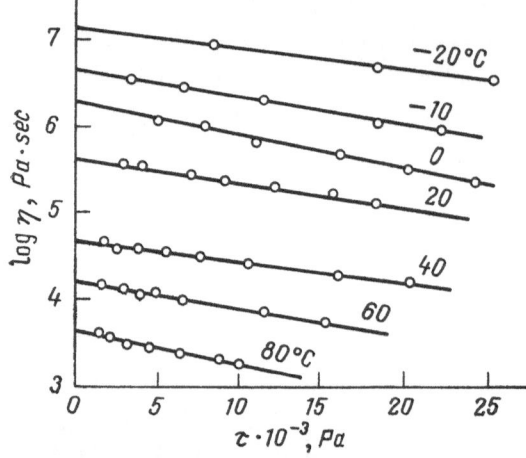

Fig. 2.23. Determining the values of initial (Newtonian) viscosity of polyisobutylene with a molecular mass of about 1×10^5 at different temperatures.

of flow corresponding to the initial (the largest) Newtonian viscosity, in spite of the strong dependence of η on τ.

The example just considered shows that the question of whether there exists a finite limiting value of viscosity or a region of its constant values cannot be answered by experiment.

The value of initial viscosity is of great importance for the characterization of the properties of polymeric systems. Therefore, in those cases where the region of the practically constant highest viscosity is experimentally unattainable, the method of finding it by extrapolation of the dependence $\eta\,(\tau)$ to the value $\tau \to 0$ becomes important. It is exactly the dependence $\eta\,(\tau)$ [and not $\eta\,(\dot\gamma)$] that is usually chosen for this purpose. The point is that the first of the indicated functions corresponds to a stronger change of viscosity, which considerably facilitates the extrapolation. For the same reason, for the purposes of extrapolation, an exponential formula is the most convenient. If we take logarithm of the relation $\eta = \eta_0 \exp\left(-a\tau\right)$ and then consider the dependence of $\log \eta$ on τ, the extrapolation of this dependence to $\tau = 0$ will enable the value of η_0 to be determined. A typical example of determination of the largest Newtonian viscosities of polyisobutylene at various temperatures, which is made by this method, is shown in Fig. 2.23. Direct measurement of values of η_0 has shown that the results of the determination of η_0 values by the extrapolation method lie within the range limited by the measurement error.

A method similar in meaning to the one just described is to determine the values of η_0 by extrapolating experimental data presented in the coordinates η^{-1} and τ. Indeed, the exponential relation between viscosity and stress at small τ may be written in the form

$$\eta^{-1} = \eta_0^{-1} \exp(a\tau) \approx \eta_0^{-1}(1 + a\tau) \approx \eta_0^{-1} + (a/\eta_0)\,\tau$$

which accounts for the use of the coordinates η^{-1} and τ for finding η_0. Good results are also obtained* when η_0 is determined by extrapolation in the coordinates η^{-1} and $\dot{\gamma}^n$, in which case the constant n is chosen so as to linearize the graphical representation of experimental data. The replacement of τ by $\dot{\gamma}^n$ is associated with the supposition that at sufficiently low stresses there is fulfilled the power law which changes at $\dot{\gamma} \to 0$ to the condition $\eta = \eta_0$.

All these methods of finding η_0 are purely empirical and are not associated with the true form of the flow curve at low τ values. One must not therefore be surprised that there exist cases where the application of this or that method does not allow for the linearization of the observed dependence $\eta(\tau)$ or $\eta(\dot{\gamma})$ and does not therefore make it possible to reliably determine the initial viscosity.

2.6. Relation Between Flow Properties, Molecular Mass and Branching of Polymers

2.6.1. Introduction

While considering the temperature dependence of the viscosity of linear polymers, we introduced the concept of the macromolecular segment as a molecular-kinetic unit performing elementary acts of translation in space from one equilibrium state to another. If the size of the segment is much smaller than that of the macromolecule, these transfers—the elementary acts of flow—are independent of molecular mass. However, for the macromolecule to be transferred irreversibly, it is necessary that the centre of gravity of the entire molecule be shifted as a result of the displacement of its constituent segments. But the higher the molecular mass of the polymer, i.e., the greater the number of segments in the macromolecule, the larger is the number of cooperative segment motions that must be effected for its centre of gravity to be shifted and the higher must be the viscosity.

Study of the effect of molecular mass on the flow properties of polymers is supposed to provide answers to a number of questions. How does molecular mass affect the initial viscosity and non-Newtonian flow behaviour (anomalous viscosity) of polymers? How can one compare the flow properties of polymers with the different structures of the macromolecular chain, considering that at one and the same molecular mass the chain length and flexibility may strongly differ for polymers of different nature? How does the molecular mass distribution affect the dependence of the Newtonian (initial) viscosity on molecular mass and how does the non-Newtonian

* Rudin, A. and K. K. Chee, *Macromolecules*, 6, 613 (1973).

flow behaviour change? In evaluating the effect of molecular-mass distribution (MMD) on the flow properties of polymers there also arises the most important question: What characteristics of MMD and what values of molecular mass of polydisperse polymers should be used to compare the flow properties of various polymers.

Experimental investigations into the entire range of questions regarding the effect of molecular mass on the flow properties of polymers are associated with the preparation of samples of very narrow MMD (monodisperse, in the ideal case) and with a reliable determination of their molecular masses. Only in such a case can the effects of molecular mass and MMD be separated. This accounts for the fact that the major achievements have been made only in recent years and the most complicated problem of the quantitative estimation of the effect of MMD on the flow properties of polymers is still far from solution.

The very setting-up of the problem of the dependence of the viscosity of a polymer on its molecular parameters implies the correctness of the assumption of the unambiguity of the values of viscosity for a given sample, independently of its previous history. This idea is associated with the concept of the existence of the equilibrium state of the polymeric system, which is attained rather rapidly after the preceding loading processes (if, of course, they do not alter the molecular mass or chemical nature of the polymer) due to a sufficiently high rate of relaxation processes taking place in melts and concentrated solutions, in spite of their high viscosity. In very many cases, such a supposition proves valid. But this is not always the case.

The most obvious case of the unambiguity of the values of viscosity of samples having identical molecular characteristics is a consequence of the retainment in the melt of residues of the crystalline structure which may be varied. This is especially typical, for example, of polyvinyl chloride since, because of the very low degree of crystallinity and high imperfection of the crystals, it can flow at temperatures below the equilibrium melting point of the crystalline phase.* Another, very peculiar case has been observed by Andrianova** who has found that the viscosity of polystyrene depends on the concentration and "quality" of the solvent from which the sample is produced by sublimation. This fact may be interpreted as a consequence of the retainment in the melt of a certain structure which existed in the polymer in solution and which was dependent on the nature of the solvent and the solution concentration. This structure was found to be quite stable to the subsequent thermomechanical effects imposed on the material. In this connection, it should also be noted that the structural rearrangements in the melt occur, in general, much more slowly than mechanical relaxation.

2.6.2. Dependence of Viscosity on Molecular Mass for Polymers of Narrow Molecular Mass Distribution

The relation between the viscosity and molecular characteristics of a polymer can be used as a physico-chemical method of investigation of high-molecular-mass compounds. The main index of the flow properties of polymers is the intrinsic viscosity $[\eta]$. The problem of the relationship

---

* Berens, A. R. and V. L. Folt, *Polymer Eng. Sci.*, 9, 27 (1969); Collins, E. A., *Pure and Appl. Chem.*, 49, 581 (1977).

** Andrianova, G. P., *J. Polymer Sci.*, *Polymer Phys. Ed.*, 13, 95 (1975).

between [η] and molecular mass M is tackled in detail in numerous specialized monographs. We shall therefore confine ourselves to the minimum information on this question, which is of value for the rheology of polymers.

Generally adopted is the relation between [η] and molecular mass called the Mark-Kuhn-Houwink-Sakurada equation, which is convenient for practical purposes:

$$[\eta] = KM^a \tag{2.60}$$

where K and a are empirical constants characteristic of a given polymer-solvent pair. For flexible-chain polymers $0.5 \leqslant a < 1$. For rigid-chain (rod-like) polymers a may exceed unity. The values of K and a are found through an independent calibration of the method, using samples of known molecular mass, i.e., formula (2.60) provides a comparative method of determining molecular masses on the basis of the results of measurements of intrinsic viscosity.

An important special case of formula (2.60) is its application to solutions in a theta-solvent. In this case, formula (2.60) assumes a simpler form:

$$[\eta] = K_\theta M^{0.5} \tag{2.61}$$

where K_θ is the value of the constant K when [η] is measured in a theta-solvent; a in this case is apparently equal to 0.5.

At present the values of K and a are known for a great number of various polymer-solvent pairs, so that the use of formula (2.60) is a universal convenient and sufficiently simple method of characterizing a polymer. A large number of empirical correlations are known between values of the constants K, K_θ, and a, on the one hand, and the detailed specific features of the macromolecular structure, on the other.[*] Moreover, methods are also known, which establish a correlation between the constants K and a.[**] All this substantially facilitates practical calculations carried out with the aid of formula (2.60) and makes the measurement of [η] a reliable tool of physico-chemical analysis of polymers.

Other empirical formulas which relate [η] and M are also known. Thus, the following empirical equation is apparently a rather convenient method[***]:

$$[\eta] = -A_2 + A_1 M^{-0.5} \tag{2.62}$$

where A_1 and A_2 are constants; as has been shown by the authors of the work cited, A_1 is close to K_θ^{-1}.

An important advantage of formula (2.62) as compared with Eq. (2.60) is that the relationship between [η] and M, plotted in the coordinates $[\eta]^{-1}$ and $M^{-0.5}$, is found to be linear over a wider range of molecular masses (and especially in the low-molecular-mass region) than when use is made of the coordinates $\log [\eta]$ and $\log M$.

A common feature of all known relationships between [η] and M is the relatively slight effect of molecular mass of the flow properties of solutions. This is associated with the fact that in the region of dilute solutions the decisive role is played by the individual properties of macromolecules and no cooperative effects connected with the interaction of polymer chains and dependent substantially on their length are observed.

---

* Krevelen, D. W. van, *Properties of Polymers. Correlations with Chemical Structure*, Elsevier Publ. Co., Amsterdam-London-New York, 2nd ed., 1976.
** Aharoni, S. M., *J. Appl. Polymer Sci.*, 21, 1323 (1977).
*** Dondos, A. and H. Benoit, *Polymer*, 18, 1161 (1977); 19, 523 (1978); Dondos, A., *Polymer*, 18, 1250 (1977).

The situation sharply changes when dealing with concentrated polymer solutions and molten polymers. The first reliable measurements of the viscosity of solutions of polymers of narrow MMD were carried out by Fox and Flory who used polyisobutylene and polystyrene for the purpose. Later, such viscosity measurements were extended and carried out for many polymers of various structure.

The main results of several dozens of works devoted to the experimental study of the dependence of initial viscosity on the molecular mass of polymers may be formulated in the following way. There are two regions of molecular masses, which are separated by the critical molecular mass M_c characteristic of each polymer-homologous series. In both regions of molecular masses, the relationship between η_0 and M can be described by a power law, namely:

$$\eta_0 (M) = \begin{cases} aM^\alpha & \text{at } M < M_c \\ bM^\beta & \text{at } M \geqslant M_c \end{cases} \tag{2.63}$$

Here a and b are the individual constants of the polymer-homologous series; the value of α is of the order of unity (though usually greater than 1); β has a value close to 3.4-3.5. The substantially higher values of α observed in some cases are associated with the failure to realize a true molecular flow of polymers. This is characteristic, for example, of polyvinyl chloride which at $T < 220°C$ is liable to form stable aggregates which are retained during flow.* Such phenomena can also be observed during the formation of strong intermolecular bonds, say, during the formation of hydrogen bonds, the cross-linking of chains containing acid groups by metal ions, or during the interaction of macromolecules at the site of the functional end groups of the chains.

A typical example of the relationship between η and M that corresponds to formula (2.63) is given in Fig. 2.24 for polymethylenes, the argument being the number Z of carbon atoms in the chain of principal valences. The solid lines drawn in this figure correspond to regions of molecular masses below and above the critical value M_c (or, what is the same thing, below Z_c), and the slopes of these lines are, respectively, equal to 1.0 and 3.5.

The question of the upper limit of applicability of formula (2.60) is controversial. The difficulty in providing an answer to this question is associated with the problem of reliable measurements of very high values of the viscosity of high-molecular samples. An example of the dependence of η_0 on the mass-average molecular mass $\overline{M}_w$ of polyethylene, shown in Fig. 2.25, points to the possibility of the unlimited extension of formula (2.60) to the entire accessible region of molecular masses. But the value of β in Fig. 2.25 is equal to 3.0, which is somewhat lower than the usual value of β. According to the author of the work cited here, this may be associated with the fact that Fig. 2.25 gives the values not of η_0 but of the apparent viscosity in the region of high stresses, which is rather far away from the region of Newtonian flow.

If use is made of a large amount of experimental data, the transition from the lower to the upper region is found to be not so sharp as in Fig. 2.24. This is clearly seen from the experimental data obtained by a number of

---

* Iyngaae-Jorgensen, J., *J. Appl. Polymer Sci.*, **20**, 2497 (1976).

authors for polydimethyl siloxanes (Fig. 2.26). Here too the existence of
two regions differing in the character of the dependence η_0 (M) is obvious,
but the transition between them is smooth. Therefore, the position of the
point corresponding to the critical molecular mass becomes indefinite.
Nonetheless, the concept, according to which there exists a critical value of
molecular mass for each polymer-homologous series, is extremely important
for the generalization of experimental data on the dependence not only
of the viscosity but also of many other characteristics of polymers on molecu-

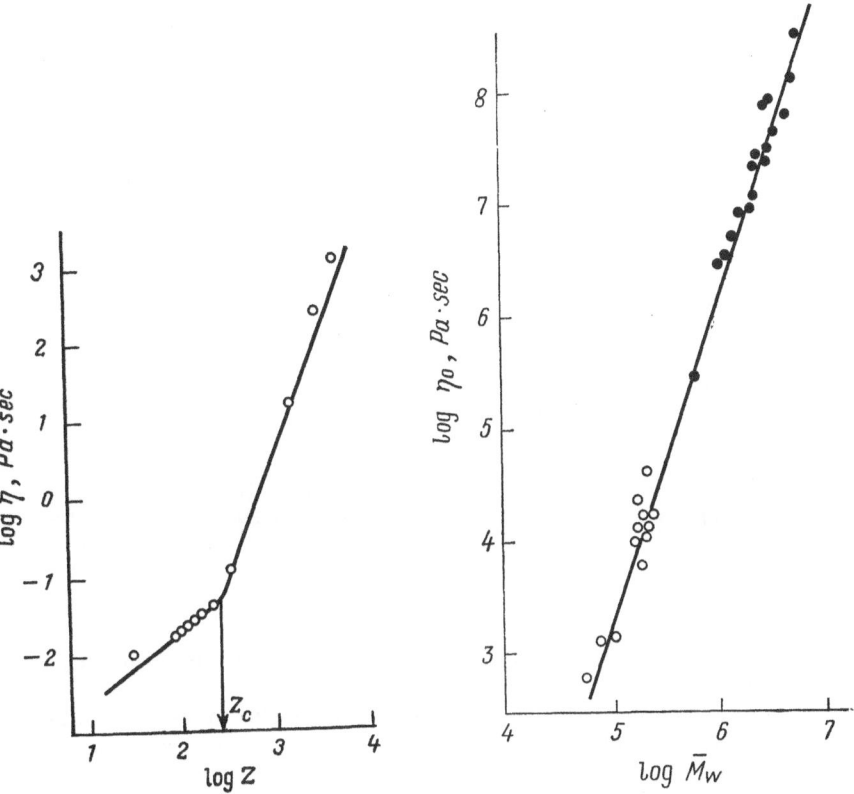

Fig. 2.24. The Newtonian viscosity of
molten polymethylenes versus the number of carbon atoms in the polymeric
chain. [Berry, G. C. and T. G. Fox,
Adv. Polymer Sci., 5, 261 (1968).]

Fig. 2.25. Viscosity-molecular mass relationship for polyethylenes. [Shaw,
M. T., *Polymer Eng. Sci.*, 17, 266
(1977).] The unfilled circles refer to data taken from the literature; the filled
circles refer to the original measurements of one of the authors.

lar masses. Therefore, M_c can be found approximately, for example, from
the point of intersection of the extensions of the two straight-line branches
of the curve of η_0 versus M plotted on a log-log scale.

When the critical molecular mass is reached, the set of properties inherent
in polymeric systems are changed. The attainment of the critical molecular
mass leads to the change of the set of properties inherent in polymeric systems.
A compound may be legitimately regarded as belonging to the class of polymers only when the critical molecular mass is attained. Therefore, for the

critical molecular mass of various polymers to be estimated, use has been made of different methods. The question of the consistency between the results of determination of M_c by different methods will be taken up in Chapter 3 devoted to the analysis of the dynamic properties of polymers, which offers wider possibilities for a comparison of the results obtained by using different approaches. At this point we confine ourselves to the consideration of the most frequently encountered values of M_c determined on the basis of the results of measurement of the dependence of viscosity within polymer-homologous series on molecular mass: polyethylene, 4000; polybutadiene, 5600; polyisobutylene, 17,000; polystyrene, 35,000; poly-

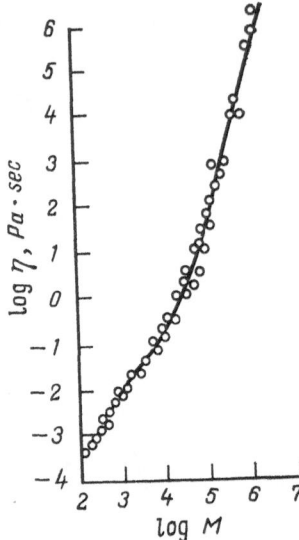

Fig. 2.26. The Newtonian viscosity of polydimethyl siloxanes versus molecular mass. [Kataoka, T. and S. Ueda, *J. Polymer Sci.*, A-1, 5, 3071 (1967).]

dimethyl siloxane, 29,000; polyvinyl acetate, 22,500; polymethyl methacrylate, 27,500, etc.

The relationship between M_c and the size of the macromolecular coil is traced out with the aid of the empirical relation deduced by van Krevelen,* who has found that

$$M_c = (13/K_\theta)^2 \qquad (2.64)$$

where K_θ is the value of the constant figuring in formula (2.61) and characterizing the behaviour of the polymer in an infinitely dilute solution in a theta-solvent. The quantity K_θ is expressed here in $cm^3 \cdot mole^{1/2}/g^{3/2}$.

It is necessary, however, to stress that the values of M_c given should only be regarded as approximate estimates since even for a single polymer there can be found in the literature the slightly different values of M_c. Sometimes, different values of M_c are given on the basis of the same experimental data, which is attributed to the spread of the transition from the lower to the upper portion of the curve of η_0 versus M.

The attainment of the critical molecular mass M_c is usually tied up to the appearance in the polymer of a spatial network of junctions of the type

_____

* van Krevelen, D. W., *Properties of Polymers*, 2nd Ed., Elsevier, Amsterdam, 1976, page 338.

of macromolecular entanglements, which are characterized by weak inter-molecular interactions and, accordingly, by a low potential energy barrier. This accounts for the very slight dependence of M_c on temperature.

2.6.3. The Fox Theory*

The value of the critical molecular mass determines the sharp change of the properties of polymers in a given homologous series. Therefore, M_c should be considered a characteristic constant of the polymer-homologous series. The most successful results in this direction have been obtained by Fox and coworkers, who proposed the following formula for the dependence of viscosity on molecular mass:

$$\eta_0 = \frac{A}{6} X_c \left(\frac{X}{X_c} \right) \zeta(\rho) \tag{2.65}$$

where the parameter X is expressed in the following way:

$$X = Z(\langle s^2 \rangle_0 / M)(\varphi_2/v_2) = \langle s^2 \rangle_0 (\varphi_2/v_a)$$

and, accordingly,

$$X_c = Z_c(\langle s^2 \rangle_0 / M)(\varphi_2/v_2)$$

The quantities appearing in this formula have the following meanings: A is Avogadro's number; a is a coefficient equal to 1.0 at $X < X_c$ and to 3.4 at $X > X_c$; $\zeta(\rho)$ is a correction factor which depends on the density (ρ) of the polymeric system; $\langle s^2 \rangle_0$ is the root-mean-square radius of the inertia of the macromolecular coil in a theta-solvent when the polymer-solvent interaction is equivalent to the intermolecular interaction of polymeric chains; φ_2 is the volume content of the polymer in the system; $v_a = v_2 (M/Z)$ is the volume per atom of the chain of principal valences; v_2 is the specific volume of the system. Under isothermal conditions, for polymers for which $X > X_c$ the density is independent of molecular mass.

According to formula (2.65), under isothermal conditions the viscosity of high polymers is determined only by the value of the parameter X. At $X < X_c$ the density depends on molecular mass, and therefore when the viscosity of polymers is compared with molecular masses below M_c, it is necessary to introduce a correction factor $\zeta(\rho)$. It is also essential that in the case of solutions account must be taken of the change of their density with dilution.

The most significant result obtained by Fox and his coworkers is the inference regarding the universality of the critical value of the parameter X_c for a wide range of polymers. This critical value is of the order of 4.0×10^{-15} with a scatter of up to ± 20 per cent, though data are known for some polymers, which differ much from this averaged value. Nonetheless, the existence of the universal value of X_c for many polymers is thought to be a very important experimental fact. This is illustrated by Fig. 2.27. In plotting this figure the viscosity values were shifted along the ordinate by a certain arbitrary value C in order to avoid the superposition of experimental points referring to different polymers. From Fig. 2.27 (after Berry and Fox) it is seen that for all the polymers concerned, to the change of the slope of the

* Fox, T. G., *J. Polymer Sci.*, C, 9, 35 (1965); Berry, G. C. and T. G. Fox, *Adv. Polymer Sci.*, **21**, 261 (1968).

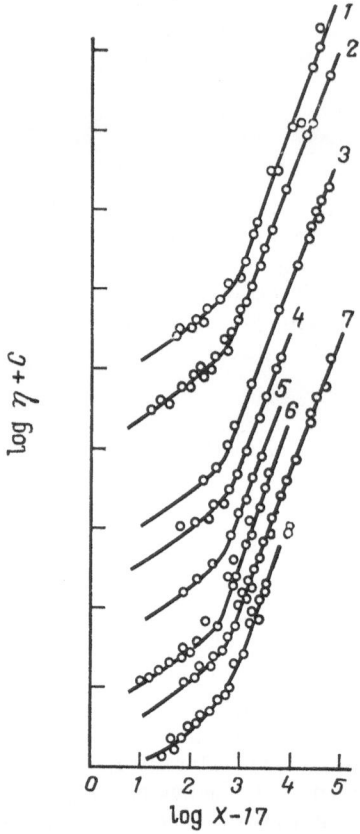

Fig. 2.27. The dependence of Newtonian viscosity on the parameter X for various polymers (after G. Berry and T. Fox):

1—polydimethyl siloxanes; 2—polyisobutylenes; 3—polybutadienes (50 per cent of cis-1,4-units); 4—polytetramethyl-p-silphenyl siloxanes; 5—polymethyl methacrylates; 6—polyethylene glycols; 7—polyvinyl acetates; 8—polystyrenes. The slopes of the lower branches of the curves are 1.0 and those of the upper branches, 3.4.

η_0 (M) curve there practically corresponds one and the same critical value of the parameter $X = X_c$. The more detailed experimental data obtained by Berry and Fox for a number of polymers are presented in Table 2.1.

Thus, on the basis of the concepts worked out by Fox, the viscosity of linear flexible-chain polymer melts and in solution is primarily determined by a characteristic of macromolecular coils. It is assumed that this characteristic is the quantity $\langle s_0 \rangle$, i.e., the size of macromolecular coils in a theta-solvent, more exactly the ratio $\langle s_0 \rangle / M$. From this there has been drawn a conclusion of fundamental importance: no ordered arrangement of macromolecules is present in melts of thermoplasts and rubbers and in concentrated solutions. These inferences have been confirmed by experimental investigations of the structure of amorphous polymers by means of small-angle neutron scattering*.

The above-considered results, which have been obtained by Fox, may be considered a generalization of experimental data but not a molecular theory. This follows even from the fact that this generalization is based on empirical relations (2.63).

---

* Benoit, H. et al., J. Polymer Sci., 14, 2119 (1978); de Gennes, P. G., J. Polymer Sci., Polymer Letters Ed., 15, 623 (1977); Fischer, E. W. et al., J. Macromol. Sci.—Phys., B12, 41 (1976).

Table 2.1.   Values of Constants That Determine, According to the Fox Theory, the Dependence of Viscosity on Molecular Mass for a Number of Polymers

Polymer	t, °C	$\dfrac{M}{Z}$	v_2	$\dfrac{\langle s^2 \rangle_0}{M} \cdot 10^{18}$	Z_c	$X_c \cdot 10^{15}$
Polyvinyl acetate	155	43	0.880	5.7	570	3.7
Polymethyl methacrylate (commercial)	217	50	0.880	6.2	550	3.9
Polymethyl methacrylate (strictly atactic)	217	50	0.880	6.8	630	4.9
Polystyrene	217	52	1.038	7.6	600	4.35
Polyisobutylene	217	28	1.123	8.7	540	4.20
Polyethylene glycol	80	14.7	0.925	9.8	300	3.20
Polydimethyl siloxane	25	37	1.04	7.2	660	4.60
Polytetramethylene-p-phenylene siloxane	218	69	1.0	7.3	582	4.50
Polybutadiene (90% cis-1,4) . . .	50	18	1.10	—	1670	8.26
Polybutadiene (80% cis-1,4) . . .	50	18	1.11	—	1200	7.20
Polybutadiene (65% cis-1,4) . . .	50	18	1.14	—	740	5.25
Polybutadiene (50% cis-1,4) . . .	27	18	1.11	12.6	330	3.6
Polyethylene	150	14	1.307	17.0	270	3.5
Poly-ε-capramide	253	16.15	1.0	10.0	310	3.1
Polydiethylene adipate	110	16.7	0.89	10.5	310	3.6
Polydecamethylene adipate . . .	110	15.8	1.027	11.5	320	3.6
Polydecamethylene succinate . .	110	16.0	0.99	12.5	230	2.9
Polydecamethylene sebacate . . .	110	15.4	1.067	14.7	270	2.9
Polyneopentyl succinate	100	23.3	1.0	16.0	710	11.4

2.6.4. The Bueche Theory*

The principal idea underlying the Bueche theory is that the movement of molecules (and of the molecular-kinetic units of flow—segments) must be cooperative. This means that the motion of one macromolecule cannot occur if other macromolecules are not involved in this motion, i.e., a moving molecule inevitably makes the other molecules move. The retarding force that arises during motion and determines the viscosity depends therefore not only on the resistance of a single macromolecule to motion but also on the sum of friction losses of all the macromolecules participating in the motion.

The initial underlying proposition of the theory is the formula derived by Debye:

$$\eta_0 = A \langle L^2 \rangle \, m\rho f / 36M \tag{2.66}$$

where $\langle L^2 \rangle$ is the root-mean-square distance between the ends of the coiled polymeric chain, which is proportional to M; A is Avogadro's number; ρ is the density; m is the number of segments in the molecule, which is proportional to the molecular mass M; f is the effective segmental friction factor.

Since for a given polymer $(\langle L^2 \rangle / M) = \text{const}$, then from formula (2.66) it follows that $\eta_0 \propto mf$. Formula (2.66) describes the dependence of the viscosity on molecular mass in an implicit form since it is not known in advance how the molecular mass affects the quantity f.

** Bueche, F., *J. Chem. Phys.*, **20**, 1959 (1952); **25**, 599 (1956).

The principal problem solved by Bueche is the elucidation of the effect of molecular mass on the quantity f.

It is assumed that each macromolecular chain is coupled with other chains at certain entanglement points. Suppose that a chain forms C_1 points of first-order interactions with the other macromolecules. The latter, in their turn, are connected, through C_2 points of second-order interactions, with other chains, etc. If now the primary macromolecule starts moving, it will pull along all macromolecules connected with it by interactions of various orders. But the junctions in the spatial entanglement network are not rigid; there exists the possibility of slippage of macromolecules past an entanglement. Then the lag in the motion of chains from the "primary" chain is the greater the higher is the bond order between the chains. Let the velocity of the first chain be v, then the chains entangled with it by first order will have the velocity vs, where s is the slippage factor; the velocity of macromolecules entangled with the primary chain by second-order interactions will be vs^2, etc. According to the estimate obtained by Bueche, s is of the order of 0.1-0.5.

The friction factor between a macromolecule and its surroundings is f_0 (called the segmental friction factor); then, evidently, the friction force will be given, without taking account of the macromolecular interaction, by

$$F_0 = vmf_0$$

If we now take into account the cooperative motion, i.e., if we consider not only the friction of the primary macromolecule but the friction of all the others entangled with it by interactions of various orders, then the total force that must be applied to make the macromolecule move through the surrounding material will be

$$F = vmf_0 (1 + C_1 s + C_2 s^2 + C_3 s^3 + \ldots)$$

Each term of the sum is the product of the number of macromolecules involved in the motion through an interaction of the corresponding order by the relative velocity v and the segmental friction factor f_0.

Next we calculate the number of points of interaction, C_1, C_2, etc. To a first approximation, it may be assumed that $C_1 = KM$, $C_2 = (KM)^2$, etc., where K is a certain constant coefficient which has the meaning of the frequency of appearance of interaction nodes in the macromolecule. But this assumption is incorrect since there may always exist a polymer with such a large molecular mass M_0 that $M_0 > (Ks)^{-1}$ and, consequently, the product KsM will be greater than unity. Then the force F becomes infinitely large; this means that the flow of polymers of very high molecular mass is impossible. Since this conclusion contradicts the available experimental data, another assumption is introduced, namely, it is assumed that the interaction of macromolecules is effective not at all entanglement points.

Further calculations lead to the final expression for f, which contains two terms.

If we consider a polymer with a relatively low molecular mass, namely, one for which $KM < 1$, i.e., if the molecular mass is so low that entanglement junctions are encountered, on the average, less frequently than one per molecule, then the second term of the formula for f is small as compared with unity and $f \approx f_0$. This means that for low-molecular-mass polymers the viscosity must be proportional to the molecular mass since $m \propto M$. But if $KM \gg 1$, which is equivalent to the formation of a network structure in the polymeric system (a large number of entanglement points per mole-cule), then the character of the dependence of viscosity on molecular mass will be sharply changed. In this case, the second term of the expression for f becomes much larger than unity, and the dependence of η_0 on M assumes the form

$$\eta_0 \propto m f_0 M^{3/2}$$

or

$$\eta_0 \propto M^{5/2}$$

This formula is the statistical law of resistance to motion of very long entangled ropes that slip over each other.

Thus, the Bueche theory provides a good explanation for the most impor-tant experimental fact: the change of the character of the dependence of viscosity on molecular mass at a certain critical value of molecular mass, the existence of M_c being associated with the beginning of the formation of a secondary structure (entanglement nodes) in the polymer. The shortcom-ings of the theory are thought to be a large number of assumptions and rather crude estimations required for the final formulas and conclusions to be arrived at.

To eliminate the contradiction between the theoretical (2.5) and experi-mental (3.5) values of the coefficient β in Eq. (2.63), Bueche presumed that additional energy losses for internal friction arise due to the deformation of the macromolecules themselves since the segments in each macromolecule are shifted relative to each other; the effective friction factor in this process is the same as during the displacement of the centres of gravity of the mole-cules relative to the surrounding material. It then turns out that to formula (2.66) there must be added a cofactor of the type $(1 + m/8q)$, where q is the length of the molecular chain between two adjacent entanglement points. The value of (m/q) is equal to the number of entanglement points per one macromolecule, on an average. If $M \gg M_c$, then $(m/q) \gg 1$, there-fore the viscosity is found to be proportional to $(m^2 f)$ and not to $(m f)$. It is thought that q is independent of M. And since at $M > M_c$ it has been found that $f \propto M^{3/2}$, then it follows that at $M \gg M_c$ the following law must really be satisfied:

$$\eta_0 \propto M^{7/2}$$

Thus, not only a qualitative but also a quantitative explanation is provided for the basic facts observed. The theoretical conceptions developed by Bueche have been further finalized by a more detailed examination of the character of the relationship between viscosity and molecular mass at $M <$ $< M_c$. The theory predicts the linear dependence of viscosity on molecular mass, whereas actually the dependence $\eta_0 (M)$ is found to be somewhat stronger and the exponent α in formula (2.63) is usually greater than unity. This discrepancy is accounted for by the fact that in the region of molecular masses below M_c it is necessary to take account of the correction factor $\zeta (\rho)$ [see formula (2.65)], which reflects the effect of the free ends of macro-molecular chains on the free volume of the polymer. Only at $M \gg M_c$ when the macromolecules are sufficiently large, does the role of the free ends become negligibly small and the density reaches a value which is limiting for a given polymer-homologous series. This is illustrated by the data presented in Fig. 2.28, where the arrow indicates the value of M_c determined from the relation between η_0 and M.

The variation of the density with molecular mass may be taken into account if we compare the values of viscosity for a series of samples with different M at different temperatures chosen so that the condition $\rho = \text{const}$ and, hence, $\zeta\,(\rho) = \text{const}$ is fulfilled. Experimental testing of this method has confirmed that the relationship between η_0 and M is strictly linear on the comparison condition $\zeta\,(\rho) = \text{const}$.

In addition to what has been said about the relative correlation of the critical molecular masses determined from viscosity and density measurements, it should be noted that the same value of M_c in polymer-homologous

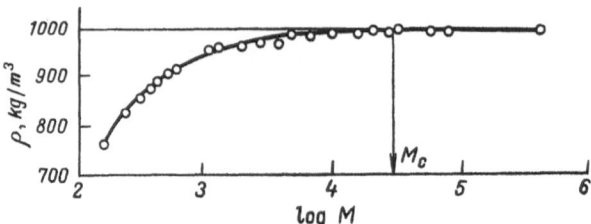

Fig. 2.28. The dependence of the density of polydimethyl siloxanes on molecular mass (see the reference to Fig. 2.8a).

series corresponds to the attainment of the limiting value of the activation energy of flow. This means that to the value of M_c or, more exactly, to a certain narrow range of molecular masses there corresponds the variation of many properties of polymers.

In conclusion, we once again emphasize the fact of the strong dependence of viscosity on molecular mass for high polymers as described by a power law with an exponent close to 3.5, which is undoubtedly associated with the cooperative motion of macromolecules.

2.6.5. Effect of Molecular-Mass Distribution on the Newtonian Viscosity

Formula (2.63) has been derived for the viscosity of monodisperse polymers. For polydisperse polymers there immediately arises the following question: What value of the averaged molecular mass should be used for the relations in formula (2.63)? An answer to the question about the correct choice of the averaged molecular mass is provided by the measurement of the viscosity of polymers of known molecular-mass distribution (MMD). Investigations of this kind have been carried out for polystyrene, polyvinyl acetate, and some other polymers. They have shown that when narrow fractions or monodisperse polymers, whose molecular masses are higher than M_c, are mixed, the viscosity of the resulting blends can be calculated from formula (2.63) into which the mass-average molecular mass $\dot{M}_w$ should be introduced. This is illustrated by the data presented in Fig. 2.29 for narrow fractions (unfilled circles) and polydisperse samples of polystyrenes of known MMD.

Though the use of M_w instead of M in formula (2.63) is the most frequently employed method of estimating the viscosity of polymers of broad MMD, it is occasionally reported in the literature that other moments of molecular-mass distribution are more exact, for example, the number-average molecular mass for polymethyl methacrylates; it has also been pointed out that MMD has an effect on the viscosity of samples of identical $\bar{M}_w$.

Some authors believe* that η_0 cannot be compared, in a general case, with the one and the same value (or values) of the average molecular mass since, depending on the nature of MMD, the value of η_0 must be determined

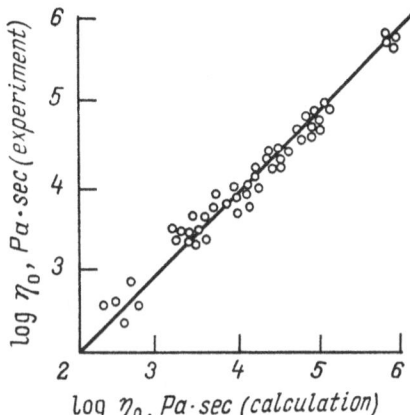

Fig. 2.29. Comparing experimental and theoretical values of Newtonian viscosity of polystyrenes. [Cantow, H.-J., *Plast. Inst. Trans. J.*, **31**, 96, 141 (1963).]

by the various moments of MMD; the longer the high-molecular-mass tail, the greater is the extent of averaging on which η_0 depends.

In the case of polyethylene (and, possibly, of other polymers as well), when considering the dependence of viscosity on MMD it is found to be also necessary to take into account the degree of chain branching**, so that the final correlative relationships must represent the dependences of a number of rheological parameters on a set of molecular characteristics—the average molecular mass, the degree of branching, MMD characteristics.

If a polydisperse polymer contains fractions with molecular masses below M_c, the dependence of viscosity on MMD and the averages of molecular mass becomes substantially complicated. In these cases, there is no simple relationship between η_0 and the mass-average molecular mass, as seen from Fig. 2.30 which presents the results of experimental investigations of the viscosity of blends of polyethylenes and low-molecular paraffin. Here the curves *1-3* refer to blends consisting of paraffin and three samples of polyethylene having different molecular masses. The curve *4* gives the relationship between η_0 and $\bar{M}_w$ plotted in accordance with formula (2.63). From Fig. 2.30 it is seen that the presence of fractions with molecular masses below M_c lowers the viscosity down to values lying much well below those calculated by formula (2.63), i.e., the viscosity is not determined by the mass-average molecular mass. The average of molecular mass, which has

---

* Budtov, V. P., Yu. L. Vagin, and E. L. Vinogradov, *Mekhanika Polimerov*, 3, 514 1978).

** Brauer, E., I. P. Briedis, V. I. Bukhgalter, L. L. Sulzhenko, L. A. Faitelson, and M. Fidler, *Mekhanika Polymerov*, 2, 283 (1977).

been found to be applicable for the calculation of the viscosity of blends of polyethylene and paraffins by means of formula (2.63), is given by the following formula

$$\bar{M}_\alpha = \left(\sum_i w_i M_i^\alpha\right)^{1/\alpha}$$

where M_i are the molecular masses of the components; w_i is their mass fraction in the mixture.

In contrast to cases where $\bar{M}_\alpha = \bar{M}_w$ and $\alpha = 1$, for the blends under consideration, which contain fractions with molecular masses both lower and higher than M_c, the value of $\alpha = 0.75$. The scope of generality of such

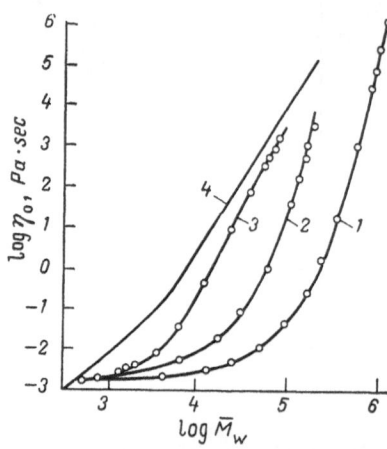

Fig. 2.30. The dependence of Newtonian viscosity on the mass-average molecular mass for blends of polyethylenes with different molecular masses and low-molecular paraffins [Longworth, R. and W. Busse, *Trans. Soc. Rheol.*, 6, 179 (1962).]:

1-3—original polymers of various molecular mass; 4—polymer blends.

a method of averaging molecular masses for estimating the initial viscosity of polymers containing low-molecular-mass fractions is, however, unknown.

A general method of taking into account the effect of molecular-mass distribution on the viscosity η_0 must be based on the introduction of various MMD averages into the corresponding calculation formula. Thus, for a number of polymers of different MMD, which contain low-molecular fractions, a satisfactory accuracy of estimation is attained when use is made of the following empirical formula*:

$$\eta_0 = (A_w \bar{M}_w + A_0 - A_n M_n^{-1})^{1/a}$$

where A_0, A_n, and A_w are empirical constants which are dependent on the nature of the polymeric chain; $\bar{M}_n$ and $\bar{M}_w$ are, respectively, the number-average and mass-average molecular masses. The constant a varies within narrow limits (obviously, $1/a$ must be close to 3.5). Its value is the exponent in the formula: $\eta_{bl}^a = \sum w_i \eta_i^a$, with the aid of which the viscosity of a blend of fractions, η_{bl}, can be calculated from the viscosity values, η_i, of the components, whose mass fraction in the blend is w_i.

The introduction of a large number of constants into this formula robs generality of it. However, for each concrete polymer-homologous series the

---

* Chistov, S. F., I. I. Skorokhodov, Ya. I. Vilenkin, and M. E. Erlykina, *Vysokomol. Soedin.*, B20, 299 (1978); B21, 364 (1979).

use of this formula may prove to be a convenient and sufficiently accurate method of calculating the viscosity of a polydisperse polymer.

2.6.6. Critical Molecular Mass and Non-Newtonian Flow

Hoffmann and Rother*, who worked with narrow fractions of polystyrenes, have shown that non-Newtonian viscosity reveals itself when $M > M_c$. The explanation of this fact can be based on the idea that at molecular masses higher than M_c there appears a structure in the polymer which can be modelled by a network of fluctuating macromolecular entanglements. The non-Newtonian flow behaviour is associated with the decrease of the entanglement network density under the influence of the deformation of the polymer. These conceptions lead to very important consequences, which lend themselves most readily to interpretation in the case of linear polymers of narrow MMD**.

At $M > M_c$ with increasing molecular mass, as the shear rate and stress increase, the entanglement network is more likely to be destroyed—the non-Newtonian behaviour becomes noticeable.

However, if M increases considerably (at any rate, if it exceeds $10M_c$), the situation is substantially changed, so that over a wide range of shear stresses polymers behave as Newtonian liquids. It should be stressed that high-molecular-mass flexible-chain polymers are only similar to Newtonian liquids but they are not Newtonian liquids proper since they are capable of exhibiting large recoverable deformations, i.e., are viscoelastic materials. Thus, the non-Newtonian behaviour of the polymers considered here is most sharply pronounced at values of M/M_c lying in the range from 1 to 10. This feature of the behaviour of high-molecular-mass polymers is accounted for in the following way. At $M/M_c > 10$ the fluctuating entanglement network becomes homogeneous and is highly dense, and it is no longer influenced by the free ends of macromolecules. The stress acting on the polymer is uniformly distributed over the network junctions. As long as the stress is below a certain critical value, which may nevertheless be considerable (it usually ranges from 0.1 to 1 MPa) each network junction is influenced by a small stress and the "structure" of the polymer remains unchanged, since there is chance for the separate junctions broken down to be compensated for by the formation of new junctions and under the action of thermal motion the state of dynamic equilibrium is maintained. When the stress attains a critical value, the rate of formation of junctions begins to lag behind the rate of destruction. This leads to a sharp increase in the load acting on the existing junctions and the destruction of the "structure" becomes avalanche-like. This means that narrow-distribution high polymers are either flowing like Newtonian liquids or are broken down like cured (cross-linked) elastomers. From the foregoing it follows that for each polymer-homologous series there must exist a single critical value of shear stress, which will describe the behaviour of all its high-molecular-mass representatives. On the other hand, it is evident that the critical rates of deformation must strongly depend on the molecular mass of each particular

---

* Hoffmann, M. and R. Rother, *Makromol. Chem.*, 80, 95 (1964).
** Vinogradov, G. V., *Pure and Appl. Chem.*, 26, 423 (1971); *Rheol. Acta*, 12, 357 (1973).

polymer, decreasing with increasing molecular mass. It will be shown at a later time that experiment not only supports what has been said above but it also enables the effects observed to be described quantitatively.

The fact that at certain shear stresses and rates polymers are broken down like cured elastomers allows us to regard such a process as being equivalent to their transition from the fluid to the rubbery state. Indeed, the transition of the polymer from the fluid to the rubber-like state (under isothermal conditions at $T > T_g$) is detectable with increase of the rate of deformation when there have not yet taken place numerous displacements of the centres of gravity of macromolecules, which are responsible for the accumulation of recoverable deformations, and the polymer behaves as a quasi-cured material in which large recoverable deformations may develop, and the fluidity is suppressed.

The ability of narrow-distribution polymers to pass from the fluid to the rubbery state is usually detected at M/M_c greater than 5.

At $M/M_c > 10$ the transition from the fluid to the rubber-like state is sufficiently sharp. Therefore, in such cases at deformation rates exceeding the critical values the fluidity of polymers should be **expected** (and **this** is actually observed experimentally) to be suppressed to such an extent that steady flow conditions will be impossible to realize.

The value of M_c may vary by a factor of 10-20, depending on the nature of the polymeric chain. But it is important to note that the conditions under which the non-Newtonian behaviour manifests itself and the transition of polymers from the fluid to the rubbery state takes place and, hence, the loss of fluidity, are determined by the ratio M/M_c and not by the absolute value of molecular mass.

The transition from the fluid to the rubber-like state is governed by the ratio of the rate of deformation to the rate of relaxation, which can be used to characterize the flow properties of the polymer. This ratio is expressed by the dimensionless product $(\dot{\gamma}\theta)$, where θ is a certain characteristic relaxation time (the reciprocal of the rate of relaxation). The transition in question must take place at a certain value of $\dot{\gamma}_s\theta = $ const. For linear polymers, the quantity θ is directly associated with the initial viscosity of the polymer (for more detail, see Chapter 3). Therefore, the critical rate of deformation $\dot{\gamma}_s$ is inversely proportional to the initial viscosity of the polymer and, accordingly, depends on temperature. But $\dot{\gamma}_s\theta = (\theta/\eta_0)\,\tau_s$, where (θ/η_0) (at $M \gg M_c$) must not depend on molecular mass and temperature, i.e., the critical shear stress τ_s is a constant quantity. It is necessary to emphasize that the condition $\tau_s = $ const applies not only to the case of the clear-cut transition of monodisperse polymers to the rubbery state; with polydisperse polymers exhibiting non-Newtonian flow behaviour there exists the relation between θ and η_0 as well. But if the transition to the rubbery state occurs in the region of non-Newtonian flow, the critical deformation conditions are attained gradually.

The quantity τ_s has a characteristic value for each polymer, although in general it depends relatively weakly on the nature of the polymeric chain; it varies in extreme cases by about a factor of up to 10-20. For carbon-chain polymers, the critical shear stress increases with increasing kinetic flexibility of macromolecules. The highest critical shear stress is exhibited by linear polyethylenes and polybutadienes, for which it equals $(3-4) \times 10^5$ Pa.

From what has been said above one arrives at the general conclusion that it is impossible to accomplish a steady-state flow of high-molecular-mass polymers at shear stresses exceeding the values indicated.

Of importance in considering the effect of deformation on the state of the polymer is the ratio of the rate of decrease of the entanglement network density under the influence of deformation to the rate of its recovery under the action of the thermal motion of macromolecules. At high values of M/M_c and high temperatures, which considerably exceed the glass-transition temperature, the rate of formation of network junctions is high and up to

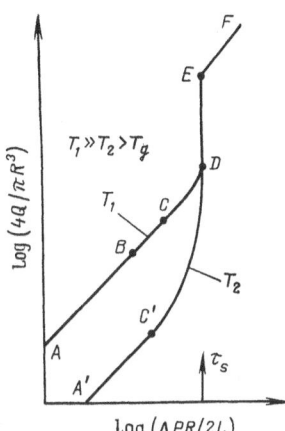

Fig. 2.31. Flow curves and the dependences of volume rates of flow on the pressure gradients in capillaries for linear high-molecular-mass polymers of narrow MMD $(T_1 > T_2)$.

shear rates close to the critical rate the deformation has a slight effect on the network density. This means that the flow of monodisperse polymers occurs under the conditions when the network density is practically constant and the polymer exhibits weakly pronounced non-Newtonian flow behaviour.

The situation changes when the temperature falls considerably. Then the rate of formation of the network decreases. The dynamic equilibrium of the processes involving the decrease of the entanglement network density under the influence of deformation and its formation under the action of thermal motion is shifted to the side of the decrease of the network density. This means that at high values of M/M_c the conditions of transition of the polymer from the fluid to the rubbery state depend on temperature, more exactly, on the remoteness of the state of the polymer from the glass-transition temperature: at higher temperatures the transition of the polymer to the rubbery state takes place only under the conditions of weakly pronounced non-Newtonian behaviour; when the temperature falls off this transition may be preceded by considerable non-Newtonian behaviour.*

What has been said above is illustrated by a graph showing the character of the dependence of the volume rate of flow Q on the pressure drop ΔP when the polymer is extruded through a cylindrical capillary (Fig. 2.31). The two quantities are standardized with respect to the dimensions of the capillary—its radius R and length L, so as to obtain parameters coincident

---

* Vinogradov, G. V., *Pure and Appl. Chem.*, **26**, 423 (1971); *Pure and Appl. Chem.*, *Macromol. Chem.*, **8**, 413 (1973); **9**, 115 (1973); *Rheol. Acta*, **12**, 357 (1973); Vinogradov, G. V., A. Ya. Malkin *et al.*, *J. Polymer Sci.*, A-2, **10**, 1061 (1972); *Plaste und Kautschuk*, **19**, 907 (1972).

in dimensions with the shear rate and stress. The portion of the graph AB corresponds to the laminary flow of the liquid, which is similar to a Newtonian liquid judging by the proportionality between the shear rate and the shear stress. The reservation regarding the similarity and not the identity is essential because the material in question is a viscoelastic liquid. Beginning from the point B, at which $\tau/\tau_s \approx 0.2$, there are observed oscillating (non-steady) flow conditions, which are known as elastic turbulence or the "melt fracture". It is enhanced with increasing shear stress. Elastic turbulence manifests itself under the influence of the development of large recoverable deformations which under flow conditions corresponding to the portion of the curve BCD may reach many tens of a per cent.

The portion of the flow curve CD represents a more or less appreciable non-Newtonian behaviour. It is the less strongly pronounced, the narrower is the MMD of the polymer and the higher is the temperature. At point D there are attained the critical values of shear stress and rate at which the transition of the polymer to the rubbery state occurs. This is accompanied by a sharp decrease of its fluidity and a decrease of the adhesive interaction of the polymer with the capillary wall. As a result, the resistance of the polymer to motion in the duct decreases and, accordingly, the volume rate of flow increases in a jumpwise manner—this is known as the flow spurt, which is represented by the vertical branch of the curve DE. To the spurt there may correspond an enormous jump of the flow rate. When a spurt takes place, the laminar flow of the polymer is replaced with the slippage of the polymer over the duct walls. Under these flow conditions the calculation of the shear rate becomes impossible. Hence, what is usually called the flow curve is the portion $ABCD$ of the curve in question and by no means the portion DE. In formal calculations (but devoid of any physical meaning) of "viscosity" on the portion DE it is found to be inversely proportional to the fictitious value of the shear rate, though in reality there is no flow in this region and neither the shear rate nor the viscosity of the polymer can be calculated.

As a result of the action of high stresses in the entrance zone of the duct, there appear breaks in the continuity of the polymer as in the case of an elastically stressed body. Therefore the portion of the curve EF describes the motion of a rubber-like body broken down into pieces. It thus follows that the branch EF resembles only outwardly the upper branch of the flow curve, i.e., that portion of it which describes the flow of polymer systems with the lowest Newtonian viscosity. As a matter of fact, the branch EF characterizes the motion of polymers, which has nothing in common with the Newtonian flow.

The position of the point E depends slightly on the molecular mass of the polymer (within a given polymer-homologous series). But the position of the branch AD depends very strongly on molecular mass since the critical shear rate decreases by about one decimal order when the molecular mass increases by 2 times. Therefore, the spurt effect is extremely enhanced with increasing molecular mass. At M/M_c of the order of 10^2 the jump of the flow rate during the spurt may be tens of thousands.

When the temperature decreases greatly in the region of high shear stresses that precede the spurt, there is observed a sufficiently sharply pronounced non-Newtonian behaviour, though the value of τ_s changes little and the spurt mechanism remains the same as at higher temperatures. Hence, at low temperatures the transition from Newtonian flow to the spurt may be smooth and the critical point D implicitly expressed (the branch $C'D$ of the lower curve in Fig. 2.31). So extreme care must be exercised in interpreting the results of a viscometric investigation of polymers at high shear stresses.

2.6.7. Non-Newtonian Flow and Polydispersity of Polymers

Molecular-mass distribution has a great effect on the non-Newtonian behaviour; moreover, in the case of high-molecular polymers it is predetermined by their polydispersity. This is because the non-Newtonian behaviour is a relaxation effect and depends therefore on the form of the relaxation spectrum, which in its turn is determined by the molecular-mass distribution (MMD) of the polymer. That the non-Newtonian behaviour is a relaxation effect in the first place and the geometric factor—large recoverable deformations and the associated nonlinear effects—is of secondary importance, follows from the weak manifestation of non-Newtonian behaviour during the flow of monodisperse polymers in the region of high shear stresses, when recoverable deformations may reach 100-200 per cent.

The non-Newtonian flow behaviour as a relaxation effect specific to polydisperse polymers becomes especially spectacular when considering the flow properties of a blend of monodisperse polymers (in the simplest case, it consists of two monodisperse polymers).* If the shear rates and stresses are sufficiently low, the components of the blend behave like Newtonian liquids. When the shear rate increases, a critical shear rate $\dot{\gamma}_s$ is attained for the higher-molecular-mass component, which is equivalent to its transition to the rubbery state. In this state it behaves as a rubber-like filler. The dissipation losses are low since at $\dot{\gamma} > \dot{\gamma}_s$ they are no longer associated with the displacement of the centres of gravity of its macromolecules and are only caused by rapid conformational motions of the macromolecular chain between the entanglement points and by the "streamlining" of these macromolecules by components that have not yet passed into the rubbery state. A decrease of dissipation losses means a decrease in the apparent viscosity: with increasing stress the rate of shear increases non-proportionately rapidly. In this case, in the high-molecular-mass component of the blend the stored recoverable deformation increases under the influence of increasing stress, which is quite typical of polymers in the rubbery state. Hence, the large reversible deformations of the blend appear to be greater than those of the high-molecular-mass component since in the pure state the latter could not flow, having passed into the rubbery state. Because of this, in polydisperse polymers containing high-molecular-mass components, at high shear stresses and rates all the effects due to large recoverable deformations are more strongly pronounced, say, the development of normal stresses and the swelling of the polymer extrudate emerging from the die. While increasing all nonlinear effects, large recoverable deformations amplify their influence on the flow properties of polymers and increase their contribution to the development of non-Newtonian behaviour (anomalous viscosity).

---

* Malkin, A. Ya., N. K. Blinova, G. V. Vinogradov, M. P. Zabugina, O. Yu. Sabsai, V. G. Shalganova, I. Yu. Kirchevskaya, and V. P. Shatalov, *Europ. Polymer J.*, 10, 445 (1974).

As the shear rate increases the components of the blend with increasingly lower molecular masses pass successively to the rubbery state, which causes the viscosity to decrease progressively. If the lowest-molecular-mass component of the blend is also capable of passing to the rubbery state, then upon attainment of a critical shear rate characteristic of it the blend will undergo a spurt. What has just been said is illustrated by Fig. 2.32a for a binary blend containing equal fractions of both components. So far as the shear rate is lower than the critical rate corresponding to the spurt of the higher-molecular-mass component, the viscosity of the blend flowing like a Newtonian liquid will be simply given by the equation

$$\eta_0 = A_4^i (\sum_i w_i M_i^\alpha)^{1/\alpha}$$

where w_i is the mass fraction of the component with molecular mass M_i; α is the exponent in Eq. (2.63) for fractions with $M_i > M_c$.

At rates exceeding the critical shear rate for the high-molecular-mass component the dissipation losses caused by it diminish and the deformation

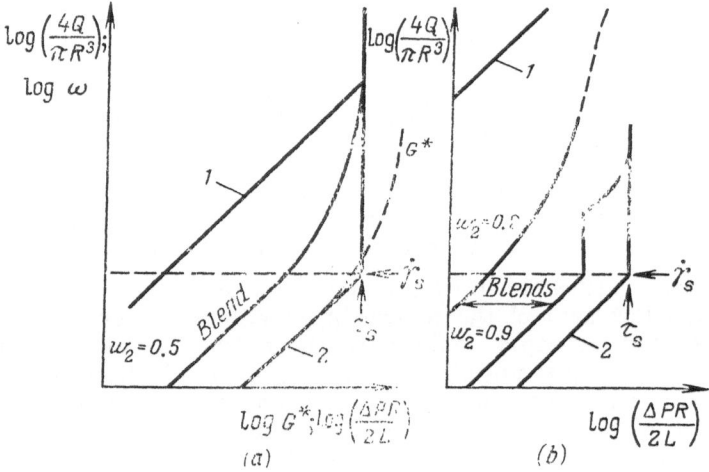

Fig. 2.32. Flow curves and the dependences of volume rates of flow on pressure gradients in capillaries for blends of monodisperse polymers:

a—blends of components 1 and 2 having close molecular masses; b—blends of polymers 1 and 2 having strongly differing molecular masses; w_2 is the mass fraction of the high-molecular-mass component.

resistance of the blend begins to increase slowly. The flow curve becomes concave towards the shear rate axis. Attention should be paid here to the asymptotic character of the attainment of the spurt of the entire blend until the critical shear rate of the less viscous component of the blend is exceeded.

The apparent viscosity of a mixture of monodisperse polymers is calculated in a simple way when $M_i > 10 \ M_c$ for all the components of the blend. The calculation is based on the empirically established correlation between the shear stress in steady-state flow and the complex shear modulus G^* determined under the conditions of small-amplitude deformation (for more detail, see Sec. 3.4). It is assumed that the circular frequency, ω, of small-amplitude deformation under harmonic conditions is numerically equal to the shear rate. It thus follows that it is possible to evaluate the flow properties of polymer systems from the complex viscosity ($\eta^* = G^*/\omega$).

of the components of the blend, which are in a state of transition to the rubbery state. Accordingly, the values of the circular frequency and complex shear modulus are also plotted along the coordinates in Fig. 2.32, and the relationship between G^* and ω for the high-molecular-mass component is represented by the dashed line. The possibility of calculating the dynamic viscosity of the blend by summing up the contributions of all the components of the blend has also been considered in the literature.*

The existence of a correlation between the complex and apparent viscosities enables one to calculate the apparent viscosity of the blend in the same way as was done for the Newtonian viscosity, namely,

$$\eta\,(\dot{\gamma}) = A\{\textstyle\sum w_i\,[\eta^*\,(\omega)]^{1/\alpha}\}^{\alpha}_{\dot{\gamma}=\omega} \qquad\qquad (2.67)$$

The simplicity of the calculation of the dependence $\eta\,(\dot{\gamma})$ for high-molecular-mass polymers with $(M/M_c) > 10$ at $T \gg T_g$ results from the fact that the density of the fluctuating network is of decisive importance and the free-volume factor does not play any significant part. The latter becomes essential when with decreasing molecular mass the free ends of the macromolecules begin to exert an ever increasing effect, which results in a decrease of the spatial homogeneity of the fluctuating entanglement network. A specific effect can also be exerted by a large difference between the molecular masses and, accordingly, between the viscosities of the components of the blend, as seen from Fig. 2.32b. In this case, when the concentration of the high-molecular-mass component is high, there is observed a "two-step" dependence of the flow rate on the pressure drop. When the critical shear rate of the high-molecular-mass component is attained, it passes to the rubbery state. The specificity of the phenomenon in this case is determined by the fact that this transition is found to be facilitated owing to the considerable inhomogeneity of the fluctuating entanglement network. The specific features of the flow behaviour of mixtures of polymer samples having sharply different molecular masses, the polymers used belonging to the same homologous series, may be attributed to the possibly incomplete compatibility of the fractions. As a result, the dissipation losses decrease in a jumpwise manner and the spurt effect is observed. The spurt, however, occurs at the stress $\tau < \tau_s$. Therefore, it has a small amplitude and as the shear stress increases further the non-Newtonian flow conditions are developing until the τ_s value typical of a given polymer-homologous series is attained.

If the concentration of the high-molecular-mass component is considerably decreased, the transition to the rubbery state cannot be so sharp as described above, and the dependence of the flow rate on the pressure gradient (the flow curve) is found to be typical of non-Newtonian systems.

With a considerable broadening of MMD, when the spatial inhomogeneity of the entanglement network increases, the increase of non-Newtonian viscosity is accompanied by the increase of the compliance of the high-molecular-mass component; the rubber-like deformations and normal stresses rapidly increase.

What has been said above leads to the following general conclusions. First, the non-Newtonian behaviour of polymers is due to their relaxation properties. Second, at least for high-molecular polymers ($M > 10\,M_c$) the non-Newtonian flow behaviour is determined by their polydispersity. Third, it is easy to account for the fact that with broadening of MMD to the side

---

* Montfort, I. P., G. Marin, J. Arman, and Ph. Monge, *Polymer*, 19, 277 (1978).

of high molecular masses the shear rates and stresses at which non-Newtonian behaviour is observed are decreased. Fourth, as the non-Newtonian behaviour is enhanced the rubbery deformations increase and the manifestation of nonlinear effects (normal stresses and the associated phenomena) is intensified.

The attempts to establish the relationship between the shape of the flow curve, i.e., the dependence of η on $\dot{\gamma}$, and the MMD function, $f(M)$, have been based on the use of a theoretically calculated dependence of η on $\dot{\gamma}$ for monodisperse polymers. It is assumed that the dependence $\eta(\dot{\gamma})$ for a polydisperse polymer is determined by summing up the contributions of all fractions, the effect of each of which is proportional to $f(M)\,dM$. This approach has been used, for example, by Middleman*, who proposed the following formula:

$$\eta(\dot{\gamma}) = \frac{1}{\overline{M}_n \cdot \overline{M}_w} \int\limits_0^\infty M^2 F[\dot{\gamma}\theta(M)]\,f(M)\,dM \qquad (2.68)$$

where $F(\dot{\gamma}\theta)$ is the function which describes the effect of the shear rate on the viscosity of a polymer; it includes the parameter θ which itself depends on the molecular mass.

Other relationships between $\eta(\dot{\gamma})$ and $f(M)$ are also known, which presuppose the possibility of calculating the dependence $\eta(\dot{\gamma})$ if the function $f(M)$ is known, and also of determining MMD on the basis of the experimentally measured function $\eta(\dot{\gamma})$.

It should be noted that this approach based on the summation of the contributions of individual fractions, each of which is calculated from the theoretical dependence $F(\dot{\gamma}\theta)$, is not reliable for a number of reasons. The function $F(\dot{\gamma}\theta)$ derived theoretically for a monodisperse polymer describes experimental data only qualitatively. Besides, the determination of the relationship between the characteristic relaxation time θ contained in the function $F(\dot{\gamma}\theta)$ and the molecular mass is itself a complicated task.

More substantiated is the approach based on the generalization not of the theoretically specified functions $F(\dot{\gamma}\theta)$ for monodisperse polymers but of experimental data, i.e., the calculation of the dependence $\eta(\dot{\gamma})$ for polymers of known MMD, which relies on the results of measurements of the viscosity of monodisperse polymers within a single homologous series. Middleman points out that such calculations based on Eq. (2.68) have been carried out by Robinson for polystyrene. The dependence $F(\dot{\gamma}\theta)$ for monodisperse polymers was described by the empirical equation:

$$F(\dot{\gamma}\theta) = \frac{\eta}{\eta_0} = \frac{1}{1 + 2.68\,(\dot{\gamma}\theta)^{0.89}}$$

---

* Middleman, S., *The Flow of High Polymers*, Interscience Publ., New York, 1968; see Chapter IV, Sec. IIIB.

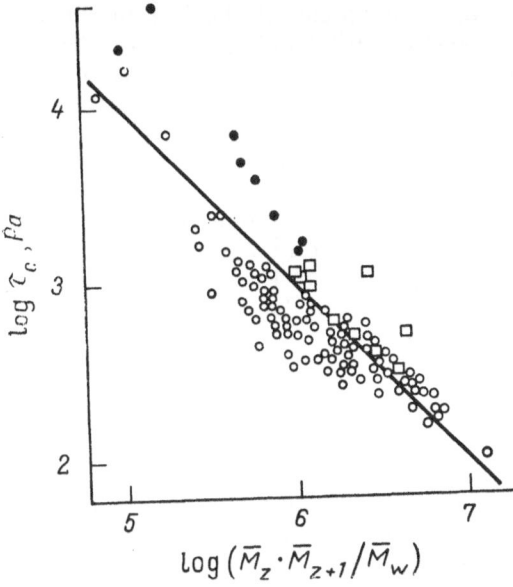

Fig. 2.33. Dependence of the critical value of shear stress corresponding to the appearance of non-Newtonian behaviour on the higher moments of MMD for polydimethyl siloxanes (○), polyethylenes (●), and polyisobutylenes (□).

where the connection between the relaxation time θ and the molecular mass of a monodisperse polymer was determined by the following relation:

$$\theta = \frac{12\eta_0 M^{0.75}}{\pi^2 \rho RT}$$

Integration of Eq. (2.68) for polymers of known MMD enables us to obtain the dependence $\eta\,(\gamma\bar{\theta}_w)$, where $\bar{\theta}_w$ is calculated by the following formula:

$$\bar{\theta}_w = \frac{12\eta_0 \bar{M}_w^{0.75}}{\pi^2 \rho RT}$$

The results of a comparison of the calculated dependences of (η/η_0) on $(\dot{\gamma}\bar{\theta}_w)$ with experiment for two series of polystyrenes of different MMD have shown that the calculations are in good agreement with experiment. It may be supposed that formula (2.68) satisfactorily describes the effect of MMD on viscosity but it requires a knowledge of the dependence $F\,(\dot{\gamma}\theta)$ for monodisperse polymers in order to carry out reliable calculations.

Attempts to describe the experimental results of an investigation into the effect of MMD on the flow curve have led to the use of various averaged molecular masses for a quantitative characterization of MMD. Of primary interest for this estimation are two characteristics of flow curves: the beginning of the region of non-Newtonian flow behaviour and the curvature of the flow curve. Thus, Kataoka* assumed that the stress at which there begins the deviation from the region of Newtonian flow, τ_c, is determined by the complex quantity $(\bar{M}_{z+1}\bar{M}_z)/\bar{M}_w$. Experimental data illustrating this supposition only point (Fig. 2.33) to a qualitative correlation between τ_c

* Kataoka, T., *J. Polymer Sci.*, B5, 1063 (1967).

and $(\overline{M}_{z+1}\overline{M}_z/\overline{M}_w)$ since the spread of the data in Fig. 2.33 is rather great.

Often the molecular-mass distribution is evaluated by using the ratio of the mass-average molecular mass to the number-average molecular mass, $Q = (\overline{M}_w/\overline{M}_n)$. This parameter can be used to characterize polymers within a single homologous series if their MMD's are expected to be similar. In a general case, however, to one and the same value of Q there may correspond as many various MMD as desirable.

As an example of the use of the parameter Q for establishing the relationship between MMD and non-Newtonian behaviour may be cited the work of Schreiber.* This author has found that for linear polyethylenes the degree of non-Newtonian behaviour can be determined from the formula

$$\log s = -1.0 + 0.78 \log Q$$

where $s = -d \log \eta/d \log \tau$. Often in order to characterize the shape of the flow curve that correlates with MMD or with parameters characterizing MMD, use is also made of the second derivative $d^2 \log \eta/d (\log \dot{\gamma})^2$.

The effect of the factor Q on the shape of the flow curve has also been demonstrated for molten polymethyl methacrylates of various degrees of polydispersity.** The flow curves were represented by the formula:

$$\eta/\eta_0 = [1 + b \ (\dot{\gamma}\eta_0)^{0.8}]^{-1}$$

and the factor b that characterizes the beginning of the region of non-Newtonian flow was found to be related with Q. The empirical formula expressing this relationship has the form

$$b = 2 \times 10^{-5} Q^{1.35}$$

On the other hand, according to Cottam***, the degree of non-Newtonian behaviour is determined by M_z, and there is not any other correlation between the degree of non-Newtonian behaviour and the ratio $(\overline{M}_w/\overline{M}_n)$. According to Ballman and Simon****, as the shear rate increases the flow curve is determined by the moments of MMD which are the lower, the higher is the shear rate, so that in the region of high shear rates the determining role is played by the number-average molecular mass. In general, the question as to which MMD averages determine the behaviour of the polymer in the region of high shear rates remains open at present; one can find contradicting information in the literature.

Incidentally, the existence of the relation between MMD and the shape (curvature) of the flow curve enables one to propose some simple methods of evaluating MMD from a limited number of viscosity measurements. Thus, it may be assumed that the ratio of the values of the standard fluidity index of a melt—the melt index—at two loads can characterize the MMD of polymers since such a ratio is a measure of the curvature of the flow curve.*****

* Schreiber, H. P., *J. Appl. Polymer Sci.*, 9, 2101 (1965).
** Panova, G. D., L. I. Myasnikova, D. N. Emelyanov, and A. V. Ryabov, *Vysokomol. Soedin.*, 18B, 273 (1976).
*** Cottam, B. J., *J. Appl. Polymer Sci.*, 9, 1853 (1965).
**** Ballman, R. L. and R. Simon, *J. Polymer Sci.*, A2, 3557 (1964).
***** Malkin, A. Ya., and M. L. Friedman, *Plastmassy*, 8, 23 (1976).

Indeed, for a linear polyethylene the following empirical formula has been obtained*:

$$Q = -5.9 + 1.51 \, (I_5/I_{1.2}) + 3.21 \, [CH_3]$$

where I_5 and $I_{1.2}$ are the values of the melt index at two loads indicated as the subscripts, and $[CH_3]$ is the content of methyl groups (i.e., a characteristic of the branching of the polyethylene molecules).

Analogous (with respect to the meaning) relationships can be obtained for other polymers as well.

The measurement of one characteristic of the flow curve (in this case, $I_5/I_{1.2}$) allows us to obtain only one characteristic of MMD (Q in this case). It has however been noted** that for real commercial polymers (but not for artificially prepared mixtures of narrow fractions) Q satisfactorily represents the other characteristics of MMD, for example: $\overline{M}_z/\overline{M}_w \quad Q^{0.75}$ and $\overline{M}_z \cdot \overline{M}_{z+1}/\overline{M}_w^2 = Q^{2.06}$, where $\overline{M}_z$ and $\overline{M}_{z+1}$ are the highest MMD averages. Therefore, knowing the $\overline{M}_w$ of a polymer and having determined Q, it is possible to find the various MMD averages.

But in general it may be noted that the calculation of a flow curve from a known MMD and the converse problem, the determination of MMD from the measured function $\eta \, (\dot{\gamma})$, are not equivalent. The first problem can be solved with a sufficiently high accuracy by summation of the contributions of the individual fractions to the viscosity of a polydisperse polymer, using various formulas proposed by various authors. In this case, rather significant changes of MMD do not strongly affect the form of the dependence $\eta \, (\dot{\gamma})$, which is approximately the same for different polydisperse polymers. The latter circumstance accounts for the low accuracy of the converse calculation; since the flow curves of polydisperse polymers are close to each other, it is difficult to derive from them any information on the differences in the MMD of the polymers being compared. Therefore, from a change in the form of flow curves it is easy to trace out the changeover from a monodisperse to a polydisperse polymer, but the MMD of a polydisperse polymer is difficult to calculate reliably.

2.6.8. Effect of Shear Rate on the Viscosity-Molecular-Mass Relationship

The form of the function $\eta \, (M)$ for high-molecular-mass compounds depends on the shear rate, as is schematically shown in Fig. 2.34, where M should be taken to mean $\overline{M}_w$ for polydisperse high-molecular compounds.

The character of the variation of the functional dependence $\eta \, (M)$ with increasing $\dot{\gamma}$ presented in Fig. 2.34 can be predicted*** by using formulas for the function $\eta \, (\dot{\gamma})$ that give a power dependence of η on $\dot{\gamma}$ with increasing shear rate. Let, for example,

$$\eta \, (\dot{\gamma})/\eta_0 = [1 + (\dot{\gamma}\theta)^a]^{-1}$$

* Shekhtmeister, I. E., S. S. Mnatsakanov, V. V. Nesterov, and B. G. Belenky, *Plastmassy*, 9, 58 (1977).
** Krevelen, D. W. van, D. J. Goedhart, and P. J. Hoftyzer, *Polymer*, 18, 750 (1977).
*** Malkin, A. Ya. and G. V. Vinogradov, *J. Polymer Sci.*, B2, 671 (1964).

where θ is a constant having the meaning of the relaxation time; $\theta \propto \eta_0 \propto$ $\propto M^{3.5}$, and a is a constant.

Then, at high shear rates $\eta \, (\dot{\gamma}) \propto (\dot{\gamma}\theta)^{-a}$ and the ratio of the viscosities of two samples having different molecular masses is

$$\frac{\eta_1 \, (\dot{\gamma}, \, M_1)}{\eta_2 \, (\dot{\gamma}, \, M_2)} \cdot \frac{\eta_0 \, (M_2)}{\eta_0 \, (M_1)} = \frac{(\dot{\gamma}\theta_2)^a}{(\dot{\gamma}\theta_1)^a} = \left(\frac{\theta_2}{\theta_1} \right)^a = \left[\frac{\eta_0 \, (M_2)}{\eta_0 \, (M_1)} \right]^a$$

or

$$\frac{\eta_1 \, (\dot{\gamma}, \, M_1)}{\eta_2 \, (\dot{\gamma}, \, M_2)} = \left[\frac{\eta_0 \, (M_1)}{\eta_0 \, (M_2)} \right]^{1-a} = \left(\frac{M_1}{M_2} \right)^{3.5(1-a)}$$

where η_1 and η_2 are the viscosity values; θ_1 and θ_2 are the relaxation times of polymers with molecular masses M_1 and M_2, respectively.

It thus follows that

$$\eta \propto M^{3.5(1-a)}$$

If it is assumed that $a = 0.71$, then at high shear rates the viscosity must be proportional to the molecular mass:

$$\eta_{\dot{\gamma} \to \infty} \propto M$$

Similar results are obtained by using other most widely used values of the exponent a. For example, η is proportional to $M^{0.85}$ at $a = 0.75$ and η is

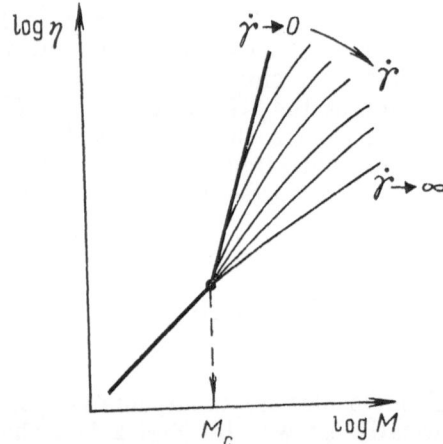

Fig. 2.34. Variation of the dependence of viscosity on molecular mass with increasing rate of shear.

proportional to $M^{1.15}$ at $a = 0.66$, i.e., at high shear rates the dependence $\eta \, (M)$ is found to be practically linear.

2.7. Viscosity of Branched Polymers

Viscosity measurements carried out on samples of branched polymers with known shapes of macromolecules have shown that the macromolecular structure has a substantial effect on the initial viscosity and on the degree of non-Newtonian behaviour.

The effect of the degree of branching on the initial viscosity was first observed by Charlesby* on polysiloxanes subjected to radioactive radiation. He found that the viscosity of branched polymers, η_{br}, is lower than the viscosity of linear polymers, η_{lin}, with the same molecular mass. The same regularity has been observed by Wyman and coworkers** for monodisperse polystyrenes. The fact that this work was conducted on monodisperse polymers makes these results especially important.

The decrease of viscosity with increasing degree of branching of macromolecules has also been described for polyesters, polyethylenes, and other

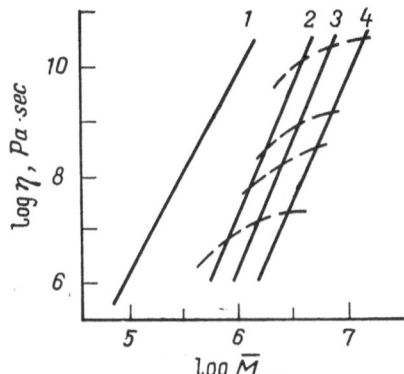

Fig. 2.35. Dependence of viscosity on the mass-average molecular mass for linear (1) and comb-shaped (2-4) polystyrenes of narrow MMD. The dashed lines correspond to polymers with the same length of side branches.

polymers. The effect of side branches in macromolecules on the viscosity of polymers may be so considerable that the ratio η_{lin}/η_{br} sometimes reaches several tens of units at given values of average molecular mass and temperature.

The foregoing can be illustrated by the data presented in Fig. 2.35, which compares the dependences of viscosity on molecular mass for monodisperse linear*** and comb-shaped**** polymers. The solid straight line represents the dependence of η on $\overline{M}_w$ for linear polystyrenes. The straight lines AA, BB, and CC give the dependence of η on $\overline{M}_w$ for comb-shaped polystyrenes containing 10, 20, and 40 side chains, respectively. These side branches are of the same length for all the polymers involved. The dashed lines show the variation of the viscosity as a function of the number of side chains, whose molecular masses are equal to 6.5, 13, 18, and 36 thousand. It is seen that at a given molecular mass, the viscosity of polystyrenes falls off abruptly with increasing degree of branching. The increase of viscosity with increasing molecular mass is especially slow as the number of side chains increases. Although the viscosity of branched polystyrenes is low as compared with linear polystyrenes, the dependence of viscosity on molecular mass for samples with a constant number of side chains is found to be steeper than for linear polystyrenes [the exponent α in Eq. (2.63) for branched polymers is equal to 4.5].

* Charlesby, A. J., *J. Polymer Sci.*, **17**, 370 (1955).
** Wyman, D. P., L. G. Elyash, and W. J. Frazer, *J. Polymer Sci.*, A3, 681 (1965).
*** Akovali, G., *J. Polymer Sci.*, A-2, 5, 875 (1967).
**** Fujimoto, T., H. Narukawa, and M. Nagasawa, *Macromolecules*, **8**, 19 (1972).

The effect of branching on the viscosity of macromolecules can be understood if one takes into account the manner in which the mean radius of
gyration of an unperturbed macromolecular coil varies under the influence
of the branching. This variation is expressed with the aid of the conforma-
tional parameter $g^{1/2} = \langle s \rangle_{br}/\langle s \rangle_{lin}$, where $\langle s \rangle_{br}$ and $\langle s \rangle_{lin}$ are, respectively, the root-mean-square radii of gyration of branched and linear macro-
molecules of the same molecular mass.

The conformational parameter g can be found directly by measuring the
characteristics of macromolecules in dilute solutions (in the theta-state).
For branched macromolecules of regular structure—star-shaped and comb-
branched polymers with branches of the same length—the g-factor is determined by calculations. For star-shaped polymers* $g = 3/p - 2/p^2$, where p
is the number of branches. For comb-branched polymers the g-factor is
expressed in a more complicated way** since it depends not only on the
length of branches but also on the ratio of the molecular masses of the
branches in the main chain.

The replacement of M with the quantity (gM) in formulas (2.63) enables
one to describe correctly, in a number of cases, the flow properties of branched
polymers. This cannot however be regarded as a general rule. Even from
the data of Fig. 2.35 it follows that the rate of increase of the viscosity of
branched polymers with increasing molecular mass is greater than in the
case of linear polymers. Besides, when the g-factor is used, the viscosity
as a function of (gM) must be independent of the number of branches. This
is not always found to be the case.

The lack of a quantitative agreement between experimental evidence and
theoretical conceptions based on the evaluation of the sizes of the macromolecular coil may be due to various causes. In the first place, it may be due
to the fact that the conditions under which an entanglement network is
formed depend on the characteristics of the branching of the polymer chain.
Indeed, the decrease of the viscosity of branched polymers as compared with
linear polymers of the same molecular mass is observed in all cases when the
length of side branches does not exceed a certain critical value. In a general
ease, the ratio of the viscosities of branched and linear polymers depends
on the length of branches. Thus, for polyvinyl acetates it has been shown
by Long and coworkers*** that, depending on the length of side branches, the
ratio (η_{lin}/η_{br}) may either be greater than unity or decrease by a factor
of up to 10. This is confirmed by the data obtained by comparing linear and
branched polyisoprenes with different types of branching****, the data obtained
by Kraus and Gruver***** for star-shaped polybutadienes of narrow MMD,
and the results obtained by Suzuki****** for polystyrenes (Fig. 2.36). In this
figure, the quantity $(A/2.3)$, which is directly proportional to viscosity
and is the pre-exponential factor in Eq. (2.14), is plotted along the ordinate
on a logarithmic scale. From the data presented in Fig. 2.36 it follows that
there exist two critical values of molecular mass. The first (M_c) corresponds
to the repeatedly discussed value at which one dependence changes to another
in Eqs. (2.59); the second (M'_c) is about 5 times higher than the first. It

---

     * Orofino, T. A., *Polymer*, 2, 305 (1961).
    ** Berry, G. C., *J. Polymer Sci.*, A-2, 6, 1551 (1968).
   *** Long, V. C., G. C. Berry, and L. M. Hobbs, *Polymer*, 5, 517 (1964).
  **** Graessley, W. W., *Accounts Chem. Res.*, 10, 332 (1977).
 ***** Kraus, G. and J. T. Gruver, *J. Polymer Sci.*, A3, 105 (1965).
****** Suzuki, T., Ph.D. Thesis, Strasbourg, France, 1970.

corresponds to the sharp increase of the steepness of the dependence of viscosity on molecular mass. The solid intersecting straight lines characterize, throughout their entire course, the dependence of viscosity on molecular mass for linear polystyrenes and also of the viscosity of branched polystyrenes up to the second critical molecular mass M_c'. Thus, here at $M < M_c'$ it is possible to arrive at a unified description of the flow properties of linear and branched polymers provided that the product (gM) is used instead of the quantity M. But at $M > M_c'$ the viscosity of branched polymers (the dashed line) appears to be higher than that of their linear counterparts.

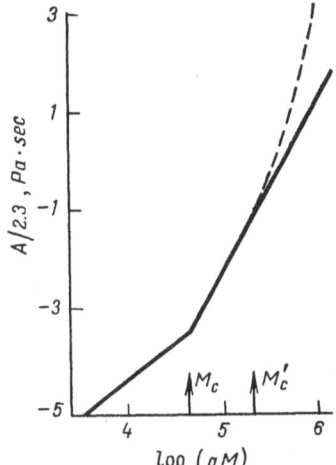

Fig. 2.36. Dependence of the pre-exponential factor (which is directly proportional to viscosity) on molecular mass for polystyrenes of narrow MMD (linear and star-like).

The effect of branching on the rheological properties of polymer melts has been studied most frequently for polyethylene. Depending on the technology of its production, macromolecules of various structure are naturally formed, which has a noticeable effect on the processing and performance properties of the material. It is well known that the presence of branches in polyethylene leads to a very strong (two-fold) increase of the activation energy of viscous flow. Branching has the same strong effect on the other flow properties of polymer melts. Thus, it has been noted[*] that the viscosity of branched polyethylene containing only 10-25 ethyl groups per 1000 atoms in the main chain may be 3 times greater than the viscosity of linear polyethylene with the same value of $\overline{M}_w$.

The general approach to the explanation of the role played by branching in the manifestation of the rheological properties of polyethylene melts is associated with the introduction of the g-factor. In a number of works it has been pointed out[**] that the representation of the dependence of viscosity on the generalized argument (gM) enables one to obtain a unified characteristic of flow properties for linear and branched samples of polyethylene. The presence of long-chain branching, however, causes an abnormally sharp increase of viscosity as compared with that predicted from the magnitude of the argument (gM). This is associated with the fact that the principal

* Saeda, S., T. Suzuki, and K. Yamaguchi, *J. Soc. Mater. Sci. Japan*, **20**, 621 (1971).

** Mendelson, R. A., W. A. Bowles, and F. L. Finger, *J. Polymer Sci.*, A-2, **8**, 105 and 127 (1970); Ram, A. and J. Miltz, *Intern. J. Polymer Mater.*, **2**, 39 (1972).

factor here is not the change of the size of the macromolecular coil but the enhanced intermolecular interactions caused by entanglements and side branches.*

Attempts to construct the rheological criterion of the branching of polyethylene have led to the introduction of a complex parameter**:

$$K = \eta_0 / [\eta]^\alpha \theta_0^\beta$$

where α and β are constants; $[\eta]$ is the intrinsic viscosity; η_0 is the initial viscosity of the melt; and θ_0 is the relaxation time determined by the position of the flow curves.

It has been found that the complex parameter K, which takes account, to various degrees, of the variation of the molecular and relaxation characteristics of the melt, is most sensitive to the branching of polyethylene macromolecules.

The discussion of the effect of the branching of macromolecules on the viscosity of polymers referred to deformation regimes involving low shear stresses and rates. As the shear stresses and rates increase the effect of the branching of macromolecules on the apparent viscosity of polymers diminishes and at high shear rates it may sometimes be insignificant.

2.8. Pressure Dependence of Viscosity

The effect of pressure on the viscosity of polymer melts and solutions belongs to the category of the most important theoretical and practical problems in the rheology of polymers. In practice (during the processing of plastics and elastomers) one deals with pressures of the order of hundreds (in extrusion) and even of thousands (in pressure moulding) of atmospheres, which may have a certain effect on viscosity and cause deviations of the parameters of technological processes from the values calculated from the apparent viscosity measured at low pressures. Nonetheless, the effect of pressure on the viscosity of polymer systems has been studied little.

In considering the effect of pressure on viscosity it is necessary to take into account the fact that the physical state of a polymer can be changed under the action of pressure: there may take place vitrification and crystallization. The change of the glass-transition temperature with increasing pressure is the more significant that the entire set of the viscoelastic characteristics are determined by how far a given state of the polymer is from the region of its transition to the glassy state.

Equation (2.6) given earlier can be generalized with account being taken of the effect not only of temperature but also of pressure on the free volume. Then

$$v_f = v_{fg} [1 + \alpha_0 (T - T_g) + \beta_0 (p - p_g)] \tag{2.69}$$

where α_0 and β_0 are the thermal expansion and compressibility coefficients for free volume; p_g is the pressure at which vitrification occurs.

If we assume that the coefficients α_0 and β_0 do not depend on the pressure and that vitrification takes place at a constant value of free volume, then

---

* Wild, L., R. Ranganath, and D. C. Knobeloch, *Polymer Eng. Sci.*, 16, 811 (1976).
** Locati, G. and L. Gargani, *Proc. 7th Intern. Congr. Rheology*, Gothenburg, 1976, p. 520.

it is easy to arrive at the following variant of the Ehrenfest equation:

$$\left(\frac{\partial p}{\partial T}\right)_{T=T_g} = \frac{\Delta\alpha_0}{\Delta\beta_0}$$

Bianchi* suggested taking into account the effect of pressure on the occupied volume as well. Then

$$\left(\frac{\partial p}{\partial T}\right)_{T=T_g} = \frac{\Delta\alpha_0}{\Delta\beta_0 - \frac{1}{v_g}\left(\frac{\partial v_g}{\partial T_g}\right)}$$

The quantity $(\partial p/\partial T)_{T=T_g}$ varies in polymers within the limits of 0.015-0.045. This gives an idea of the scope of the possible effect of pressure on their glass-transition temperature under isothermal conditions, and, as a consequence, on the viscosity of polymers at increased pressures.

The dependence of the initial viscosity on pressure in the region of not too high pressures is satisfactorily described by an exponential relation, just as in the case of low-molecular-mass liquids:

$$\eta_p = \eta^0 e^{\delta p} \tag{2.70}$$

where η_p and η^0 are the viscosity values at pressure p and atmospheric pressure, respectively; δ is a constant.

The value of the viscosity piezo-coefficient δ for polymers is several thousandths of bar^{-1}. As the temperature rises the quantity δ decreases and the correlation between δ and the p-v-T diagram of the polymer can be established. This relation is expressed by the following formula for a number of vinyl polymers**:

$$-K = (\partial \ln \mu/\partial T)_p \, (\partial v/\partial p)_T \, (\partial \ln \mu/\partial p)_T$$

where the subscripts signify that the derivatives are calculated at constant values of the corresponding quantities; the parameter K has been found to be constant for a number of polymers; it is equal to $(17 \pm 1) \times 10^{-4}$ cm^3/g·deg; the parameter μ in this formula is equal to the viscosity η divided by the pre-exponential factor A in formula (2.14).

Equation (2.70) can be arrived at in different ways: on the basis of the Eyring theory, assuming that the quantity T_g in the WLF equation increases linearly with pressure, or on the basis of the supposition made by Ferry and Stratton*** that the free volume is decreased under the action of pressure, the coefficient β_0 being independent of pressure, i.e.,

$$f_p = f - \beta_0 p$$

where f_p and f are the relative free volumes at pressure p and at atmospheric pressure.

By analogy with the effect of temperature on viscosity,

$$\log (\eta_p/\eta^0) = \frac{1}{f_p} - \frac{1}{f} \tag{2.71}$$

which corresponds to the above-given exponential dependence if it is assumed that the right-hand side of Eq. (2.71) is equal to $(\delta/2.3)$.

In estimating the effect of pressure on viscosity one has to make use of two viscosity piezo-coefficients (just as the effect of temperature on viscosity

---

* Bianchi, U., *J. Phys. Chem.*, **69**, 1497 (1965).
** Miller, A. A., *Amer. Chem. Soc. Polymer Preprints*, **12**, 517 (1971); *Polymer*, **19**, 899 (1978).
*** Ferry, J. D. and R. A. Stratton, *Koll.-Z.*, **171**, 107 (1960).

can be characterized by the "activation energies of flow" at constant values of shear stress or rate). The viscosity piezo-coefficients are written as follows by definition:

$$\alpha_{\dot{\gamma}} = \frac{1}{\eta}\left(\frac{\partial \eta}{\partial p}\right)_{\dot{\gamma},\,T} = \frac{1}{\tau}\left(\frac{\partial \tau}{\partial p}\right)_{\dot{\gamma},\,T}$$

$$\alpha_{\tau} = \frac{1}{\eta}\left(\frac{\partial \eta}{\partial p}\right)_{\tau,\,T} = \frac{1}{\dot{\gamma}}\left(\frac{\partial \dot{\gamma}}{\partial p}\right)_{\tau,\,T}$$

They are interrelated by the following equation:

$$\alpha_{\tau} = \frac{\tau}{\dot{\gamma}} \cdot \frac{\partial \dot{\gamma}}{\partial \tau}\,\alpha_{\dot{\gamma}} = \frac{d \log \dot{\gamma}}{d \log \tau}\,\alpha_{\dot{\gamma}}$$

Here $\alpha_{\tau} \geqslant \alpha_{\dot{\gamma}}$ since $(d \log \dot{\gamma}/d \log \tau) \geqslant 1$. The effect of pressure on viscosity when measured at constant shear stress or rate can be illustrated by the data obtained by Semjonow[*]: for polystyrene at 180°C the increase of pressure up to 1 kbar at $\tau = $ const caused a 55-fold increase of viscosity and at $\dot{\gamma} =$ = const the increase was two-fold.

It has been pointed out above that polymers can be vitrified and crystallized when the pressure is considerably increased. In capillary viscometers and, in general, during the flow in ducts this may be accompanied by a very great increase of the resistance to flow. To this there corresponds a fictitious increase of viscosity; the meaning of the word "fictitious" is that the calculation of viscosity in such cases is in general nonlegitimate.

In the case of flexible-chain high-molecular-mass crystallizable polymers the effect of pressure may be superimposed by the specific effect observed during their flow in dies. In the entrance zone of the die there arise considerable tensile stresses. This means that under the conditions of the action of high stresses there is created a strong orientation effect, which by itself exceedingly facilitates the crystallization of polymers. Thus, it has been shown[**] that when forced through the capillary, natural rubber is capable of crystallizing at temperatures of up to 146°C. A thorough investigation of this kind of effect for linear polyethylene has been carried out by Porter and coworkers.[***] It has been shown that linear polyethylene while flowing in the duct under the action of high pressure near the melting temperature is capable of forming a specific crystalline structure which possesses an extremely high elastic modulus and strength.

What has been said points to the necessity to differentiate between the direct effect of pressure and the orienting effect on the properties and structure of polymers.

    [*] Semjonow, V., *Adv. Polymer Sci.*, **5**, 387 (1968).
   [**] Volt, V. R., C. E. Smith, and C. E. Wilkes, *Rubb. Chem. Techn.*, **44**, 1, 1 (1971).
  [***] Southern, J. and R. S. Porter, *Am. Chem. Soc. Polymer Preprints*, **10**, 1028 (1969); **11**, 266 (1970); Niikuni, T. and R. S. Porter, *J. Materials Sci.*, **9**, 389 (1974); Perkins, W. G. and R. S. Porter, *J. Materials Sci.*, **12**, 2355 (1977).

2.9. Viscosity of Concentrated Polymer Solutions

2.9.1. Introduction

The data on the viscosity of polymer solutions has been extensively reported in the literature. Here we shall consider only the general approach to the description of the flow properties of concentrated solutions and also of plastisized polymers, which are also in fact highly concentrated solutions.

The concept of concentrated solutions is associated with the idea of the interaction between macromolecules; this interaction is governed by the content of macromolecules in solution, the characteristic of macromolecular chains and the nature of the solvent. A convenient measure of the concentration from the standpoint of the problem in question is the product $(c\,[\eta])$, where the polymer concentration c and the intrinsic viscosity $[\eta]$ have dimensions which are inverse to each other. For dilute solutions the quantity c characterizes the filling of their volume with macromolecules. The dimensionless product $\tilde{c} = (c\,[\eta])$ may be regarded as the reduced concentration, i.e., the concentration standardized in a definite way with respect to the molecular mass of the polymer since $[\eta]$ is related to the molecular mass of the polymer.

The transition from dilute to concentrated solutions corresponds to $\tilde{c}$ values of the order of several units. Depending on the quantity $[\eta]$ (and, hence, the molecular mass of the polymer), this transition, which occurs in the region of the critical concentration c_c, is accompanied by a qualitative change of the properties of solutions.

It has been repeatedly pointed out in the previous sections that at molecular masses exceeding the critical value $(M > M_c)$ of polymers in bulk there is formed a fluctuating entanglement network. At $M > M_c$ and $c > c_c$ this occurs also in polymer solutions. Therefore, at $M < M_c$ and $M > M_c$ concentrated polymer solutions may have different properties, just as in the case of polymer melts.

2.9.2. Empirical Correlations

The initial viscosity of concentrated polymer solutions depends on the concentration and molecular mass in the following manner:

$$\eta_0 = K c^\alpha M^\beta \tag{2.72}$$

where K, α, and β are constants.

Formulas of this type are widely used to represent the dependence of the viscosity of solutions on the concentration and molecular mass of the dissolved polymer for diverse polymer-solvent pairs.

The effect of the nature of the solvent on the flow properties of polymer solutions depends on the concentration range concerned.[*] In the region of low concentrations the viscosity of polymer solutions in poor solvents is lower but it changes more strongly with concentration. Therefore, as the

---

[*] Tager, A. A., V. E. Dreval, G. O. Botvinnik, S. B. Kenina, V. I. Novitskaya, L. K. Sidorova, and T. A. Usoltseva, *Vysokomol. Soedin.*, A14, 1381 (1972).

concentration increases the viscosity of polymer solutions in a poor solvent may be found to be higher than in a good solvent. The nature of the solvent has a slight effect on the character of the dependence of viscosity on molecular mass. For nonpolar and weakly polar polymers, whose macromolecules are highly flexible, the thermodynamic quality of the solvent exerts a very slight influence on the viscosity of their solutions. At a specified volume concentration of such polymers the difference in viscosity between solutions is largely determined by the difference in the viscosity values of the solvents used. The quality of the solvent bears an enormous influence on the viscosity of solutions of rigid-chain polymers, the direction of this effect being substantially different in regions of dilute and concentrated solutions.

The coefficients α and β in formula (2.72) are, in a general case, functions of concentration. They increase upon approach to a polymer in bulk. For linear flexible-chain polymers the coefficients α and β may reach 5-7 and 3.5, respectively. However, up to certain, rather often high concentrations the coefficients α and β vary in the same direction, so that their ratio remains constant; in such a case this ratio is close to the exponent a in formula (2.60). The discrepancy between (β/α) and a does not usually exceed 10-15 per cent. For the theta conditions the ratio (β/α) is close to 0.5 and increases to unity when good solvents are used. For rigid-chain polymers the value of α increases rapidly with concentration (at a constant temperature), reaching values of the order of 15-17, while β does not usually exceed 3.5. Therefore, for such solutions the indicated correlation between (β/α) and a is valid only in the region of low concentrations.

If $(\beta/\alpha) = a$, then formula (2.72) can be rewritten in the form:

$$\eta_0 = K\,(cM^{\beta/\alpha})^{\alpha} = K\,(cM^a)^{\alpha}$$

that is, η_0 is a function of cM^a, which is equivalent to the use of the dimensionless concentration $\tilde{c}$ as the argument. Thus, either the dimensionless concentration $c\,[\eta]$ or the quantity (cM^a) may be used as the argument in describing the concentration dependence of viscosity. Use is often made of a certain averaged value of a, in which case a single argument, e.g. $cM^{5/8}$, is introduced for a group of polymers.[*]

Determination of the critical concentration c_c may depend on the method of representation of experimental data, which is illustrated by the example given in Fig. 2.37. Here the relationship between $\log \eta$ and $\log c$ appears to be expressed by a smooth curve, while the concentration dependence of the difference $(\eta_0 - \eta_s)$ (where η_s is the viscosity of the solvent) is reliably approximated by two straight lines. To their point of intersection there corresponds the critical concentration indicated by an arrow on the upper abscissa scale.

The closeness of the values of the ratio (β/α) and the exponent a in the Mark-Houwink equation allows one to presume that the parameter $\tilde{c}$ must predetermine the flow properties of polymer solutions over a wide range of values of c and M—from infinitely dilute solutions up to $c \gg c_c$.

For experimental data on the relationships between η_0 and c measured for polymers of different molecular masses to be generalized, Simha and

---

* Chitrangad, B., H. R. Osmers, and S. Middleman, *Polymer Eng. Sci.*, 17, 806 (1977).

coworkers* have introduced the concept of reduced viscosity:

$$\tilde{\eta} = \eta_{sp}/\tilde{c} = (\eta_0 - \eta_s)/\eta_s \tilde{c}$$

The quantity $\tilde{\eta}$ determines the thickening action of the polymer referred to the volume filling of the solution with macromolecular chains. At $c \to 0$ this quantity is equal to unity, and therefore the function $\tilde{\eta}(\tilde{c})$ has a common initial point for all curves for the concentration dependences of viscosity, which are obtained for different polymers or at different temperatures.

The use of the function $\tilde{\eta}(\tilde{c})$ enables one to arrive at the characteristic of the flow properties of polymer solutions in each given solvent in a form

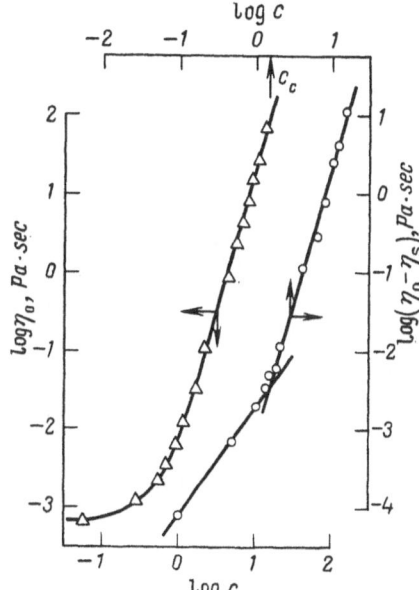

Fig. 2.37. The viscosity versus the polymer concentration for solutions of emulsion polystyrene ($\overline{M}_v = 4.9 \times \times 10^5$) in benzene. [Vinogradov, G. V. and L. V. Titkova, *Rheol. Acta*, **7**, 297 (1968).]

invariant to the concentration, molecular mass (at $M > M_c$) and temperature. In order to obtain a more complete generalization of the flow properties of solutions with respect to molecular masses of polymers, Simha and coworkers make use not of the quantity $\tilde{c}$ but of the reduced concentration in the form of $c_r = c (M_0/M)^{a_1}$, where M_0 is the molecular mass of the polymer used as the normalizing parameter. Depending on the nature of the polymer and solvent, the value of a_1 appears to be slightly higher or lower than the value of a in the Mark-Houwink formula, though the differences between these values are not great.

To obtain a temperature-invariant characteristic of the flow properties of polymer solutions, use may also be made of the shift factor which is determined by the superposition of the curves of $\tilde{\eta}$ versus $\tilde{c}$ for different temperatures in double logarithmic coordinates. The effect of temperature on the flow properties of solutions becomes especially significant upon approach to the region of phase separation and vitrification.

---

* Utracki, L. and R. Simha, *J. Polymer Sci.*, **A1**, 1089 (1963); Simha, R. and F. S. Chan, *J. Phys. Chem.*, **75**, 256 (1971).

Using the function $\tilde{\eta}\,(\tilde{c})$, it is possible, in many cases, to obtain invariant flow characteristics of solutions for a wide range of polymers.* For each solvent and polymer-homologues of various molecular masses there is obtained an invariant, master curve of the flow properties of solutions. For flexible-chain polymers it may cover the region from infinitely dilute solutions to polymer melts. With rigid-chain polymers the master curve reaches a region

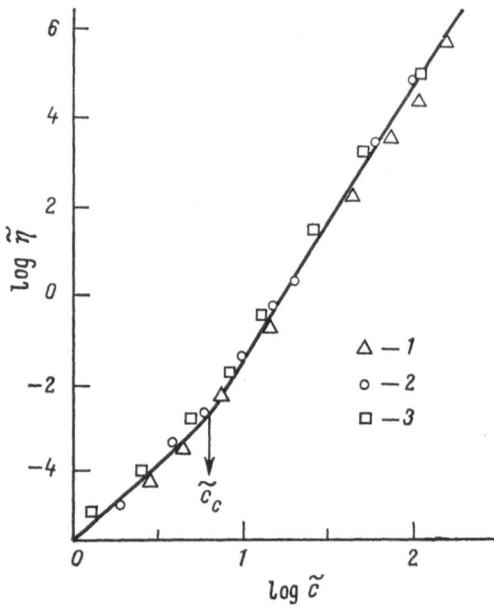

Fig. 2.38. Master curve of the viscosity of solutions of polybutadienes of narrow MMD in methylnaphthalene for polymers having different molecular masses [Malkin, A. Ya., *Rheol. Acta*, **12**, 486 (1973).]: *1*—1.5 × 10⁵; *2*—2.4 × 10⁵; *3*—3.6 × 10⁵.

of concentrations at which the viscosity increases especially rapidly because of the transition to the glassy state.

Examples of the characteristic of the flow properties of solutions of flexible-chain polymers, which is invariant to molecular mass and holds over a very wide range of concentrations, are the data for solutions of polybutadienes of narrow MMD presented in Fig. 2.38. The generalized parameter $c\,[\eta]$ for the description of the flow properties of solutions is also applicable to rigid-chain polymers.**

In going from one solvent to another the invariance of the flow characteristic is attained by introducing additionally the parameter K_M into the argument of the function; this parameter is equal to $(K'_M/2.3)$, where the quantity K'_M corresponds to the representation of the initial region of the curve of $\tilde{\eta}$ versus $\tilde{c}$ by the Martin exponential equation:

$$\tilde{\eta} = \exp\,(K'_M\tilde{c}) \tag{2.73}$$

* Dreval, V. E., A. Ya. Malkin, and G. O. Botvinnik, *J. Polymer Sci., Polymer Phys. Ed.*, **11**, 1055 (1973); *Mekhanika Polimerov*, 6, 1110 (1972); Malkin, A. Ya., *Rheol. Acta*, **12**, 486 (1973); Zakin, J. L., R. Wu, H. Luh, and K. G. Mayhan, *J. Polymer Sci., Polymer Phys. Ed.*, **14**, 299 (1976).

** Berry, G. C. and S. M. Liwak, *J. Polymer Sci., Polymer Phys. Ed.*, **14**, 1717 (1976).

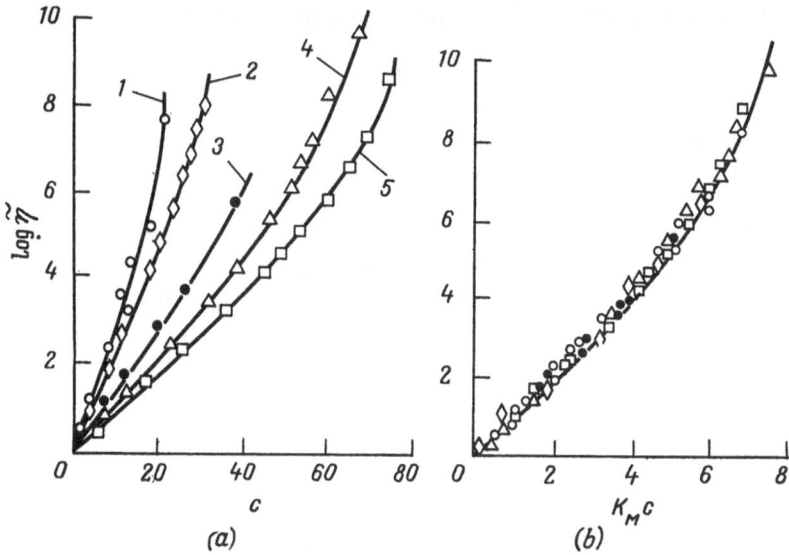

Fig. 2.39. The viscous properties of solutions of polystyrene in various solvents:
a—the dependences of the reduced viscosity on the reduced concentration for solutions of polystyrene·
in decalin (*1*), toluene (*2*), a mixture of ethylbenzene and decalin (25 : 75) (*3*), carbon tetrachloride (*4*),.
and cyclohexane (*5*); *b*—master curve for the reduced viscosity versus the concentration of polymer
solutions in various solvents. The symbols are the same as in figure *a*. [Dreval, V. E., A. Ya. Malkin,.
and G. O. Botvinnik, *J. Polymer Sci., Polymer Phys. Ed.*, **11**, 1110 (1973).]

An example of the use of the function $\tilde{\eta} = \tilde{\eta}\,(K_M\tilde{c})$ is shown in Fig. 2.39,.
and Fig. 2.40 presents the functions $\tilde{\eta}\,(K_M\tilde{c})$ for various polymers produced.
by using a wide gamut of solvents.

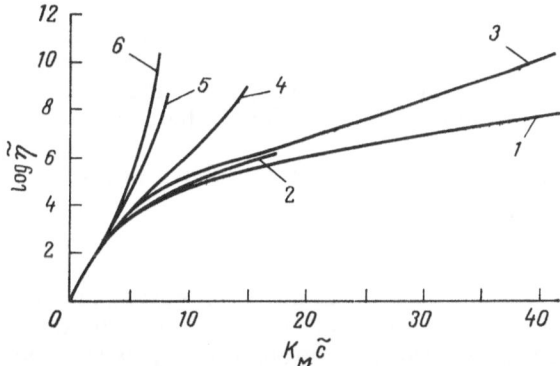

Fig. 2.40. Master curves for the flow properties of polymer solutions:
1—polybutadiene; *2*—polydimethyl siloxane; *3*—polyisobutylene; *4*—cellulose acetate; *5*—polyvinyl
acetate; *6*—polystyrene. See the reference to Fig. 2.39.

An approach to the estimation of the flow properties of concentrated polymer solutions, which is different in form from the one considered above but
similar in meaning, has been suggested by Linow and Philipp*. It is based
on the parameters characterizing the viscosity of dilute solutions and makes

* Linow, K. J. and B. Philipp, *Plaste und Kautschuk*, **18**, 721 (1971).

use of the Schultze-Blaschke formula for dilute solutions:

$$\tilde{\eta} = 1 + K_{SB}\eta_{sp}$$

where K_{SB} is a constant.

The upper limit of applicability of this linear formula is restricted to the value of $\eta_{sp} = 1.104\eta_{sp}^*/K_{SB}$, where η_{sp}^* is the characteristic (normalizing) value of η_{sp}. If now we normalize η_{sp} by η_{sp}^* and consider the dependence of $\tilde{\eta}$ on the reduced parameter $\tilde{\eta}_{sp} = \eta_{sp}/\eta_{sp}^*$, it will turn out that for a wide range of concentrated polymer solutions there exists a universal power dependence of $(\tilde{\eta} - 1)$ on $\tilde{\eta}_{sp}$ that generalizes the above-given relationship between $\tilde{\eta}$ and η_{sp}, which applies only to the region of dilute solutions.

The question of an analytical description of the dependence of $\tilde{\eta}$ on $\tilde{c}$ over a wide range of compositions is very complicated and no general solution has yet been found. The Lyons-Tobolsky empirical equation*, which is a generalized form of Eq. (2.73), is most convenient for this purpose:

$$\tilde{\eta} = \exp\left[K_M'\tilde{c}/(1 - q\tilde{c})\right] \tag{2.74}$$

The introduction of a new, additional constant, q, is associated with the necessity to correlate the viscosity value calculated by formula (2.74) with the limiting case of the viscosity of polymer melts and elastomers, i.e., to take into account the divergence of the curves in the fan shown in Fig. 2.40. Formula (2.74) provides, in a number of cases, a satisfactory agreement with experimental data over a wide range of compositions.**

A somewhat different approach to the description of the dependence of η on $(c\,[\eta])$ for dilute and moderately concentrated solutions has been proposed by Budtov***, who derived the following formula:

$$\eta = \eta_s\,(1 + \gamma c\,[\eta])^{1/\gamma}\exp\left(\frac{H_s}{RT}\frac{a\varphi}{1 - a\varphi}\right)$$

where η_s and H_s are the viscosity and activation energy of the viscous flow of the solvent; φ is the volume concentration of the polymer; and the constant a, which depends on temperature, lies within the limits 0.4-0.8. The quantity γ is a constant equal to 0 for a theta-solvent and to 0.3-0.4 for good solvents.

In the region of low concentrations the above formula takes the following form:

$$\eta = \eta_s\,[1 + c\,([\eta] + A) + \ldots]$$

The constant A is equal to several units (in cm³/g) and plays an important part only for oligomeric products. For high-molecular-mass polymers the Huggins constant is determined through γ: $k_H = (1 - \gamma)/2$.

The above relationships may be regarded as one of the possible quantitative descriptions of the dependence of η on c with the generalized parameter $(c\,[\eta])$ or $(K_M c\,[\eta])$ being used as the argument. From the relations derived in this theory it follows that it is possible to analyse the effect of the concentration of the solution and of the nature of the solvent on the value of

* Lyons, P. F. and A. V. Tobolsky, *Polymer Eng. Sci.*, 10, 1 (1970).
** Quadrat, O., *Collect. Czech. Chem. Communs.*, 42, 1520 (1977); Rink, M., A. Pavan, and S. Roccasalvo, *Polymer Eng. Sci.*, 18, 755 (1978).
*** Budtov, V. P., *Vysokomol. Soedin.*, A12, 1355 (1970); B17, 339 and A17, 1353 (1976); *Mekhanika Polimerov*, 1, 172 (1976).

the apparent activation energy of viscous flow. An essential point here is
the decrease of the activation energy with increasing concentration (predicted for the region of low concentrations) when poor solvents are used. As
the concentration is further increased the activation energy increases with
increasing solution concentration.

The experimental data considered above point to the existence of an intimate relation between the characteristics of infinitely dilute solutions and
the flow properties of concentrated polymer solutions. This means that the
conformational characteristic of an isolated macromolecule, with account
taken of the specific effect of the solvent on it, determines the conditions

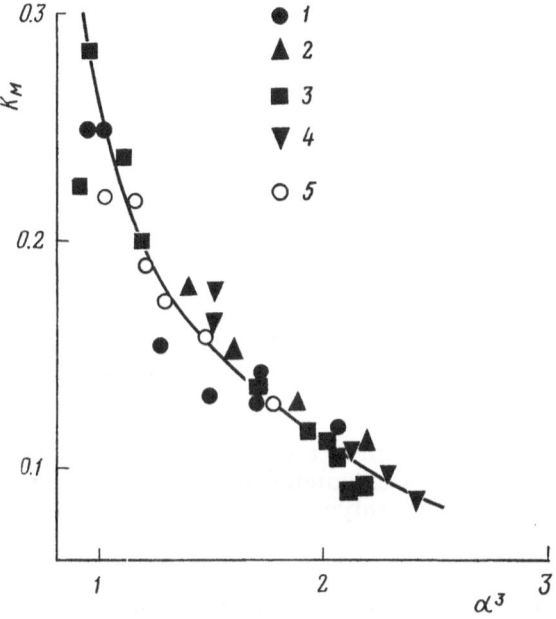

Fig. 2.41. The constant K_M plotted against the expansion factor α of a macromolecular
coil for polydimethyl siloxane (1), polyisobutylene (2), polystyrene (3), and polyvinyl
acetate (4). For comparison, the experimental data obtained by Bohdanecký are given (5).
[Bohdanecký, M., Coll. Czech. Chem. Commun., 35, 1972 (1970).] See the reference to
Fig. 2.39.

of formation of an entanglement network of macromolecules and the character
of the influence of this network on the viscosity of polymer solutions.

The successful use of the function $\tilde{\eta}$ ($K_M \tilde{c}$) for constructing the invariant
characteristics of the concentration dependences of the viscosity of solutions
of polymers of various homologous series in different solvents may be explained
by the fact that the introduction of the coefficient K_M allows one to take
into account the flexibility of the macromolecular chain and the polymer-
solvent interaction.

An overall measure of such interactions is the value of the expansion factor
α for the macromolecular coil. Indeed, as seen from Fig. 2.41, there exists
a correlation between K_M and α. The quantity α includes both the entropy
and the energy effects of intermolecular interactions. The effects of these
two factors can be separated and it is possible in this way to find the general
form of the dependence of K_M on the thermodynamic characteristics of the

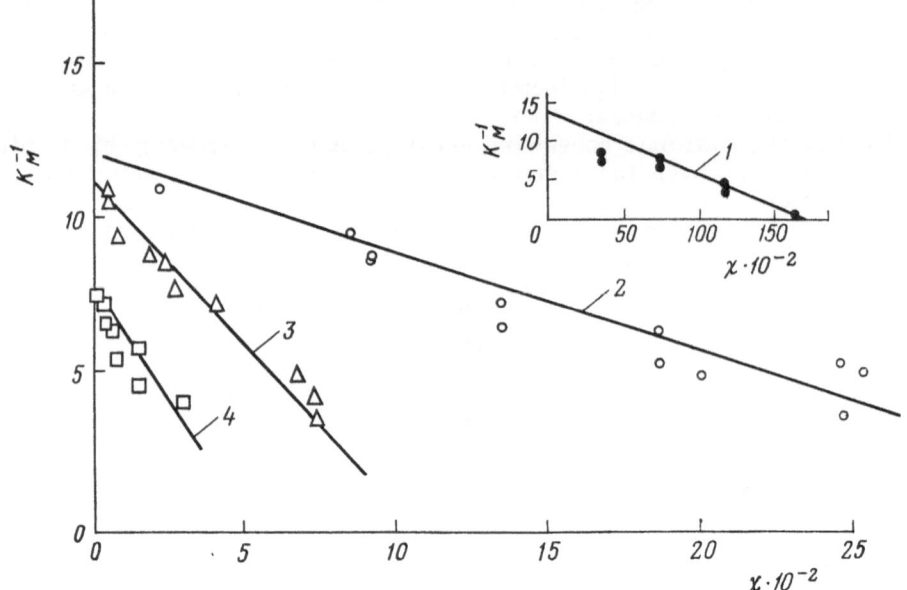

Fig. 2.42. The constant K_M versus the parameter χ for solutions of polydimethyl siloxanes (1), polyisobutylene (2), polystyrene (3), and cellulose acetate (4). See the reference to Fig. 2.39.

solvent and polymer. A measure of the energetic interactions in the solution is the parameter χ, which is calculated through the difference between the solubility parameters of the polymer, δ_2, and the solvent, δ_1:

$$\chi = V_1 (\delta_2 - \delta_1)^2 / RT$$

where V_1 is the molar volume of the solvent, R is the gas constant, and T is the absolute temperature.

The examination of experimental data on various polymers and solvents has shown that the dependence of K_M on χ is hyperbolic in character, i.e., it straightens out in the coordinates K_M^{-1} and χ, as shown in Fig. 2.42. The difference in the position of these lines in this figure is associated with the role played by the entropy factor. The position of the straight lines is characterized by two parameters: the point of intersection with the ordinate axis, which gives the value of $K_{M,0}$ and the slope of the straight lines of

$$\gamma = -dK_M^{-1}/d\chi.$$

The quantity $K_{M,0}$ corresponds to the condition $\delta_1 = \delta_2$, i.e., to the absence of energetic interactions in the solution. Therefore, it is completely determined by entropy effects. Indeed, it can be compared with unperturbed sizes of the macromolecular coil, as shown in Fig. 2.43, where the argument is the value of the ratio $(\overline{h_\theta^2}/\overline{h_0^2})^{0.5}$, where $(\overline{h_\theta^2})^{0.5}$ is the root-mean-square distance between the chain ends in a theta-solvent, and $(\overline{h_0^2})^{0.5}$ is the distance between the ends of the freely jointed chain. The ratio $(\overline{h_\theta^2}/\overline{h_0^2})^{0.5}$ may be regarded as a measure of the internal rigidity of the macromolecule. As the ratio $(\overline{h_\theta^2}/\overline{h_0^2})^{0.5}$ increases, i.e., as the skeleton rigidity of the chain increases,

$K_{M,0}$ increases too (or $K_{M,0}^{-1}$ decreases linearly, as shown in Fig. 2.43). Thus, the dependence of $K_{M,0}$ on $(\overline{h_\theta^2}/\overline{h_0^2})^{0.5}$ describes the correlation between the structural factor and the entropy contribution to the value of the constant K_M with the energy effects being excluded.

The quantity γ, which reflects the contribution of the energy effects, correlates with the cohesive energy density of the polymer, δ_2, namely, as the energy of intermolecular interaction in the polymer increases the sensitivity

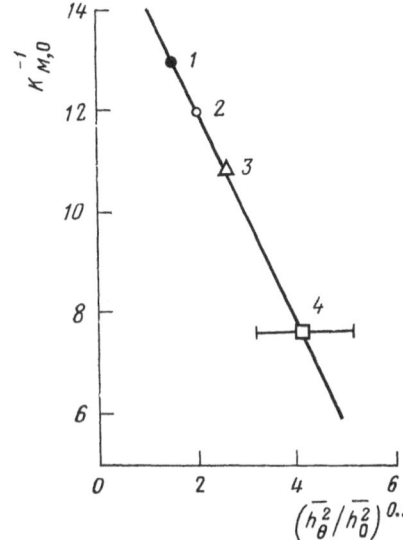

Fig. 2.43. The constant $K_{M,0}$ versus the skeleton rigidity of the macromolecular chain for polydimethyl siloxane (1), polyisobutylene (2), polystyrene (3), and cellulose acetate (4). See the reference to Fig.2.39.

of the constant K_M to the variation of χ also increases, the dependence of γ on δ_2 being linear and γ being equal to 0 at $\delta_2 = 7.15$.

Thus, the Martin constant, which shows the rate of increase of the viscosity of the solution with increasing concentration of the polymer, depends on the following quantities: the difference between the solubility parameters of the polymer and the solvent, the molar volume of the solvent, the skeleton rigidity of the chain, and the cohesive energy density of the polymer. This result can be expressed by the following formula, which includes the parameters indicated above:

$$K_M = [a - b\,(\overline{h_\theta^2}/\overline{h_0^2})^{0.5} - k\,(\delta_2 - 7.15)\,\chi]^{-1}$$

The meaning and numerical values of the coefficients a, b, and k are well seen from the figures discussed above.

The quantities figuring in this formula can be found in many reference books since they are the objective characteristics of various polymers and low-molecular-mass solvents. Therefore, the use of the approach under consideration offers a quantitative method of estimating the viscosity of polymer solutions in various solvents.

The experimental findings considered above and their generalization based on the introduction of various parameters characterizing the individual properties of the components in solution have shown how the viscosity of the solvent varies upon addition of a polymer to it. The data referring to the other limit of the concentration range and showing the variation of the

viscosity of the polymer upon addition of a low-molecular-mass liquid are
not so numerous and systematic. But, in any case, it may be stated that the
viscosity of the low-molecular-mass component introduced into the melt
plays an important part. An illustration of this statement is Fig. 2.44 which
shows the intensity of the effect of the solvent on the viscosity of the polymer. The following parameter is used as a measure of the phenomenon under
consideration:

$$\beta = |d \log \eta / d \log \varphi_2|$$

where φ_2 is the volume fraction of the polymer in solution, the values of β
being calculated at $\varphi_2 \to 1$, i.e., in the limiting case of addition of small
amounts of the solvent (a plastisizer) to the polymer.

From Fig. 2.44 it is clearly seen that as the viscosity of the low-molecular-
mass agent increases its plastisizing effect on the melt is decreased.

This does not, however, cover all possible cases involving the effect of the
low-molecular-mass component on the viscosity of the melt. The point is

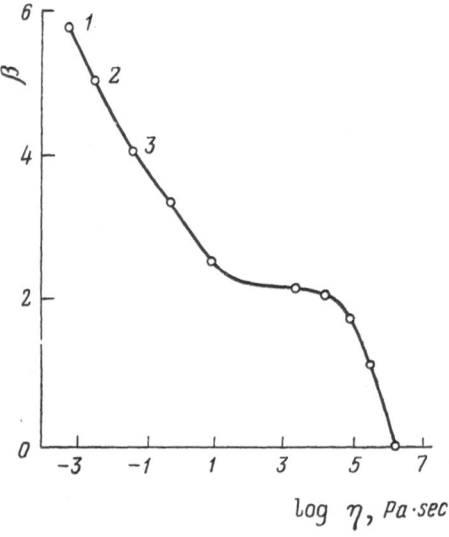

Fig. 2.44. Effect of addition of the
low-molecular-mass component on the
viscosity of polybutadiene with molecular mass 2.4×10^{5}. The abscissa
shows the viscosity of the component
added: toluene (1); α-methylnaphtha-
lene (2); diheptyl phthalate (3); the
remaining points refer to low-molecular-mass polybutadienes added to
the base high-molecular-mass polymer.
[Malkin, A. Ya., G. Zh. Zhangereeva, and M. P. Zabugina, Vysokomol.
Soedin., A-16, 2360 (1974).]

that the monotonous influence of the composition on the properties of the
polymer is characteristic of cases where the homogeneity of the blend is
retained with change of the composition. This is reflected in the fact that
the parameters of the individual chain play the determining role in the
manifestation of the flow properties of solutions over the entire range of
compositions and that the variation of β in Fig. 2.44 is monotonic. A spectacular example that illustrates the possibility of a quite different behaviour
of the melt upon addition of low-molecular agents to it is given in Fig. 2.45.
The exceedingly sharply pronounced extremal course of the concentration
dependence of the viscosity of the melt (which has been described for a number of compositions) to which there are added micro amounts of certain
low-molecular-mass (or oligomeric) agents, is evidence that the effects associated with the structure-formation within the amorphous state may have
a strong effect on the flow properties of polymer melts*.

* Andrianova, G. P., N. F. Bakeev, and P. V. Kozlov, Vysokomol. Soedin., A13,
266 (1971).

The tendency towards specific interactions in polymer solutions is enhanced with increasing rigidity of the macromolecular chain. If for flexible-chain polymers (of the type of polydimethyl siloxane, polybutadiene, etc.) the formation of a random coil in solution is thought to be undoubtful, as the rigidity of the chain increases the number of segments in the chain becomes relatively small, so that the macromolecule is no longer capable of assuming the shape of the Gaussian coil with a statistical distribution of segments relative to the centre of gravity of the chain. Therefore, in considering the concentration dependence of the viscosity of rigid-chain polymers containing, for example, benzene rings in the main chain (polyphenylene-oxadiazo-

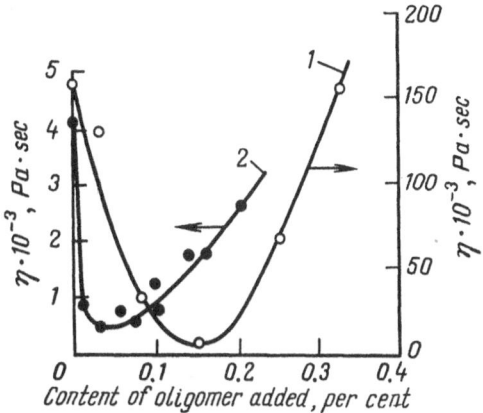

Fig. 2.45. Apparent viscosity of molten polypropylene of various molecular mass versus the content of organosilicone at 194°C [Andrianova, G. P. and V. A. Kargin, *Vysokomol. Soedin.*, **A13**, 1564 (1971).]:

1—polypropylene of grade Moplen at shear stress 1×10^4 Pa; *2*—polypropylene of grade ICI with a lower molecular mass at shear stress 5×10^3 Pa.

les, polyterephthalamides, etc.) there arise certain anomalies due to the specific interactions in such solutions.* Rather high values of a in formula (2.60) are typical for such solutions in the region of low concentrations. For the region of concentrated solutions the exponent α in formula (2.72) is found to be lower than for solutions of flexible-chain polymers. Furthermore, when generalizing the concentration dependences of viscosity it appears to be necessary to take into account the possibility of specific interactions in solutions when the critical concentration is exceeded, by introducing the factor (c/c_c) into the argument of the generalized concentration dependence of viscosity. This is not required for flexible-chain polymers, which points to the monotonic character of variation of the properties (and, possibly, of the structure) of solutions of flexible-chain polymers, in contrast to rigid-chain polymers.

The most striking effect of specific interactions in polymer solutions is associated with the transition of an isotropic solution of extremely rigid-chain (rod-like) polymers to the liquid-crystalline state upon attainment of a certain concentration c^*. This phenomenon is typical of polymers of both

* Kulichikhin, V. G., A. S. Semenova, V. I. Vasiliev, E. G. Kogan, A. V. Volokhina, and A. Ya. Malkin, *Vysokomol. Soedin.*, **B19**, 594 (1977); Kulichikhin, V. G., A. Ya. Malkin, E. G. Kogan, and A. V. Volokhina, *Khim. Volokna*, 6, 26 (1978).

biological and artificial origin. In this case, the dependence of η on c passes
through a very sharply pronounced maximum at $c = c^*$, the value of c^*
being decreased with increasing molecular mass of the polymer.* An example
of this phenomenon is shown in Fig. 2.46. The value of c^* exceeds c_c at
which the exponent α in the formula for the concentration dependence of
viscosity is changed, so that the transition to the liquid-crystalline state
occurs in the region of concentrated solutions. Over the entire range of com-

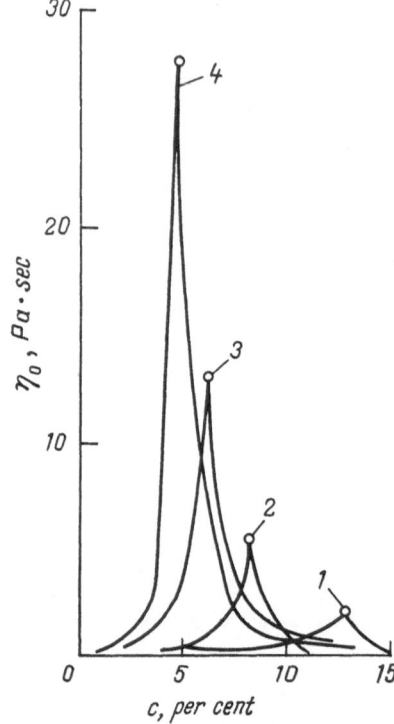

Fig. 2.46. Dependence of viscosity on
the concentration of a polymer having
various molecular masses for solutions
of poly-p-benzamide in dimethyl acetamide at 20°C:
1—10×10^3; 2—16×10^3; 3—24×10^3;
4—69×10^3.

positions at $c > c^*$ the viscosity of solutions decreases with increasing concentration of the polymer.

In generalizing the concentration dependence of the viscosity of such
solutions it is necessary to take account of the existence of a phase transition. Therefore, if we normalize the ordinate scale by the viscosity value η_m
at the maximum point and make use of the product of the concentration by
the molecular mass as the argument, we shall obtain a generalized concentration dependence of the flow properties of solutions of rigid-chain polymers
passing to the liquid-crystalline state (Fig. 2.47). It is natural that such
a dependence is fundamentally different from the concentration dependence
of the viscosity of flexible-chain polymers.

* Kulichikhin, V. G., A. Ya. Malkin, S. P. Papkov, O. N. Korolkova, V. D. Kalmykova, A. V. Volokhina, and O. B. Semenov, *Vysokomol. Soedin.*, **A16**, 169 (1974);
Papkov, S. P., V. G. Kulichikhin, V. D. Kalmykova, and A. Ya. Malkin, *J. Polymer
Sci., Polymer Phys. Ed.*, **12**, 1753 (1974); Lukasheva, N. V., A. V. Volokhina, and S. P. Papkov, *Vysokomol. Soedin.*, **B20**, 151 (1978); Kiss, G. and R. S. Porter, *Amer. Chem. Soc.
Polymer Preprints*, **18**, 185 (1977); **20**, 45 (1979); *J. Polymer Sci.: Polym. Symp.*, **65**,
193 (1978).

A further fundamental specific feature of the flow properties of liquid-crystalline polymer solutions is the appearance of viscosity anisotropy in the various directions of shear measurement.* This phenomenon is associated with the strong orientation of anisodiametric particles in the extrudate. The degree of orientation depends on the shear rate since there is a factor that counteracts the stresses—this is Brownian motion. The degree of orientation attained (equilibrium for a given shear rate) is determined by the relation

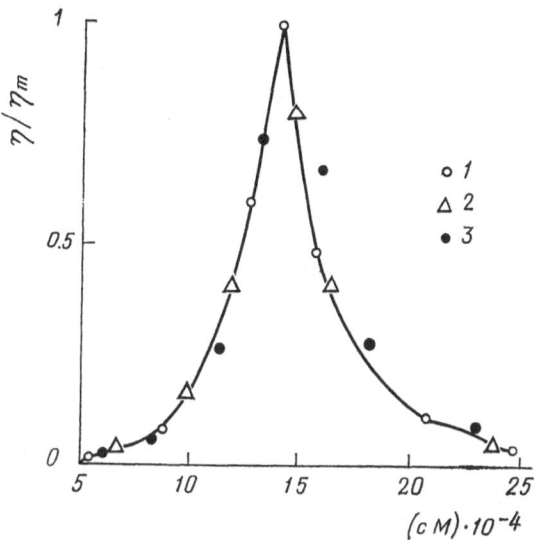

Fig. 2.47. Master concentration curve of the viscosity of solutions of poly-p-benzamide in dimethyl acetamide at 20°C. The molecular masses of the polymer are as follows:
$\overline{M}w \times 10^{-3}$: 16 (1); 24 (2); 69 (3). [(Kulichikhin, V. G., A. Ya. Malkin, S. P. Papkov et al., Vysokomol. Soedin., A16, 169 (1974).]

between the effect of deformation and the random molecular motion. The resistance to motion depends on whether the motion is directed along or across the predominating orientation in the flow. Accordingly, the viscosity coefficients, which constitute a measure of this resistance, depend on whether the force inducing the motion along or across the direction of flow of the solution is measured. Although the existence of viscosity anisotropy is the most characteristic feature of liquid-crystalline polymer solutions, this phenomenon has not yet been studied.

The transition to the liquid-crystalline state, which gives rise to an extremal dependence of viscosity on solution concentration, can be initiated by deformation. As has been demonstrated for polyamide-hydrazine (known under the tradename X-500)**, its solutions are isotropic in a state of rest or at low rates of shear. Therefore, the η vs. c curve is monotonic. It consists of two portions with different slopes; the curve is quite similar to analogous dependences that have been repeatedly observed for solutions of the various flexible-chain polymers. But in the region of high shear rates the form of the η vs. c curve is sharply changed: there appears a maximum on it and the entire curve becomes quite analogous to the curves shown in Fig. 2.45. This

---

* Malkin, A. Ya., N. V. Vasilieva, T. A. Belousova, and V. G. Kulichikhin, Kolloid. Zhurn., 41, 200 (1979).
** Valenti, B. and A. Ciferri, J. Polymer Sci., Polymer Lett. Ed., 16, 657 (1978).

may be regarded as a rheological proof of the formation of an ordered (liquid-crystalline) phase in solutions of X-500 at sufficiently high shear rates. Hence the possibility of producing highly oriented high-modulus fibres from this polymer if the process is effected at appropriate rates.*

2.9.3. Concentration Dependence of the Newtonian Viscosity (the Free Volume Theory)

The effect of free volume on the flow properties of solutions is only manifested in the region of high polymer concentrations.

Whereas the free volume of a polymer melt is a function of temperature alone, as was the case in Sec. 2.2, with solutions it is determined, in addition, by the concentrations of the components. It is reasonable to begin the discussion of this problem with the case where $M < M_c$ and no fluctuating network exists.

The relative free volume of a solution of concentration c_1 at temperature T is expressed as $f_c = f(T, c_1)$; for a polymer in bulk $f_c = f(T, 0)$. Here c_1 is the solvent concentration. The use of Eq. (2.5), in which $B_0 \approx 1$, leads to the following expression for the shift factor a_c with respect to concentration:

$$\ln a_c = \ln (\eta_{0c}/\eta_0) = f_c^{-1} - f^{-1}$$

where η_{0c} and η_0 are the initial viscosities of the solution and the polymer.

Fujita and Kishimoto ** suggested using the linear form of the dependence of the relative free volume on the solvent concentration:

$$f_c = f + \beta c_1 \tag{2.75}$$

where the parameter β depends on the nature of the polymer and solvent. Hence,

$$\frac{f^2 \ln a_c}{1 + f \ln a_c} = -\beta c_1 \tag{2.76}$$

Expressions (2.75) and (2.76) and the conclusions that follow from them are only valid for solutions of high concentrations. Using expressions (2.75) and (2.76), one can arrive at concentration dependences of the viscosity and glass-transition temperatures of highly concentrated solutions.

If the concentration of the solvent is low, then from Eq. (2.76) it follows that

$$\ln a_c = -\beta c_1/f^2$$

and therefore

$$\eta_{0c} = \eta_0 \exp(-\beta c_1/f^2) = \eta_0 \exp(-Kc_1) \tag{2.77}$$

where $K = \beta/f^2$ is a constant for a given polymer-solvent system.

The physical significance of the constant β was elucidated by Ferry and Stratton***, who proceeded from the idea that when a polymer is mixed with a solvent, the occupied volumes are additive. Then, the free volumes of

---

* Valenti, B. and A. Ciferri, *J. Polymer Sci., Polymer Lett. Ed.*, 16, 657 (1978).
** Fujita, H. and A. Kishimoto, *J. Polymer Sci.*, 28, 547 (1958); *J. Chem. Phys.*, 34, 393 (1961).
*** Ferry, J. D. and R. Stratton, *Koll.-Z.*, 171, 107 (1960).

the polymer and solvent appear to be non-additive. This has to be considered in a general case when the dissolution is accompanied by the decrease of the total volume of the system as compared with the sum of the volumes of the components. It is evident that the decrease of the total volume of the system can only occur due to the free volumes of the components. Ferry and Stratton have shown that if the occupied volumes of the components of the solution are assumed to be additive, the following relation holds true:

$$\beta = f_s v_2 \qquad (2.78)$$

where f_s is the relative free volume of the solvent; v_2 is the specific volume of the polymer.

Usually, f_s is of the order of 0.15-0.30, which agrees with the values of the parameter β determined independently from the variation of the glass-transition temperature of the polymer when a solvent is introduced into it; this will be discussed below. Then, for concentrated solutions the following relation is fulfilled:

$$\eta_{oc} = \eta_0 \exp\left[-(f_s v_2/f^2)\, c_1\right] \qquad (2.79)$$

Formulas (2.77) and (2.79) are the basic quantitative outcome of the theory of the viscosity of concentrated polymer solutions, which enables one to calculate the viscosity of solutions with a high content of polymers on the basis of the independently determined values of the parameters f_s, v_2, and f.

From relation (2.79) it follows that the decrease of the viscosity of the polymer upon addition of a low-molecular-mass solvent to it is primarily determined by two factors—the free volume of the solvent and the remoteness of the temperature of solutions from the glass-transition temperature of the polymer. The larger the value of f_s and the lower the value of f, the more intensive is the decrease of the viscosity of the solution at the same content of the solvent in the system. Besides, since f decreases with fall of temperature, the strongest effect produced by the addition of the solvent is observed near the glass-transition temperature of the polymer.

The theoretical results formulated in terms of the final formulas (2.77) through (2.79) apply equally well to solutions, which may be regarded as homogeneous mixtures of a polymer and a solvent. But if the solvent has only a limited compatibility with the polymer, this leads to various anomalies of the flow properties. This circumstance, for example, may be responsible for the extremely strong plasticizing effect of micro additions with the subsequent sharp increase of the viscosity upon further increase of the concentration of the agent added (see Fig. 2.44 and the relevant discussion). Then there may take place specific polymer-solvent interactions (for example, the formation of hydrogen bonds). In this case too, there are observed anomalies of the concentration dependence of viscosity. This mechanism accounts, in particular, for a very strong decrease of the viscosity of molten polymers containing small amounts of water.[*]

Basing on expression (2.75), one can obtain the dependence of the glass-transition temperature of the solution, T_{gc}, on the concentration of the solvent if it is taken into account that the relative free volume of the polymer, f_0, is expressed in terms of the volume at the glass-transition temperature (f_g), and the thermal expansion coefficient of the free volume (α_0) is

_____

* Malkin, A. Ya. and B. M. Rendar, *Plastmassy*, 5, 43 (1977).

given by the formula $f_0 = f_g + \alpha_0 (T - T_g)$. Then

$$f_c = f_g + \beta c_1 + \alpha_0 (T - T_g) \tag{2.80}$$

Assuming that at the glass-transition temperature of the solution, $T = T_{gc}$, its relative free volume, just as that of the polymer melt, is f_g (i.e., the condition $f = f_g$ determines the attainment of the glassy state), it is not difficult to obtain a simple relation between the glass-transition temperatures of the solution and the polymer melt:

$$T_{gc} = T_g - (\beta/\alpha_0) c_1 \tag{2.81}$$

Addition of a low-molecular-mass component to a polymer causes a decrease in the glass-transition temperature, this effect being proportional to the amount of the solvent (plasticizer) added (the Kargin-Malinsky rule).

The analysis of experimental data carried out by Fujita and Kishimoto has shown that β varies from 0.13 to 0.28 for solutions of polystyrene in 12 solvents; from 0.14 to 0.25 for solutions of polymethyl methacrylate in 9 solvents; for the system polymethyl methacrylate-diethyl phthalate $\beta = 0.07$.

As a general rule, the value of β decreases with increasing molecular mass of the solvent, just as in the case shown in Fig. 2.43 for a polymer plasticized by low-molecular-mass compounds of different viscosity. In some cases, however, the value of β may be influenced by the polymer-solvent interaction. Thus, for polystyrene solutions in various phthalates the course of the dependence of T_{gc} on the molecular mass of phthalates is extremal.[*]

A further supposition, on the basis of which there can be deduced the concentration dependence of the viscosity of polymer solutions, is the hypothesis of the additivity not only of the occupied but also of the free volumes. In this case, however, no account is taken of the possible effect of the change of the volume of the solution during mixing. The basic equation of this theory is a quantitative formulation of the assumption of the additivity of the free volumes of the components at temperatures above the region of their vitrification.[**] This means that

$$f_c = \varphi_2 f_2 + (1 - \varphi_2) f_s$$

where φ_2 is the volume fraction of the polymer in the solution and f_2 is its free volume.

It is also presumed that the condition for the vitrification of the polymer and solvent is the equality of the corresponding relative free volumes to 0.025. Besides, as usual, $\alpha_0 = 4.8 \times 10^{-4}$ K^{-1}. Then

$$f_c (T, c) = c [0.025 + 4.8 \times 10^{-4} (T - T_g)] + (1 - c) [0.025 + \alpha_s (T - T_{gs})] \tag{2.82}$$

where α_s and T_{gs} are, respectively, the thermal expansion coefficient of the free volume of the solvent and its glass temperature; here and afterwards c is the concentration of the polymer in solution.

The first term in the sum under consideration represents the relative free volume of the polymer, and the second term expresses the same for the solvent. It is important to note that expression (2.82) is not subject to any limitations concerning the concentration of the solutions. From this standpoint, the Kelley-Bueche theory is more general than the above-discussed Fujita-Kishimoto theory.

---

   * Tager, A. A. and A. I. Suvorova, *Vysokomol. Soedin.*, **8**, 1698 (1966).
  ** Kelley, F. N. and F. Bueche, *J. Polymer Sci.*, **50**, 549 (1961).

Equating the left-hand side of Eq. (2.82) to 0.025, we can find the glass-transition temperature of the solution:

$$T_{gc} = \frac{4.8 \times 10^{-4} c T_g + \alpha_s (1-c) T_{gs}}{4.8 \times 10^{-4} c + \alpha_s (1-c)} \qquad (2.83)$$

An experimental testing of this formula has shown that it correctly describes the variation of the glass-transition temperature of the system upon addition of a low-molecular-mass solvent over a wide range of compositions. In such cases, in the region of small amounts of solvent the departures from the linear dependence, which corresponds to the rule of "volume fractions", are negligibly small. A typical example that illustrates the correla-

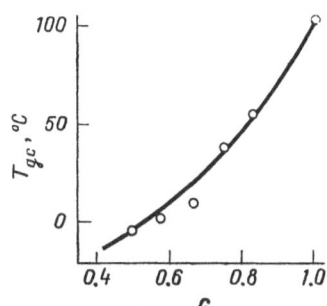

Fig. 2.48. The glass-transition temperature of the system polymethyl methacrylate-diethyl phthalate [the curve has been calculated by formula (2.83)].

tion between the values of T_{gc} calculated from formula (2.83) and the experimentally determined values is presented in Fig. 2.48 for the system polymethyl methacrylate-diethyl phthalate (according to the data obtained by Kelley and Bueche). As seen from Fig. 2.48, formula (2.83) correctly reflects the variation of the glass-transition temperature of the solution, depending on the content of the polymer, c, when the value of c changes from 1.0 to 0.4. The glass temperature of the solution falls by more than 100°C.

2.9.4. The Entanglement Network and the Newtonian Viscosity

The change of T_g is the most important factor having an effect on the viscosity of the solution. The variation of viscosity upon dilution of the solution is, however, determined not only by the shifting of T_g and, hence, by the increase of the difference $(T - T_g)$, but also by the dilution effect proper. Only when the role of both these factors is taken into account can correct estimates of the viscosity of the solution be obtained.* The discussion of the effect of dilution of the polymer with the solvent boils down to the consideration of the role of the fluctuating entanglement network which leads to the existence of a critical concentration and to the change of the form of the dependence of polymer viscosity on concentration and molecular mass.

If $M < M_c$, the structure-formation effect simulated by the fluctuating entanglement network is absent. Then

$$\eta = K' M_c \xi = K'_z Z_c \xi \qquad (2.84)$$

* Hoftyzer, P. J. and D. W. van Krevelen, *Angew. Makromol. Chem.*, 56, 1 (1976).

where ξ is the monomeric friction coefficient; Z_c is the number of atoms in the main chain of the macromolecule corresponding to M_c; K' and K_Z' are constants.

When $M > M_c$, then in the presence of an entanglement network

$$\eta = KM^{3.5}c^4\xi = K_Z Z^{3.5}c^4\xi \tag{2.85}$$

The constants K and K_Z appearing in this formula are different from K' and K_Z'.

The effect of free volume on viscosity is expressed by the coefficient ξ, for which the Doolittle equation is assumed to be valid, i.e., $\ln \xi = A' + B'/f_c$ (A' and B' are constants; $B' \approx 1$).

Just as in the case of formula (2.82) for the free volume, we can write the following expressions for viscosity:

$$\eta = B'c \exp \{c [0.025+4.8 \times 10^{-4} (T-T_g)] + (1-c) [0.025+\alpha_s (T-T_{gs})]\}^{-1}$$

and

$$\eta = Bc^4 \exp \{c [0.025+4.8 \times 10^{-4} (T-T_g)] + (1-c) [0.025+\alpha_s (T-T_{gs})]\}^{-1}$$

where B and B' are constants.

In a complete form expressions (2.84) and (2.85) are written in the form proposed by Pezzin and Gligo*:

$$\ln (\eta/K_Z'Z) = \ln c + [cf + (1-c) f_s]^{-1} \tag{2.86}$$

$$\ln (\eta/K_Z Z^{3.5}) = 4 \ln c + [cf + (1-c) f_s]^{-1} \tag{2.87}$$

where f and f_s are the relative free volumes of the polymer and solvent, respectively; K_Z' and K_Z incorporate various constants contained in the corresponding expressions.

It is essential to emphasize three circumstances. First, expressions (2.84) through (2.87) can describe the flow properties of solutions of high-molecular-mass polymers of various concentrations. Second, the values of the constants K_Z' and K_Z vary within narrow limits: for various polymer systems $K_Z = -11.8 \pm 0.7$ and $K_Z' = -5.75 \pm 0.5$. Third, for rigid-chain polymers at $c \gg c_c$ the dependence of viscosity on concentration may be considerably stronger than predicted by Eq. (2.87).

A comparison of relations (2.86) and (2.87) enables one to find the critical concentration from experimental data. For a solution of polystyrene in benzene this is shown** in Fig. 2.49. Here, use is made of viscosity values in the form of $(\eta_0 - \eta_s)$, which is especially important in the region of low concentrations.

Useful qualitative considerations regarding the concentration dependence of the viscosity of solutions are offered by Berry***, who, proceeding from formulas (2.60) and (2.61) and differentiating expression (2.60) with respect to φ_2 at constant T and Z, obtained the following equation:

$$\frac{\partial \ln \eta}{\partial \ln \varphi_2} = a - \frac{1}{\alpha (T-T_0)} \left[\frac{\partial \ln (T-T_0)}{\partial \ln \varphi_2} + \frac{\partial \ln \alpha}{\partial \ln \varphi_2} \right] \tag{2.88}$$

where the quantities α and T_0 enter into the expression for the monomeric friction coefficient: $\xi = \xi_0 \exp [\alpha (T - T_0)]^{-1}$ (ξ_0 is a constant).

    * Pezzin, G. and N. Gligo, J. Appl. Polymer Sci., 10, 1, 21 and 1637 (1966).
   ** Vinogradov, G. V. and L. V. Titkova, Koll.-Z., 239, 655 (1970).
  *** Berry, G. C., J. Phys. Chem., 70, 1194 (1966).

As before, $a = 3.5$ or 1 if $X > X_c$ or $X < X_c$, respectively. The first term in brackets has a greater value at φ_2; it is close to unity and decreases with decreasing φ_2, whereas the second term remains small until φ_2 has a small value. It follows that as the solvent is introduced, η first falls off rapidly and then there is attained a region of concentrations in which η decreases

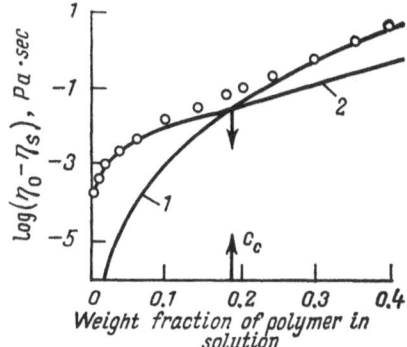

Fig. 2.49. Determination of the critical concentration of suspension polystyrene ($\overline{M}_p = 7 \times 10^4$) in benzene at 25°C. The curves 1 and 2 correspond to formulas (2.86) and (2.87).

less rapidly with further decrease of φ_2. This is also determined by the quantity a, which in its turn depends on the ratio X/X_c.

2.9.5. Non-Newtonian Viscosity

This problem has been thoroughly studied by many investigators. We shall confine ourselves to the consideration of those questions which are associated with the construction of complete flow curves for systems exhibiting non-Newtonian flow behaviour.

The most important feature of polymer solutions is that it is possible to plot complete flow curves for them, which cover the regions of the highest (η_0) and lowest (η_∞) viscosities, which are independent of the shear rate and stress. These extreme values of viscosity limit the region of non-Newtonian flow behaviour ("structural" viscosity) which varies more or less strongly with shear stress and rate.

The region of polymer concentrations, in which it is possible to obtain complete flow curves, is determined by the nature of the polymer and solvent (and also by temperature).

True solutions of flexible-chain polymers can exist over the entire concentration range. An example of the flow curves of such solutions is illustrated by the data* of Fig. 2.50 for solutions of high-molecular-mass polybutadiene of narrow MMD in methylnaphthalene and diheptyl phthalate. It has been pointed out above that in the case of high-molecular-mass linear polymers of narrow MMD the non-Newtonian flow behaviour up to the spurt is weakly pronounced, and the vertical spurt branch and the upper inclined branch have nothing to do with laminary flow and, hence, cannot be regarded as manifestations of non-Newtonian behaviour. As the polymer is diluted

---

* Vinogradov, G. V., A. Ya. Malkin, N. K. Blinova, S. I. Sergeenkov, M. P. Zabugina, L. V. Titkova, Yu. G. Yanovsky, and V. G. Shalganova, *Europ. Polymer J.*, 9, 1231 (1973).

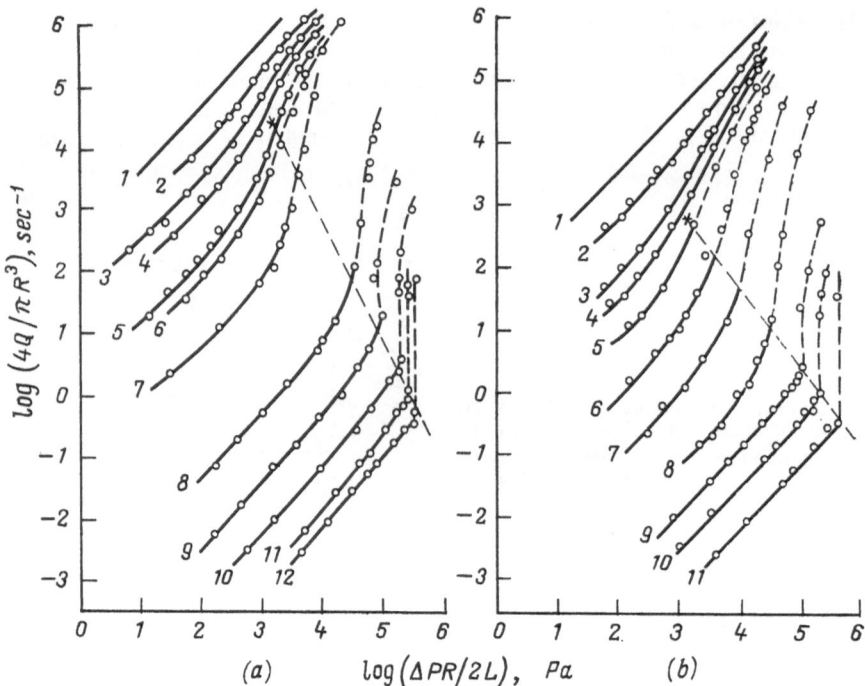

Fig. 2.50. Flow curves for solutions of polybutadiene ($\overline{M}_v = 2.4 \times 10^5$; $\overline{M}_w/\overline{M}_n \approx 1.1$) in methylnaphthalene (*a*) and diheptyl phthalate (*b*) at 25°C:
a—curves *1-12* correspond to the following concentrations [per cent (by mass)]: 0; 1; 2; 3; 5; 6; 10; 20; 30; 50; 70; 100; the critical concentration cc $\approx$ 4 per cent; *b*—curves *1-11* correspond to the following concentrations: 0; 1.5; 3; 4; 7; 10; 20; 30; 50; 70; 100 per cent; the critical concentration $c_c \approx$ 4-6 per cent.

with the solvent the critical parameters corresponding to spurt conditions vary in a manner shown by dashed lines. Here the extent of the vertical branches of the curve of flow rate against pressure gradient in the capillary is shortened and the flow rate vs. pressure gradient curves gradually become S-shaped, a form typical of complete flow curves. It is, however, very difficult to find out at what concentrations there are really obtained true flow curves and not the curves that show the masked manifestation of the spurt effect with all the characteristic features caused by non-steady flow conditions. The lower the concentration of the flexible-chain polymer, the more rapidly are the relaxation processes taking place in it. Therefore, for the non-steady flow conditions to be detected, higher and higher speeds of recording are required, which complicates the correct solution of the problem of the nature of flow.

In the case of linear polymers that have no highly flexible chains, such as, for example, polystyrene, complete flow curves can be obtained when use is made of thermodynamically "poor" solvents over a rather wide range of concentrations.*

Complete flow curves for rigid-chain polymers can be obtained at low concentrations in relatively poor solvents. It is exactly for such systems—

* Tager, A. A. and V. E. Dreval, *Rheol. Acta*, 9, 517 (1970); Ito, Y. and S. Shishido, *J. Polymer Sci., Polymer Phys. Ed.*, 16, 725 (1978).

solutions of nitrocellulose—that Philippoff and Hess obtained nearly complete flow curves for the first time in the thirties. The upper region of concentrations, up to which complete flow curves are realized, is determined by the conditions of gelation of the system. If no gelation takes place, complete flow curves for solutions of rigid-chain polymers can be observed up to rather high concentrations.

Macromolecules of the most rigid-chain polymers, such as aromatic polyamides or polypeptides in the form of α-helices, are similar to rigid rods. Their solutions behave as Newtonian liquids at low concentrations and exhib-

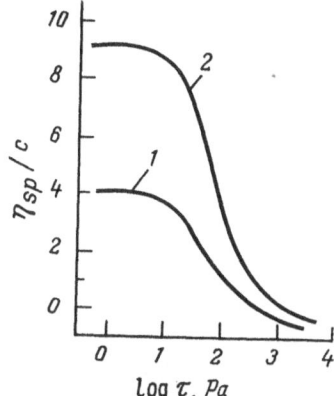

Fig. 2.51. Flow curves for solutions of poly-γ-benzyl-L-glutamate of various concentrations in cresol:

1—0.51 per cent (by mass); 2—1.41 per cent (by mass). [Yang, J. T., J. Amer. Chem. Soc., 8, 1783 (1958).]

it non-Newtonian behaviour at higher concentrations. At a certain critical concentration a tactoid structure* may form in them and they become liquid-crystalline systems with all the features characteristic of such liquids. In the region of low concentrations there can be obtained complete flow curves for rigid-chain polymers with a well-developed structural branch, as shown in Fig. 2.51.

The existence of a flow region with constant (initial, Newtonian) viscosity η_0 is undoubtful for isotropic solutions and melts, and for such liquids η_0 is the most important reference characteristic of their rheological properties. However, for anisotropic polymer solutions (solutions forming a liquid-crystalline structure) no flow region with constant viscosity exists. As the shear stress decreases the viscosity increases indefinitely, so that in such solutions there exists a yield value corresponding to the structural strength of the solution.** The appearance of the yield value is characteristic of materials having a certain inherent structure that hinders the development of flow. Viscous flow becomes possible only after this structure is destroyed.

A distinctive feature of complete flow curves is their symmetry relative to regions of constant viscosity. This leads to the following simple relationship between the highest and lowest viscosities and the viscosity at the inflection point (η_i) of the flow curve plotted in the coordinates $\log \dot{\gamma}$ and

---

* Flory, P. J., Proc. Royal Soc., A234, 73 (1956).
** Papkov, S. P., V. G. Kulichikhin, A. Ya. Malkin, V. D. Kalmykova, A. V. Volokhina, and L. I. Gudim, Vysokomol. Soedin, 14B, 244 (1972); Papkov, S. P., V. G. Kulichikhin, V. D. Kalmykova, and A. Ya. Malkin, J. Polymer Sci., Polymer Phys. Ed., 12, 1753 (1974); Wong, C.-P., H. Ohnuma, and G. C. Berry, Amer. Chem. Soc., Polymer Preprints, 18, 167 (1977).

$\log \tau$:

$$\log \eta_i = \frac{1}{2} (\log \eta_0 + \log \eta_\infty)$$

This relation enables one to determine η_∞ from η_0 and η_i, and second, to estimate the reliability of determination of η_0 and η_∞ since the above formula must be satisfied.

Umstätter* noticed that complete flow curves having an S-shape are similar to curves of normal distribution of probabilities restricted by the values of the initial and upper Newtonian viscosity. These curves can be linearized if use is made of an appropriate coordinate scale. The parameters

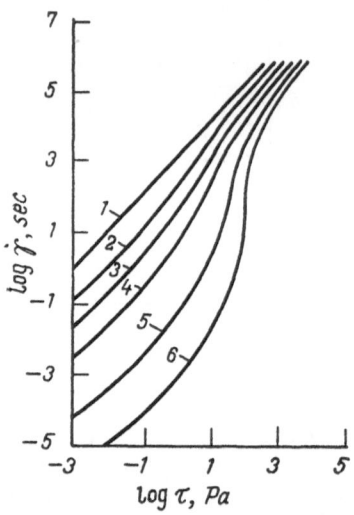

Fig. 2.52. Complete flow curves for solutions of cellulose nitrate in n-butyl acetate; curves *1-6* correspond to the following polymer concentrations [per cent (by mass)]:

0; 0.0575; 0.115; 0.2; 0.5; 1.0. [Philippoff, W., F. H. Gaskins, J. G. Brodnyan, J. *Appl. Phys.*, 28, 1118 (1957).]

that are used in such a graphical representation of flow curves are the product $\tau\dot{\gamma}$ (the internal friction power) and the function

$$\Phi = \frac{\log \eta - \log \eta_\infty}{\log \eta_0 - \log \eta_\infty}$$

The choice of the internal friction power as the parameter that defines the state of a deformed polymer solution is justified by the above-indicated symmetry of the curves.

What has been said above is clarified by a comparison of the left part of Fig. 2.18 which gives complete flow curves for polyisobutylene solutions with Fig. 2.52, which presents analogous data for solutions of cellulose nitrate, and with Fig. 2.53, where all these data are presented (after Wright and Crouse**) in the form of the dependence of $(\tau\dot{\gamma})$ on Φ in coordinates corresponding to the normal law of probability distribution. Experimental data for solutions of various concentrations fall within the hatched narrow bands in Fig. 2.53. This shows that it is really possible to linearize the complete S-shaped flow curves by using the Umstätter method. The use of Fig. 2.53

* Umstätter, H., *Einfürung in die Viskosimetrie.* Berlin, Springer, 1952, p. 138.
** Wright, W. A. and W. W. Crouse, *ASME-ASLE Intern. Lubrication Conf.*, Washington, October, 1964, preprint No. 64LC-11, New York, Academic Press, 1964.

makes it possible to perform a careful extrapolation of experimental data with respect to the solution concentration or deformation conditions. In all cases, however, it is necessary to know the initial (Newtonian) viscosity and at least one value of viscosity in the region of non-Newtonian flow. It is obvious that from the graphs constructed by using the normal law of

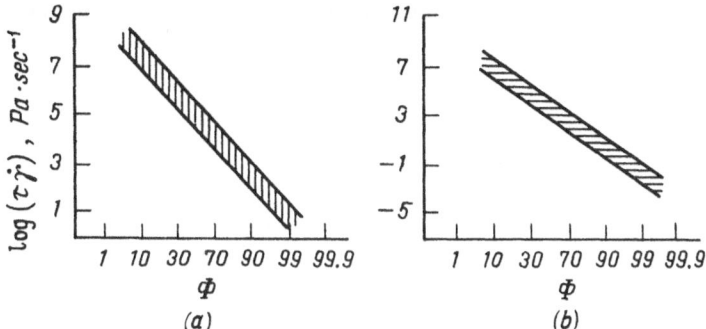

Fig. 2.53. Data of Figs. 2.18a and 2.51 in coordinates corresponding to the normal law of probability distribution:
a—solutions of polyisobutylene in decalin; b—solutions of cellulose nitrate in n-butyl acetate.

probability distribution for the dependence of $(\tau\dot{\gamma})$ on Φ it is easy to find the value of η_∞ if η_0 and one value of η $(\dot{\gamma})$ are known. Unfortunately, the above-considered method of linearization of experimental data is justified not in all cases and the limits of its applicability are unknown.

2.10. Generalized Description of the Flow Properties of Polymers by Master Curves

The variation of the viscosity of polymer systems under the influence of deformation can be conveniently estimated by the ratio of the viscosity under given flow conditions to its highest value: (η/η_0). The ratio (η/η_0) quantitatively characterizes the departure of the properties of the system from the limiting state corresponding to the flow properties of the simplest (Newtonian) liquid. If the flow properties of the system are described by a complete flow curve with regions of the highest and lowest viscosities (η_0 and η_∞), the state of the system under each given set of deformation conditions is determined* by the quantity $(\eta - \eta_\infty)/(\eta_0 - \eta_\infty)$.

The simplest possibility of comparing and, accordingly, of generalizing the flow characteristics of polymer systems by means of master curves over a wide range of temperatures, molecular masses, concentrations, shear rates and stresses, is based on the use of the parameter characterizing the ratio of the shear rate to the rate of relaxation processes, the latter being the inverse of the relaxation time. Such a normalization leads to the concept of the reduced shear rate. The concept of the relaxation time θ is commonly used as a characteristic of the rate of relaxation processes. Then, the reduced

---

* Malkin, A. Ya. and V. E. Dreval, *Vysokomol. Soedin.*, A10, 1907 (1968).

shear rate is expressed as $(\dot{\gamma}\theta)$ and the description of the flow properties of polymer systems boils down to that of the dependence $\eta/\eta_0 = f(\dot{\gamma}\theta)$.

In various molecular-kinetic theories of polymer systems (see Chapter 3) it is pointed out that within each relaxation-time spectrum the various relaxation times are unambiguously interrelated. Then, it is convenient to use, in the expression for the reduced shear rate, the longest relaxation time, θ_m, which is given, within a certain constant, by the following expression:

$$\theta_m = \frac{\eta_0 M^a}{T^b c^k} \tag{2.89}$$

where c is the concentration; the values of the constants a, b, and k depend on the choice of the model of the polymer system. Thus, in the Bueche theory all the constants are each assumed to be equal to unity, whereas in the Williams theory $a = 0$, $b = 1$, $k = 2$.

If the temperature dependence of viscosity for polymer melts is considered, in which case $M = $ const and $c = $ const, then in the expression for θ_m only the ratio η_0/T^b is essential. But T^b does not practically change by more than 1.5 times, whereas η_0 may vary by decimal orders. Therefore, at a first and often sufficient approximation it may be assumed that θ_m is proportional to η_0 and the reduced shear rate is equal to $(\dot{\gamma}\eta_0)$. This expression is easily transformed in the following manner: $\dot{\gamma}\eta_0 = (\tau/\eta)\,\eta_0 = \tau/(\eta/\eta_0)$. Accordingly the ratio (η/η_0) is found to be a unique function of τ. A specific feature of the dependence of (η/η_0) on τ is that it gives a more steeply descending viscosity curve and therefore proves less convenient than the dependence of (η/η_0) on $(\dot{\gamma}\eta_0)$.

Since $\eta/\eta_0 = F(\tau)$, it follows that for each given polymer system the temperature dependence (η/η_0) (this is also valid for η at $\tau = $ const) is unambiguously determined by the temperature dependence of η_0. The form of the function $f(\dot{\gamma}\eta_0)$ does not depend on temperature[*], which allows us to consider it temperature-invariant. An example of such dependence is presented in Fig. 2.54. In the case under consideration T varies by 1.4 times, and viscosity η_0 varies by 2300 times.

The method of constructing a master curve for the temperature-invariant description of viscosity enables one to determine the value of η at various $\dot{\gamma}$ for various temperatures if the temperature dependence of η_0 and dependence of apparent viscosity on the shear rate at any one temperature are known.

The above-described method of generalization of experimental data on the dependences of viscosity on shear rate and temperature may be extended to the case where the viscosity depends also on pressure.[**] An example is shown in Fig. 2.55. The effect of pressure and temperature brings about a change in the value of η_0 but not in the form of the flow curve, which is only shifted parallel to itself along the coordinate axes of log η and log $\dot{\gamma}$ by an amount dependent on temperature and pressure.

---

[*] Vinogradov, G. V., A. Ya. Malkin, N. V. Prozorovskaya, and V. A. Kargin, *Doklady Akad. Nauk SSSR*, 150, 574 (1963); Vinogradov, G. V. and A. Ya. Malkin, *J. Polymer Sci.*, A2, 2357 (1964).
[**] Semjonow, V., *Adv. Polymer Sci.*, 5, 387 (1968).

The problem of the generalized representation of the flow properties of polymer systems is considerably complicated when comparing polymers of different nature. This is seen even from examples of polymers within a single homologous series, which have different MMD. It has already been indicated (see Sec. 2.6.7) that the viscosity of high-molecular-mass linear polymers of narrow MMD is but slightly dependent on the shear stresses,

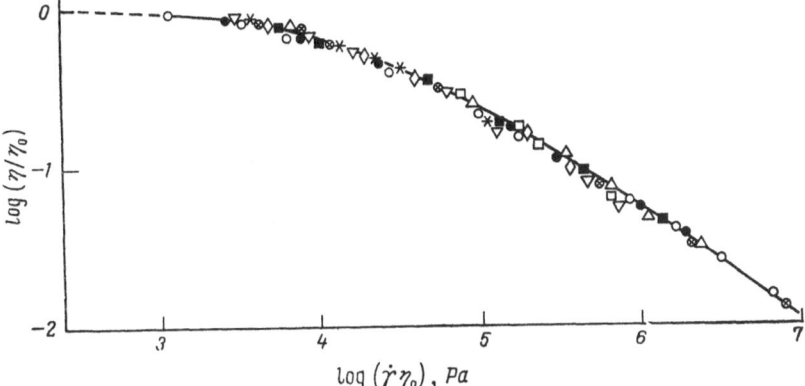

Fig. 2.54. Temperature-invariant curve of the viscosity of polyisobutylene. The different symbols correspond to different temperatures in the range from —20 to 80°C. [Mustafaev, E., A. Ya. Malkin, E. P. Plotnikova, and G. V. Vinogradov, *Vysokomol. Soedin.*, 6, 1515 (1964).]

up to their high values, whereas in the case of polydisperse polymers the non-Newtonian behaviour may be very sharply pronounced. It is clear that the flow curves of narrow- and wide-distribution polymers cannot be similar

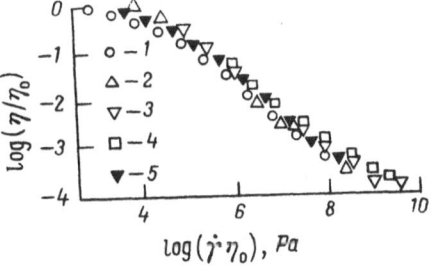

Fig. 2.55. Temperature- and piezo-invariant curve of the viscosity of molten polystyrene:

1-4 correspond to the temperature 165°C and the pressures (in Pa): 1×10^5, 5×10^6, 1×10^8, 1.5×10^8; *5*—temperature 200°C and pressure 5×10^7 Pa. [Hellwege, K. H. *et al.*, *Rheol. Acta*, 6, 105 (1967).]

and, hence, do not converge when shifted parallel to themselves in the coordinate axes of $\log \dot{\gamma}$ and $\log \tau$ or $\log \eta$ and $\log \dot{\gamma}$. This is illustrated by experimental data referring to polybutadienes of narrow and wide MMD (Fig. 2.56). What has been said can be explained by comparing the relaxation characteristics of polymer systems of different MMD. The point is that the above-given expression for θ_m is valid for monodisperse polymers and it may not be extended to wide-distribution polymers the more so that it is unknown what quantity characterizing MMD should be substituted into the expression for θ_m.

The flow curves must be similar for polymer systems with similar relaxation spectra. This requirement is satisfied, to a crude approximation, by many of the commercially available polydisperse polymers. Indeed, in the coordinates (η/η_0) and $(\dot{\gamma}\eta_0)$ on a log-log scale the flow properties of commercial polydisperse polymers are found to be represented by a narrow band* (Fig. 2.57). The solid lines delineating this band give departures from the averaged curve (the dashed line), which are maximally two-fold. The dashed

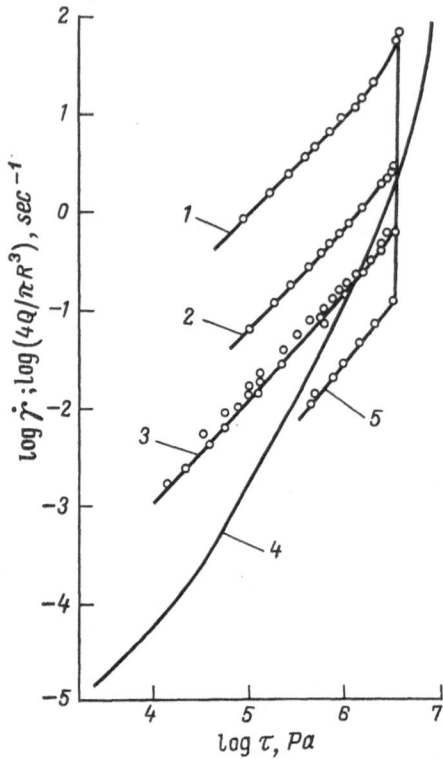

Fig. 2.56. Plots of rate of shear or volume rate of flow versus shear stress for polybutadienes of narrow and wide MMD:

1, 2, 3, 5—polybutadienes of narrow MMD and molecular masses ($M \times 10^{-4}$), respectively: 6.7; 15; 20; 24; 4—commercial polybutadiene.

line in Fig. 2.57 describes the properties of melts of many thermoplasts and elastomers. It is approximated by the following empirical formula:

$$\eta/\eta_0 = [1 + 0.4\,(\dot{\gamma}\eta_0/1.32 \times 10^4)^{0.355} + (\dot{\gamma}\eta_0/1.32 \times 10^4)^{0.71}]^{-1} \qquad (2.90)$$

where $(\dot{\gamma}\eta_0)$ has the dimensions of Pa. In the region of high shear rates this formula corresponds to a power dependence of viscosity on the parameter $(\dot{\gamma}\eta_0)$ with an exponent equal to -0.71. If it is necessary to finalize this formula for any particular polymer one must use** the values of the constants that deviate somewhat from the "universal" quantities contained in formula (2.90).

Using the dashed line in Fig. 2.57, which determines the universal characteristic of the viscosity of a polymer in bulk, one can obtain the function

* Vinogradov, G. V., A. Ya. Malkin, N. V. Prozorovskaya, and V. A. Kargin, *Doklady Akad. Nauk SSSR*, 154, 890 (1964); Vinogradov, G. V. and A. Ya. Malkin, *Zhurn. Prikl. Mekh. i Tekhn. Fiz.*, 5, 66 (1964); *J. Polymer Sci.*, A2, 4, 135 (1966).

** Kalinchev, E. L. and M. B. Sakovtseva, *Plastmassy*, 11, 30 (1975); 5, 32 and 7, 13 (1977).

$\eta/\eta_0 = F(\tau)$, given in Fig. 2.58, which is universal to the same approximation as the function $\eta/\eta_0 = f(\dot{\gamma}\tau_0)$.

The possibility of representing the flow properties of polydisperse polymers in a temperature-invariant form has been repeatedly tested for the last 15 years by many authors for a large variety of polymers. This method has proved useful for the engineering evaluation of apparent viscosity at various

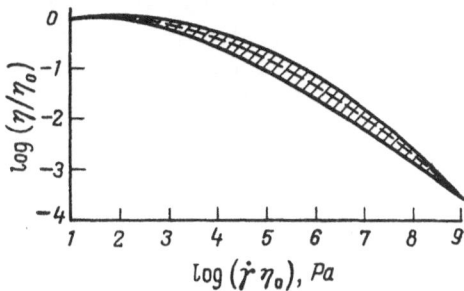

Fig. 2.57. Master curve for the flow properties of melts of commercial thermoplasts and rubbers of wide MMD. The hatched area is the spread of experimental data.

shear rates or for the determination of the shear rate at which the viscosity reaches a specified level.

The quantity η_0 plays an important part in the use of the temperature-invariant viscosity curve. If use is made of the universal dependence $\eta/\eta_0 = F(\tau)$, then on the basis of a single measurement of apparent viscosity it is possible to approximately evaluate η_0, which is often done with satis-

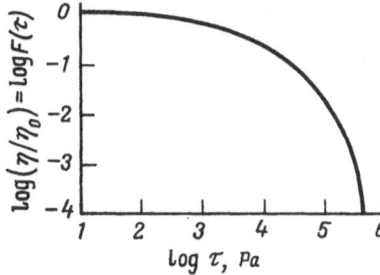

Fig. 2.58. The averaged master curve for the flow properties of melts of thermoplasts and rubbers of wide MMD, represented in the form of the function $F(\tau)$ (corresponds to the dashed line in Fig. 2.57).

factory reliability.* It should be noted, in this connection, that the use of the function $F(\tau)$ makes it possible to estimate the temperature dependence of η_0 if $\eta(\tau)$ has been measured at different temperatures.

The phenomenological approach to the solution of many problems of polymer rheology remains to be efficient up to the present time and this applies fully to the question of the generalized flow characteristics of polymer solutions and melts considered here. Thus, a useful procedure is the use of the concept of the initial relaxation time. It can be determined for the initial state of the polymer when $\dot{\gamma} \to 0$. It is equal to the ratio of the initial values of viscosity η_0 and the rubbery modulus G_0 (see Chapter 5). Such an estimation of the initial relaxation time θ_0, which characterizes the properties of a polymer system, is convenient** since it does not require the use of

* Malkin, A. Ya. and G. V. Vinogradov, *J. Appl. Polymer Sci.*, **10**, 767 (1966).
** Vinogradov, G. V., A. Ya. Malkin, and G. V. Berezhnaya, *Vysokomol. Soedin.*, A13, 2793 (1971).

any additional theoretical conceptions for this purpose. Therefore, the quantity θ_0 can be found from the following relations:

$$\theta_0 = \eta_0/G_0 = \zeta_0/\eta_0 \tag{2.91}$$

Here ζ_0 is the initial value of the normal stress coefficient (see Chapter 4).

The representation of the reduced shear rate in the form of $(\dot{\gamma}\theta_0)$ must be of greater generality than that of $(\dot{\gamma}\eta_0)$. Indeed θ_0 is expressed in terms of two independently determined parameters η_0 and G_0, which are differently related with such fundamental characteristics as molecular mass, molecular-mass distribution (MMD), concentration and temperature. It should be kept in mind that the modulus G_0 is rather sensitive to changes of MMD and concentration. Only in those cases when molten polymers of wide MMD and high average molecular mass are dealt with, does the quantity G_0 vary within narrow limits. This justifies the use of the reduced shear rate in the form of $(\dot{\gamma}\eta_0)$ for constructing master curves for commercial polymers. Moreover, the value of G_0 for polymer melts depends very slightly on temperature. Hence, for their temperature-invariant curves to be obtained, use may be made of the quantity $(\dot{\gamma}\eta_0)$. It is most expedient to use the general expression for the reduced shear rate in the form of $(\dot{\gamma}\theta_0)$ for comparing the characteristics of the flow properties of polymers of different MMD and polymer solutions. The latter case is illustrated by the data presented in Fig. 2.59.

The method of representation of the flow properties of polymer systems in a generalized form proves also applicable, in some cases, to solutions of rigid-chain polymers.* This method, however, has not yet been sufficiently tested. For instance, for polymer solutions there are no data on quantitative comparison of the quantity θ_0 with the concentration of the solution, which could limit the applicability of the method since it is necessary to measure not only η_0 but also the elastic modulus G_0, which is not always easy to determine. In practice, it would be more preferable to make use of formula (2.89), but the question of the values of the constants a, b, and k remains to be controversial.

The possibility of using different approaches to the estimation of the quantity θ_0 and also the fact that the use of the Newtonian viscosity for the evaluation of the reduced shear rate does not always allow one to construct master curves for the flow properties of polymer systems explain why the "relaxation time" is found by purely empirical means: graphs of (η/η_0) versus $\dot{\gamma}$ are plotted in logarithmic coordinates and then the curves are shifted along the log $\dot{\gamma}$ axis until a superposition is achieved. The use of such an approach for linear and branched polyethylenes of different MMD has shown** that it is possible to achieve a superposition of flow curves for linear polymers. But master curves for the flow properties of linear and branched polyethylenes are found to be different. This demonstrates the limited effectiveness of the procedure under consideration for polymers that differ in the structure of the macromolecular chain.

In connection with the discussion of the master curve of the viscosity of polymers, it should be emphasized that the possibility of generalization of

---

* Lobanova, G. A. and E. A. Pakshver, *Khim. Volokna*, 3, 29 (1977).
** Mendelson, R. A., W. A. Bowles, and F. L. Finger, *J. Polymer Sci.*, A-2, 8, 127 (1970).

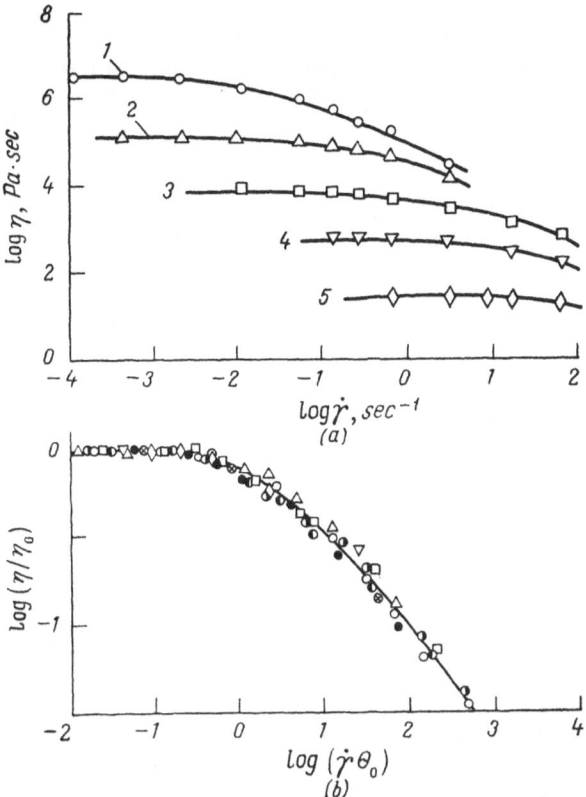

Fig. 2.59. Viscosity plotted against rate of shear for solutions of polystyrene in decalin in an ordinary form (a) and in generalized coordinates (b):
curves *1-5* correspond to the polymer contents in solution, φ_2, at 25°C (volume fractions): 0.573, 0.466, 0.380, 0.290, 0.184.

data on the dependences of apparent viscosity on temperature, concentration, and shear rate rests, in an implicit way, on the idea that the variation of apparent viscosity is predetermined by the properties of the polymer in the corresponding initial state, i.e., by the quantities η_0 and θ_0. Therefore, the generalization of the flow characteristics of polymer solutions and melts by means of the procedures considered above is possible when their relaxation spectra are similar in the initial state and vary on deformation in an analogous manner.

2.11. The Rule of Logarithmic Additivity

The possibility of using the reduced coordinates (η/η_0) and $(\dot{\gamma}\eta_0)$ for generalizing the flow properties of polymer melts leads to the useful rule of separate estimation of the effect of shear stress, temperature, and molecular mass on viscosity. Indeed, it has been pointed out above that

$$\eta = \eta_0 F(\tau)$$

$$\eta = \eta_0 F\,(\tau)$$

At the same time, Fox and coworkers found long ago that the dependence of η_0 on molecular mass and temperature may be represented by the product of two independent functions:

$$\eta_0 = f_1^!(T) \cdot f_2^!(M)$$

This is accounted for by the fact that the activation energy of viscous flow is independent of molecular mass. Then,

$$\eta = f_1\,(T) \cdot f_2\,(M) \cdot F\,(\tau) \qquad\qquad\qquad (2.92)$$

The last expression may be called, for obvious reasons, the rule of logarithmic additivity.*

What has been said above concerning the generalization of the dependences of viscosity on the shear rate at various pressures allows us to write the expression for η in a more general form:

$$\eta = f_1\,(T) \cdot f_2\,(M) \cdot F\,(\tau) \cdot F_1\,(p)$$

The rule of logarithmic additivity may be given a concrete form if use is made of some of the results obtained above and the functions $f_1\,(T)$, $f_2\,(M)$, and $F\,(\tau)$ involved are represented in the form of those universal expressions which, according to the existing theoretical concepts and available experimental data, are applicable to a wide range of industrially important polymers. Indeed, the temperature dependence of viscosity is described in many cases by means of the WLF formula with universal values of the constants, which is expressed, within a constant term, by the formula:

$$\log\!f_1\,(T) = \frac{C_{1g}\,(T - T_g)}{C_{2g} + T - T_g}$$

The dependence of viscosity on molecular mass is written in the following form, also within a constant term:

$$\log f_2\,(M) = 3.4 \log\,(M/M_c)$$

The function $F\,(\tau)$ contained in formula (2.92) is represented by the dependence shown in Fig. 2.58. Combination of what has been said above regarding the functions $f_1\,(T)$, $f_2\,(M)$, and $F\,(\tau)$ with the analytical expression of the rule of logarithmic additivity leads to the following formula**:

$$\log \eta = \frac{C_{1g}\,(T - T_g)}{C_{2g} + T - T_g} + 3.4 \log\,(M/M_c) + \log F\,(\tau) + \log \eta_g \qquad (2.93)$$

The constant η_g figuring in this formula includes the terms omitted in the expressions for $f_1\,(T)$ and $f_2\,(M)$. The physical significance of this constant becomes clear from the following considerations: let $T = T_g$ and $M = M_c$; then it is seen that η_g represents the Newtonian viscosity of a polymer of the corresponding structure and molecular mass $M = M_c$ at its glass-transition temperature. If we now take into account that the condition for the vitrification of a liquid with molecular mass M_c, which has not yet acquired the characteristic properties of high-molecular-mass compounds, may be assumed to be the attainment of a viscosity of the order of 10^{12} Pa·sec, it

---

  * Bartenev, G. M. and L. A. Vishnitskaya, *Vysokomol. Soedin.*, **6**, 751 (1964).
 ** Malkin, A. Ya. and G. V. Vinogradov, *Vysokomol. Soedin.*, **7**, 1134 (1965).

will then be found that the constant η_g in formula (2.93) has a universal value.

Formula (2.93) shows the relations among the various factors determining the viscosity of a given polymer under chosen measurement conditions: the viscosity depends on the difference between the temperature of the experiment and the glass-transition temperature, on the length of the macromolecular chain, expressed by the number of equivalent segments (where the length of the segment corresponds to the critical molecular mass), and the level of acting stresses. With the aid of this formula it is possible to take into account the dependence of viscosity on hydrostatic pressure, for which purpose it will suffice to consider the effect of this factor on the glass-transition temperature T_g contained in Eq. (2.93).

The above-discussed results of the generalization of available experimental evidence and existing theoretical conceptions concerning the flow properties of various polymeric systems describe, in a general form, the effect of various factors on the viscosity of polymers, though formula (2.93) can hardly be used for practical estimations of viscosity because of the approximate and averaged character of the constants involved.

CHAPTER 3

Viscoelastic Properties of Polymer Melts and Solutions

3.1. Viscoelastic Properties of Dilute Solutions

3.1.1. Introduction

This chapter (primarily Sections 3.1 and 3.2) is mainly concerned with the use of mechanical models as analogues of the behaviour of dilute and concentrated polymer solutions or molten polymers, which predict the relaxation phenomena observed on deformation and their regularities. The theory of the relaxation properties of the material in this approach is based on the solution of hydrodynamic problems, which include the scrutiny of the behaviour of mechanical models in the simplest mechanical fields— on shear, oscillatory loading, extension, etc.—with account taken of kinetic (or statistical) phenomena. Therefore, such models may be called kinetic. The simplest, though far from realistic, model of the polymer molecule in solution is a bead. A more complex model is an ellipsoid; even more complex is the dumbbell model (two beads connected by a rigid rod or spring). Further finalizing is achieved through the use of a model in the form of a chain of interconnected beads. Rigid-chain macromolecules are modelled by means of rods, etc. A large number of diverse variants of such mechanical models whose component parts are variously connected are possible. It is evident that the behaviour of such systems is varied. Accordingly, the predictions made on the basis of various kinetic models of polymers and |their solutions are different.

The general task of kinetic theories based on mechanical models of polymers is a comparison of the actually observed behaviour with predictions following from the analysis of the behaviour of a model and the estimation of the role of the constants introduced to characterize the elements of the models, which are supposed to correspond to certain molecular parameters of the real polymeric chain. Introduction of the various new elements into the model immediately leads to the prediction of new mechanical phenomena, which may be either present or absent on deformation of a particular material. A model is made more and more complex (or is finalized) until one arrives at an undoubted qualitative and suitable quantitative concordance between the phenomena involved and the regularities following from a theoretical analysis of the behaviour of the mechanical model and the properties of the polymer.

This chapter deals chiefly with the physical ideas and conceptions used to formulate and construct kinetic models simulating the behaviour of polymers and their solutions; the calculating methods employed are not practically discussed here; they may be highly sophisticated and therefore comprise the subject of special investigations*.

---

* These problems are treated in greater detail in Bird, R. B., O. Hassager, R. C. Armstrong, and C. F. Curtiss, *Dynamics of Polymeric Liquids*, vol. 2, *Kinetic Theory*, John Wiley and Sons, New York, 1977, which is considered by the authors themselves to be

Diverse manifestations of the relaxation (viscoelastic) properties are the most important feature of the behaviour of polymeric systems. Therefore, since the thirties numerous attempts have been made to account for and to describe quantitatively the relaxation properties of polymeric systems on the basis of the fundamental features of their structure and the most general considerations pertaining to the regularities of the development of deformation.

The total deformation γ is made up of the instantaneous elastic component γ_0, the delayed (rubbery or high-elastic) component γ_e and the plastic (viscous-flow) component γ_f:

$$\gamma = \gamma_0 + \gamma_e + \gamma_f \tag{3.1}$$

This is represented by a mechanical model (Fig. 3.1) which is known as the Burgers-Frenkel model. Here the spring simulates the instantaneous elastic deformation; the element consisting of a spring and a dashpot connected in parallel represents the delayed deformation; and the lower dashpot simulates the viscous resistance to deformation. The displacement of each of the component parts simulates the relative deformation and the force required for this corresponds to the stress. A special feature of the Burgers-Frenkel model is that each component of the total deformation is related linearly to the stress:

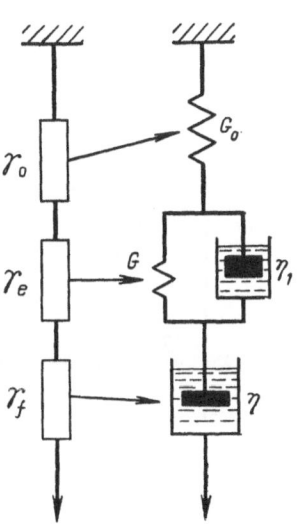

Fig. 3.1. The Burgers-Frenkel model.

$$\gamma_0 = \frac{\tau}{G_0}; \quad \eta_1 \dot{\gamma}_e + G\gamma_e = \tau; \quad \dot{\gamma}_f = \frac{\tau}{\eta}$$

In order to derive a rheological equation of state (a constitutive equation) corresponding to the Burgers-Frenkel model, it is necessary to replace the deformation components in Eq. (3.1) with stresses. Appropriate transformations lead to the following differential equation which describes the properties of a viscoelastic liquid:

$$\tau + \left(\frac{G_0}{\eta_1} + \frac{G}{\eta_1} + \frac{G_0}{\eta}\right)\dot{\tau} + \frac{G_0 G}{\eta \eta_1}\tau = G_0\dot{\gamma} + \frac{G_0 G}{\eta_1}\dot{\gamma} \tag{3.2}$$

A quite analogous equation is obtained when considering a model built up of two Maxwell elements connected in parallel. If we equate the coefficients before the terms of the same order of magnitude in the two models being compared, one series of constants may be expressed in terms of another. Thus, the Burgers-Frenkel model is mathematically quite equivalent to the model consisting of two Maxwell elements connected in parallel. In particular, both models have two relaxation times and one retardation time. But the Burgers-Frenkel model is more preferable since it provides a more spectacular modelling of the mechanical processes taking place during deformation, and it is for this reason that this model was used as the basis for further generalizations which led to the idea of the relaxation spectrum arising during the deformation of polymeric systems.

"an introduction to the kinetic theory of polymers" or "an introductory textbook on this subject". This once again stresses the complexity of the problems of molecular-kinetic (statistical) interpretation and prediction of the behaviour of real polymeric materials.

3.1.2. The Bead Model (the Kargin-Slonimsky-Rouse Model)

The basic supposition that enables one to quantitatively describe the relaxation properties of a polymer is the idea of the possibility of regarding the polymer molecule as a set of identical elements connected in series— segments, each of which is deformed independently of the others, and the condition of the continuity of the chain is provided by the connection of the segments. The viscoelastic properties of all segments are assumed to be identical and it is presumed that the specificity of the behaviour of each segment is reflected by the Burgers-Frenkel model. The division of the real polymeric chain into segments is provisional and is associated with the assumptions that are not always valid and can be verified, to a certain ex-

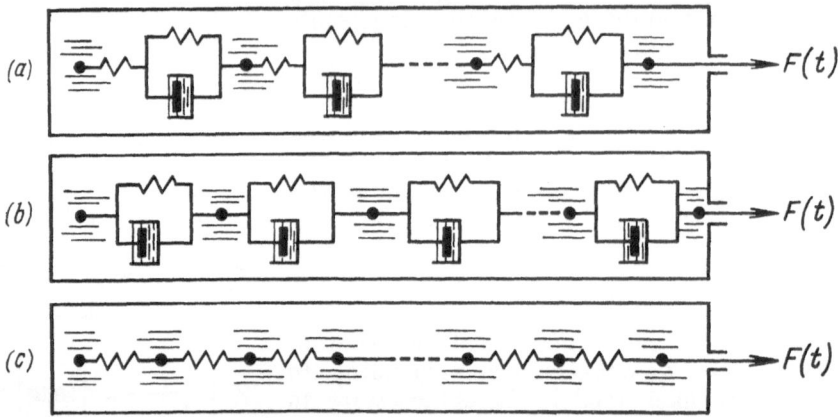

Fig. 3.2. The Kargin-Slonimsky-Rouse (KSR) model:
a—the complete model with account taken of instantaneous elastic deformations and inner viscosity; b—the model without account of instantaneous deformation; c—the bead model without account of inner viscosity.

tent, only by the agreement between theoretical predictions of the properties of the model used and experimental data.

The viscous resistance to the displacement of the dashpot is identified with the interaction between the macromolecule and the surrounding medium, which hinders its movement relative to the solvent or other polymer molecules in the system. This is schematically shown in Fig. 3.2a. If we assume that the modulus G_0, which is responsible for the instantaneous elastic deformation, is large as compared with the modulus G expressing rubber-like deformations (which corresponds to the actual case), then a simpler model results, which is shown in Fig. 3.2b. The elements of the models a and b in Fig. 3.2 describe the delayed strain associated with the imperfect elasticity of the polymeric chain. It may be assumed that the energy dissipation associated with the spring-dashpot pair connected in parallel as shown in Fig. 3.2a and b characterizes the internal viscosity of the polymer molecule. If we disregard the resistance to deformation due to the internal viscosity of the chain as compared with the interaction with the medium, the model will be further simplified (Fig. 3.2c).

All these variants of the model of the polymer chain have been proposed by Kargin and Slonimsky* and examined in detail by Gotlib and Volkenshtein.** An analogous model was considered by Rouse***, who brought the relevant calculations up to the computation of the case of harmonic loading, which is especially important in practice. This model is called the bead model. Further in the text, any of the variants of the model of the polymeric chain shown in Fig. 3.2 will be called the Kargin-Slonimsky-Rouse model (abbreviated to the KSR model).

Figure 3.2 shows the unidimensional model. Therefore, it must reflect the specific features of the behaviour of the projection of the polymer chain onto the chosen direction in space. The extension of this model to a three-dimensional case leads to the prediction of the existence of normal stresses on shear.

Suppose that an external force dependent on the time $F(t)$ is applied to one of the ends of the chainlike model shown in Fig. 3.2b. The response of the KSR model to the externally applied force is progressing in the following manner (as shown in the works cited here).

The resistance to the movement of an ith bead in the KSR model is expressed by the product of the viscosity of the medium, η, times the velocity of that bead, i.e., by the quantity $\eta \, (du_i/dt)$, where (du_i/dt) is the time derivative of its coordinate. The force acting on the spring with modulus G is equal to the product of G by the difference in the coordinates of the spring ends since this difference is the amount of relative deformation (extension) of the spring. The force acting on the dashpot with viscosity η_1 is equal to the product of the viscosity η_1 by the difference in the rates of displacement of the beads surrounding the dashpot.

The equilibrium condition for the first bead subjected to the action of the indicated forces may be written in the form of the following equation:

$$\eta \frac{du_1}{dt} + \eta_1 \left(\frac{du_1}{dt} - \frac{du_2}{dt} \right) + G\,(u_1 - u_2) = 0 \tag{3.3}$$

The ith bead is acted on by the medium and the spring-dashpot pair both from the left and right sides. Therefore, the equilibrium condition for the ith bead assumes the form

$$\eta \frac{du_i}{dt} + \eta_1 \left[\left(\frac{du_i}{dt} - \frac{du_{i+1}}{dt} \right) - \left(\frac{du_{i-1}}{dt} - \frac{du_i}{dt} \right) \right] +$$
$$+ G\,[(u_i - u_{i+1}) - (u_{i-1} - u_i)] = 0; \quad i = 2,\ 3,\ \dots,\ n$$

or

$$\eta \frac{du_i}{dt} + \eta_1 \left(2\frac{du_i}{dt} - \frac{du_{i-1}}{dt} - \frac{du_{i+1}}{dt} \right) + G\,(2u_i - u_{i-1} - u_{i+1}) = 0 \tag{3.4}$$

$$i = 2,\ 3,\ \dots,\ n$$

Finally, the last $(n+1)$st bead. apart from being influenced by the medium and the ith spring-dashpot pair, is also subjected to the action of an external force, and therefore the equilibrium condition for it has the form

$$\eta \frac{du_{n+1}}{dt} + \eta_1 \left(\frac{du_{n+1}}{dt} - \frac{du_n}{dt} \right) + G\,(u_{n+1} - u_n) = -F(t) \tag{3.5}$$

* Kargin, V. A. and G. L. Slonimsky, *Doklady Akad. Nau ₒSSR*, **62,** 239 (1948); *Zhurn. Fiz. Khim.*, **23,** 563 (1949).
** Gotlib, Yu. Ya. and M. V. Volkenshtein, *Zhurn. Tekh. Fiz.*, **23,** 1936 (1953).
*** Rouse, P., *J. Chem. Phys.*, **21,** 1272 (1953).

In a simplified variant of the KSR model, which is shown in Fig. 3.2 c, it is assumed that $\eta_1 = 0$, in which case the terms of the above equations that contain η_1 are dropped.

The system of equations (3.3) through (3.5) define the law of motion of the polymer chain under the action of a specified force. If we now subtract, from each of the equations given above, the subsequent one (of course, with the exception of the last one), the ith equation will be written in the following form:

$$\eta \frac{d\,(u_i - u_{i+1})}{dt} + \eta_1 \left[2 \frac{d\,(u_i - u_{i+1})}{dt} - \frac{d\,(u_i - u_{i+2})}{dt} - \frac{d\,(u_{i-1} - u_i)}{dt} \right] +$$

$$+ G\,[2\,(u_i - u_{i+1}) - (u_{i+1} - u_{i+2}) - (u_{i-1} - u_i)] = 0$$

Instead of u_i, new variables may be introduced:

$$y_i = u_i - u_{i+1}$$

which are called normal coordinates. In normal coordinates the system of equations (3.3) through (3.5) becomes

$$
\begin{cases}
\eta \dot{y}_1 + \eta_1\,(2\dot{y}_1 - \dot{y}_2) + G\,(2y_1 - y_2) = 0 \\
\eta \dot{y}_i + \eta_i\,(2\dot{y}_i - \dot{y}_{i+1} - \dot{y}_{i-1}) + G\,(2y_1 - y_{i+1} - y_{i-1}) = 0 \\
\qquad i = 2,\ 3,\ \ldots,\ (n-1) \\
\eta \dot{y}_n + \eta_1\,(2\dot{y}_n - \dot{y}_{n-1}) + G\,(2y_n - y_{n-1}) = F\,(t)
\end{cases}
\qquad (3.6)
$$

The above n linear differential first-order equations form a system of unknown functions y_i for n. The solution of the corresponding homogeneous system of equations (minus the right-hand side) is well known. It has the form

$$y_i = \sum_{\mu=1}^{n} C_{\mu i} e^{-s_\mu t}$$

where $C_{\mu i}$ and s_μ are the constants to be determined.

Substitution of the expressions for y_i into the system of homogeneous differential equations under consideration leads to a system of algebraic equations for s_μ, whose solution gives the sought-for values:

$$s_\mu = \frac{4G \left[\sin \dfrac{\mu\pi}{2\,(n+1)} \right]^2}{\eta + 4\eta_1 \left[\sin \dfrac{\mu\pi}{2\,(n+1)} \right]^2} \qquad (3.7)$$

At $\eta_1 = 0$ the values of s_μ are expressed as

$$s_\mu = \frac{4G}{\eta} \left[\sin \frac{\mu\pi}{2\,(n+1)} \right]^2 \qquad (3.7a)$$

We can now write the final solution of the above-formulated problem of finding y_i:

$$y_i = \sum_{\mu=1,\,3,\,5}^{n} \frac{4G \sin \dfrac{\mu\pi}{n+1} \cdot \sin \dfrac{i\mu\pi}{n+1}}{(n+1) \left\{ \eta + 4\eta_1 \left[\sin \dfrac{\mu\pi}{n+1} \right]^2 \right\}} \left[\int_0^t -e^{-s_\mu(t-t')} F\,(t')\,dt' \right] \qquad (3.8)$$

With a constantly acting force $F\,(t) = F_0 = $ const the time dependence of the displacement of the chain end, i.e., of the $(n+1)$st bead, will be given by

$$u_{n+1}(t) = \frac{F_0 t}{\eta(n+1)} + F_0 \sum_{\mu=1}^{n} B_\mu (1 - e^{-t/\lambda_\mu}) \tag{3.9}$$

where the structure of the quantities $\lambda_\mu = s_\mu^{-1}$ and B_μ can easily be obtained from Eqs. (3.7) and (3.8).

From the form of the resulting relation (3.9) it is seen that the total displacement is made up of the first term of formula (3.9), which depends linearly on time, and the sum of the terms which depend on time exponentially. The displacement of the chain end is composed of the displacement of the centre of gravity (irrecoverable flow) and the movement of the chain end relative to the centre of gravity u_e (elastic displacement). After the external force is removed the chain end returns by the amount u_e, while the centre of gravity of the chain stays in the position reached during the period of the action of the external force. It is easy to see that the sum in formula (3.9) corresponds to recoverable deformation, and the constants λ_μ represent the retardation times of the system.

For the simplified variant of the model shown in Fig. 3.2c

$$\lambda_\mu = \frac{\eta}{4G} \left[\sin \frac{\mu\pi}{2(n+1)} \right]^{-2} \tag{3.10}$$

$$B_\mu = \frac{1}{2G(n+1)} \left[\cot \frac{\mu\pi}{2(n+1)} \right]^2 \tag{3.10a}$$

Thus, the presence of a set of identical elements ("segments") having identical properties, which are connected into a chain, leads by itself to the appearance of the retardation time spectrum λ_μ, the number of lines in the spectrum being determined by the number, n, of segments in the chain. Thus, if $n = 2$ (the so-called "dumbbell" model), then there exists only one retardation time λ_0 equal to $\eta/2G$.

For a chain consisting of $(n + 1)$ segments (beads connected by springs or a spring-dashpot pair), the number of retardation times is n. The range of distribution of retardation times for a long chain containing a large number of segments $(n \gg 1)$ is determined by the extreme values:

$$\lambda_{\min} = \frac{\eta}{4G} \left[\sin \frac{n\pi}{2(n+1)} \right]^{-2} \approx \frac{\eta}{4G} = \frac{\lambda_0}{2}$$

and

$$\lambda_{\max} = \frac{\eta}{4G} \left[\sin \frac{\pi}{2(n+1)} \right]^{-2} \approx \frac{(n+1)^2}{\pi^2} \frac{\eta}{G} = \lambda_0 \frac{2(n+1)^2}{\pi^2} \tag{3.11}$$

where $\lambda_0 = \eta/2G$.

Thus, the minimum retardation time $\lambda_0/2$ is almost independent of the total chain length; this result is quite natural since the most rapid relaxation processes occur due to the motion (displacement) of a separate segment. The maximum value of the retardation time increases as the square of the number of segments in the chain, i.e., as the square of the chain length; this is associated with the fact that the slowest relaxation process takes place during the displacement of the entire molecular chain.

An important feature of the spectrum λ_μ obtained for large n is that, according to Eq. (3.10), all the values of the retardation time can be expressed in terms of the maximum retardation time $\lambda_{\max}$ as follows:

$$\lambda_\mu = \lambda_{\max} \mu^{-2} \tag{3.12}$$

This formula holds for the retardation time λ_μ with the subscript less than $n/5$.

The retardation time spectrum expressed by a set of values of λ_μ and the corresponding constants B_μ is discrete. If n is sufficiently large, the change-over to a continuous spectrum can be made by replacing the summation in the above-given formulas by integration. In the region of short times or high frequencies, in which the replacement of a discrete (line) spectrum with a continuous spectrum is correct, the function $\Phi(\lambda)$ may be represented in the following manner:

$$\Phi(\lambda) \propto \lambda^{-1/2}$$

Or for the function $L(\lambda)$

$$L(\lambda) \propto \lambda^{3/2}$$

Hence, the spectral functions of relaxation time distribution $F(\theta)$ and $H(\theta)$ [see formula (1.86)] are found to be proportional:

$$F(\theta) \propto \theta^{-3/2}; \quad H(\theta) \propto \theta^{1/2}$$

The proportionality constant is found from any integral property of the spectrum; integration should be carried out from λ_{min} to λ_{max}.

The segmental friction factor is related in a simple way to the viscosities of the solution, η_0, and the solvent, η_s:

$$\eta(n+1) = \eta_0 - \eta_s$$

or

$$\eta = (\eta_0 - \eta_s)/(n+1) \tag{3.13}$$

When the deformation regime $F_0 = $ const is specified, irreto erable deformations increase infinitely in time. Elastic (rubbery) deformations also increase, approaching the asymptotic value of $u_e(\infty)$, the value of which at $t \to \infty$ is as follows according to formula (3.9):

$$u_e(\infty) = F_0 n/3G$$

Hence, rubbery modulus $G_e = \lim_{t \to \infty}(F_0/u_e)$ is expressed in the following manner:

$$G_e = F_0/u_e(\infty) = 3G/n$$

The rubber elasticity of cross-linked (cured) elastomers (rubbers) is associated by its nature with the Brownian motion of separate portions of the chain. The same mechanism is responsible for the elasticity of polymer chains that are not united into a network by chemical bonds. Therefore, in both cases the modulus G_e is proportional to the kinetic factor (kT), where k is Boltzmann's constant, T is the absolute temperature, and N is the number of chains per unit volume of the polymeric system. Therefore, within a certain constant factor a

$$G_e = akTN$$

and

$$G = \frac{1}{3}akTNn \tag{3.14}$$

Substitution of formulas (3.13) and (3.14) into the expression for the maximum retardation time λ_{max} gives according to Eq. (3.11):

$$\lambda_{max} = \frac{(n+1)^2}{\pi^2} \cdot \frac{\eta}{G} = \frac{3(\eta_0 - \eta_s)}{a\pi^2 kTN} \propto \frac{3(\eta_0 - \eta_s)^{a}}{\pi^2 NkT}$$

And, as before, the remaining values of the retardation times λ_μ are expressed in terms of λ_{max} by formula (3.12). The last expression exactly corresponds to the longest relaxation time* θ_m if we assume the proportionality constant a to be equal to 1/2.

Thus, in the KSR model the maximum relaxation time θ_m written in terms of the macroscopically measured characteristics of polymeric systems—the viscosity η_0 and the content of macromolecules in the solution, N—is expressed in the following way:

$$\theta_m = \frac{6(\eta_0 - \eta_s)}{\pi^2 NkT} = 0.608 \frac{\eta_0 - \eta_s}{NkT} \tag{3.15}$$

and the remaining relaxation times θ_p (at $p < n/5$), just like λ_μ n Eq. (3.12), are expressed in terms of θ_m as follows:

$$\theta_p = \theta_m p^{-2} \tag{3.16}$$

On the basis of the results obtained in Chapter 1, the relaxation spectrum predicted by the KSR model may be given in the form

$$F(\theta) = NkT \sum_{p=1}^{n} \delta(\theta - \theta_p)$$

or

$$H(\theta) = NkT \sum_{p=1}^{n} \theta_p \delta(\theta - \theta_p)$$

The components of the complex modulus are written in the following manner:

$$\left\{ \begin{array}{l} G'(\omega) = NkT \sum_{p=1}^{n} \dfrac{(\omega\theta_p)^2}{1 + (\omega\theta_p)^2} \\[4mm] G''(\omega) = NkT \sum_{p=1}^{n} \dfrac{(\omega\theta_p)}{1 + (\omega\theta_p)^2} + \omega\eta_s \end{array} \right. \tag{3.17}$$

or

$$\left\{ \begin{array}{l} G'(\omega) = NkT \sum_{p=1}^{n} \dfrac{(\omega\theta_m)^2}{p^2 + (\omega\theta_m)^2} \\[4mm] G''(\omega) = NkT \sum_{p=1}^{n} \dfrac{p(\omega\theta_m)^2}{p^2 + (\omega\theta_m)^2} + \omega\eta_s \end{array} \right. \tag{3.18}$$

The last equations show that the dependences of the components of the complex modulus of the KSR model on frequency may be represented in

---

* In the KSR model, the relaxation and retardation times are expressed in the same manner, the only difference being in the numerical values of the constants.

dimensionless form:

$$
\begin{cases}
\dfrac{G'(\omega)}{NkT} = \displaystyle\sum_{p=1}^{n} \dfrac{(\omega\theta_m)^2}{p^2 + (\omega\theta_m)^2} \\[4mm]
\dfrac{G''(\omega) - \omega\eta_s}{NkT} = \displaystyle\sum_{p=1}^{n} \dfrac{p\,(\omega\theta_m)}{p^2 + (\omega\theta_m)^2}
\end{cases}
\tag{3.18a}
$$

The KSR model is "linear" by its nature, i.e., it gives results that are within the framework of the linear theory of viscoelasticity.

This model has been suggested to account for the behaviour of long macro-molecules in infinitely dilute solutions. The basic concepts underlying this model are valid, in many respects, for concentrated polymer solutions and molten polymers. In such cases, however, the value of the segmental friction factor η is due to the resistance to the movement of the chain offered not only by the low-molecular-mass solvent but also by the other macromole-cules contained in the system.

Certain difficulties arise when the KSR model or similar models are applied to dilute solutions of relatively short chains. In such a case, some detailing is required, which is associated with the reconsideration of the basic assump-tions used in constructing the KSR model. Naturally, this leads to different final results.*

The KSR model played historically an important part in the physics of polymers since it demonstrated the possibilities of the molecular-statistical approach to the analysis of the relaxational properties of polymeric systems. Of special importance is the fact that for the existence of the relaxation spectrum of a polymer to be predicted, it proved to be sufficient to assume the development, during deformations, of various modes of motion of iden-tical structural units (called "submolecules" or "segments") united into a chain. This has clearly demonstrated that the presence of a relaxation spectrum, even a broad one, is not yet evidence of the complexity of the chemical structure of the chain or of the variety of the nature (mechanism) of its molecular-kinetic motions.

3.1.3. The Free-Draining Coil (the Bueche model**)

The KSR model deals with the behaviour of an extended chain subjected to the action of longitudinal forces. The Bueche model rests on the analysis of the behaviour of the macromolecule modelled by a random coil which is placed in the field of shearing forces. This model is shown in Fig. 3.3, where the black circles represent the points of interaction between the segments and the medium. The motion of these segments relative to the centre of mass is studied. A deviation of the position of a given segment from the mean statistical position gives rise to a force tending to return it to the most probable position. This force is directly proportional to the deviation from the equilibrium position, which is simulated by a spring. Thus, the motion of separate segments relative to the centre of mass is simulated by a series

---

* Freire, J. J. and A. Horta, *J. Chem. Phys.*, 65, 2860 (1976).
** Bueche, F., *Physical Properties of Polymers*, Interscience Publishers, |New York, 1962.

of Maxwell elements, one of the ends of which is rigidly connected to the centre of mass and the other end carries a bead immersed in a viscous liquid and connected with the centre of mass by a spring.

The treatment of the macromolecule as a series of segments, each of which is a Maxwell element, is close to the ideas underlying the KSR model. The main difference between these models is, however, that the Bueche model takes account of the effect of the rotational motion of the macromolecular coil on the behaviour of the segments. The total displacement of the coil is composed of the translational motion of the centre of mass and the rotation of the coil relative to this point.

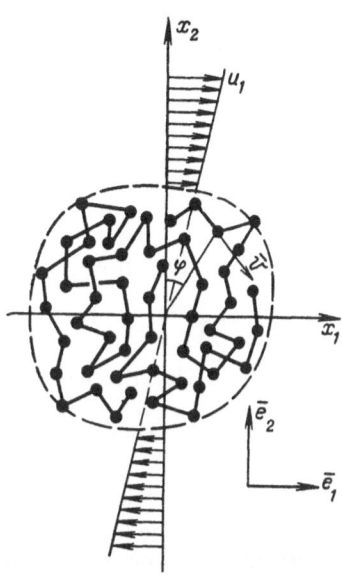

Suppose that the coordinate origin is the centre of mass of the macromolecule, and the motion of a certain arbitrarily chosen segment relative to the coordinate origin (the centre of mass) is to be analysed. The velocity profile is also shown in Fig. 3.3. It is presumed that the macromolecular coil does not disturb the velocity distribution that would have existed in the absence of the macromolecule. This means that the coil is absolutely permeable to the stream and does not introduce perturbation into the motion of the medium. An analogous supposition in an implicit form has been used in the analysis of the motion of beads in the KSR model.

Fig. 3.3. A random macromolecular coil in a hydrodynamic field.

At the chosen point, the position of which is determined by the radius-vector $\mathbf{r}$, the velocity of the stream $\mathbf{V}$ relative to the coordinate origin is expressed in the following manner:

$$\mathbf{V} = \dot{\gamma} \mathbf{e}_1 \cdot (\mathbf{r} \cdot \mathbf{e}_2)$$

where $\mathbf{e}_1$ and $\mathbf{e}_2$ are the unit vectors (shown in Fig. 3.3) directed along the coordinate axes, and $\dot{\gamma}$ is the velocity gradient, $\dot{\gamma} = du_1/dx_2$. The last equality may be represented in a form devoid of vector notations:

$$V = \dot{\gamma} r \cos \varphi$$

The angle φ is shown in Fig. 3.3. The circular component of the velocity V_φ is equal to

$$V_\varphi = V \cos \varphi = \dot{\gamma} r \cos^2 \varphi$$

The angular velocity ω of the segment residing at the chosen point in space is calculated as

$$\omega = \frac{V_\varphi}{r} = \dot{\gamma} \cos^2 \varphi \tag{3.19}$$

Here it is assumed that the shape of the macromolecule does not change during motion, i.e., it is the steady-state and not transient flow conditions that are considered.

The angular velocity of the segment is changed on rotation of the macromolecule since it depends on the angle φ. The time-averaged angular velocity

$\langle \omega \rangle$ corresponds to the value of ω averaged with respect to the angle since the rotation occurs periodically. Therefore,

$$\langle \omega \rangle = \frac{1}{2\pi} \int\limits_0^{2\pi} \omega \, d\varphi = \frac{\dot{\gamma}}{2} \qquad (3.20)$$

Thus, on motion (flow) of the polymeric system in the shearing flow with the velocity gradient $\dot{\gamma}$ the segments and, hence, the macromolecule itself, rotate with an angular velocity equal, on an average, to $\dot{\gamma}/2$.

During the rotation of the macromolecular coil there arise oscillations of the segments relative to the centre of mass with a frequency equal to $\dot{\gamma}/2$. As a result of this, there appear friction losses, the intensity of which depends on the frequency of oscillations, i.e., on the velocity gradient in the long run. The viscous resistance to deformation is composed of two components: the energy dissipation during viscoelastic oscillations and during the flow of the solvent. Therefore, the viscosity η of the system can be defined as

$$\eta = \eta_s + D/\dot{\gamma}^2$$

where D is the intensity of energy dissipation during viscoelastic oscillations.

The Bueche theory leads to two final formulas: the dependence of the apparent viscosity on the shear rate $\eta \, (\dot{\gamma})$ due to the mechanisms described, and the outwardly analogous dependence of the dynamic viscosity η' on the frequency ω. The first of these formulas was given in Chapter 2 [see Eq. (2.44)] in discussing the plausible explanations for the effect of non-Newtonian behaviour. The second is automatically obtained from Eq. (2.44) through the replacement of η with η' and of $\dot{\gamma}$ with ω.

The longest relaxation time in the Bueche theory, θ_m^B, is equal to

$$\theta_m^B = \frac{12 \, (\eta_0 - \eta_s)}{\pi^2 N k T} = 2\theta_m \qquad (3.21)$$

Just as in the KSR theory, the remaining relaxation times θ_p^B may be expressed in terms of θ_m^B with the aid of the simple relation $\theta_p^B = \theta_m^B \cdot p^{-2}$. This enables the dependence of η' on ω to be expressed in terms of θ_m^B or of a set of values of θ_p^B in an equivalent way:

$$\frac{\eta' \, (\omega) - \eta_s}{\eta_0 - \eta_s} = 1 - \frac{6}{\pi^2} \sum_{p=1}^n \frac{(\omega \theta_m^B)^2}{p^2 \, [p^4 + (\omega \theta_m^B)^2]} \left[2 - \frac{(\omega \theta_m^B)^2}{p^4 + (\omega \theta_m^B)^2} \right] =$$

$$= 1 - \frac{6}{\pi^2} \sum_{p=1}^n \frac{(\omega \theta_p^B)^2}{p^4 + p^2 \, (\omega \theta_p^B)^2} \left[2 - \frac{(\omega \theta_p^B)^2}{p^4 + (\omega \theta_p^B)^2} \right] \qquad (3.22)$$

The prediction of the agreement between the dependences $\eta \, (\dot{\gamma})$ and $\eta' \, (\omega)$ is one of the main results of the Bueche theory, which is of interest to the analysis of the relaxation properties of fluid polymer systems. As regards the quantitative calculation of the relaxation characteristics, the Bueche model does not provide a satisfactory agreement between theory and experimental observations, especially in the region of high frequencies (short times) since the Bueche theory predicts a considerably stronger dependence of the components of the complex modulus on frequency than is actually observed. According to Bueche, in the region of short retardation times, when the line spectrum predicted by theory can be approximated by a continuous function, $\Phi(\lambda)$ is found to be proportional to $\lambda^{1/2}$ and $L(\lambda) \propto \lambda^{3/2}$.

The Bueche model is important primarily in connection with the necessity to consider the rotation of the macromolecular coil on shear deformations and, as a consequence, in connection with the qualitative explanation of the known experimental fact of the analogy between the forms of the dependences of the apparent viscosity on the shear rate and of the dynamic viscosity of fluid polymer systems on frequency (for more detail, see Sec. 3.4).

3.1.4. The Nondraining Coil (the Kirkwood-Riseman-Zimm Model)

A further step in theoretical conceptions of the viscoelastic properties of polymeric chains is the Zimm model,* which is based on the works of Kirkwood and Riseman. This model (called briefly the KRZ model) rests on the analysis of the behaviour of the same macromolecular chain as in the KSR model. But a substantially new point in the KRZ theory is the consideration of the hydrodynamic interaction between separate segments in the chain. This means that, in contrast to the KSR and Bueche models considered above, in calculations involved in the KRZ model account is taken of the fact of the perturbation of the field of the flow rates of the liquid caused by the presence in it of foreign particles.

In principle, two extreme cases are possible: the first is the case where the polymeric chain does not give rise to perturbations of the rate of flow, i.e., no hydrodynamic interaction is present; this limiting case reduces to the KSR model; and the second is the case where the space occupied by the macromolecule is found to be impermeable to the solvent (the nondraining coil), to which there corresponds the maximum possible hydrodynamic effect of a given segment on the motion of the surroundings.

In the latter case, in describing the effect of the medium on the motion of the bead, which simulates segmental motion, the magnitude of the force F_i acting on the ith bead must be calculated as $\eta(V_i - W_i)$ and not as $\eta(V_i - W_i^0)$, as was done in the KSR model. The difference is that account must be taken here of the difference between the velocity of the bead, V_i, and the flow rate of the stream perturbed by the presence of all segments, W_i, and not of the flow rate of the unperturbed stream, W_i^0, as it would have been at a given point if there were no macromolecules in the solution. The quantity W_i is expressed in terms of W_i^0 and the sum of interactions in the system:

$$W_i = W_i^0 + \sum_{k=}^{n} T_{ik} F_k$$

in which expression each term of the sum represents the perturbation component due to the kth segment at a point in which the ith segment is situated. The quantity F_k is the force with which the kth segment interacts with the medium, and the parameter T_{ik} defines the intensity of perturbation expressed in terms of the numbers of the segments.

---

* Zimm, B., *J. Chem. Phys.*, **24**, 269 (1956); Zimm, B., G. Roe, and L. Epstein, *J. Chem. Phys.*, **24**, 279 (1956).

The quantitative calculations carried out by Zimm apply to the case of a completely nondraining coil, though his work contained a method of calculation of the effect of the partial permeability of the coil on its relaxation properties.

The dynamic components of the complex shear modulus $G'(\omega)$ and $G''(\omega)$, by means of which theory can be most conveniently and spectacularly compared with experiment, are expressed in any theory in the same way as written in formulas (3.17), but, depending on the model adopted, the relaxation times θ_p are calculated in different ways. Thus, in the KSR model, θ_m defined in terms of the concentration, c, and the molecular mass, M, of the polymer, is written [compare with Eq. (3.15)], as follows:

$$\theta_m = 0.608 \frac{(\eta_0 - \eta_s) M}{cRT} \qquad (3.23)$$

where R is the universal gas constant.

In the KRZ model, the maximum relaxation time θ_m^Z differs somewhat from θ_m, so that for the nondraining coil

$$\theta_m^Z = 0.422 \frac{(\eta_0 - \eta_s) M}{cRT} \qquad (3.24)$$

The most essential difference between the KSR and KRZ theories is, however, in the form of the relaxation time distribution: for the KRZ model the relax-

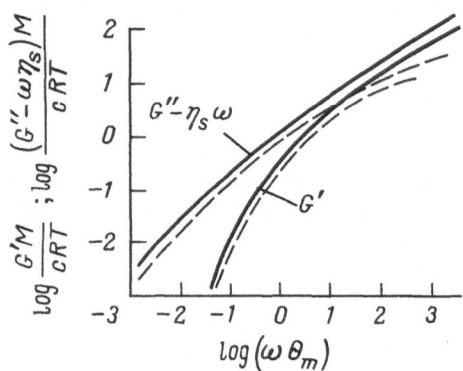

Fig. 3.4. The dynamic functions predicted by the KSR theory (dashed lines) and the KRZ theory (solid lines).

ation time distribution is narrower than in the KSR model. This is seen from a comparison of the first seven values of relaxation times in the KSR model (θ_p) and in the KRZ model (θ_p^Z), which are compared below in the form of the ratios of the maximum values of θ_m and θ_m^Z to θ_p and θ_p^Z, respectively:

$p =$	1	2	3	4	5	6	7
KSR θ_m/θ_p	1	4	9	16	25	36	49
KRZ θ_m^Z/θ_p^Z	1	3.17	5.99	9.38	13.24	17.50	22.13

The difference in the distribution of relaxation times leads to various predictions concerning the form of the functions $G'(\omega)$ and $G''(\omega)$, but in both cases, they can be represented in dimensionless form. This substantially facilitates the comparison of experimental results with theoretical predictions. Figure 3.4 shows the frequency dependences of the dynamic functions for the KSR and KRZ models, which are reduced to the dimensionless form. For this purpose, the dependences $G'(\omega)$ and $G''(\omega)$ in the KSR model were calculated from formulas (3.18), and in the KRZ model the argument was reduced to the variable $(\omega\theta_m)$ and the calculation was made with the aid of the formulas

$$
\begin{cases}
\dfrac{G'(\omega)\,M}{cRT} = \displaystyle\sum_{p=1}^{n} (\theta'_p)^2\, \dfrac{(\omega\theta_m)^2}{1+(\theta'_p\omega\theta_m)^2} \\[4mm]
\dfrac{[G''(\omega)-\omega\eta_s]\,M}{cRT} = \displaystyle\sum_{p=1}^{n} \theta'_p\, \dfrac{(\omega\theta_m)}{1+(\theta'_p\omega\theta_m)^2}
\end{cases}
\tag{3.25}
$$

where the reduced relaxation times θ'_p are expressed in terms of θ^Z_p as follows:

$$\theta'_p = \theta^Z_p/\theta_m = (\theta^Z_p/\theta^Z_m)\,(\theta^Z_m/\theta_m) = 0.695\,(\theta^Z_p/\theta^Z_m)$$

This allows us to construct the dependences of G' and $(G'' - \omega\eta_s)$ on the reduced frequency in the same coordinates in both theories.

In order to test theoretical predictions it is necessary to make any of the points on the experimental curve coincide with the chosen point on the graph of the calculated dependence and then to compare the courses of the experimental and theoretical curves. The decisive criterion by which the agreement between theory and experiment should be estimated is the region of high frequencies because at low frequencies, approximately up to $\omega\theta_m \approx 1$, both models give practically concordant results. But in the region of high frequencies the KSR theory predicts that G' must coincide with $(G'' - \omega\eta_s)$; the slope of the graph in doubly logarithmic coordinates must be equal to 1/2 for both functions. According to the KRZ theory, $(G'' - \omega\eta_s)$ exceeds G' by $\sqrt{3}$ times and the slope of both graphs is 2/3.

3.1.5. The Partially Draining Coil

For cases intermediate between the above-considered extreme models [the model of the absolutely free-draining coil (the KSR model) and the model of the completely nondraining coil (the KRZ model)], the relative permeability is expressed by the hydrodynamic interaction parameter h, whose extreme values, 0 and ∞, correspond to the KSR and KRZ theories, respectively. At any intermediate values of h the resulting values of the pth relaxation time θ^h_p lie between the corresponding pair of values, θ_p and θ^Z_p. The dependences $G'(\omega)$ and $G''(\omega)$ calculated from formula (3.17) by using the values of the relaxation times θ^h_p, which correspond to the various values of h, lie between the curves shown in Fig. 3.4.

The ideal statistical (Gaussian) distribution of the positions of the segments of the macromolecular chain in space is only realized in a θ-solvent, in which the polymer-polymer interaction is equivalent to the polymer-solvent interaction. Depending on the nature (or "quality") of the solvent used, this distribution may be different from the Gaussian distribution, which leads to additional perturbation effects that change the values of relaxation times of the system.

In the case of the Gaussian distribution, the root-mean-square distance $\langle r^2 \rangle^{0.5}$ between the ith and kth segments of length L is expressed, in accordance with the theory of the freely jointed chain, as follows:

$$\langle r^2 \rangle = |k-i|L^2$$

Deviations from the Gaussian distribution are manifested in the change of the form of the dependence of the root-mean-square distance $\langle r^2 \rangle^{0.5}$ on the difference in the serial numbers of the segments, namely: it is assumed that for the non-Gaussian distribution

$$\langle r^2 \rangle = |k-i|^{1+\varepsilon} L^2$$

The value of the parameter ε serves as a quantitative measure of the deviation from the Gaussian distribution and, hence, characterizes the specific features of the polymer-solvent interaction. The parameter ε has an influence on the volume of the macromolecular coil in the solution and may be associated with the relation between the molecular mass of the polymer and the intrinsic viscosity of the infinitely dilute solution. Therefore, in investigating the viscoelastic properties of polymer solutions in "good" solvents it is necessary to take into account the effect of the parameter ε on the relaxation-time spectrum and, hence, on the form of the $G'(\omega)$ and $G''(\omega)$ curves.

Tschoegl[*] calculated the frequency dependences of the components of the complex modulus for various values of h and ε (the latter may vary from 0 for a θ-solvent to 0.20). By matching the experimentally found functions $G'(\omega)$ and $G''(\omega)$ with the graphs obtained by theoretical calculations, it is possible to find h and ε, for which there is attained the best agreement with experiment for the system under investigation. An independent method of testing the results obtained by this method is the estimation of the quantity ε from the results of measurement of the dependence of the apparent viscosity on molecular mass for the polymer-solvent system chosen. For experimental testing of the theory, solutions in a θ-solvent have obvious advantages since it is known in advance that $\varepsilon = 0$ for them. Some experimental results of testing of the partially draining coil will be considered below.

3.1.6. Viscoelastic Properties of Solutions of Short and Rigid Macromolecules

The theoretical results discussed above referred to the behaviour of coil-like macromolecules, in which the spatial distribution of segments obeyed the Gaussian law or, at least, did not deviate significantly from the "ideal" statistical distribution. In these cases, the macromolecular chain with a large number (n) of segments is considered, which leads to the appearance of a spectrum consisting of a wide set of relaxation times. It may often be assumed that $n \to \infty$. The value of n may however be small. This results in the appearance of the dependence of the form of the dynamic functions on n, so that the dimensionless dependences $G'(\omega\theta_m)$ and $G''(\omega\theta_m)$ are not "universal", in contrast to cases involving large n values. This is illustrated by Fig. 3.5 which presents the results of calculations of the frequency dependence of the dynamic viscosity η' in the coordinates $(\eta' - \eta_\infty)/(\eta_0 - \eta_\infty)$ vs. $(\omega\theta_m)$ (η_0 and η_∞ are the limiting values of η' corresponding to the con-

---

[*] Tschoegl, N. W., *J. Chem. Phys.*, 39, 149; 40, 473 (1963); Tschoegl, N. W. and J. D. Ferry, *J. Phys. Chem.*, 68, 867 (1964); Frederick, J. E., N. W. Tschoegl, and J. D. Ferry, *J. Phys. Chem.*, 68, 1974 (1964). Analogous calculations of the Zimm model with account taken of the possibility of the partial permeability of the coil to chains of limited length have been made by A. Peterlin: Peterlin, A., *J. Polymer Sci.*, A-2, 5, 179 (1967). The most complete results of the corresponding calculations are presented in: Lodge, A. S. and Y.-J. Wu, *Rheol. Acta*, 10, 539 (1971).

ditions $\omega \to 0$ and $\omega \to \infty$, respectively). The extreme cases ($n = 1$ and $n \to \infty$) describe the relaxation of the Maxwell model and the KRZ model for a chain consisting of an infinitely large number of segments. Obviously, as the relaxation time spectrum is broadened the dynamic viscosity decreases with frequency over a frequency range which is the wider the larger is the value of n and, hence, the larger is the number of characteristic relaxation times of the system.

If the chain becomes very short in the sense that it cannot be regarded as a series of statistical segments, the entire apparatus of the above-considered theories becomes, in principle, inapplicable to the calculation of the visco-

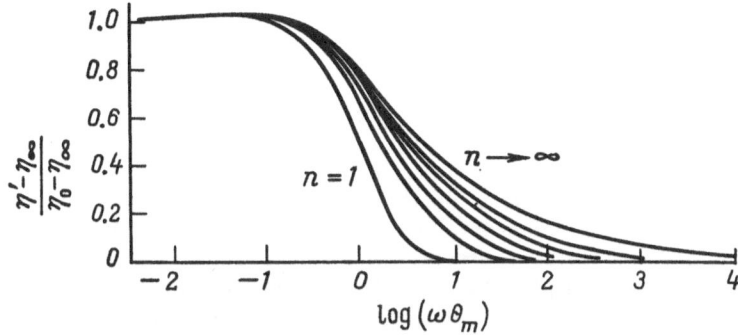

Fig. 3.5. Frequency dependences of dynamic viscosity according to the KRZ theory for different numbers (n) of segments in the chain. For the curves from left to right n equals 1, 5, 10, 15, 50, 100, ∞. [Ferry, J. D., *J. Polymer Sci.*, C, **15**, 307 (1966).]

elastic properties of the polymer system. In this case, a different mechanical model is required to describe the relaxation (viscoelastic) properties of macromolecules. Such a model proposed by Kirkwood and Auer* (the KA

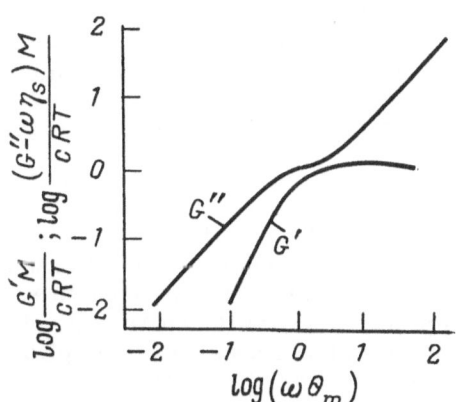

Fig. 3.6. The dynamic functions predicted by the KA theory. [Ferry, J. D., *J. Polymer Sci.*, C, **15**, 307 (1966.)]

model) is based on the representation of the macromolecule in the solution as a rigid rod. Very close results are obtained if a suspension of rigid rods is replaced with a suspension of ellipsoid-like particles of elongated form. An example of the dynamic characteristics of such a system according to calculations based on the KA model is shown in Fig. 3.6. If we compare Figs. 3.4 and 3.6, the difference between the predictions of the theories of statistical

---

* Kirkwood, J. G. and P. L. Auer, *J. Chem. Phys.*, **19**, 281 (1951).

coils (the KSR and KRZ models) and those of the theory of rigid particles (the KA model) becomes quite obvious.

The viscoelastic properties of a suspension of rigid macromolecules depend on their shape. The extreme cases here are the spherical non-deformed particles, whose solution does not at all exhibit viscoelastic properties different from the relaxation properties of the solvent, and rod-like rigid particles, whose behaviour is described by the KA theory and by a number of other theories which give close results (Cerf and Scheraga). Solutions of rigid-chain molecules are also characterized by definite regularities of the theoretically predicted dependence of the apparent (and intrinsic) viscosity on the rate of shear.*

The investigation of the viscoelastic properties of solutions of rigid macromolecules of various forms, which are considered by the KA theory and the related theories of other authors, is especially important for materials of biological origin: in virtue of the specific features of intramolecular bonds such macromolecules retain the rigid conformation of chains, which is responsible for the specificity of the viscoelastic properties of their solutions.

The overwhelming majority of theories of the rheological properties of solutions of rigid-chain macromolecules have been concerned with infinitely dilute solutions, in which there are no hydrodynamic interactions between macromolecules. An important exception is liquid-crystalline solutions of rigid-chain polymers, whose rheological properties have been described in Chapter 2. Attempts have, however, been made lately to further develop the kinetic theory of solutions of rod-like macromolecules in concentrated solutions.** Such a theory predicts an exceedingly strong dependence of the initial (Newtonian) viscosity on molecular mass: $\eta_0 \propto M^6$ and an even stronger dependence of the normal-stress coefficient (see Chapter 4): $\zeta_0 \propto$ $\propto M^{13}$ with a relatively weak concentration dependence of these coefficients: $\eta_0 \propto c^3$ and $\zeta_0 \propto c^5$. Moreover, the theory also predicts that the dependence of the shear stress on the shear rate is extremal. This must serve as an indication of the existence of a region of unstable flow. These theoretical predictions, however, have not been compared with experimental data.

3.1.7. Experimental Results

The theoretical predictions of the frequency dependences of the components of the complex modulus, which have been calculated according to the above-discussed models of the viscoelastic properties of polymer systems, have repeatedly been subjected to extensive experimental tests.

The most important circumstance that should be taken into account in comparing theory with experiment is the fact that all these theoretical results have been obtained on the assumption of the homogeneity of the lengths of macromolecules in the polymer sample under study. The consideration of the molecular-mass distribution introduces complications which are difficult to take into account. Therefore, the testing of theory can provide sufficiently reliable results only when the viscoelastic properties of solutions of monodisperse polymers are investigated. In experimental practice, monodisperse polymers are considered to be high-molecular-mass compounds characterized by the ratio of the mass-average molecular mass to the number-average molecular mass, which is close to 1.05-1.1.

---

* Mele, D., *Europ. Polymer J.*, 14, 623 (1978).
** Doi, M. and S. S. F. Edwards, *J. Chem. Soc. Faraday Trans.*, 74, 918 (1978).

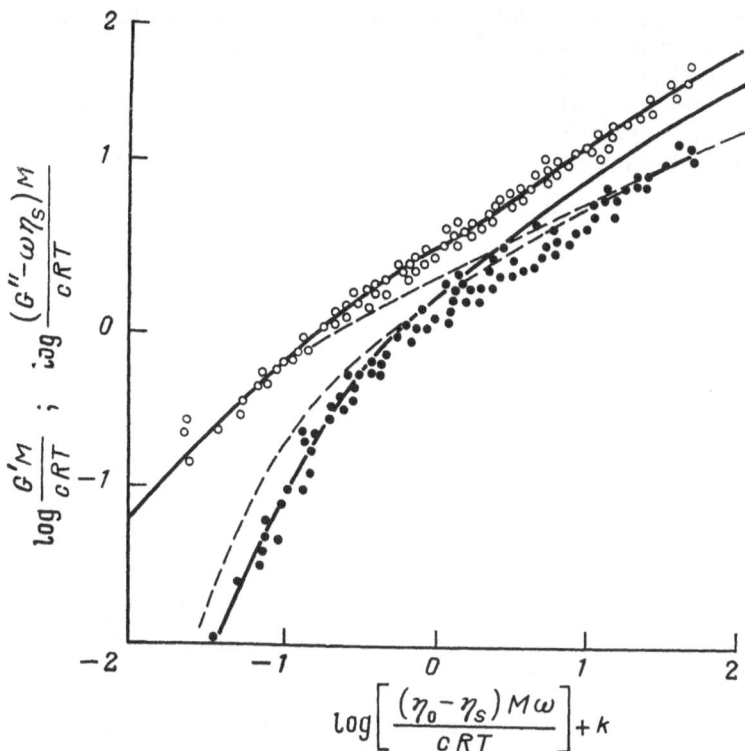

Fig. 3.7. Experimental results obtained by investigating the dynamic properties of solutions of monodisperse polystyrenes in toluene. The range of molecular masses studied is from 4.77×10^4 to 1.2×10^6; the range of concentrations from 2.71 to 0.49 g/dl; the concentration decreases with increasing molecular mass of the polymer.
Dashed lines correspond to the KSR theory; solid lines, the KRZ theory; $\bigcirc = G'$; $\bullet = G''$. [Harrison, J. C., J. Lamb, and A. J. Matheson, *J. Phys. Chem.*, 68, 1072 (1964).]

A comparison of theory with experiment is conveniently made by using dimensionless values of the modulus and frequency. A fit of experimental curves with theoretical curves is provided by shifting them along the coordinate axes; if the form of the dependences G' (ω) and G'' (ω) rather than the estimation of the values of the longest relaxation time is of interest, the values of the argument are determined only within an arbitrary constant. A typical example of such treatment of experimental data is given in Fig. 3.7 for a series of solutions of monodisperse polystyrenes in toluene over a wide range of solution compositions and molecular masses of the polymer. The experimental data given are, on the whole, in good agreement with the Kirkwood-Riseman-Zimm theory; this refers to many other data on the viscoelastic properties of solutions of flexible-chain polymers cited in the literature.

Other cases are, however, also possible. Thus, Fig. 3.8 shows the frequency dependences of G' and G'' for solutions of polyisobutylene, which are in satisfactory agreement with the predictions of the KSR theory. The entire body of available experimental data on the viscoelastic properties of dilute solutions of flexible-chain polymers lie in a region whose boundaries are outlined by the KSR and KRZ theories. Therefore, when choosing the parameters h and ε in the model of the partially draining coil, a reasonable concordance between theory and experiment was obtained.

The general regularity of the effect of concentration on the character of the viscoelastic properties of solutions of flexible-chain polymers boils down to the following: as the content of the polymer in the system increases the behaviour of the polymer predicted by the KRZ theory changes to that de-

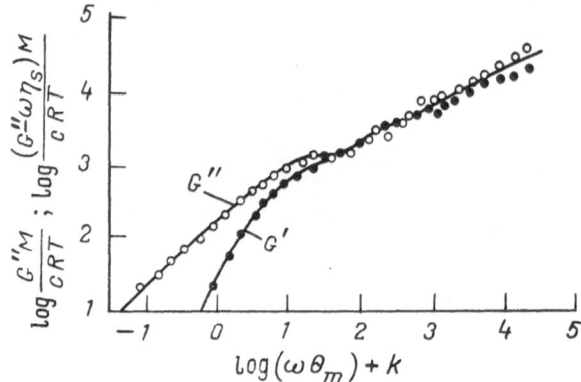

Fig. 3.8. Experimental results obtained by investigating the dynamic properties of a 4-percent solution of polyisobutylene in a mineral oil. [Tschoegl, N. W. and J. D. Ferry, *Koll.-Z.*, **189**, 37 (1963).]

scribed by the KSR model. Analogous consequences are brought about by the increase of the molecular mass of the polymer. This is shown in Fig. 3.9

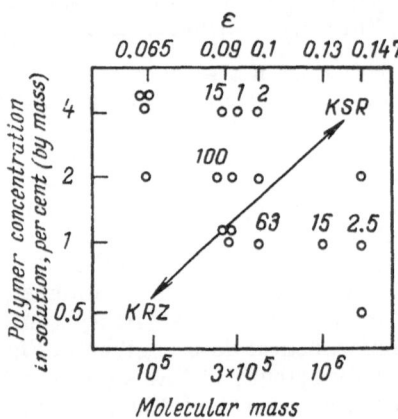

Fig. 3.9. Values of the hydrodynamic interaction parameter h (indicated at circles) for solutions of monodisperse polystyrenes of different molecular masses in Aroclor-1248. The arrows indicate the direction of variation of the values of the parameter corresponding to transitions to the extreme cases—the KRZ and KSR models.

in the form of a "map" of the values of the hydrodynamic interaction parameter h as a function of the molecular mass and the concentration of the solution for the polystyrene-Aroclor* system. Analogous "maps" are also

---

* Aroclor, a partially chlorinated diphenyl, is extensively used in investigations of the properties of solutions of polystyrene and some other polymers as a "good" solvent of high viscosity. Depending on the extent of chlorination, the viscosities of Aroclor of grades 1232, 1248, and 1254 are, respectively, 0.014, 0.27, and 9.0 Pa·sec (at 25°C). Aroclor was widely used in the sixties, but it was later found to be strongly cancerogenic. At present, its production and utilization are forbidden.

known to exist for other polymer-solvent systems. The parameter ε is chosen as one of the axes of the map. This allows one to take account of the role of the molecular mass of the polymer and the value of ε associated with the size of the molecular coil in the solvent chosen.

The consideration of the effect of concentration and molecular mass on the character of the frequency dependences of the components of the complex shear modulus in terms of the KSR and KRZ theories is only valid until at high concentrations and/or high values of molecular mass of the polymer there takes place a qualitative change of the dependences in question, which is caused by the appearance of the region of the rubbery state. This corresponds to a qualitative change in the mechanism of manifestation of the viscoelastic properties of polymer systems, and the above-discussed theoretical conceptions that attribute the observed dependences $G'(\omega)$ and $G''(\omega)$ to the viscoelastic properties of an individual flexible-chain macromolecule become incorrect because the viscoelasticity of the structural network becomes dominant. The viscoelastic properties of polymer systems of this type will be discussed in Sec. 3.3.

Experimental investigations of the viscoelastic properties of solutions of rigid-chain macromolecules have shown that even in those cases when the macromolecule in the solvent chosen retains a strictly helical conformation (for example, poly-γ-benzyl-L-glutamate in m-methoxyphenol), the behaviour of such a solution is satisfactorily described by the rigid-rod model only in the region of relatively low frequencies.* As the frequency increases there are observed ever increasing deviations from the predictions of the Kirkwood-Auer theory, and the frequency dependences of the components of the complex shear modulus approach the theoretical predictions arrived at for the statistical coil model. When the content of rigid macromolecules in the solution increases, their intermolecular interaction is intensified, which favours the departure of the chain conformations from the strictly helical conformation, as a result of which the deviations of the properties of solutions from the predictions of the KA theory become larger. In investigations of solutions of the same polymers but in solvents in which they adopt the form of the statistical random coil or if the transformation from the helical to the coil-like conformation is caused by other factors, the regularities of the manifestation of the viscoelastic properties of such systems correspond to the predictions of the theories of statistical random coils.

3.1.8. Viscoelastic Characteristics of Infinitely Dilute Solutions

The most reliable testing of theoretical concepts of the regularities of manifestations of the flexibility of the polymeric chain is associated with the measurements of the viscoelastic characteristics of dilute solutions in the region of very low concentrations c with the subsequent extrapolation of the experimental data to $c \to 0$. Such investigations have become possible with the advent** of new highly sensitive instruments for measuring the dynamic properties of low-modulus solutions (with the storage modulus G' up to 1 Pa); these instruments have been used for determining the functions

---

* Tschoegl, N. W. and J. D. Ferry, *J. Am. Chem. Soc.*, 86, 1474 (1964).
** Schrag, J. L. and R. M. Johnson, *Rev. Sci. Instr.*, 42. 224 (1971).

G' (ω) and G'' (ω) for a wide range of materials with the polymer content in solution not exceeding several fractions of a per cent.* The importance of these measurements is the same as that of reliable determinations of the viscosity of very dilute solutions, which has enabled one to deal with the concept of the intrinsic viscosity of infinitely dilute solutions. What we are

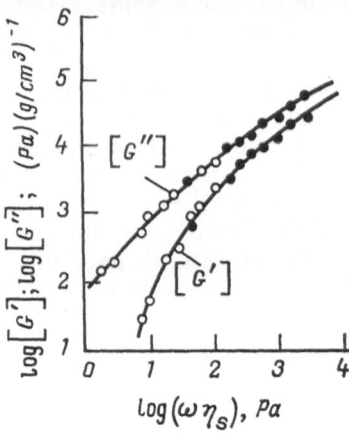

Fig. 3.10. Comparing theoretically calculated frequency dependence of the components of the dynamic modulus for the partially draining coil (the solid line corresponding to the interaction parameter $h = 0.25$) with experimental data referring to infinitely dilute solutions of polystyrene (with $M = 8.6 \times 10^5$) in two theta-solvents—decalin (○) and bis-2-ethylhexyl phthalate (●). [Johnson, R. M., J. L. Schrag, and J. D. Ferry, *Polymer J.*, **1**, 742 (1970).]

speaking about here is the concepts analogous to the intrinsic viscosity of solutions, namely the characteristic values of the components of the complex shear modulus $[G']$ and $[G'']$; these quantities are defined as follows:

$$[G'] = \lim_{c \to 0} (G'/c); \quad [G''] = \lim_{c \to 0} [(G'' - \omega\eta_s)/c]$$

The determination of $[G']$ and $[G'']$ consists of the construction of graphs showing the dependences of $(G'/c)^2$ (G'/c in the earlier works, but the use of the quadratic dependences has proved more reliable) and $[(G'' - \omega\eta_s)/c]$ on c, which are straight lines, and their extrapolation to $c = 0$.

The experimental results obtained for various polymers—polydimethyl siloxane, polybutadiene, polystyrene, etc.—have shown that, on the whole, they all behave as partially draining coils, so that the appropriate choice of the interaction parameter h provides a satisfactory agreement between theory and experiment and the chemical nature of the flexible polymeric chain plays no part at all. An example of the results of a comparison of theory with experimental evidence referring to infinitely dilute solutions is presented in Fig. 3.10 for solutions of polystyrene in two different theta-solvents. The agreement between theory and experiment is also preserved for solutions in "good" solvents. Further investigations have also shown that the examination of the frequency dependences of $[G']$ and $[G'']$ enables us

---

* A large number of works by J. D. Ferry and his school have been devoted to this subject: *Macromolecules*, 4, 210 (1971); 5, 17 (1972); 6, 541 (1973); 8, 539 (1975); *Polymer J.*, 1, 742 (1970); 2, 541 (1971); 4, 24 and 668 (1973); 7, 195 (1975); *J. Polymer Sci.*, *Polymer Phys. Ed.*, 16, 1031 (1978). Of the latest works in this field, special mention should be made of the following article: Rosser, R. W., J. L. Schrag, and J. D. Ferry, *Macromolecules*, 11, 1060 (1978). In this work, use was made of samples of polystyrene of exceedingly high molecular mass. It was shown that the viscoelastic properties of a solution in a good solvent at infinite dilution are well described by the Rouse theory and in a theta-solvent, by the Zimm theory. Reviews of investigations in this direction have been written by J. D. Ferry: *Accounts of Chem. Research*, 6, 60 (1973); *Pure and Appl. Chem.*, 50, 299 (1978).

to confirm the existing theoretical conceptions of the viscoelastic properties not only of linear but also of branched macromolecules, the behaviour of whose solutions may be treated in terms of the Zimm model provided that account is taken of the partial permeability of the macromolecular coil. Here, however, the value of h for branched polymers is found to be somewhat lower than for linear macromolecules of the same chemical constitution.

As regards infinitely dilute solutions of rigid-chain polymers, for them in the high-frequency region there are observed sharp discrepancies between theoretical predictions based on the rigid-rod model and experimental data*, just as in the earlier investigations carried out in the region of relatively high concentrations. The frequency dependences of $[G']$ and $[G'']$ at high ω values are well described by the model of the partially draining coil. The authors of the original work note, however, that this fact should not be regarded as evidence of the actual existence of the coil-like conformation of macromolecules. This result is rather characteristic in the sense that it clearly points to the impossibility of an unambiguous comparison of the experimental results of measurements of the rheological properties of the polymeric system with its structural characteristics, so that the agreement between the observed dependences and the predictions of the various theories is not yet a sufficient proof of the reality of the physical model used for appropriate calculations.

3.1.9. Dynamic Properties at Very High Frequencies

The problem of determination of the limiting dynamic properties of polymer systems in the region of very high frequencies is an independent task in the theory. An answer to the question of the actual behaviour of a polymer

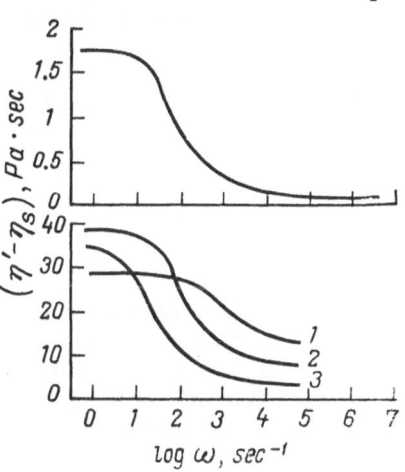

Fig. 3.11. Frequency dependences of the dynamic viscosities of solutions of polystyrene at 25°C. The characteristics of the solutions:

$1-M = 19.8 \times 10^3$, $c = 0.075$ g/cm³; $2-M = 82 \times 10^3$, $c = 0.0507$ g/cm³; $3-M = 267 \times 10^3$, $c = 0.0274$ g/cm³; for the upper curve $M = 860 \times 10^3$, $c = 0.0154$ g/cm³. The solvent is Aroclor-1248. [Massa, D. J., J. L. Schrag, and J. D. Ferry, *Macromolecules*, 4, 210 (1971).]

system at very high frequencies can be obtained by means of direct experimental measurements.

A typical example of the results obtained is shown in Fig. 3.11. Judging from these data, the limiting dynamic (the lowest Newtonian) viscosity of the solution, η'_∞, appears to be substantially higher than the viscosity of the

* Warren, T. C., J. L. Schrag, and J. D. Ferry, *Biopolymers*, 12, 1905 (1973).

solvent, η_s. The existence of the inequality $\eta'_\infty > \eta_s$ can be accounted for by the fact that at high frequencies the macromolecular coils (while not being actually deformed and giving no contribution to the viscoelastic losses) behave as an inert filler with respect to the solvent; it is well known that the viscosity of a liquid containing a solid filler is higher than its intrinsic viscosity. In this respect, it is characteristic that in the limiting region of high frequencies, where $\eta' = \eta'_\infty = $ const, for a series of solutions of various concentrations and polymers of various molecular masses the intrinsic viscosity $(\eta'_\infty - \eta_s)/\eta_s c$ remains practically constant (according to the data presented in Fig. 3.11), i.e., the additional contribution to viscous losses at high frequencies, which is due to the presence of the polymer, is only proportional to the volume content of macromolecules in the solution.*

When the region of investigated frequencies was extended to the megahertz region, there were detected new experimental facts, which lead one to the reconsideration of the question of the limiting properties of the polymer system. The point is that at $\omega > 10^6$ Hz the dependence of dynamic viscosity on frequency is exhibited by such compounds which at lower frequencies behave as Newtonian liquids, in particular low-molecular solvents. Even more spectacular in this respect are the results of an investigation of the frequency dependence of dynamic viscosity of solutions of monodisperse polystyrene in n-dibutyl phthalate**, which at 25°C is similar to a theta-solvent by its attitude towards polystyrene. For this system, over a rather wide frequency range, reaching 100 MHz, there is observed the existence of a practically constant quasilimiting value of dynamic viscosity. But, as the frequency increases further, up to 300 MHz, the viscosity again begins to decrease. This effect is due to the fact that at such high frequencies the viscosity of the solvent itself begins to fall. Control measurements of the dynamic viscosity of the solvent have shown that at frequencies greater than 100 MHz the value of η' becomes lower than that of η_s. It is apparently for this reason that the dependence of η' on ω shown in Fig. 3.11 is only a portion of the complete curve of η' versus ω, so that at still higher frequencies the dynamic viscosity must continue to decrease.

3.2. The Superposition Principle

The form of the frequency dependences of the dynamic functions is varied, depending on the model of the polymer system chosen. But, for any model it proves possible to represent the dependences $G'(\omega)$ and $G''(\omega)$ in dimensionless form, using the "reduced" frequency $(\omega\theta_m)$ as the argument and the "reduced" normalized values of the components of the modulus as the

---

* A theory that takes into account the additional contribution of the "Einstein" (due to the presence of particles that behave in the limiting high-frequency case as hard beads) component to the viscoelastic properties of dilute polymer solutions has been proposed in the following work: Peterlin, A., *Polymer*, 18, 747 (1977). The general approach to the theoretical treatment of this problem is associated with the use of the concept of the internal viscosity (or kinetic stiffness) of the chain: Peterlin, A., *J. Polymer Sci.*, A-2, 5, 179 (1967); Volkov, V. S. and V. N. Pokrovsky, *Vysokomol. Soedin.*, 20B, 834 (1978).
** Moore, R. S. et al., *J. Chem. Phys.*, 50, 5088 (1969).

variables:

$$\begin{cases} G'_r = \dfrac{G'}{NkT} = \dfrac{G'M}{cRT} \\[3mm] G''_r = (G'' - \omega\eta_s)\dfrac{1}{NkT} = (G'' - \omega\eta_s)\dfrac{M}{cRT} \end{cases} \qquad (3.26)$$

The dimensions of the quantities contained in these formulas are chosen so as to render the reduced variables dimensionless.

For a polymer containing no solvent:

$$G'_r = \frac{G'M}{\rho RT}; \qquad G''_r = \frac{G''M}{\rho RT} \qquad (3.26a)$$

where ρ is the density.

From this it follows that the dependences $G'_r(\omega\theta_m)$ and $G''_r(\omega\theta_m)$ must be "universal" for a group of polymers and their solutions whose relaxation time spectra are similar in form or whose behaviour is described by the same kinetic model.

Then, the use of the reduced variables allows one to achieve a superposition of the $G'(\omega)$ and $G''(\omega)$ curves obtained for different systems or for different measurement conditions if the relaxation spectra are similar in form for the cases involved. This method of treatment (generalization) of experimental data referring to different temperatures, concentrations, and molecular masses of polymers is known as the superposition principle.

3.2.1. Temperature-Frequency Superposition

The superposition principle was first advanced by Aleksandrov and Lazurkin in 1939. They noticed that the dynamic functions corresponding to various temperatures are similar to one another in form and are shifted along the frequency axis. The distance between the curves corresponding to different temperatures was, later termed the shift factor, a_T. With the product ωa_T chosen as the argument one can construct master curves for the components of the complex shear modulus, $G'(\omega a_T)$ and $G''(\omega a_T)$, which are invariant to temperature. In practice, this is accomplished by the parallel shifting of the experimentally obtained $G'(\omega)$ and $G''(\omega)$ curves by the distance $\log a_T$ along the $\log \omega$ axis. In this way there are obtained the curves of G' and G'' vs. (ωa_T) reduced to the reference temperature T_0. The shift factor a_T is dimensionless. For the temperature T_0 chosen as an arbitrary reference temperature, $a_T = 1$; for all the other temperatures studied a_T is determined empirically. The reduction procedure boils down to determining the temperature dependence of the shift factor, $a_T(T)$, and the result is a temperature-invariant master curve.

Comparison of the argument (ωa_T) with the dimensionless parameter $(\omega\theta_m)$ allows one to establish that by its physical significance a_T is the ratio of maximum relaxation times at different temperatures to the maximum relaxation time at the reference temperature T_0, to which the reduction is made, since the function $a_T(T)$ characterizes the temperature dependence of the maximum relaxation times.

The variation of temperature must also be taken into account in calculating G'_r and G''_r by introducing a temperature-density correction [see formula (3.26a)]. If at the reference temperature the density is ρ_0, then the temperature-density correction is found as $(\rho_0 T_0/\rho T)$. For solutions, the correction is calculated in an analogous manner since the temperature coefficients of the densities of the polymer and solvent are different in a general case. In practice, the temperature-density correction is usually not great and may be neglected.

The choice of the reference temperature is, of course, arbitrary and a matter of convenience; the replacement of the temperature T_0 leads to the entire temperature-invariant curve being shifted along the abscissa by an amount determined by the ratio of the values of a_T at these temperatures. Therefore the argument (ωa_T) is determined within an arbitrary constant.

Comparison of a_T with θ_m allows one to express $a_T (T)$ through the temperature dependence of viscosity, a quantity that can be measured directly and independently. Indeed, θ_m can be represented in the form of $\theta_m = = K (\eta_0 - \eta_s)/\rho T$, where the constant K includes the quantities that remain constant when the temperature superposition is being accomplished. If the constant K is excluded, a_T may be given by

$$a_T = \frac{\eta_0 (T) - \eta_s (T)}{\rho (T) T} \cdot \frac{\rho_0 T_0}{\eta_0 (T_0) - \eta_s (T_0)} \tag{3.27}$$

For concentrated polymer solutions and polymer melts $\eta_s \ll \eta_0$, and therefore

$$a_T = \frac{\eta_0 (T)}{\eta_0 (T_0)} \cdot \frac{\rho_0 T_0}{\rho T} \tag{3.27a}$$

If the temperature-density correction factor is neglected as compared with the variation of viscosity, then

$$a_T = \eta_0 (T)/\eta_0 (T_0) \tag{3.27b}$$

From the considerations given above it follows that the temperature dependence of the shift factor a_T can be determined from the temperature dependence of the viscosity of the system. Indeed, viscosity may be regarded as an integral characteristic of the viscoelastic properties of the material. Therefore, the change of the frequency axis with temperature by a_T times will have the same effect on the variation of the viscosity of the system.

The identity of viscosity- and shift factor-temperature dependences permits us to extend what has been said in Chapter 2 in respect to the form and general properties of the function $\eta (T)$ for polymer systems to the analysis of the temperature dependences of viscoelastic functions. Thus, in the literature, extensive use is made of the WLF equation (see page 119) in the analysis of viscoelastic functions for describing the temperature dependence of relaxation times, although the temperature superposition principle itself is far from being restricted to any one particular form of the function $a_T (T)$ and requires only the equivalency of the forms of the dependences of the viscoelastic functions at various temperatures.

The temperature superposition principle is especially important in those cases when at different temperatures there are experimentally obtained curves which are only partly overlapped when shifted along the log ω axis. The superposition of such curves allows one to significantly extend the range of variation of the argument and to cover such wide regions of reduced frequencies which are very difficult, if not impossible at all, to cover by direct methods of measuring the moduli at a single temperature, because one would have to carry out measurements with the frequency varying by a factor of up to 10-16 decimal orders.

The temperature-superposition principle was formulated above in application to the analysis of the temperature dependences of the components of

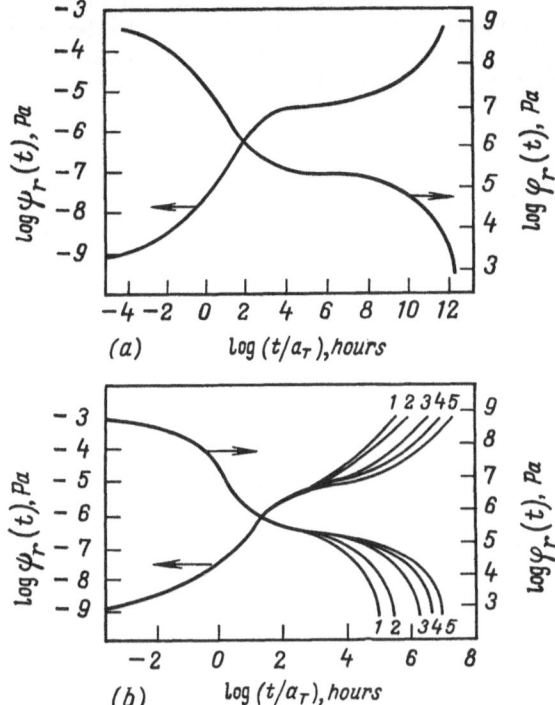

Fig. 3.12. Temperature-invariant creep and relaxation functions for polyisobutylene of wide MMD (a) and monodisperse polystyrenes (b) with molecular masses ($M \times 10^{-5}$): 1—0.8; 2—1.25; 3—1.93; 4—2.39; 5—2.67. [Tobolsky, A. V., J. Polymer Sci., C, 9, 157 (1965).]

the complex shear modulus. However, in virtue of the existence of the relations of linear theory of viscoelasticity the variation of the argument (frequency) by a_T times in one of the viscoelastic functions corresponds to the identical change of the frequency scale in considering the relaxation and creep function. Consequently, the temperature-time or temperature-frequency superposition principle may generally be regarded as a method of curve fitting for the viscoelastic properties of polymer systems by shifting the initial time or frequency dependences of the corresponding functions along the log ω or log t axis by an amount equal to log a_T which depends on temperature. Here the reduced time scale is expressed in the following manner:

$$t_r = \omega_r^{-1} = (\omega a_T)^{-1} = t/a_T$$

The reduced (temperature-invariant) relaxation and creep functions (φ_r and ψ_r, respectively) are expressed in terms of the initial functions with the aid of the temperature-density correction factor:

$$\varphi_r = \varphi \, (T_0 \rho_0 / T\rho)$$

$$\psi_r = \psi \, (T\rho / T_0 \rho_0)$$

The **potentialities** of the temperature superposition principle are illustrated by Fig. 3.12, which presents the creep and relaxation functions for polyisobutylene and a series of monodisperse polystyrenes of various molecular masses with the argument varying over a range of up to 16 decimal orders, and also by Fig. 3.13 which presents the frequency dependences of the dynamic functions of polystyrenes. Master (temperature-invariant) curves of the type shown in Fig. 3.12 have been obtained for a large variety of polymers and their solutions, which has made it possible to establish the regularities governing the manifestation of the relaxation (viscoelastic) properties of various polymers and to find the characteristic parameters which can be used to classify the states of polymers according to their viscoelastic properties.

Common to a wide range of amorphous polymers and melts of crystalline polymers is first of all the form of the relaxation and creep functions over a wide region of values of $t_r = t/a_T$. Proceeding from this form of the visco-

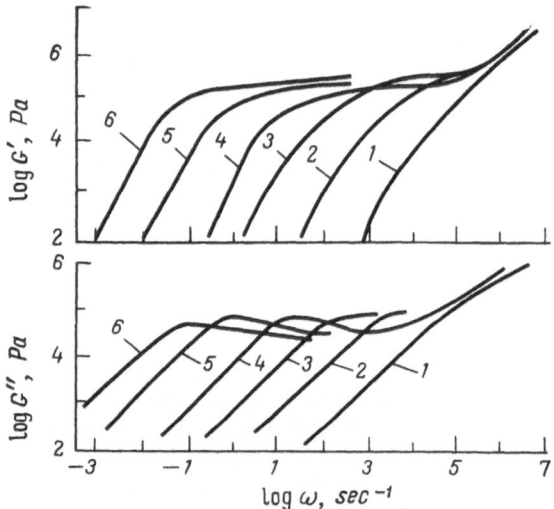

Fig. 3.13. Dynamic characteristics of polystyrenes of narrow MMD at 190°C The molecular masses of the samples used ($M \times 10^{-5}$): 1—0.20; 2—0.59; 3—1.80; 4—2.15; 5—5.10; 6—8.60.

elastic functions, one can determine the characteristic regions of frequencies (or the corresponding temperatures), in which the behaviour of the system shows certain specific features. These are a region of nearly constant, very high values (of the order of 3×10^9 Pa for extension) of the relaxation modulus, which is associated with the leathery state of the material; a region of decreasing values of the relaxation function, which varies from about 3×10^9 to about 10^6 Pa; this region may be called the transition region or the region of the leathery state of the polymer; a more or less wide plateau of constant values of the relaxation function, which is identified with the high-elastic (rubbery) state of the polymer system; and, finally, a region of rapid relaxation at large values of t_r, which is identified with the fluid state of the polymer. Of course, in all these cases, one is speaking of the same material, and the change of its behaviour, which is characterized by the values of the relaxation function, occurs due to the variation of the loading frequency or temperature.

3.2.2. Concentration-Invariant Characteristics of the Viscoelastic Properties of Solutions

The empirical approach to the construction of concentration-invariant curves for the viscoelastic properties of polymer systems is based on observations which are quite analogous to those described above in discussing the temperature superposition principle. If the dependences $G'(\omega)$ and $G''(\omega)$ for solutions of different concentrations are similar in form, they can be superimposed by shifting them horizontally along the log ω axis by an amount equal to the concentration shift factor a_c, which depends on concentration. But here one has to take into account the necessity of vertical shifting along the modulus axis. Therefore, when plotting the concentration-invariant curves for viscoelastic properties it is convenient to proceed from the preliminarily normalized functions, as described above, in order to find empirically only the value of the concentration shift factor.

Such a characteristic of the viscoelastic properties of solutions is, for example, the dynamic viscosity since it can conveniently be considered in the form of the dependence of η'/η'_0 on ωa_c, where η'_0 is the initial value of dynamic viscosity (corresponding to the limiting case $\omega \to 0$), or, what is the same thing, the initial Newtonian viscosity η_0. In normalizing the values of η' by η_0, the reduction along the ordinate is automatically provided since at $\omega = 0$ the ratio η'/η_0 is equal to unity.

A comparison of the frequency reduced to concentration, $\omega_r = \omega a_c$, with the value of the argument ($\omega\theta_m$), allows us to conclude that the concentration shift factor a_c is given by two cofactors:

$$a_c = \eta_0(c)/c$$

Experiment shows, however, that if the concentration in a_c is used in a first power it does not provide, in a general case, a concentration-invariant characteristic of the dynamic viscosity. Therefore, using the empirical approach to the concentration-invariant master curves for the viscoelastic properties of polymer systems, the following function should be used as the argument of these dependences:

$$a_c = \eta_0(c)/f_1(c) \tag{3.28}$$

where the dependence $f_1(c)$ is nonlinear in a general case and its precise form must be found empirically.

A typical example of the concentration-invariant master curve of dynamic viscosity is shown in Fig. 3.14 for solutions of polyisobutylene in decalin when the content of the polymer in solution varies from 3 to 100 per cent. In this particular case, the function $f_1(c)$ was given as c^α, where the exponent α varied from 1 to 3. Figure 3.14 also shows a master (concentration-invariant) curve for the storage modulus G'_r, which was plotted with the use of the same function, $f_1(c)$. The example given points to the possibility of constructing concentration-invariant dynamic functions over a wide range (about 9 decimal orders) of variation of the argument with the shift factor a_c being appropriately chosen.

The function $f_1(c)$ has the meaning of the concentration dependence of a certain characteristic modulus of the solution. Indeed, in discussing the question of plotting concentration-invariant characteristics of the flow properties of solutions (see Chapter 2) it was pointed out that the dimensionless argument must have the form $(\dot{\gamma}\eta_0/G_0)$. An analogous role for the dynamic functions is played by the argument $[\omega\eta_0/f_1\,(c)]$ It is obvious that the functions $G_0\,(c)$ and $f_1\,(c)$ are identical. It is easy to show that this function

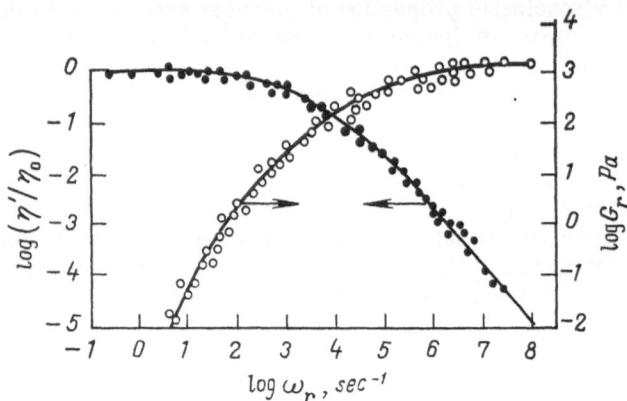

Fig. 3.14. Concentration-invariant characteristics of the dynamic viscosity and dynamic modulus of solutions of polyisobutylenes with $M = 1 \times 10^6$ in decalin. The original data were obtained by investigating 3-100 percent solutions of the polymer. [De Witt, T., H. Markovitz, F. J. Padden, Jr., and L. J. Zapas, *J. Coll. Sci.*, 10, 174 (1955).]

must be used as the normalizing parameter for G' and G'' in calculating the reduced values of these quantities, and also in normalizing the relaxation

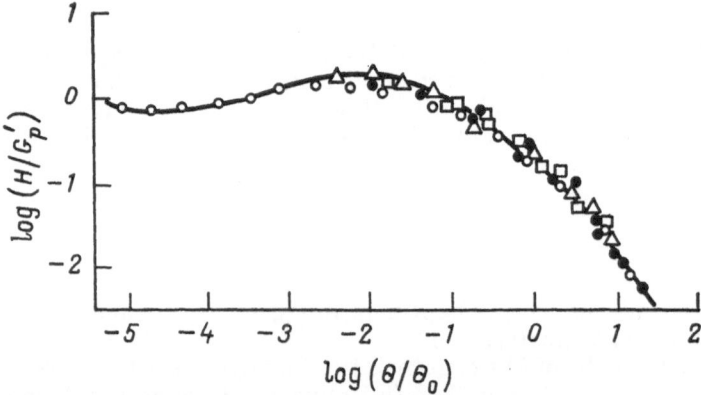

Fig. 3.15. The relaxation spectrum in dimensionless coordinates [Dreval, V. E., A. Ya. Malkin, G. V. Vinogradov, and A. A. Tager, *Europ. Polymer J.*, 9, 85 (1973).]:
$\bigcirc$ —solutions of polyisobutylene in toluene and decalin (the volume fraction of the polymer φ_2 from 0.338 to 1.0); $\triangle$ = solution of polystyrene in ethylbenzene ($\varphi_2 = 0.67$) and decalin ($\varphi_2 = 0.65$); $\bullet$ = = solution of polymethyl methacrylate in toluene ($\varphi_2 = 0.65$) and dihexyl phthalate ($\varphi_2 = 0.4$-0.65); $\square$, = solution of cellulose acetate in cyclohexanone ($\varphi_2 = 0.25$) and dioxane ($\varphi_2 = 0.25$).

spectrum which has the dimensions as the modulus. The validity of such an approach is illustrated in Fig. 3.15, which is a generalization of the experimental data on the viscoelastic properties of a large number of various polymer solutions, which were obtained by varying the concentration over wide ranges.

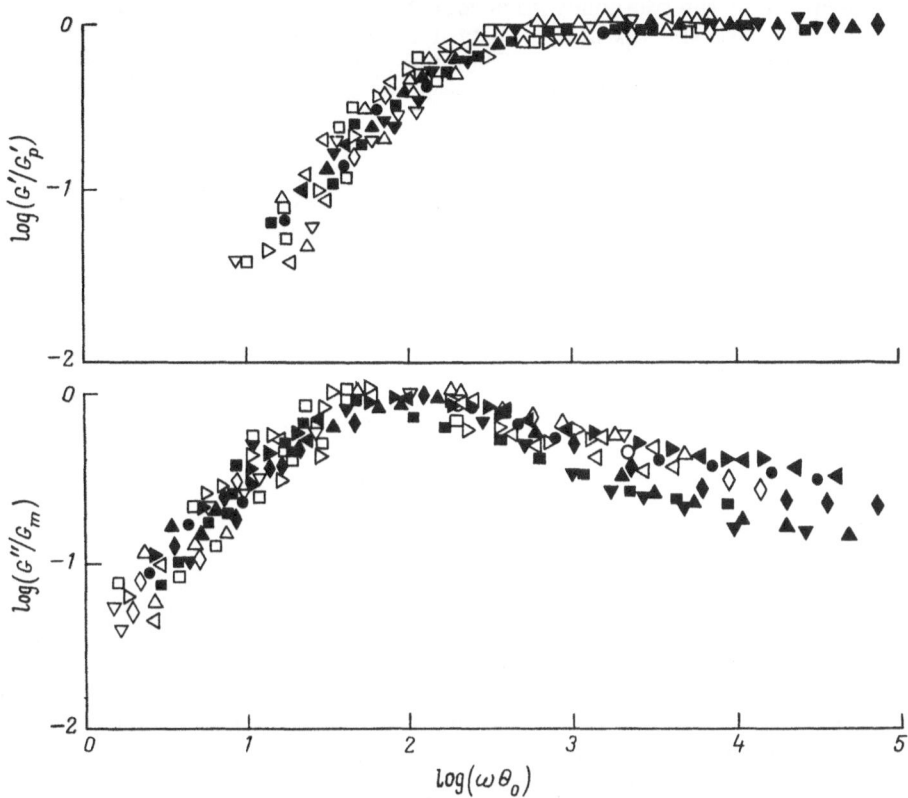

Fig. 3.16. Concentration-invariant characteristics of the dynamic properties of solutions of polybutadienes of various molecular mass in different solvents. [Blinova, N. K., S. I. Sergeenkov, A. Ya. Malkin, Yu. G. Yanovsky, and G. V. Vinogradov, *Mekhanika Polimerov*, **1**, 132 (1973).]

In considering the dynamic functions of polymer solutions it is always convenient to normalize the dependences $G'(\omega)$ and $G''(\omega)$ by the characteristic points of these functions: G' by the value of G'_p corresponding to the rubbery plateau, and G'' by its value at a maximum, G''_m, corresponding to the transition from the fluid to the rubbery state. The role of a_c can be played in this case by a certain characteristic relaxation time θ_0, say, one at which there is attained the maximum of the function $G''(\omega)$. An example that illustrates this method of constructing generalized (concentration-invariant) dynamic functions of polymer solutions is given in Fig. 3.16.

3.2.3. Reduction by Molecular Mass

Consideration of methods of constructing master curves for the viscoelastic properties of polymer systems, which are invariant with respect to the molecular mass of the polymer, must basically be a repetition of what has been said above concerning the method of plotting concentration-invariant master curves. Just as with concentration, the effect of molecular mass on the dis

placement along the frequency axis manifests itself in two ways: through the
value of the viscosity η_0 and a certain function of molecular mass, which
is not linear in a general case. Therefore, the shift factor appropriate to
molecular mass, a_M, should be expressed by the relation

$$a_M = \eta_0 (M)/f_2 (M) \qquad (3.29)$$

where $f_2 (M)$ is a certain a priori unknown function.

According to the theories discussed in Sec. 3.1 this function is equal
to M^{-1}; but when dealing with polymers having sufficiently high molecular
masses it appears that this function is equal to unity and the reduction along
the log ω axis is effected by shifting by an amount log η_0. This is due to the
fact that $f_2 (M)$, like $f_1 (c)$, has the meaning of the rubbery modulus (G_0),
and for high-molecular-mass polymers G_0 does not depend on molecular
mass. This greatly simplifies the generalization of experimental data on
the flow and viscoelastic properties of polymers having different molecular
masses; in particular, there is no vertical shift of the functions $G' (\omega)$ and
$G'' (\omega)$.

3.2.4. Limitations of the Superposition Principle

. As follows from the foregoing, for polymers or their solutions with relaxation spectra similar in form there can be constructed master curves of their
viscoelastic properties. They are obtained if the viscoelastic functions and
their arguments are normalized by independent parameters—temperature,
concentration, molecular mass. The reduced frequency ω must be expressed
as follows in a general case:

$$\omega_r = \omega\theta_m = \omega \left[\eta_0 (T, M, c) - \eta_s (T) \right] \frac{1}{\rho T \cdot f_1 (c) \cdot f_2 (M)} \qquad (3.30)$$

The factor θ_m reflects the effect of the indicated parameters of the system
on the value of its longest (terminal) relaxation time. In an analogous way,
the reduced time t_r is given by

$$t_r = \frac{t}{\theta_m} = \frac{\rho T f_1 (c) f_2 (M)}{\eta_0 (T, M, c) - \eta_s (T)} \qquad (3.30a)$$

In the simplest cases, the functions $f_1 (c) = c$ and $f_2 (M) = M^{-1}$; then ω_r
reduces to the dimensionless parameter $\omega\theta_m$, which appears in various
theories of the mechanical properties of macromolecules treated as statistical
coils. In a more general case, this is not so, and the problem of the form of
the functions $f_1 (c)$ and $f_2 (M)$ has no general solution.

The superposition principle in its diverse versions has found wide application for the treatment of experimental data obtained by measuring various
functions characterizing the viscoelastic properties of polymers. Therefore,
it is important to outline the limits of its applicability. In the most general
case, the range of validity of this method is determined by the initial requirement of the similarity of the forms of the relaxation spectra of the systems
being compared or (for a given system) of the spectra when the external
parameters are changed (temperature and pressure, though the latter factor
has not been thoroughly studied). The fulfillment of this requirement automatically provides the applicability of the superposition principle, although
a particular form of the dependences of shift factors on temperature, concentration and molecular mass cannot be found by fulfilling this requirement.

In very many cases, the viscoelastic properties of polymers are really found to be similar in form. We may, however, indicate, with certainty, several typical cases where the basic condition that provides superposition is not fulfilled. This is first of all the variation of the solution concentration and the transition from concentrated to dilute solutions. Even in the region of relatively low concentrations there occurs a gradual change in the form of the spectrum, depending on concentration. The variation of the hydrodynamic interaction parameter (in the theory of the partially draining coil) also leads to the impossibility of superposition of experimental data. A similar effect is exerted by the change of the nature of the solvent if the character of the polymer-solvent interaction is strongly altered, say, if there takes

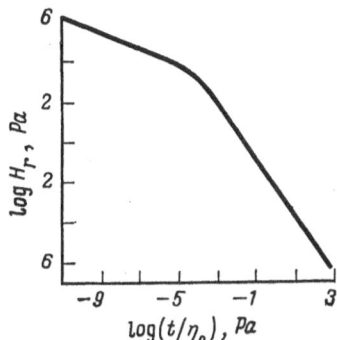

Fig. 3.17. The "universal" logarithmic spectrum characterizing the relaxation time distribution for melts of poly-disperse polymers.

place a coil-helix transition and the viscoelastic properties of the polymer chain are substantially changed.

The last case, which actually limits the possibility of application of the temperature superposition principle, is associated with the existence of several various relaxation mechanisms, each of which can be characterized by its own temperature dependence (or activation energy). Then, at different temperatures the relaxation times referring to various molecular mechanisms make different contributions to the observed viscoelastic functions. This makes impossible the superposition of experimental data into a single master curve.

The question of the existence of "universal" relaxation spectra for polymer systems of similar composition in analogous states is important. It is often possible to obtain the universal form of relaxation functions and relaxation spectra, as, for example, in the case of melts and concentrated solutions of polydisperse polymers. An example demonstrating the validity of this approach is given in Fig. 3.15: the relaxation spectrum constructed in normalized coordinates is found to be common to a very wide range of polymer solutions; the individual properties of the system are described here with the aid of two parameters: θ_m and G_0 (or η_0 and G_0). This problem is even more simply solved for melts of polydisperse polymers*, whose viscoelastic properties are described by means of a "universal" relaxation spectrum (Fig. 3.17) and the individual properties of the polymer are represented by a single parameter, η_0. This parameter is used to normalize the relaxation

---

* Vinogradov, G. V., A. Ya. Malkin, N. V. Prozorovskaya, and V. A. Kargin, *Doklady Akad. Nauk SSSR*, **154**, 1421 (1964); Vinogradov, G. V. and A. Ya. Malkin, *Zhurn. Prikl. Mekh. i Tekhn. Fiz.*, 5, 66 (1964); *J. Polymer Sci.*, A-2, **4**, 135 (1966).

times in the spectrum (θ/η_0) and frequencies $(\omega\eta_0)$ in considering the dynamic functions.

The use of "universal" spectra, which describe the viscoelastic properties of a wide range of polymer systems, is important for practical purposes, especially if no other information concerning the behaviour of a particular material is available. Of course, there are polymeric materials that differ in their relaxation spectra. But it seems to be important that polymers can be classified into rather extensive groups covering a large number of materials, according to the form of relaxation-time distribution. For them the position of the spectrum is determined by the characteristic relaxation time, θ_m, because other relaxation times in the spectrum are expressed in terms of θ_m, and the role of temperature, concentration, and other factors in the manifestation of the viscoelastic properties of the material is expressed through their effect on θ_m.

3.3. Viscoelastic Properties of Polymer Melts and Concentrated Solutions

The mechanism of viscoelasticity of polymer systems described in Sec. 3.1 is based on the consideration of the deformation of individual polymeric chains, each of which is thought to be a series of independent segments (subchains). When such a chain is deformed, there arise various modes of motion, which leads to the appearance of a discrete spectrum of interrelated relaxation times. This model most directly corresponds to dilute polymer solutions. The general principles of the theory of viscoelasticity of polymeric chains remain valid for concentrated solutions and melts because for these systems the original cause of viscoelastic properties is the ability of an individual chain to experience various modes of motion. But for these concentrated systems the theory must be reconsidered and modified, primarily with account taken of the interaction among the chains, which leads to the non-linearity of the dependence of the total resistance to motion on the chain length of the macromolecule.

3.3.1. Viscoelasticity of Low-Molecular-Mass Liquids

In the range of frequencies amounting to values of the order of 10^6 sec^{-1}, low-molecular-mass liquids, as a general rule, do not exhibit viscoelastic properties, i.e., in other words, their relaxation times lie in the region of values of $\theta < 10^{-6}$ sec, but at higher frequencies reaching 3×10^8 Hz in low-molecular-mass liquids too there can be detected relaxation phenomena. In accordance with what has been said in Sec. 3.2, the fall of temperature must bring about effects equivalent to the increase of frequency. In practice, however, the reduced frequency $\omega_r = \omega a_T$ cannot be increased substantially in this way because the viscosity of low-molecular liquids depends relatively weakly on temperature and therefore the value of the shift factor a_T is not high. Therefore measurements carried out over a wide temperature range do not lead to a substantial increase of ω_r. For example, the viscosity of water when cooled from the boiling to the freezing point increases only 6 times, which is equivalent to the same change of the reduced frequency.

There exist, however, a group of simple (non-polymeric) liquids which undergo vitrification when overcooled. This allows us to substantially broaden the concepts of the viscoelastic properties of liquids by comparing them with the properties of polymer systems proper. A typical example is provided by measurements of the creep and relaxation of rosin overcooled by rapid freezing. These measurements have shown* that rosin is characterized by a very narrow relaxation time distribution, so that it can be approximated, to a sufficiently good accuracy, by means of a Maxwell model with a single relaxation time.

Detailed investigations of a large number of organic vitrifiable liquids, including low-molecular-mass polymers, have shown** that they are also

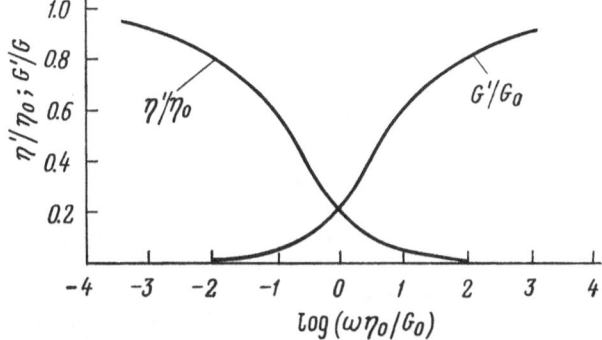

Fig. 3.18. Dependences of dynamic functions on frequency (in dimensionless form) for vitrifiable liquids (after J. Lamb).

characterized by a narrow (though non-Maxwellian) relaxation-time spectrum. A quantitative picture of this empirically found spectrum can be drawn if use is made of the frequency dependences of the components of the complex compliance and the modulus, which are obtained on the basis of calculations from this spectrum. These relations have the following form:

$$G' = \frac{4G_\infty (x/2)^{3/2} [1+(x/2)]^{1/2}}{\{[1+(x/2)^{1/2}]^2+(x/2)\}^2} \; ; \quad G'' = \frac{2G_\infty (x/2) [1+(2x)^{1/2}]}{\{[1+(x/2)^{1/2}]^2+(x/2)\}^2} \tag{3.31}$$

$$I' = \frac{1}{G_\infty} + \frac{1}{(x/2)^{1/2}} \; ; \quad I'' = \frac{1}{\omega\eta_0} + \frac{1}{G_\infty (x/2)^{1/2}} \tag{3.31a}$$

$$x = \omega\eta_0/G_\infty$$

where η_0 is the viscosity and G_∞ is the instantaneous elastic modulus. The frequency dependences of the dynamic viscosity and the dynamic modulus as predicted by these functions are shown in Fig. 3.18 in dimensionless variables. The quantity (η_0/G_∞) plays the role of the characteristic relaxation time of the system.

   * Tobolsky, A. V. and R. B. Taylor, J. Phys. Chem., 67, 2439 (1963).
  ** Barlow, A. J. et al., Proc. Royal Soc., A298, 467 and 481, A300, 356 (1967); A309, 497 (1969); Lamb, J., Rheol. Acta, 8, 428 (1969); Proc. Fifth Intern. Congr. Rheol., 4, 325 (1970).

Relaxation phenomena taking place in vitrifiable organic liquids develop in the range of values of the argument $(\omega \eta_0 / G_\infty)$ from 10^{-1} to 10^3, i.e., when the reduced frequency varies up to four decimal orders. This is not, of course, a very narrow range of relaxation times, but it is at least markedly narrower than the relaxation-time spectrum of high-molecular-mass polymers, for which there are often observed relaxation phenomena with the variation of frequency in the range of up to 14 decimal orders.

The results described above are applicable to vitrifiable liquids in the region of temperatures close to the glass-transition temperature. As the temperature mounts the relaxation spectrum is narrowed because molecular motion ceases to be cooperative, and at very high temperatures (with respect to the glass-transition temperature) the spectrum may degenerate to a single relaxation time, as has been observed with rosin.

3.3.2. General Concepts of the Viscoelastic Properties of Polymer Melts and Concentrated Solutions

In going from dilute solutions to concentrated systems and polymers in bulk and also from low-molecular-mass compounds to high polymers the character of the manifestation of viscoelastic properties is gradually changed, which is reflected in the shape of the frequency dependences of the dynamic functions $G'(\omega)$ and $G''(\omega)$. This is clearly seen from a comparison of Fig. 3.7 or 3.10 with Figs. 3.13 and 3.16.

In earlier investigations of the dynamic properties of concentrated solutions and melts the measurements were made on samples with a wide molecular-mass distribution (MMD). This complicated the analysis of experimental data and made it impossible to obtain unambiguous results on the relations between the molecular parameters of the polymer and the specific features of its chemical constitution and the viscoelastic properties of the material. In this case the main features of the viscoelastic properties of a high polymer are found to be smoothed out since the resulting dependences are arrived at by the superimposition of a large number of functions, each of which has its own characteristic features and points.

The dynamic characteristics of concentrated polymer solutions and polymer melts exhibit certain typical features (see Fig. 3.13). When a certain chain length is attained, on the $G'(\omega)$ curve there appears a plateau. Its level G'_p is practically independent of the molecular mass of the polymer and temperature. When the molecular mass of the polymer is about $5M_c$ (it will be recalled that M_c is such a critical value of molecular mass at which the dependence of the initial viscosity on molecular mass is sharply changed; see Chapter 2), a maximum appears on the $G''(\omega)$ curve, the position of which approximately corresponds to the appearance of a plateau on the curve of G' versus ω, and at higher frequencies the function $G''(\omega)$ passes through a minimum. To the minimum of the $G''(\omega)$ graph there evidently corresponds the minimum on the frequency dependence of the loss factor (loss tangent).

The existence of a rubbery plateau on the frequency dependence of $G'(\omega)$ is the most characteristic feature of the mechanical properties of cured elastomers. It is found to be due to the three-dimensional network of chemical bonds. Here the value of the modulus on the plateau, G'_p, is associated with the molecular mass M_e of the chain portion between the two effective cross links. The relation between them is expressed thus:

$$G'_p = \frac{\rho RT}{M_e} \tag{3.32}$$

where ρ is the density of the polymer; R is the gas constant; T is the absolute temperature at which the modulus G'_p is measured.

This formula is valid within the front-factor, which has somewhat different values in different theories of the mechanical behaviour of rubbers and which possibly depends on the features of the network topology. Using formula (3.32), one can determine the quantity M_e which quantitatively characterizes the density of the network of chemical bonds and may be called the molecular mass of a "dynamic segment" or simply "segment" of the chain.

If a cross-linked elastomer is compared with a linear polymer, an adequate physical model should be found, which would account for the existence in the fluid polymer of properties characteristic of the rubber with its network of permanent chemical bonds. Such a model is a network with temporary (fluctuating) junctions formed both by purely mechanical macromolecular entanglements and by any types of physical interactions localized in a number of points along the chain length. The concept of the entanglement network is used as a means of modelling the properties of a system. This method is useful when the detailed structure of the material is unknown; it allows the properties of the material being measured to be expressed in terms of the parameters associated with the specific molecular structure.

The most important parameter of the fluctuating entanglement network is the average lifetime of a junction or the relaxation time when the elements forming the network are subjected to mechanical action. If this time is infinitely long and comparable with the time of existence of chemical bonds, then the stresses in the network do not relax, except for the mechanism of chemical relaxation due to the rupture of chemical bonds. Then the polymer is capable of retaining strains and stresses for an indefinitely long time. This case corresponds to cross-linked rubbers or, in general, to polymers with a three-dimensional structural framework formed by covalent bonds. If the relaxation time is very short, or at least substantially shorter than the time of observation, the structural units appear to be absolutely nonlinked to the observer; they slip freely at the junctions and the system behaves as a typical liquid. In all intermediate cases, there occur a wide range of relaxation phenomena associated with the existence of a set (spectrum) of relaxation times of the motions of the polymeric chain. Thus, the spectrum may be simply divided into two parts: the region of slow relaxation processes which are completed more slowly than the fluctuating junctions of the network are broken down, and the region of rapid relaxation processes which develop more rapidly than the relaxation in the network junctions.

Rapid relaxation processes take place in fluid polymers in the same manner as they occur in cross-linked rubbers or in separate nonbonded chains because they are caused by the motion of short segments. Therefore, the region of rapid relaxation processes may be similar for all systems, from infinitely dilute to concentrated solutions of linear polymers and cured rubbers. The region of slow relaxation processes is varied in a general case and the description of the relaxation (viscoelastic) properties of the material in this region is the principal task of the theory of concentrated polymer systems.

In fact, both types of relaxation processes overlap because of the existence of a distribution of distances between the network junctions and the difference in their nature in a real material. This leads, in particular, to the following: during flow which occurs at a certain rate some of the junctions are retained in the form of a quasistable network and some are broken down rapidly enough and their presence is not essential to the resistance of the material to flow. Therefore, depending on the rate of deformation, the response of the system may prove to be quite different.

The main structural parameter of the fluctuating entanglement network, just as in the case of a network of chemical bonds, is its density characterized by the average statistical molecular mass (M_e) of the chain length between two successive network junctions. The quantity M_e, by analogy with what

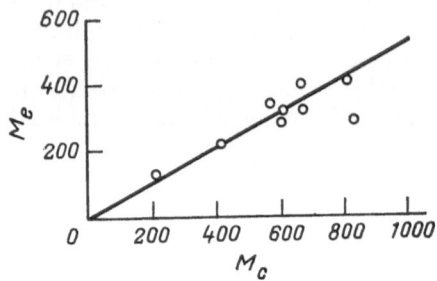

Fig. 3.19. Relation between M_e and M_c for polymethyl methacrylate, polybutadiene, polyvinyl acetate, polyisobutylene, polystyrene, and natural rubber. [Porter, R. S. and J. F. Johnson, $Rheol. Acta$, 7, 332 (1968).]

is done for cross-linked rubbers, can be determined from the experimentally measured value of the modulus G'_p with the aid of formula (3.32).

When considering the dependence of viscosity on molecular mass the critical molecular mass M_c was used as the main molecular characteristic determined by the same factors. Apparently, a correlation between M_e and M_c should be expected to exist. Indeed, independent determinations of these quantities for a number of polymers have shown* (Fig. 3.19) that

$$M_c \approx 2 M_e \qquad\qquad (3.33)$$

Then, either the number of dynamic segments equal to (M/M_e) or the value of the ratio $M/M_c = M/2M_e$ may be used as the dimensionless characteristic of the length of the molecular chain.

The extent of the plateau along the frequency axis of $\Delta \log \omega$ depends on molecular mass. This important experimental result may be represented in a generalized form if the number of dynamic segments is used as a measure of the length of the molecular chain of various polymers. The treatment of experimental data obtained by investigating the dynamic properties of a number of polymers in bulk has made it possible to construct a general dependence of the extent of the plateau, $\Delta \log \omega$, on the number of dynamic segments (Fig. 3.20). This dependence is described by the formula

$$\Delta \log \omega = \alpha \log (M/M_e) + B \qquad\qquad (3.34)$$

The value of the slope α is close to the exponent 3.4 in the relation between viscosity and molecular mass (see Chapter 2). This is due to the fact that the high-frequency limit of the plateau does not depend on the molecular mass of the chain and the low-frequency limit of the plateau, just as the entire region of slow relaxation processes, is displaced along the frequency axis in proportion to the viscosity of the polymer.

* Porter, R. S. and J. F. Johnson, $Rheol. Acta$, 7, 332 (1968); Malkin, A. Ya., E. A. Dzura, and G. V. Vinogradov, $Doklady Akad. Nauk SSSR$, 188, 1328 (1969).

The constant B in formula (3.34) is associated with the evaluation of the number of dynamic segments in the chain required for a plateau to appear. The point of intersection of the straight line in Fig. 3.20 with the abscissa exactly corresponds to the appearance of the plateau (since for this point $\Delta \log \omega = 0$). From Fig. 3.20 it follows that here $M = 4M_e$, i.e., the molecular mass M_p, at which the plateau appears on the frequency dependence

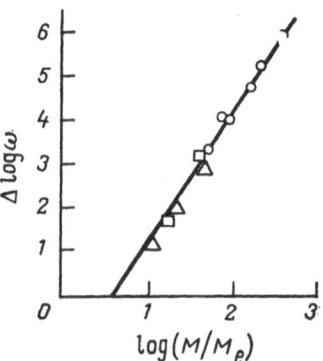

Fig. 3.20. Dependence of the extent of the rubbery plateau, $\Delta \log \omega$, on the number of dynamic segments in the polymeric chain for polybutadienes ($\bigcirc$), polystyrenes ($\triangle$), and polyvinyl acetate ($\square$) with different molecular masses. [Vinogradov, G. V., E. A. Dzura, A. Ya. Malkin, and V. A. Grechanovsky, *J. Polymer Sci.*, A-2, 9, 1153 (1971).]

$G'(\omega)$, is equal to the four-fold molecular mass of the dynamic segment. Then, formula (3.34) may be rewritten in the following manner:

$$\Delta \log \omega = \alpha \log (M/4M_e) = \alpha \log (M/2M_c) = \alpha \log (M/M_p) \qquad (3.34a)$$

With a low-molecular-mass solvent introduced into the polymer (but if the system still retains the fluctuating entanglement network) the following characteristic changes of the relaxation (viscoelastic) properties are observed*: the decrease of the values of storage modulus G'_p and the loss modulus G''_m at the point of the maximum of the curve of G'' versus ω; the gradual degeneration of the plateau on the G' versus ω dependence; and the disappearance of the extrema on the $G''(\omega)$ curve. In the region of high concentrations, for flexible polymers the dependences of G'_p and G''_m on the concentration of the polymer in the solution, φ_2, are close to quadratic relations: $G'_p \propto \varphi_2^2$ and $G''_m \propto \varphi_2^2$ and they are not affected by the quality of the solvent (Fig. 3.21). As the concentration of the polymers decreases further it is difficult to determine the values of G'_p and G''_m reliably and therefore it becomes impossible to speak of the concentration dependences of these parameters for solutions.

For a series of polymers belonging to the same polymer-homologous series or concentrated solutions at a sufficiently high content of the polymer in the solution the shape of the frequency dependences of the functions $G'(\omega)$ and $G''(\omega)$ in the region of low frequencies is the same until the plateau is attained. Therefore, the only individual characteristic of the system is still the value of the characteristic relaxation time, which depends on the molecular mass of the polymer and its content in the solution.

On the basis of what has been said above, the shape of the relaxation spectra, corresponding to the region of slow relaxation phenomena (until

---

* Blinova, N. K., S. I. Sergeenkov, A. Ya. Malkin, Yu. G. Yanovsky, and G. V. Vinogradov, *Mekhanika Polimerov*, 1, 132 (1973); Vinogradov, G. V. and A. Ya. Malkin, *Europ. Polymer J.* 9, 1231 (1973).

there appears a plateau on the G' (ω) curve and a maximum on the G'' (ω) dependence is attained), is insensitive to the length of the polymeric chain and the concentration of the solution. The form of the spectrum responsible for a set of relaxation phenomena in the region of the plateau and loss minima must also be the same provided that polymers with the same number of dynamic segments in the chain are compared. The latter proposition is illustrated by Fig. 3.22, which presents the frequency dependences of the loss tangent for two different polymers—polybutadiene and polystyrene. The molecular masses of the samples being compared are chosen so that the reduced chain length is nearly the same. The normalization with respect to the frequency axis was done by the choice of the characteristic relaxation time θ_0 such that the positions of the minima coincide. As seen from Fig. 3.22,

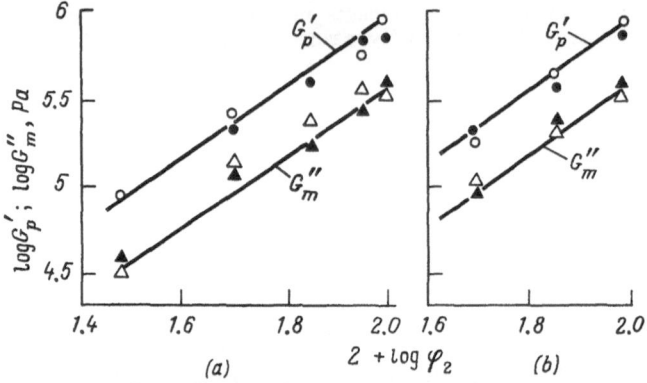

Fig. 3.21. Concentration dependences of the characteristic parameters of the dynamic functions for solutions of polybutadienes in methylnaphthalene (a) and diheptyl phthalate (b). The molecular masses of the polymers ($M \times 10^{-5}$): open symbols = 2.4; black symbols = 1.5.

this provides a practical identity of the frequency dependences of tan δ for different polymers.

The general regularities formulated above for the manifestation of the viscoelastic properties of concentrated polymer solution and molten polymers are sufficiently clearly observed for monodisperse or nearly monodisperse polymers. As the molecular-mass distribution is broadened such a clearcut picture disappears and there arises the problem (that has not yet found a general solution) of summing up all the contributions made by different fractions to the viscoelastic properties of the material.

Consideration of the dynamic characteristics of polymers enables us to work out a general classification of their relaxation states at temperatures above T_g. Thus, the region of low frequencies corresponds to the fluid state of linear polymers; the plateau region on the frequency dependence of the modulus, G' (ω), corresponds to the rubber-like state of linear polymers when they are no longer capable of flowing and behave, at high rates of loading, as quasicured systems. The transition from one state to another takes place in the region of frequencies corresponding to the maximum of the loss modulus or the loss factor. At still higher frequencies when the storage modulus G' begins to increase as compared with the value of G_p', there takes place the transition to the glassy state, but in such cases the specific features of fluid polymer systems are lost since, as has been noted above, the behaviour of the material in this region is independent of the molecular mass of the polymer and also of whether the macromolecules are still linear, or are bonded by chemical cross links.

In small-amplitude oscillating shear polymers may undergo indefinitely prolonged adestructive deformation over a very wide frequency range, covering many decimal orders. This makes it possible to attain, at a single temperature, the fluid, rubbery, leathery, and glassy states. The question often arises as to the legitimacy of this kind of estimation of the states of polymers when they are at temperatures considerably higher than the glass-transition temperature. What is meant here is that a polymer, which is a viscoelastic

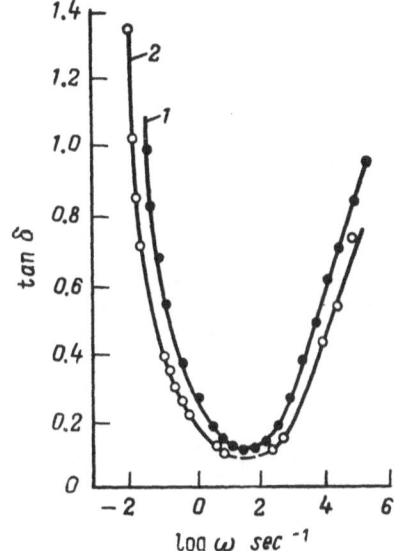

Fig. 3.22. Frequency dependences of loss tangent for polybutadiene with $M/M_e = 44$ (1) and polystyrene with $M/M_e = 51$ (2). [Vinogradov, G. V., E. A. Dzura, and A. Ya. Malkin, *Inzhenerno-Fiz. Zhurn.*, **18**, 965 (1970).]

liquid, behaves like cured rubber-like elastomers and glass-like bodies in the region of high frequencies. In answering the above question one should rely on the existence of the equivalency of the thermomechanical (determined under a certain deformation regime) and velocity (measured at a certain constant temperature) characteristics of polymeric materials, which are determined by the relaxation nature of their behaviour.

The dynamic method of studying the relaxation properties of polymers makes it possible to determine the frequency regions corresponding to the transition of polymers from the fluid to the rubbery, leathery, and glassy states. These transition frequencies depend on the temperature (they decrease with fall of temperature) and diminish with increasing molecular mass of the polymer within a polymer-homologous series.

The measurement of the relaxation (viscoelastic) properties of polymer-homologous series, beginning from monomers and oligomers and up to typical high polymers, makes it possible to deduce the following general pattern of variation of the viscoelastic properties at characteristic values of molecular masses. If M_k is the molecular mass of a kinetic segment that participates in a single act of transport on viscous flow, upon attainment of molecular masses of the order of (5-10) M_k there appear entanglements and the value of $M_e = $ (5-10) M_k corresponds to the distance between the junctions of the

Table 3.1.   Classification of Low-Molecular-Mass Substances, Oligomers and Polymers According to Their Viscoelastic Properties

Low-viscosity liquids	Oligomers	Polymers	High-Polymers	
M_k = kinetic flow segment	$M > M_e = (5\text{-}10)\,M_k$	$M > M_c = (2\text{-}6)\,M_e$	$M \gg 10\,M_e$	$M \gg 20\,M_e$
	M_e = kinetic unit in the entanglement network—a dynamic segment according to oscillatory measurements.	M_c = critical value of M corresponding to an increase in the slope of the viscosity vs. molecular mass curve.	Increase in the uniformity of segmental density of the entanglement network.	Uniformity of segmental density in the entanglement network is attained.
			The long-time part of the relaxation spectrum is developed.	Separation of the short- and long-time parts is completed.
	Appearance of an entanglement network.	An entanglement network is developed.	Appearance of the symptoms of the rubbery state.	All the specific features of the rubbery state become prominent.
	The long-time part of the relaxation spectrum appears.	Appearance of large recoverable (rubbery) deformations.	The first symptoms of transition from the fluid to rubbery state become noticeable.	The transition from the fluid to the rubbery state becomes sharp with increasing deformation.
		Non-Newtonian behaviour appears.	The inflection point on the curves of G' and G'' vs. frequencies and the weakly pronounced loss maximum on the G'' vs. ω curve appear.	The loss maximum and the rubbery plateau are sharply pronounced on the G' and G'' vs. ω curves.
			The zero shear modulus decreases.	Loss of fluidity with increasing rate of deformation.
			The glass transition temperature is stabilized.	Flow behaviour resembles the behaviour of Newtonian fluids over a wide range of stresses.
				The zero shear modulus is stabilized at a low value level.
				The stress optical rule is obeyed.

fluctuating entanglement network [see formula (3.32)]. In the region of molecular masses M_c exceeding $2M_e$ there is formed an entanglement network, which leads to the change of the character of the dependence of viscosity on molecular mass and to the possibility of development of large recoverable (rubber-like) deformations. As the chain length increases further, when $M = M_p \approx$ (4-5) M_e, there appears an inflection point on the $G''(\omega)$ curve, which corresponds to the attainment of the rubbery state. In the same region of molecular masses the value of the glass-transition temperature is stabilized in a given polymer-homologous series. The completion of the changes of the properties and the transition to the characteristics typical of high polymers take place at molecular masses of the order of $10M_c$ or $20M_e$. Here, the maxima in the relaxation spectrum are clearly pronounced and the plateau on the curve of G' versus ω is stabilized and the compound acquires all the features of a polymer (see Table 3.1).

3.3.3. Molecular Theories of Viscoelasticity of Polymer Melts and Concentrated Solutions*

During the flow of concentrated polymer solutions and molten polymers the motion of macromolecules is subject to rather complicated restrictions imposed by similar surrounding macromolecules and by the solvent. In accordance with the conceptions of the mechanics of constrained systems, from the consideration of the motion of an ensemble of interconnected chains we may turn to the analysis of an equivalent system of free chains. In such an approach, when setting up equations of motion for separate chains, the restrictions imposed on the motion are taken into account by introducing additional forces, the action of which is equivalent to the action of the real environment. These forces are unknown, and therefore for the redetermination of the motion of polymer chains there are required additional assumptions. The indicated scheme of consideration is common to all the proposed theories of the viscoelasticity of polymers, which are based on the analysis of macromolecular motion. The distinction between different molecular theories is in the manner in which they solve the basic problem—the modelling of the effect of the environment on the motion of macromolecules.

The attempt to take into account this effect in the simplest way, by introducing a viscous liquid instead of environmental macromolecules, has proved successful only for polymers of not high molecular mass, in which case the complications introduced by chain entanglements may be neglected. For polymers of high molecular mass the KSR model of subchains which has been considered earlier proves to be insufficient. It is not able to account for the experimentally observed plateau on the frequency dependence of the storage modulus and also for the local extrema (the maximum and the minimum) on the frequency dependence of the loss modulus. Therefore, attempts have been made to develop the theory further, which has been done in two directions—either local entanglements were supposed to exist or it was assumed that the surrounding continuum has certain, more complex properties than a Newtonian liquid.

* Sections 3.3.3-3.3.6 have been written in collaboration with V. S. Volkov.

3.3.4. The Concept of Entanglements

An important stage in the development of the theory of viscoelasticity of concentrated polymer systems was the work of Bueche*, which laid the foundation of the concept of entanglements. According to its basic assumption, the interaction between the macromolecule and the environment is localized at isolated points along the chain—at the entanglements with other macromolecules. Bueche supposed that the effect of entanglements can be taken into account with the aid of an increased value of the friction factor at points of contact between the entangled macromolecules as compared with the friction factor of non-entangled macromolecules. Based on these conceptions, Bueche determined the dependence of viscosity on molecular mass for concentrated polymer solutions and polymer melts.

The Ferry-Landel-Williams Model. This scheme was then extended by Ferry and his coworkers** to the calculation of the effect of entanglements on the viscoelastic properties of concentrated polymer systems on the basis of the KSR model. They suggested that the presence of entanglements has an effect only on the friction factor of low normal modes of motion corresponding to the motions of long chain portions, as follows:

$$f = \begin{cases} f_0 & \text{at } p_e < p \leqslant n \\ f_0 p_e^{2.4} & \text{at } 1 \leqslant p \leqslant p_e \end{cases}$$

where f_0 is the friction factor of the bead in a "segmental" liquid; n is the number of segments; $p_e = M/2M_e = M/M_c$, and $M > M_c$.

In accordance with this, for polymers having a molecular mass greater than M_c, the relaxation spectrum of the system is split into two sets. The short relaxation times

$$\theta_p = f_0 \sigma^2 n^2 / 6\pi^2 p^2 kT, \quad p_e < p \leqslant n \tag{3.35}$$

are associated with the motion of short chain portions, which occurs independently of the presence of entanglements. These relaxation times are the same for polymer melts and concentrated solutions and dilute solutions.

The motion of long chain portions, which depends to a large extent on entanglements, leads to the appearance of long relaxation times:

$$\theta_p = f_0 p_e^{2.4} \sigma^2 n^2 / 6\pi^2 p^2 kT, \quad 1 \leqslant p \leqslant p_e \tag{3.35a}$$

In the region of slow relaxation processes the theory predicts a very strong dependence of relaxation times on molecular mass. According to formula (3.35a), the maximum relaxation time θ_m is proportional to molecular mass raised to a power of 4.4.

The viscosity of the system is given by

$$\eta = \frac{\rho RT}{M} \sum_{p=1}^{p_e} \theta_p = 1.645 \frac{\rho RT}{M} \theta_m \tag{3.36}$$

It is mainly associated with the contribution of slow relaxation processes, the viscosity being automatically proportional to a molecular mass raised to a power of 3.4 since the parameter of the theory, p_e, is determined so that η is proportional to $M^{3.4}$.

---

* Bueche, F., *J. Chem. Phys.*, **20**, 1959 (1952); Bueche F., *Physical Properties of Polymers*, Interscience Publishers, New York, 1962.
** Ferry, J. D., R. F. Landel, and M. C. Williams, *J. Appl. Phys.*, **26**, 359 (1955).

The equilibrium compliance on shear, in accordance with the theory under consideration, is proportional to molecular mass:

$$J_e = 0.4M/\rho RT \qquad (3.37)$$

This prediction does not agree with an important experimental fact—the compliance of high-molecular-mass polymers is independent of their molecular mass.

The prediction of the theory that the plateau level is independent of the molecular mass of the polymer,

$$G'_p = \frac{\rho RT}{M_c} \qquad (3.38)$$

is in accord with known experimental data.

Figure 3.23 compares experimental and theoretical frequency dependences of the storage modulus, $G'(\omega)$, for samples of low- and high-molecular-mass

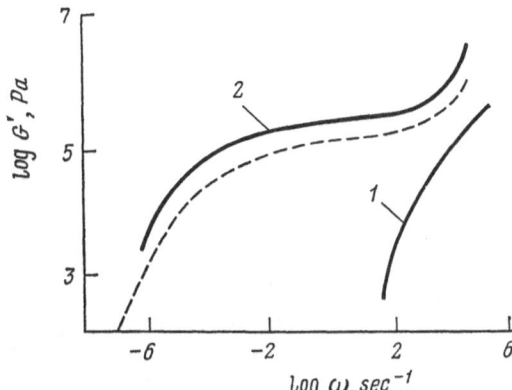

Fig. 3.23. Frequency dependences of the elastic modulus at 25°C for polyisobutylenes with molecular masses 1.09×10^4 (1) and 1.56×10^6 (2). Solid lines, experiment. Dashed line represent calculation from relaxation times modified with account taken of the existence of entanglements (Ferry J. D., R. Landell, and M. Williams)'

polyisobutylenes. The main result is that the theory in question accounts for the appearance of the rubbery plateau, which is associated with the division of the relaxation spectrum of a high polymer into two groups of relaxation times which strongly differ in magnitude.

The Chompff-Duiser-Prins Model. The ideas of Bueche and Ferry and coworkers were further developed by the works of Duiser, Chompff, and Prins*. Their model (abbreviated to the CDP model) is an extension of the model of the network formed by permanent cross-links, which was suggested earlier by Duiser and Staverman, to a network of temporary entanglement. This model is also based on the assumption that the intermolecular interaction in concentrated polymer solutions and polymers in bulk may be regarded as a consequence of the local interactions between a given macromolecule and the other macromolecules at a certain number of isolated points along the chain.

Considering a system of two chains with fixed ends at which the entanglement point is in the middle of each of the chains, the authors have shown that such a system is equivalent to two non-interconnected chains with fixed ends. Here the mobility of one of the chains at the previous contact point remains unchanged, while the mobility of the same point of the other chain

---

* Chompff, A. J. and J. A. Duiser, *J. Chem. Phys.*, **45**, 1505 (1966); Chompff, A. J. and W. Prins, *J. Chem. Phys.*, **48**, 235 (1968).

is multiplied by the slippage factor δ, which is a measure of the decrease of mobility because of the presence of entanglements.

The extension of this procedure to the entanglement network makes it possible to pass from the consideration of an ensemble of interacting macromolecules to the analysis of the behaviour of a mechanically equivalent system, which is a set of non-interconnected macromolecules. A distinctive feature of the CDP model is the assumption that the chains of an equivalent ensemble contain different numbers of slowly moving points and, hence, the study of the systems does not boil down to the consideration of the motion of a single chain. This work should be regarded as the first attempt to examine the non-unimolecular approximation by taking account of the fundamental fact of the non-equivalency of the conditions of motion of separate macromolecules.

The relaxation spectrum of the entire system is obtained in such a case by summing up the contributions of the various chains of the equivalent system with account taken of their molar fractions. In case the parameter δ of the theory, which is a measure of the relative mobility of the entangled macromolecules at the point of contact, equals zero, the model describes a network with stable junctions; for an entanglement network $0 < δ < 1$; for free macromolecules in dilute solutions $δ = 1$.

The resulting relaxation-time spectrum coincides in the region of short relaxation times with the ordinary Rouse spectrum with a slope of −1/2. This is due to the fact that the presence of junctions has no effect in this region. In the long-time region associated with the motion of chain portions, whose lengths are commensurate with or greater than the distances between the junctions, the spectrum passes through a maximum after having passed through a deep minimum. The presence of such a maximum is observed experimentally in monodisperse polymers. After the maximum is passed the slope is practically constant and equal to −1/2 up to the longest relaxation time. This part of the spectrum is the part of the Rouse spectrum shifted to the side of long relaxation times.

The above comparison of the frequency dependences of G' and G'' calculated by the CDP theory with the experimental data obtained by Ferry for poly-n-octyl methacrylate shows that the theory satisfactorily describes the plateau region of the storage modulus. But the CDP model, just like the model worked out by Ferry and coworkers, is not in harmony with the actual behaviour of the function $G'(ω)$ in the region where the plateau appears. This shortcoming is attributed to the fact that the relaxation-time spectrum corresponding to this region is the shifted part of the Rouse spectrum, which has a wider distribution than that observed experimentally.

The Models of Non-Equivalent Entanglements. The investigations carried out by Ferry and by Chompff, Duiser, and Prins have shown that the effect of entanglements on the dynamics of separate macromolecules of a high polymer within the framework of the KSR model is not sufficiently completely reflected by the assignment of equally increased values of friction factors to the beads chosen in a certain special way (as compared with the other beads).

An attempt* has been made to describe, in a more successive manner, the effect of entanglements by taking into account the possible difference between

* Pokrovsky, V. N., Yu. G. Yanovsky, G. N. Kargapolova, and G. V. Vinogradov, *Mekhanika Polimerov*, 6, 1006 (1972); Vinogradov, G. V., V. N. Pokrovsky, and Yu. G. Yanovsky, *Rheol. Acta*, **11**, 258 (1972).

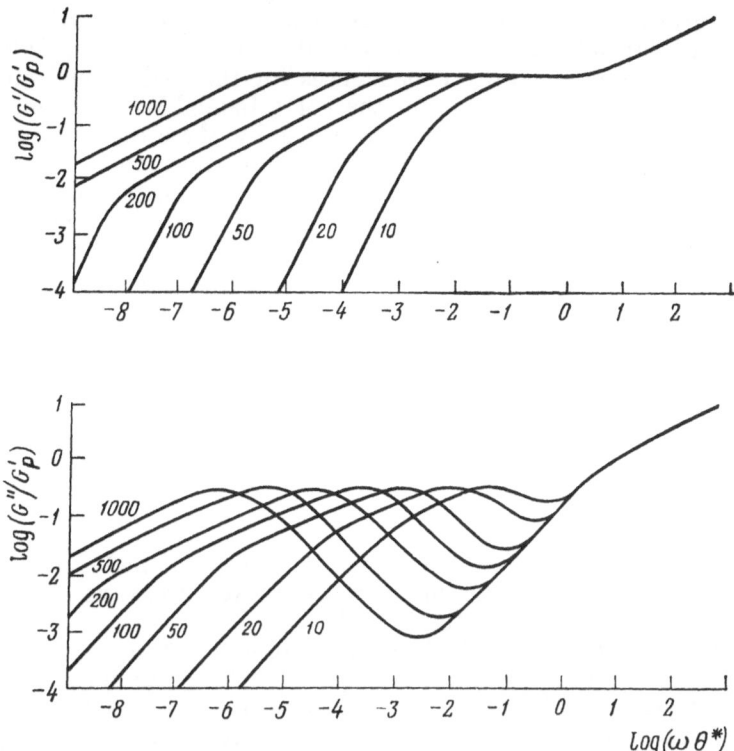

Fig. 3.24. Dependences of the real and imaginary parts of the dimensionless dynamic modulus on dimensionless frequency at the values of the parameter u indicated by numerals.

the friction factors of slow beads, depending on their position along the chain. Pokrovsky and coworkers, just as Chompff, Duiser, and Prins, believed that the friction coefficient of slow beads increases from the centre of the macromolecule to its ends. The values of the friction coefficients of slow beads, depending on their number counted off from the middle of the macromolecule towards its ends, are approximated in the following way:

$$\zeta_e = \beta_0 \frac{L}{u} (1 + | q\alpha |^\delta) \qquad (3.39)$$

where L is the length of the macromolecule, which has u slow points (beads); β_0 is the friction coefficient of the unit length of the macromolecule between the adjacent slow beads; q and δ are constants which determine the character of the increase of the friction coefficient of slow beads. These constants are unequivocally determined by the requirement of a correct description of the dependence of viscosity on molecular mass ($\eta \propto M^{3.4}$). Here $q = 2$, $\delta = 2.4$.

The dynamic functions $G'(\omega)$ and $G''(\omega)$ calculated for this model at various values of the parameter u indicated on the corresponding curves are presented in Fig. 3.24 in dimensionless form. The values of G' and G'' are normalized by the value of the modulus G'_p corresponding to the plateau:

$$G'_p = NuT$$

where N is the number of macromolecules per unit volume.

18—0489

The frequency ω is normalized by the characteristic relaxation time θ^*. The frequency dependences of the components of the dynamic modulus shown are close to the corresponding dependences for the CDP model and to the dependences obtained by Marvin for the Ferry-Landel-Williams model, i.e., the relaxation-time distributions determined by these theories are similar.

As seen from the graphs, the theory predicts that upon attainment of a certain value of molecular mass a plateau appears on the curve of G' versus ω, and a maximum and a minimum on the curve of G'' versus ω. The position of the maximum approximately coincides with the onset of the plateau. The extent of the plateau on the frequency scale of $\Delta \log \omega$ at a specified temperature is proportional to the number of entanglements per one macromolecule, raised to a power of 2.4:

$$\Delta \log \omega = 2.4 \log \left(\frac{M}{M_c} \right) \tag{3.40}$$

The value of G''_{max}, just as the value of G'_p, is inversely proportional to the molecular mass of the chain between the neighbouring entanglements, M_e:

$$G''_{max} = 0.32 \frac{\rho RT}{M_e} \tag{3.41}$$

According to the theory under discussion, the critical molecular mass M_c and the molecular mass of the chain between two successive entanglement points M_e are interrelated as follows: $M_c = 2M_e$.

A comparison of the theoretical and experimental frequency dependences of G' and G'' shows that the theory in question describes qualitatively the viscoelastic behaviour of polymers over a wide frequency range. The largest difference is observed upon transition from the fluid region to the plateau, just as in the theories discussed earlier.

Another disposition of slow beads along the chain, when the slowest beads are located not at the chain ends but in the middle of the chain*, does not eliminate the above-noted shortcoming either.

In general, the relaxation-time distribution in the region of transition from the fluid to the rubber-like behaviour cannot be described correctly by varying the friction coefficient of slow beads, as has been found by Hayashi**. As established later, this is associated with the neglect of an essential circumstance—the elastic effect of entanglement on the motion of the entangled polymer chain.

The Graessley Model.*** The physical basis for the Graessley model is the assumption that the medium exerts (via entanglements) not only a viscous drag but also an elastic resistance when the macromolecule chosen is moving through it. Each entanglement is treated as a separate bead-spring interaction between the macromolecule as a whole and the medium. In other words, it is assumed that the slow beads are interconnected by an elastic spring through the centre of gravity of the chain. The theory also makes the assumption of the dependence of the friction coefficient of slow beads and the elastic coefficient of the springs that connect them with the centre of mass on their contour position. The friction coefficient rapidly increases as the segment is moving away from the centre of mass.

Graessley examined the behaviour of the model proposed by him and obtained the basic quantitative results—the expressions for the components

* Hansen, D. R., M. C. Williams, and M. Shen, *Macromolecules*, 9, 345 (1976).
** Hayashi, S., *J. Phys. Soc. Japan*, 18, 131 and 249 (1963); 19, 101 (1964).
*** Graessley, W. W., *J. Chem. Phys.*, 54, 5143 (1971).

of the stress tensor in simple shear, the relaxation-time distribution, the functions $G'(\omega)$ and $G''(\omega)$, the time dependence of the relaxation modulus, the expression for viscosity and equilibrium compliance in terms of the parameters of the model and first of all in terms of the main parameter—the number of entanglements E per one chain. Figure 3.25 shows the dependences of G' and G'' on ω predicted by the Graessley model for $E = 200$, the components of the dynamic modulus being normalized by the value of νEkT (where ν is the number of chains per unit volume), which is equal to G_p', and ω by the value of the maximum relaxation time θ_m^G. The components of the dynamic modulus in the Graessley theory in the frequency region corresponding to the onset of the plateau are in good agreement with

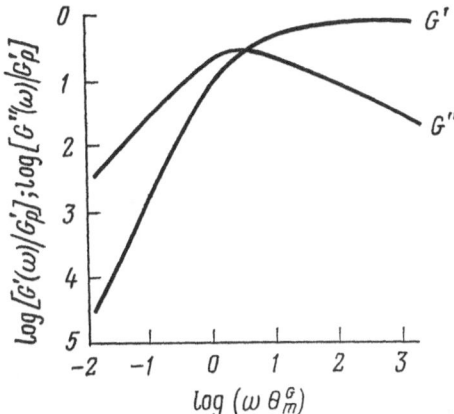

Fig. 3.25. Frequency dependences of the components of the dynamic modulus in normalized (dimensionless) form calculated for $E = 200$ (after W. W. Graessley).

experimental data, in contrast to the molecular theories proposed earlier. The Graessley theory gives a number of interesting relations between the characteristic constants of the material. For instance, the critical molecular mass M_c is related to the molecular mass of the chain portion between the neighbouring entanglements, M_e, by means of the formula $M_e = 0.397 \cdot M_c$ [compare with formula (3.33)]. The value of the loss modulus at the maximum $G''_{\max}$ is found to be equal to $0.332\,G_p'$, which is close to experimental data* and the value of the storage modulus G_s' at a frequency corresponding to the maximum of the loss modulus is given in terms of G_p' as $G_s' = 0.383\,G_p'$. The maximum of G'' is attained at a frequency substantially smaller than the frequency ω_m at which the values of G' are in the plateau region. The quantity ω_m is connected with θ_m^G by the relation $\omega_m\theta_m^G = 3.32$. An important parameter in the Graessley theory is the number of chain entanglements E. This parameter is common to all the theories based on the entanglement concept but the meaning and importance of the other parameters prove to be different. For polymer solutions the value of E is proportional to the product of the polymer concentration by the molecular mass (cM).

The maximum relaxation time is related to E in the following manner:

$$\theta_m^G = 3.575\eta/\nu EkT \qquad (3.42)$$

where η is the steady-state viscosity.

* According to the data reported, $G''_{\max} \approx 0.4G_p'$: Vinogradov, G. V., A. Ya. Malkin et al., J. Polymer Sci., A-2, 10, 1061 (1972).

According to the Graessley theory, the relationship between the equilibrium compliance J_e and E has the following form in a general case:

$$J_e = \frac{\alpha_1 E}{1 + \alpha_2 E} \qquad (3.43)$$

where α_1 and α_2 are constants. Therefore, at low values of E, just as in many other theories, J_e is proportional to E. But for polymers of high molecular mass J_e is no longer dependent on E. This corresponds to an important experimental fact which is difficult to explain by means of other theories— the compliance (or modulus of rubber elasticity) of polymers of high molecular mass is independent of their molecular mass. All these results referred to monodisperse polymers. Although J_e does not depend on molecular mass, it has been found experimentally that it is very sensitive to the molecular-mass distribution. The attempt made by Graessley to extend his theory to polydisperse polymers was not successful. The theory overexaggerates, to a very large extent, the effect of polydispersity on the value of J_e for blends of monodisperse samples.

The Model of Elastically Coupled Entanglements. The idea of the elastic response of the medium to the macromolecule that moves through it is used in a number of molecular-kinetic models*. The calculations carried out by Shen and coworkers have been especially fruitful; they arrived at a good agreement between the theoretical predictions and the observed viscoelastic behaviour of real polymers.** According to the model proposed by these authors, polymeric chains are, as usual, represented as a linear sequence of beads and springs. The effect of entanglements on the dynamics of chains is introduced as follows. Some beads chosen in a certain special manner are assigned increased (as compared with the others) values of friction coefficients and also there are introduced additional elastic constraints between pairs of slow beads. Since all the points of the chosen macromolecule are interconnected via an entanglement network, the elastic forces acting on a given entanglement are transmitted through the network and can affect the motion of the other points of the chain. It is presumed that the intensity of the elastic constraint diminishes with increasing distance between the entanglement loci. The theory also takes account of the effect of the slippage of chains over one another at an entanglement point and it is assumed that the increase of the friction coefficient brought about by the entanglements near the chain ends is less than that given by the entanglements near the centre of the chain.

The equations for the dynamics of the macromolecule in the model of entanglements exhibiting elasticity are very similar to the analytical equations of the Zimm theory, except for the fact that the hydrodynamic (viscous) interactions are replaced with viscoelastic constraints. The linear viscoelastic functions that follow from this model are calculated from the known relaxation-time distribution. Comparison of the calculated and experimental dependences of G' and G'' on ω has shown (Fig. 3.26) that they are in good agreement for monodisperse polymers.

* Forsman, W. C. and H. S. Grand, *Macromolecules*, 5, 289 (1972).
** Hansen, D. R., M. C. Williams, and M. Shen, *Macromolecules*, 9, 345 (1976); Hong, S. D., D. R. Hansen, M. C. Williams, and M. Shen, *J. Polymer Sci., Polymer Phys. Ed.*, 15, 1869 (1977); Hong, S. D., D. Soong, and M. Shen, *J. Appl. Phys.*, 48, 4019 (1977).

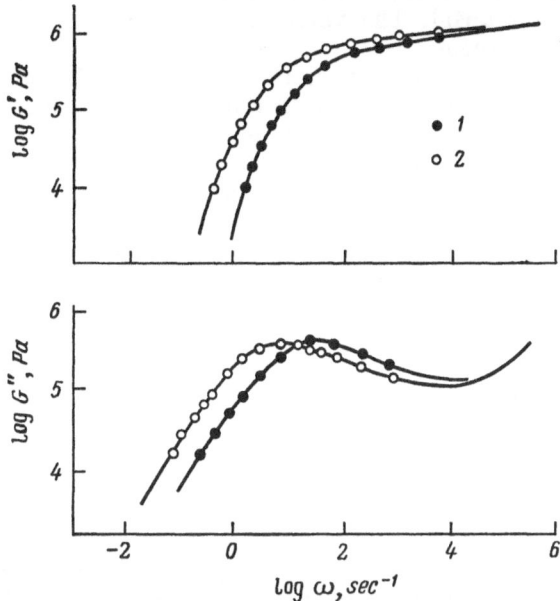

Fig. 3.26. Comparing the theoretical dependences G' (ω) and G'' (ω) (solid lines) with experimental data [Vinogradov, G. V., A. Ya. Malkin *et al.*, *J. Polymer Sci.*, A-2, **10**, 1061 (1972)] for monodisperse polybutadienes with different numbers of entanglements in the macromolecule: 34 (*1*) and 50 (*2*). [Hong, S. D., D. Soong, and M. Shen, *J. Appl. Phys.*, **48**, 4019 (1977).]

Concerning the Theory of the Rubber-Like Network. The model of a network of chains connected by fluctuating interacting points is an equivalent representation of the real structure of concentrated solutions and polymers in bulk. In the theories discussed above the mechanical properties of the network were considered by introducing the concept of the effective friction coefficient in the junctions in which the ends of dynamic segments of the chain are interacting. Lodge* analysed the model of the network itself, assuming as its main characteristics the rate of formation of a junction between two chains and its lifetime up to breakdown. Then the number of junctions retained in unit volume at time t with respect to some elapsed time t' is proportional to the value of N ($t - t'$) equal to

$$N(t-t') = \sum_{a,b} L_{a,b}\, e^{-(t-t')/\theta_{a,b}} \tag{3.44}$$

where $L_{a,b}$ is the rate of formation of a junction by the chains a and b; $\theta_{a,b}$ is a constant characterizing the rate of breakdown of the junction, which has the meaning of the relaxation time of the corresponding entanglement.

The stresses acting on the network at each instant of time are proportional to N, which is why the memory function μ included in the constitutive equation for the medium is expressed in terms of N. The stress-strain relationship is described with the aid of an integral expression which includes the memory function as a kernel, as is usual the case in the linear theory of viscoelasticity (see Chapter 1). Thus, the Lodge theory indicates that the

* Lodge, A. S., *Rheol. Acta*, **7**, 379 (1968); **10**, 539 (1971).

memory of the strain history is associated with the existence of fluctuating junctions in the material. The values of the constants $L_{a,b}$ and $\theta_{a,b}$ are not detailed in the theory.

The network theory of Lodge is an extension of the Green-Tobolsky theory of rearranging networks of chemical cross-links to concentrated polymer solutions and polymer melts, the difference being only in details.

Later, the rubber-like network model has been developed in two directions. First, the initial proposition of the theory that the distribution of the distances between the junctions of the fluctuating network is described by the Gaussian probability law has been generalized so as to introduce "non-Gaussian" terms of the distribution of the distances into the treatment (Yamamoto). This leads to the correct prediction of certain nonlinear effects observed during the flow of polymers. However, in this case too the viscoelastic properties of the model are not detailed, so that the theory provides for the possibility of the free choice of the form of the relaxation spectrum and, hence, the forms of all viscoelastic functions. Second, it has been suggested (Kaye) that the probability of the formation of junctions or their lifetime depends on the acting stress. This supposition which substantially generalizes the Lodge theory makes it possible to describe various nonlinear effects, in particular the phenomenon of non-Newtonian viscosity. Such an approach is associated with the arbitrary choice of the shape of the function that is called for to take into account the effect of stresses on the parameters characterizing the properties of fluctuating network junctions. This trend in the development of the network model, being highly versatile, does not allow us to detail the predictions concerning the form of the viscoelastic properties of the material either.

Although the rubber-like network model makes use of propositions different from those employed in the "subchain" model, a correlation can nevertheless be established between them. The basis for this is that the relations between the macroscopic stresses and strains in the subchain model are quite the same as in the network model, i.e., are represented by the equation of the theory of viscoelasticity common to both cases. The use of the subchain model has the advantage that it allows one to express, in a concrete form, the values of relaxation times in the spectrum. Then, the expression for the memory function in the network model is replaced by an equivalent expression which is more concrete:

$$\mu\,(t) = \nu kT \sum_p \theta_p^{-1} e^{-t/\theta_p}$$

where ν is the number of macromolecules per unit volume; θ_p are the values of relaxation times calculated for the necklace model in any of its generalized modifications. The conditions of the motion of the chain in the cases being compared are varied: in the rubber-like network model each chain moves affinely to the deformation of the body as a whole (just as in the case of elastomers with a network of chemical cross-links); in the subchain model, the chain moves as a whole relative to its surroundings.

The analogy between the basic relationships deduced in the network and subchain models makes it possible to relate the rate of formation of network junctions and their lifetime to the measured relaxation times of the system. The significance of this result consists also in that it allows one to limit oneself to the consideration of the behaviour of a unit chain divided into dynamic segments in constructing mechanical (or molecular-kinetic) models and theories not only for dilute but also for concentrated polymer solutions. The friction during the motion of each of these segments in a homogeneous medium surrounding the chain simulates not only the resistance of the macromolecule to motion in a low-molecular-mass solvent but also the interaction of the given chain with the other chains with which it forms a network of fluctuating contacts (physical interactions of any type). The concrete features of the structure of the system must be taken into account by a correct choice of the friction law. In the simplest case, this may be the Newton-Stokes linear law, and for concentrated solutions there may be introduced a certain operator. The concrete form of the friction law may be either given *a priori* or found from any physical considerations. But, in any case, there exists the possibility of considering the behaviour of an individual macromolecular chain in order to model the manifestation of the viscoelastic (relaxation) properties of any polymeric systems, including concentrated polymer solutions and polymer melts and blends.

3.3.5. The Concept of Microviscoelasticity*

In spite of the attractiveness of the concept of entanglements, which takes account of the polymeric nature of the medium surrounding the macromolecule, the basic proposition concerning the localized character of the dynamic interaction of macromolecules with the surroundings is far from obvious. In any case, the presence of entanglements requires explanations and direct proofs.

There has lately been observed the tendency to dismiss with the concept of the entanglement network and attempts are being made to work out a theory by imposing limitations of kinematic character on the motion of macromolecules. The KSR model as applied to concentrated polymer solutions and melts is developed further by replacing the real material surrounding the macromolecule by a certain averaged complex (non-Newtonian) continuum. It is assumed that when the macromolecule is moving among similar macromolecules and in the solvent (if there is any), the response of the surroundings is delayed, i.e., it is equivalent to the response of an elastic liquid.

The basis for the theory is the equations of the dynamics of the macromolecule simulated by a linear chain of beads and springs in a viscoelastic medium. The consideration of the after-effect is associated with exclusion of the quantities characterizing the state of the surrounding macromolecules from the equations of motion for the primary macromolecule. This, however, entails serious difficulties associated with the fact that the motion of the macromolecule becomes a stochastic non-Markovian process. If the chosen macromolecule and its surroundings are regarded as a single complex system with many degrees of freedom, the after-effect may be excluded from the consideration. However, although the introduction of parameters that characterize the surrounding macromolecules into the equations of motion for the chosen macromolecule eliminates the indicated difficulty, the problem becomes very unwieldy and practically unresolvable. Therefore, in the

* Volkov, V. S., V. N. Pokrovsky, and V. A. Roshnev, *Izv. Akad. Nauk SSSR, Mekh. Zhidk. i Gaza*, 6, 3 (1976); Pokrovsky, V. N., V. S. Volkov, and G. V. Vinogradov, *Mekhanika Polimerov*, 4, 186 (1977); Pokrovsky, V. N. and V. S. Volkov, *Vysokomol. Soedin.*, 20A, 255 (1978); Pokrovsky, V. N. and V. S. Volkov, *Vysokomol. Soedin.*, 20A, 2700 (1978).

theory under consideration use is made of the apparatus suitable for random systems involving an after-effect.

For Brownian particles, which simulate macromolecular segments, the after-effect is equivalent to the motion of the particles in a certain effective viscoelastic fluid. The characteristics of the viscoelastic medium that surrounds the primary polymer molecule define the microviscoelasticity. They are not known in advance and can be found from the self-consistent condition, which is based on a comparison of the characteristic relaxation time of the macromolecule with the unknown relaxation time of its surroundings. Thus, in the unimolecular approximation of the theory there appears the concept of microviscoelasticity which, strictly speaking, does not coincide with the observed macroviscoelasticity.

The basic theoretical equation is the dynamic equilibrium condition for a macromolecule, which is simulated by a linear bead-and-spring chain surrounded by a viscoelastic continuum. In this case, the short- and long-range interactions are taken into consideration.

The constitutive equation is set up directly from the equations of motion for separate macromolecules without using the diffusion equation for the distribution function, which is commonly employed in the traditional method of constructing a molecular theory. When the functions involved are appropriately individualized, these equations appear to be applicable to polymer solutions, melts and blends.

The dynamic modulus and relaxation times that are used in the model are determined from oscillatory shear deformations. At high values of the inherent viscosity of the molecular coil and low values of the parameter χ (which in the theory characterizes the hydrodynamic interaction between the beads in the coil) the dependence of the dynamic modulus on frequency in the low frequency range can be approximated with the aid of a single relaxation time τ_{max}. The value of G' in the plateau region will then be expressed in the following manner:

$$G'_p = 0.72 nT\chi^{-1} \tag{3.45}$$

where n is the number of macromolecules in unit volume. The initial shear viscosity of the system is expressed in terms of the second parameter utilized in the model, B, and the maximum relaxation time calculated according to Rouse, θ_m^R, as follows:

$$\eta_0 = 1.46 nTB\theta_m^R \tag{3.46}$$

Thus, the dependence of the dimensionless dynamic modulus on the dimensionless frequency is determined by two parameters, X and B. Their values depend on molecular mass and concentration and can be determined from experimental data with the aid of formulas (3.45) and (3.46). Figure 3.27 presents experimental data for polystyrene having a molecular mass of 2.15×10^5 at 160°C; it also shows theoretical curves calculated at the following values of the parameters: $B = 145$ and $\chi = 0.06$. The theory correctly reflects the main specific features of the dynamic viscoelastic functions found experimentally—the presence of a plateau on the G' vs. ω curve and of a maximum and a minimum on the curve of G'' versus ω. Moreover, it predicts the existence of a new effect—the appearance of an inflection point on the dependence $G'(\omega)$ in the low frequency range. Thus, the molecular theory based on the concept of microviscoelasticity provides an explanation for the general regularities of the manifestation of the viscoelastic properties of

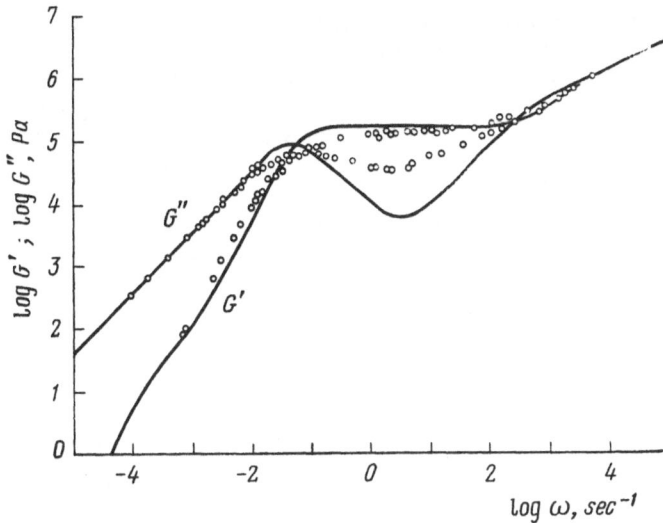

Fig. 3.27. Comparing theoretical and experimental dependences of the components of the dynamic shear modulus on frequency. The points have been reproduced from the data given in: Onogi, S., T. Masuda, and K. Kitagawa, *Macromolecules*, **3**, 109 (1970).

concentrated polymer solutions and polymers in bulk, in particular, the existence of the rubbery plateau on the curve of G' versus ω, on the basis of consideration of the motion of macromolecules among macromolecules of their kind, just as in a viscoelastic continuum, without using the concept of the entanglement network.

Based on the concept of microviscoelasticity, let us consider the interrelation between the parameters characterizing the linear viscoelastic properties of polymers, which have been determined by small-amplitude dynamic measurements. On the dependences of G', G'' and tan δ versus frequency we deal with the following characteristic points and regions. The most important for the loss modulus are the points of intersection of its frequency curve with the curve of frequency versus the storage modulus. In the region of relatively low frequencies the point of their intersection corresponds to the maximum of the loss modulus. On the frequency scale this point is conveniently taken as the beginning of the rubbery plateau. The second intersection point corresponds to the end of the plateau. In this way the extent of the plateau is determined on the frequency scale. The points of the minima of the loss modulus and tan δ are important. The corresponding frequency divides the plateau into two equal portions (to a good approximation).

In the region of the terminal zone (which corresponds to the region of fluidity) the theory predicts the presence of an inflection point on the curves of frequency against the loss modulus and tan δ. No relevant data are available at present.

The position of the point of the maximum on the frequency scale (ω_{max}), just as in the Graessley theory and in some other theories, is uniquely determined by the initial viscosity and, hence, by the temperature and the molecular mass of the elastomer. It should be pointed out that, in general, as will be seen below, most of the parameters characterizing the dependences under discussion are associated with ω_{max} and, via this quantity, with the initial

viscosity. This is natural since the initial viscosity is unambiguously determined by the longest relaxation time.

The loss modulus at the point of the maximum must amount to half
(0.4, according to experimental data) of the value of the storage modulus
on the rubbery plateau, the position of which may be assumed to be independent of frequency [more strictly, the values of the moduli in question are
proportional to $(\omega_{max})^{-0.1}$].

The position of the point of the minimum of tan δ and the loss modulus
on the frequency scale and also the values of these moduli themselves are
proportional to $(\omega_{max})^{1/2}$. The ratios of the values of loss moduli at the minimum and maximum points are proportional to $(\omega_{max})^{-3/5}$. This means that
the depth of the minimum increases with increasing molecular mass and
decreasing temperature. According to all experimental data available at
present, the depth of the minimum is smaller than that predicted by the
theory in question.

In the terminal zone the loss moduli are proportional to ω. The extent of
the region of transition from the indicated proportionality to the point of
the maximum must amount to 0.7 of a decimal order of the frequencies.
According to experimental data, the extent of this transient region is usually
not less than one decimal order of the frequencies for polymers of very narrow
MMD. As regards the storage modulus, the region of transiton from its
quadratic relation to the point of intersection with the curve of frequency
against storage modulus must be equal to 1.2 decimal orders of frequency.
In actual fact, it is somewhat higher.

3.3.6. Extension of the Theory to the Transition Zone

The molecular theories of the viscoelasticity of polymers that rest on the
model of submolecules (the KSR theory and its extensions to concentrated
systems) are inapplicable to the short-range part of the relaxation spectrum,
i.e., to the "transition" zone. This is due to the fact that the viscoelastic
behaviour of the polymer in the transition zone is caused by relaxation processes that take place in those parts of macromolecules which are shorter
than the submolecule. As has been shown by Williams, the range of applicability of the subchain model is restricted to the values of the modulus G'
less than 10^7 dynes/cm^2.

In order to extend the molecular theory to higher values of the modulus
in the transition zone, Tobolsky and Aklonis* proposed a model, according
to which the relaxation modulus (on extension) is composed of two independent Rouse-like functions due to different molecular mechanisms:

$$E_r(t) = E_1 \sum_{p=1}^{z_1} \frac{1}{z_1} e^{-\frac{tp^2}{\tau_1 z^2}} + E_2 \sum_{p=1}^{z_2} \frac{1}{z_2} e^{-\frac{tp^2}{\tau_m}} \tag{3.47}$$

The first component, which describes the behaviour of the relaxation
modulus in the transition zone, is tied up with the torsional vibrations and
internal rotation in chain molecules. In these conditions they may be regarded
as linear chains of interconnected torsional oscillators, which can sometimes

* Tobolsky, A. V. and J. J. Aklonis, *J. Phys. Chem.*, 68, 1970 (1964); Tobolsky, A. V.,
J. Polymer Sci., Part C, 9, 157 (1965).

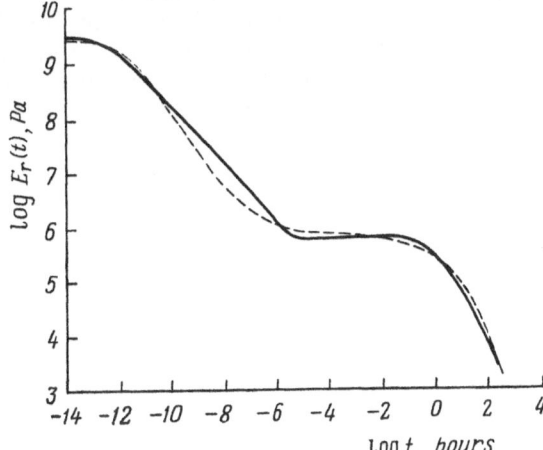

Fig. 3.28. Theoretical (solid) and experimental (dashed) curves of log E_r versus log t for polyisobutylene NBS at 25°C (after A. V. Tobolsky and J. J. Aklonis).

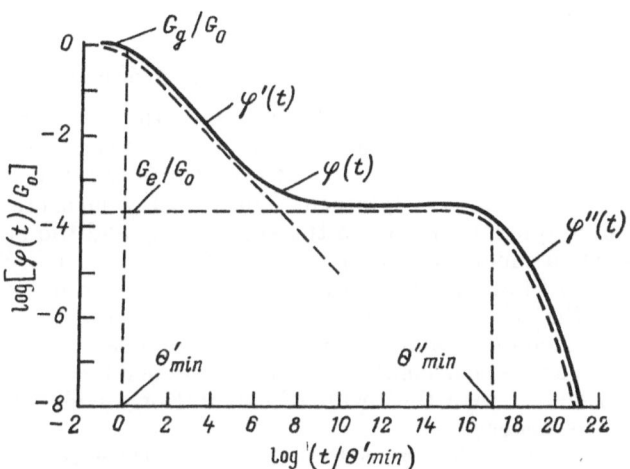

Fig. 3.29. The relaxation function consisting of two components, φ' and φ''. The function $\varphi(t)$ has been normalized by the instantaneous modulus G_0, and the time by the minimum relaxation time θ_{min} (after A. V. Tobolsky and J. J. Aklonis).

Fig. 3.30. Schematic representation of the relaxation spectrum $H(\log \theta)$ in the form of two components—the triangular $H'(\log \theta)$ (1) in the region of short relaxation times and the rectangular $H''(\log \theta)$ (2) for long relaxation times (after A. V. Tobolsky).

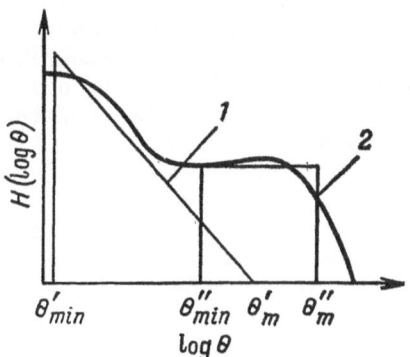

overcome the internal rotation barriers. Segments in this case are no longer Gaussian submolecules but consist of two or three atoms in the main chain. The force constant C associated with torsional oscillations (or, possibly, with the deformations of valence angles) is assigned a value of about 1000 times greater than is the force constant in the KSR theory. The value of the modulus corresponding to these conditions (at time $t = 0$) is the modulus in the glassy state and is expressed by the formula

$$E_1 = Ca^2/v$$

where a is the length of a short (non-Gaussian) segment, which is associated with torsional oscillations and internal rotation; v is the volume per torsional oscillator.

The minimum relaxation time τ_1 is associated with the relaxation time of an elementary process of internal rotation. The number of short segments, z_1, is very great, so that the first sum in Eq. (3.47) may be replaced by an integral and is given in this case by

$$R_1(t) = \frac{1}{2} E_1 \left(\frac{\tau_1}{t}\right)^{\frac{1}{2}} \left[\Gamma_{t/\tau_1}\left(\frac{1}{2}\right) - \Gamma_{t/\tau_1 z_1^2}\left(\frac{1}{2}\right)\right] \tag{3.48}$$

where $\Gamma_t\left(\frac{1}{2}\right)$ is the incomplete gamma function. Proceeding from Eq. (3.48) and relevant experimental data, we can determine the values of τ_1, E_1, and z_1.

The second component of the relaxation modulus is tied up with the entanglement network of Gaussian segments. At $t = 0$ it is equal to the modulus E_1 corresponding to the rubbery plateau. The directly measurable quantities are the maximum relaxation time of the system, τ_m, and the partial modulus E_m, the number of modes of vibration, z_2, being equal to E_2/E_m. In a concrete example considered by the authors of the theory in question, z_2 equals 20. This means that the replacement in formula (3.47) of the second sum by an integral is intolerable and it should be calculated as a discrete sum. Figure 3.28 compares experimental data on the dependence of the relaxation modulus E_r on time t with the result of calculations made according to the Tobolsky-Aklonis theory.

The Tobolsky-Aklonis theory predicts that in the transition zone the dependence of $\log E_2$ on $\log t$ must be a straight line with a slope of $-1/2$. For some polymers, however, this slope is found to be greater than $-1/2$.

The Tobolsky-Aklonis theory which permits the separation of relaxation phenomena taking place in a polymer into two groups provides a spectacular interpretation. According to this theory, the relaxation function $\varphi(t)$ is divided into two parts which may be represented schematically by simple figures (Fig. 3.29)—a trapezoid in the region of low values of t and a rectangle in the region of high values of t. When speaking of the relaxation spectrum, this is equivalent to the replacement of the function $H(\log \theta)$ shown in Fig. 3.30 by two components of the same geometric form as was used for the schematic representation of the relaxation function. These components are often called the "wedge-like" (or simply "wedge") and "box-like" relaxation spectra. Such a replacement is quite satisfactory for a wide range of times in a qualitative treatment of the problem.

The two components of the relaxation spectrum are written as follows:

$$H(\log \theta) = H'(\log \theta) + H''(\log \theta)$$

$$H' = \begin{cases} G_g \, (\theta'_{min}/\theta)^{0.5} & \text{at } \theta'_m > \theta > \theta'_{min} \\ 0 & \text{at } \theta < \theta'_{min} \text{ and } \theta > \theta'_m \end{cases}$$

$$H'' = \begin{cases} G_e & \text{at } \theta''_{min} > \theta > \theta''_{min} \\ 0 & \text{at } \theta < \theta''_m \text{ and } \theta > \theta''_m \end{cases}$$

The viscosity in steady-state flow is expressed in terms of the relaxation spectrum:

$$\eta = 2G_g\theta'_{min} \, [(\theta'_m/\theta'_{min})^{0.5} - 1] + \sum_{p=1}^{z} G_e\theta''_p = 2G_g \, (\theta'_m\theta'_{min})^{0.5} + G_e\theta''_m \sum_{p=1}^{z} p^{-2}$$

The first term reflects the contribution of short relaxation times and the second, the box-like part of the spectrum. Since the relaxation times of the wedge-like part of the spectrum θ'_p are negligibly small as compared with the relaxation times θ''_p, we may write:

$$\eta = G_e\theta''_m \sum_{p=1}^{z} p^{-2}$$

If z were very great, the sum on the right-hand side of the equation would be equal to $\pi^2/6 \approx 1.64$. At not too high z values the sum differs from $\pi/6$, but beginning even from $z = 8$ this difference is not more than 10 per cent. Therefore, with good accuracy we may assume this sum to be equal to 1.64. Then, the following formula is obtained, which relates viscosity to the other constants of the material:

$$\eta = 1.64 G_e \theta''_m \tag{3.49}$$

The quantity G_e is independent of the molecular mass of the polymer since this is the modulus of the polymer in the rubbery state, which is determined by the properties of a dynamic segment and not of the chain as a whole. Therefore, the form of the dependences of η and θ''_m on molecular mass must be identical, namely, both these quantities are proportional to M^α, where α is close to 3.4. The dependence of θ''_m on molecular mass also determines the effect of M on the extent of the plateau along the frequency axis of $\Delta \log \omega$. The beginning of the plateau in Fig. 3.29, which corresponds to the point of intersection of the functions $\varphi'(t)$ and $\varphi''(t)$, is independent of M and the long-time end of the plateau is shifted in proportion to M^α. Therefore, $\Delta \log \omega$ is proportional to M^α, as follows from many experimental data. For a large number of various polymers the value of G_g remains practically constant and is close to a value of the order of 10^9 Pa and therefore the wedge-like part of the spectrum is universal to a certain extent. As regards the box-like part of the spectrum, for it the individual property of the polymer system is the rubber-elasticity modulus G_e and, hence, the corresponding height of the box-like part of the spectrum, $H''(\log \theta)$.

The theory provides qualitatively correct predictions concerning the relaxation (viscoelastic) properties of polymer systems and the nature of the effect of various factors on it. It is however difficult to expect that the viscoelastic functions calculated by the simplified scheme of separation of the relaxation spectrum into the wedge-like and box-like regions will agree quantitatively with experiment. For instance, according to the theory, in the region of long relaxation times the ratios $\theta''_m/\theta''_{m-1}$, $\theta''_m/\theta''_{m-2}$, and $\theta''_m/\theta''_{m-3}$ will vary in the same manner as p^2, i.e., must be equal to 4, 9, and 16, respectively. But according to the measurements carried out by the authors of the theory, these ratios vary within the following ranges for monodisperse polystyrenes having different molecular masses: 4.8-5.7; 18-25, and 77-110, respectively. Although the results obtained are somewhat different, the relaxation time is found to be distributed substantially more sparsely than predicted by the theory. In general, the Tobolsky-Aklonis theory predicts a more gently sloping transition to the plateau than is experimentally observed for monodisperse polymers. This is common to all the theories, the authors of which try to describe the region of long relaxation times by means of the shifted spectrum of the KSR theory.

We should, however, emphasize here the basic idea of the theory under consideration, which consists in indicating that the high values of the modulus (in the short-time part of the transition zone) are due to the energy mechanism of elasticity, in contrast to the entropy mechanism of rubber elasticity in the region of long relaxation times.

3.3.7. The Phenomenological Approach

In constructing mechanical models, the analogues of the viscoelastic behaviour of polymer systems, various combinations of the simplest elements (the viscous dashpot and the elastic spring) are possible (see Chapter 1). In an analogous manner, in constructing mechanical analogues of the polymeric chain one may make various assumptions on how the resistance to flow and elastic deformation of the macromolecule are summed up on application of external load. Depending on the method of representation of the viscoelastic properties of the chain, various relaxation-time spectra can be obtained, which leads to the essentially different predictions concerning the expected features of the mechanical behaviour of the polymer system.

Models of the "bead" type and their generalizations to concentrated polymer solutions have been discussed; common to this group of models was the assumption of the series connection of the viscous and elastic elements making up chain segments. This section is concerned with models of the polymeric chain, which assume the possibility of the parallel connection of the viscous and elastic elements, which leads to the prediction of a different viscoelastic behaviour than in the cases considered above.

Some of the models of the viscoelastic chain with the parallel connection of the elements are shown in Fig. 3.31; scheme a which was considered by Bueche*, scheme b by Blizard** and schemes c and d by Marvin and Oser.*** Common to all these models is the representation of the macromolecule in the form of an elastic chain divided into submolecules (segments) of equal rigidity C (or C_2), to each of which there is connected in parallel a viscous (or viscoelastic) element that simulates the resistance during the motion of a segment in the medium.

The Bueche model (scheme a) takes account of the possibility of two types of resistance—one associated with the motion of a separate segment, which is characterized by the value of the monomeric friction coefficient r_1, and

---

* Bueche, F., *J. Appl. Phys.*, 26, 738 (1956); *Physical Properties of Polymers*, Interscience Publishers, New York, 1962.
** Blizard, R. B., *J. Appl. Phys.*, 22, 730 (1951).
*** Marvin, R. S. and H. Oser, *J. Res. NBS*, 66B, 4, 171 (1962); Oser, H. and R. S. Marvin, *J. Res. NBS*, 67B, 87 (1963).

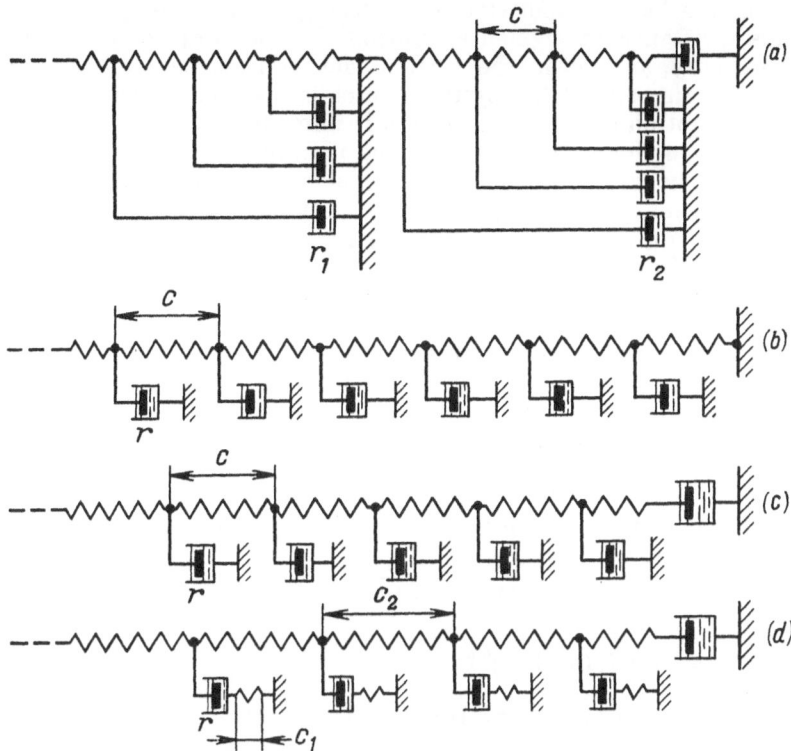

Fig. 3.31. Mechanical models of the behaviour of macromolecules in concentrated solutions, constructed by parallel connection of an elastic spring with viscous resistance:
a—the Bueche model with two groups of relaxation times; *b*—the Blizard model of a viscoelastic solid; *c*—the Marvin model of a viscoelastic liquid; *d*—the Marvin-Oser model of a viscoelastic liquid with viscous resistance of segments to motion.

the other due to the entanglements with other macromolecules, this being described by the coefficient r_2 which depends on the number of segments in the chain. The latter is necessary for the model to correctly describe the dependence of viscosity on molecular mass in the region of high molecular masses.

The Blizard model (scheme *b*) describes the behaviour of a viscoelastic solid possessing a set of relaxational properties and incapable of flowing because of the main chain being an elastic thread rigidly fixed by one end. In contrast to this, in the Marvin-Oser model (scheme *c*) a viscous element is introduced into the main chain, which corresponds to the transition to a viscoelastic liquid. Finally, in the most complete version of the Marvin-Oser model (scheme *d*) it is additionally taken into account that the resistance to segment motion may be viscoelastic. To this there corresponds the introduction of an elastic element of high rigidity C_1 ($C_1 \gg C_2$), which allows one to take into account the viscoelastic effects on deformation of the polymeric chain in the region of very high frequencies. For the region far away from the glassy state, the difference between the models shown in schemes *c* and *d* disappears.

Of greatest interest is the consideration of the mechanical properties of the model shown in Fig. 3.31*d*, which will be called further in the text the Blizard-Marvin-Oser (BMO) model since schemes *c* and *d* differ but insignificantly from scheme *b*.

An important basic proposition of the theory is the assumption of the law of viscous resistance to segmental motion. In the simplest modification of the BMO theory, just as in the KSR model, it is presumed that $r = \eta/N$, where η is the viscosity of the system, which is the sum of the viscous resistances to the motion of all the segments. It immediately follows that the viscosity η is proportional to N, i.e., to the molecular mass of the polymer in the long run. The viscoelastic properties of such a chain-like model which simulates the behaviour of dilute polymer solutions, as shown by Marvin and Oser, are very similar to those predicted by the KSR theory; in particular, the functions $G'(\omega)$ and $G''(\omega)$ for the BMO model are similar to those obtained in the KSR model. In other words, for the region of dilute solutions the method of constructing the model of a viscoelastic chain proves to be immaterial and new results cannot be obtained from the BMO model as compared with KSR theory.

An important generalization of the BMO theory is, however, achieved if use is made of the idea of separation of the relaxation-time spectrum into two regions. Then, for polymers of high molecular mass or for concentrated solutions there is obtained the following expression for the complex dynamic modulus G^* in terms of the parameters of a mechanical model:

$$G^* = \frac{\nu C_2}{[1+(C_2/C_1)\,\nu^2]^{0.5}} \cdot \frac{\tanh \nu + L \tanh (K\nu)}{1+L \tanh \nu \tanh (K\nu)} \qquad (3.50)$$

where

$$\nu = (i\omega r/C_2)^{0.5} = (1+i)\,(\omega r/2C_2)^{0.5}$$

$$L = [(l^{3.4} - 1)/(l-1)]^{0.5}; \quad K = L\,(l-1); \quad l = M/M_c$$

From this formula there can be obtained the frequency dependences of the components of the complex modulus, though this requires rather cumbersome calculations.

For concrete applications it is convenient to represent formula (3.50) in normalized form, for which purpose the reduced value of the dynamic modu-

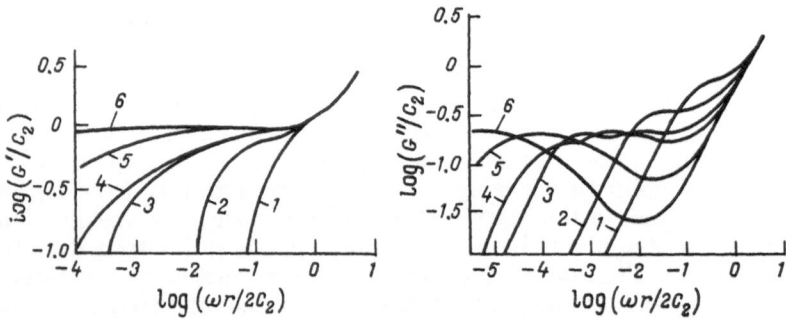

Fig. 3.32. Frequency dependences of the components of the dynamic modulus in normalized (dimensionless) form calculated according to the BMO theory for the following values of the parameter $\log l = \log (M/M_c)$:
1—0.12; *2*—0.35; *3*—0.75; *4*—0.90; *5*—1.50; *6*—2.0.

lus (G^*/C_2) is introduced. The dependences of (G'/C_2) and (G''/C_2) on $(\omega r/2C_2)$ calculated in accordance with the BMO theory at various values of the parameter l are shown in Fig. 3.32 for the region of viscous flow, the plateau region and for the beginning of the transition from the rubbery to the glassy state. The quantity C_2 has the meaning of the modulus G_p' on the plateau of polymers of high molecular mass, and the parameter $(r/2C_2)$ represents the characteristic relaxation time θ_m.

The BMO theory leads to a number of predictions concerning the specificity of the form of the dynamic functions and the effect of the length of the molecular chain on it. It is first of all seen that at $l \geqslant 10$ on the dependence $G'(\omega)$ there appears a distinctly prominent plateau corresponding to the developed region of the rubbery state, and on the G'' vs. ω curve there is detected a maximum upon transition from the fluid to the rubbery state and a minimum in the plateau region. As the molecular mass increases the extent of the plateau of the function $G'(\omega)$ and the distance between the maximum and the minimum of the function $G''(\omega)$ on the frequency axis increase. In the region of $5 \leqslant l \leqslant 10$ there is predicted the existence of two minima and two maxima of the function but they are pronounced very weakly. This effect has not been observed in practice.

For polymers of sufficiently high molecular mass (estimated by the quantity l) the theory establishes a certain relationship between the parameters characterizing the maximum on the curve of G'' versus ω and the plateau height. For example, the value of the function $G''(\omega)$ at the maximum is $0.207\,C_2$. From this it follows that

$$G''_{max} = 0.207 G_p' \tag{3.51}$$

The frequency ω_{max} at which the maximum of the function $G''(\omega)$ is attained is shifted as the chain length increases to the side of lower values. According to the BMO theory

$$\omega_{max} = C_2/rL^2$$

The frequency ω_{min} at which the function $G''(\omega)$ passes through a minimum is given by the same parameters:

$$\omega_{min} = 1.04 C_2/rL^{2/3}$$

The minimum value of the loss modulus G''_{min} is

$$G''_{min} = 1.04 C_2/L^{2/3}$$

and the value of the loss tangent $\tan \delta$ at the minimum of its frequency dependence is expressed only in terms of the parameter L:

$$(\tan \delta)_{min} = 1.04 L^{-2/3}$$

Instead of this formula, a somewhat different formula for the BMO model over the range of l values from 2 up to about 20 has been proposed*, which was derived by calculating the values of $(\tan \delta)_{min}$ by means of an electronic computer, namely:

$$(\tan \delta)_{min} = 1.02 l^{-0.80} \tag{3.52}$$

Assuming that $M_e = M_c/2$, we get

$$(\tan \delta)_{min} = 1.81 (M/M_e)^{-0.80} \tag{3.52a}$$

The results obtained by comparing experimental data with this formula show** that it reflects qualitatively correctly the character of the dependence of $(\tan \delta)_{min}$ on (M/M_e) up to $(M/M_e) \approx 150$, but the numerical coefficient

* Högberg, H., S. E. Lovell, and J. D. Ferry, *Acta Chem. Scand.*, 14, 1424 (1960).
** Vinogradov, G. V., E. A. Dzura, and A. Ya. Malkin, *Inzhenerno-Fiz. Zhurn.*, 18 975 (1970).

is found to be equal to 2.5. At high values of (M/M_e) the quantity $(\tan\delta)_{min}$ becomes practically constant, which does not contradict formula (3.52) since it has been obtained for a limited region of relatively small values of the parameter (M/M_e).

The characteristic relaxation time $\theta_0 = r/2C_2$ in the BMO model can be expressed in terms of the directly measurable parameters. By combining the above-written expressions for ω_{min}, G''_{min}, and G'_p it is easy to find that

$$\theta_0 = \frac{1}{2\omega_{min}} \left(\frac{G''_{min}}{G'_p} \right)$$

From Fig. 3.32 it is seen that when the frequency ω reaches values of the order of θ_0^{-1}, the effect of molecular mass on the viscoelastic properties of the polymer completely disappears. This is because in the region of frequencies $\omega > \theta_0^{-1}$ the viscoelastic properties of the polymeric chain are governed by motions only within a dynamic segment.

The shape of the functions $G'(\omega)$ and $G''(\omega)$ predicted by the BMO theory and also the relations between the values of the parameters at the extremal points and the dependences of the characteristic values of these functions on the length of the molecular chain that follow from this theory may serve as the basis for the experimental testing of the theory. An example that may be regarded as an experimental test of the BMO theory is the relationship between G'_p and G''_{max} found[*] for a series of polybutadienes of narrow molecular-mass distributions. This relationship has the form

$$G''_{max} \approx 0.4 G'_p \tag{3.53}$$

that is, the experimental value of the coefficient (0.4) appears to be twice as high as the theoretically predicted value (0.207).

Oser and Marvin gave an example of treatment of experimental data on the dependences $G'(\omega)$ and $G''(\omega)$, which, with the numerical constants being appropriately chosen, are in satisfactory argeement with the theoretically calculated dependences of the dynamic functions over a wide range of frequencies corresponding to the transition from the fluid to the rubbery state, the plateau and the beginning of the transition to the region of the glassy state.

3.3.8. Empirical Functions

Along with attempts to describe the viscoelastic properties of polymers with the aid of kinetic models, of importance is the approximation of experimental results by means of simple empirical expressions containing a limited number of constants. This route is fruitful because it allows one to obtain sufficiently simple analytical expressions which are convenient for further calculations, and also to relate the constants of the resulting expressions to the molecular characteristics of the polymer under investigation.

It has already been pointed out (Chapter 1) that a convenient analytical representation of the viscoelastic properties of a polymer is the Cole-Cole plot. Proceeding from this function, it has been shown[**] that for polymers

  * Vinogradov, G. V., A. Ya. Malkin, et al., J. Polymer Sci., A-2, 10, 1061 (1972).
 ** Marin, G. and W. W. Graessley, Rheol. Acta, 16, 527 (1977).

of narrow MMD over a wide range of frequencies and temperatures the following empirical relation for the dynamic compliance is fulfilled with good accuracy:

$$J^*(\omega) = J_g + \frac{1}{i\omega\eta_0} + \frac{J_p}{1+(i\omega\tau_p)^{1-\alpha}} + \frac{J_l}{1+(i\omega\tau_l)^{1-\beta}} \tag{3.54}$$

where the first term corresponds to the instantaneous compliance J_g (in the region of the glassy state), the second term corresponds to viscous flow with constant viscosity η_0, and the next two terms reflect the viscoelastic behaviour in the region of viscous flow and in the transition zone.

Estimations show that the characteristic relaxation time τ_p, just as the viscosity η_0, is proportional to $\overline{M}_w^{3.4}$. The quantities J_l, τ_l, and β, which characterize the viscoelastic behaviour in the transition zone, are independent of the mass-average molecular mass, which naturally corresponds to the mechanism of viscoelastic relaxation in this frequency range, this mechanism being associated with molecular motions within each segment. And the quantities J_p and α strongly depend on MMD and not on $\overline{M}_w$, as in the case of the dependence of apparent viscosity on the rate of shear in the region of non-Newtonian flow. The following values of the constants have been obtained (for polystyrene): $J_l = 4.5 \times 10^{-6}$ Pa^{-1} and $\beta = 0.30$, and for another relaxation region: $J_p = 7.5 \times 10^{-6}$ Pa^{-1} and $\alpha = 0.34$.

The frequency dependence of the dynamic viscosity can be described in an analogous manner*:

$$\eta'(\omega) = \eta_0/[1+(\tau_r\omega)^\alpha] \tag{3.55}$$

It is important here that the three constants figuring in this formula (η_0, τ_r, and α) can be related to the main molecular parameters of the polymer (polyethylene, for example): the average molecular mass, the polydispersity index $Q = \overline{M}_w/\overline{M}_n$, which characterizes the molecular-mass distribution of the polymer, and the degree of branching of macromolecules. It is interesting to note that such an approach enables one to establish the quantitative and readily experimentally determinable degree of branching of polyethylene, a parameter which is difficult to reliably evaluate by means of conventional physico-chemical methods.

There are known numerous attempts to relate the viscoelastic properties of polymers not with the average values of molecular mass but with MMD**. But, just as with attempts to establish correlations between the flow curve of η vs. $\dot{\gamma}$ and MMD, here too it is possible to estimate the viscoelastic properties of polymers with satisfactory accuracy from their MMD***, but the calculation of MMD from the measured relaxation characteristics of the polymer appears to be unreliable, as a general rule, because of the similarity of the relaxation spectra of polydisperse polymers of different MMD.

* Briedis, I. P. and L. A. Faitelson, *Mekhanika Polimerov*, 3, 523 (1975); 1, 120 (1976).

** Shah, B. H. and R. Darby, *Polymer Eng. Sci.*, 16, 579 (1976).

*** Malkin, A. Ya., O. Yu. Sabsai, G. Zh. Zhangereeva, M. P. Zabugina, and G. V. Vinogradov, *Mekhanika Polimerov*, 3, 519 (1977).

3.3.9. Temperature Dependence of the Extent of the Rubbery Plateau

The rubbery plateau on the frequency dependence of the storage modulus G' (ω) is limited by the frequencies ω_p and ω_f (Fig. 3.33a). The boundaries of the rubbery state of a linear polymer, which are given by the frequencies ω_p and ω_f, refer to isothermal conditions. If measurements of this type are repeated at different temperatures, the values of ω_p and ω_f will be found to be temperature-dependent. It may then be supposed that there exists a certain dependence G' (ω, T), whose unidimensional cross-section is the isothermal functions G' (ω). But it is possible to construct other cross-sections of this

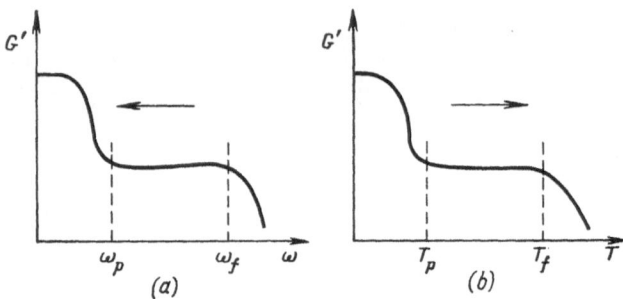

Fig. 3.33. Cross-sections of the dependence G' (ω, T) for $T = $ const (a) or $\omega = $ const (b). The arrows indicate the direction of the increase of frequency or temperature.

dependence, which would correspond to the condition $\omega = $ const, and to consider G' as a function of temperature (see Fig. 3.33b).

The measurement of the dependences of the viscoelastic functions on temperature at a constant value of the frequency chosen (or any other time factor) is a thermomechanical method of investigating polymers. By using one of the versions of this method it is possible to determine the temperature dependence of the relaxation modulus for a specified length of relaxation, and the other version is used to determine the temperature dependence of the compliance function again for a specified period of time. One may also consider the temperature dependence of the storage modulus G' at the frequency chosen.

The temperature dependence of G' can also be used to determine the extent of the plateau zone on the temperature rather than the frequency scale. To the boundaries of the plateau in Fig. 3.33b there correspond the temperatures T_p and T_f. The extent of the plateau of the rubbery state may equally well be defined as the region of frequencies in which $G' = $ const or as the region of temperatures in which the same condition is fulfilled. But of importance here is the question of the relation between $\Delta \log \omega = \log (\omega_f/\omega_p)$ and $\Delta T = T_f - T_p$.

The dependence G' (ω) does not change qualitatively with temperature, although the frequencies ω_p and ω_f are shifted and become the greater the higher is the temperature. This is shown in Fig. 3.34 where the distance between the solid lines along the vertical $\Delta \log \omega$ axis corresponds to the extent of the rubbery-state plateau on the frequency axis under isothermal conditions and the distance along the horizontal ΔT axis corresponds to the extent of the plateau along the temperature axis found at $\omega = $ const.

In order to find the quantitative agreement between the quantities ΔT and $\Delta \log \omega$ it is necessary to know the equations for the solid lines in Fig. 3.34, i.e., the temperature dependence of the relaxation times θ_p corresponding to ω_p and θ_f, the latter being responsible for relaxation upon transition to the region of viscous flow. On the basis of the considerations expounded in the preceding sections concerning the relation between the viscoelastic properties of a polymer and its viscosity it is easy to see that θ_p and θ_f vary with temperature in the same manner as the viscosity. The question of the general character of the dependence $\eta(T)$ has been considered in detail in

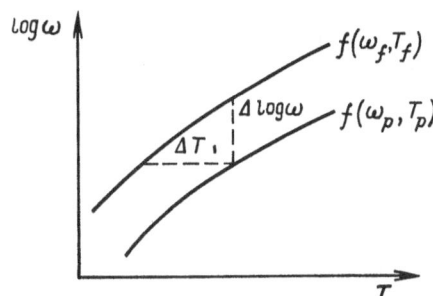

Fig. 3.34. Temperature dependences of frequencies corresponding to the boundaries of the rubbery state.

Chapter 2. At this point, for the purposes of illustration use may be made of the exponential form of the function $\eta(T)$. Then

$$\omega_f \approx \theta_f^{-1} \propto \exp(-E/RT_f); \quad \omega_p \approx \theta_p^{-1} \propto \exp(-E/RT_p) \tag{3.55}$$

where E is the activation energy of viscous flow.

It thus follows that

$$\Delta \log \omega = \log(\omega_f/\omega_p) = \frac{E}{2.3R}\left(\frac{1}{T_p} - \frac{1}{T_f}\right) = \frac{E}{2.3RT_p} \cdot \frac{\Delta T}{T_p + \Delta T} \tag{3.56}$$

Formula (3.56) is an expression for the extent of the rubbery-state plateau on a frequency scale calculated from the known width of the temperature region of the rubbery state. It is easy to write the inverse formula expressing ΔT in terms of $\Delta \log \omega$:

$$\Delta T = \frac{T_p \cdot \Delta \log \omega}{(E/2.3RT_p) - \Delta \log \omega} \tag{3.57}$$

According to formula (3.57), ΔT increases indefinitely when the value of $\Delta \log \omega$ approaches $E/2.3RT_p$. In practice, however, this corresponds to very high values of $\Delta \log \omega$. As a general rule, the inequality $\Delta \log \omega \ll E/2.3RT_p$ is fulfilled. Therefore, the relation between ΔT and $\Delta \log \omega$ may be assumed to be approximately linear:

$$\Delta \log \omega \approx (E/2.3RT_p^2) \Delta T \tag{3.58}$$

Formula (3.56) can be used to relate the extent of the temperature region corresponding to the rubbery state to the molecular mass of the polymer:

$$\log(M/M_p) = \frac{E}{2.3RT_p\alpha} \cdot \frac{\Delta T}{T_p + \Delta T}$$

or

$$\log(M/M_p) = B\,\Delta T/(C + \Delta T) \tag{3.59}$$

where B and C are constants.

Experimental data that illustrate the dependence of ΔT on the ratio (M/M_p) for various polymers are shown in Fig. 3.35; the line drawn in this figure is described by Eq. (3.59) with the following values of the constants; $B = 3.22$, $C = 122$, and $B/C = 2.6 \times 10^{-2}$. The resulting dependence of ΔT on (M/M_p) has proved to be common to the group of polymers investigated, although, generally speaking, this does not follow from the theory.

3.4. Correlation Between Dynamic and Steady-State Properties of Polymer Systems

3.4.1. Introduction

Investigations of the rheological properties of polymer solutions and melts have shown that there is an agreement (or a correlation) between the shapes of the function that characterize their properties in steady-state conditions

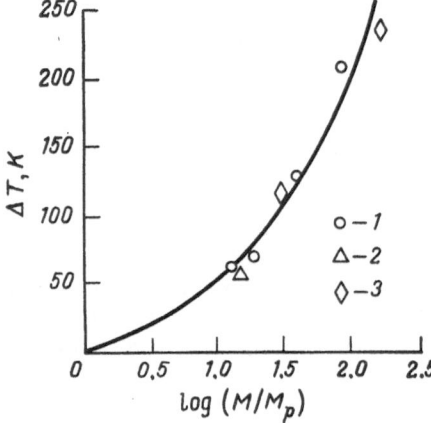

Fig. 3.35. Relation between ΔT and the ratio (M/M_p) for polybutadienes (1), polystyrenes (2), and organosilicone (3). [Malkin, A. Ya., E. A. Dzura, and G. V. Vinogradov, *Doklady Akad. Nauk SSSR*, 188, 1328 (1969).]

of shearing flow (i.e., the dependences of the shear stresses τ and the first difference of normal stresses* σ on the rate of shear $\dot{\gamma}$) and the functions describing the dynamic properties of the system [i.e., $G'(\omega)$ and $G''(\omega)$]. Instead of stresses, their coefficients may be considered: $\eta = \tau/\dot{\gamma}$ and $\zeta = \sigma/2\dot{\gamma}^2$, which are compared with the dynamic viscosity $\eta' = G''(\omega)/\omega$ and the relation $A_G = G'/\omega^2$, respectively. An essential point here is that the dependence of the dynamic functions on frequency is determined in the region of small amplitudes where these functions do not depend on the amplitude, i.e., at small deformations. The dynamic characteristics are compared with the dependences measured in steady-state flow, under the conditions in which the deformations may be indefinitely large. This means that there is established a correlation between the linear (dynamic) regimes and the regimes which may be basically nonlinear (steady-state flow).

When comparing the dynamic and steady-state properties of polymer sys-

---

* The normal-stress effect will be described in Chapter 4.

tems one should take into account that $\tau\,(\dot{\gamma})$ and $G''\,(\omega)$ define the intensity of dissipation of the work done by external forces, and $\sigma\,(\dot{\gamma})$ and $G'\,(\omega)$ characterize the elasticity of the material, i.e., its ability to store the work of external forces. It is exactly this factor that accounts, in the most general sense, for the existence of a correlation between the corresponding functions.

The problem under consideration has two interrelated aspects. On the one hand, the establishment of an analytical relationship between the functions indicated above may prove a convenient experimental procedure for the switchover from some characteristics of a polymer system to quite different ones. On the other hand, the existence of a correlation between various characteristics of the system, being a general feature of the rheological properties of polymers, can be used as an important fact against which the rheological theories, models, and constitutive equations must be tested.

3.4.2. Phenomenological Theories

The theory of linear viscoelasticity predicts the existence of certain, generally speaking, rather complicated dependences of G' and G'' on frequency; the concrete form of these dependences is determined by the salient features of the relaxation spectrum of a given system. Within the framework of this theory the shear stress is linearly dependent on the rate of shear. No correlation should therefore be expected to exist between the functions $\eta'\,(\omega)$ and $\eta\,(\dot{\gamma})$. The only exception is the limiting point

$$\eta_0 = \lim_{\omega \to 0} \eta' = \int_0^\infty \theta F\,(\theta)\,d\theta$$

Generalization of the linear theory of viscoelasticity to the case of large deformations allows us to consider the question of the possible forms of the correlation between the steady-state and dynamic characteristics of polymeric systems. As pointed out in Chapter 2, depending on the form of the differential operator used, various predictions can be made concerning the shape of the dependence of the shear stress on the rate of shear. The functions $G'\,(\omega)$ and $G''\,(\omega)$ are invariant to the method of describing the nonlinear effects in steady-state flow. Therefore, as far as the question of the correlation between the steady-state and dynamic characteristics of polymer systems is concerned the use of differential operators of complex structure enables one to modify theoretical predictions regarding the steady-state characteristics, i.e., the functions $\tau\,(\dot{\gamma})$ and $\sigma\,(\dot{\gamma})$, but it does not affect the shape of the functions $G'\,(\omega)$ and $G''\,(\omega)$ which are only determined by the choice of the values of the constants of the rheological model employed.

The use of the Oldroyd linear operator D_0 for describing the relations between the components of the stress tensor and the rate of strain in comparisons of the steady-state and dynamic properties of viscoelastic bodies has a consequence* that the dependence of τ on $\dot{\gamma}$ remains linear, but the additional geometrical effect is the appearance of normal stresses during shear flow.

---

* Leonov, A. I. and A. Ya. Malkin, *Izv. Akad. Nauk SSSR*, *Ser. Mekhanika Zhidkosti Gaza*, 3, 184 (1968).

The theory predicts the quadratic character of the dependence of σ on $\dot{\gamma}$ and the agreement between σ and $2G'$ in the limiting case at $\omega \to 0$.

In order to compare the predictions following from the use of other differential operators, it will be expedient to derive expressions for $G'(\omega)$ and $G''(\omega)$ for a discrete distribution of relaxation times, which is described by the operator equation (1.104). Appropriate calculations lead to the formulas

$$\begin{cases} G'(\omega) = \dfrac{N(\omega) \cdot Q(\omega) - M(\omega) \cdot P(\omega)}{[P(\omega)]^2 + [Q(\omega)]^2} \\[3mm] G''(\omega) = \dfrac{N(\omega) \cdot Q(\omega) + M(\omega) \cdot P(\omega)}{[P(\omega)]^2 + [Q(\omega)]^2} \end{cases} \tag{3.60}$$

where

$$\begin{cases} M(\omega) = A_2 \omega^2 - A_4 \omega^4 + A_6 \omega^6 - \ldots \\ N(\omega) = A_1 \omega - A_3 \omega^3 + A_5 \omega^5 - \ldots \\ P(\omega) = B_0 - B_2 \omega^2 + B_4 \omega^4 - \ldots \\ Q(\omega) = B_1 \omega - B_3 \omega^3 + B_5 \omega^5 - \ldots \end{cases}$$

The coefficients A_i and B_i are sets of the values of the parameters included in the operator equation (1.104).

Formulas (3.60) are easy to compare with the dependences $\tau(\dot{\gamma})$ and $\sigma(\dot{\gamma})$ obtained in Chapter 2 for various differential operators.

The application of the Jaumann operator to the operator equation for a viscoelastic liquid yields the dependences $\tau(\dot{\gamma})$ and $\sigma(\dot{\gamma})$ [see formulas (2.51) and (2.52)], which coincide with the results of calculation of $G'(\omega)$ and $G''(\omega)$ upon replacement of the argument $\dot{\gamma}$ by ω, i.e.,

$$\begin{cases} \tau(\dot{\gamma}) = G''(\omega) \\ \sigma(\dot{\gamma}) = 2G'(\omega) \end{cases} \quad \text{at } \dot{\gamma} = \omega \tag{3.61}$$

The shape of the dependences $\tau(\dot{\gamma})$ and $\sigma(\dot{\gamma})$ is governed by the viscoelastic properties of the system. The use of the Jaumann operator allows one to predict, in a natural way, the most important properties of polymer systems, such as the nonlinear character of the function $\tau(\dot{\gamma})$ and the nonquadratic nature of the function $\sigma(\dot{\gamma})$ by way of relating these phenomena with the viscoelastic properties of the material; these effects arise invariably when the system exhibits viscoelastic properties and the correlation between the dynamic and steady-state characteristics is a property inherent in all viscoelastic liquids.

The use of the Oldroyd nonlinear operator (see Chapter 1) in expressions (3.60) leads to the following relationships between the components of the complex shear modulus and stresses in steady-state shearing flow:

$$\begin{cases} \tau(\dot{\gamma}) = \sqrt{3/2}\, G''(\omega) \\ \sigma(\dot{\gamma}) = 3G'(\omega) \end{cases} \tag{3.62}$$

The condition for comparison is here $\omega = a\dot{\gamma}$, where $a = \sqrt{2/3}$. The corresponding formulas for η and ζ have the form:

$$\begin{cases} \eta(\dot{\gamma}) = \eta'(\omega) \\ \zeta(\dot{\gamma}) = G'(\omega)/\omega^2 \end{cases} \tag{3.62a}$$

Thus, the use of the Oldroyd nonlinear operator leads, in the limiting case ($\dot{\gamma} \to 0$ and $\omega \to 0$), to the relations of the linear theory of viscoelasticity, and at sufficiently high values of $\dot{\gamma}$ and ω, to the correlation between the dynamic and steady-state functions but with a shift along the axis of $\log \omega$ versus $\log \dot{\gamma}$ equal to about 0.09 unit on a logarithmic scale.

An example of the use of more complex rheological equations of state (constitutive equations) for establishing the correlation between the dynamic functions and stresses in steady-state flow of viscoelastic liquids is provided by the results obtained by Pao.* In his theory, constitutive equations are written by using the Green finite-strain tensor and the inverse Finger tensor, and the change to the fixed coordinate system is made by means of the Jaumann derivative. The introduction of the sum of two measures of large strains has led to the formulation of a constitutive equation from which there have been derived expressions for $\tau(\dot{\gamma})$ and $\sigma(\dot{\gamma})$, which are different from those considered above and which are, however, also tied up with the relaxation spectrum of the system. Pao obtained the following relations between $\tau(\dot{\gamma})$, $\sigma(\dot{\gamma})$ and the dynamic functions:

$$\begin{aligned} \tau &= \frac{(G')^2 + (G'')^2}{G''} \\ &\qquad\qquad\qquad \text{at } \omega = \dot{\gamma} \\ \sigma &= 2G' \frac{(G')^2 + (G'')^2}{(G'')^2} \end{aligned} \tag{3.63}$$

The quantity $G''/[(G')^2 + (G'')^2]$ represents the loss compliance on shear, J'', and $[(G')^2 + (G'')^2]/(G'')^2 = \sin^2 \delta$, where δ is the loss angle. Therefore, formulas (3.63) may be rewritten thus:

$$\begin{cases} \tau = G''/\sin^2 \delta = (J'')^{-1} \\ \sigma = 2G'/\sin^2 \delta \end{cases} \qquad \text{at } \omega = \dot{\gamma} \tag{3.63a}$$

As a final typical example of the use of differential operators of complex structure for establishing the correlation between the stresses on steady-state shear flow and the components of the complex shear modulus may be cited the results following from the Spriggs model.**

The dynamic functions of a viscoelastic liquid that conform to this model are expressed in the following way:

$$\begin{cases} G'(\omega) = \dfrac{\eta_0}{\theta\zeta(\alpha)} \displaystyle\sum_{p=1}^{\infty} \dfrac{(\omega\theta)^2}{p^{2\alpha} + (\omega\theta)^2} \\[4mm] G''(\omega) = \dfrac{\eta_0}{\theta\zeta(\alpha)} \displaystyle\sum_{p=1}^{\infty} \dfrac{(\omega\theta)^2}{p^{2\alpha} + (\omega\theta)^2} \, p^{\alpha} \end{cases} \tag{3.64}$$

---

* Pao, Y.-H., J. Appl. Phys., 28, 591 (1957); J. Polymer Sci., 61, 413 (1962).
** Spriggs, T. W., Chem. Eng. Sci., 20, 931 (1965); see also Chapter 1 (Sec. 1.9.3) of this book.

The shear stress and the normal stress difference in the Spriggs model have the form

$$
\left\{
\begin{aligned}
\tau\,(\dot{\gamma}) &= \frac{\eta_0}{c\theta\zeta\,(\alpha)} \sum_{p=1}^{\infty} \frac{(c\dot{\gamma}\theta)}{p^{2\alpha}+(c\dot{\gamma}\theta)^2}\cdot p^{\alpha} \\
\sigma\,(\dot{\gamma}) &= \frac{2\eta_0}{c^2\theta\zeta\,(\alpha)} \sum_{p=1}^{\infty} \frac{(c\dot{\gamma}\theta)^2}{p^{2\alpha}+(c\dot{\gamma}\theta)^2}
\end{aligned}
\right.
\tag{3.65}
$$

The constant c is expressed in terms of the independent parameter of the model, ε:

$$
c^2 = \frac{1}{3}\,(2-2\varepsilon-\varepsilon^2)\, \raisebox{-0.3ex}{\textbullet}
$$

If $\varepsilon = -1$, then $c = 1$ and the Spriggs model reduces to the model of a viscoelastic liquid with a Jaumann derivative and a known relaxation-time distribution; here $\dot{\gamma} = \omega$.

But if $\varepsilon \neq -1$, then the comparison of the resulting formulas shows that in the general case the correlation between the components of t̶ ̶complex shear modulus and the stresses τ and σ assume the form:

$$
\left\{
\begin{aligned}
\tau\,(\dot{\gamma}) &= c^{-1}G''\,(\omega) \\
\sigma\,(\dot{\gamma}) &= 2c^{-2}G'\,(\omega)
\end{aligned}
\right.
\qquad \text{at } \omega = c\dot{\gamma}
\tag{3.66}
$$

or

$$
\left\{
\begin{aligned}
\eta\,(\dot{\gamma}) &= \eta'\,(\omega) \\
\zeta\,(\dot{\gamma}) &= G'\,(\omega)/\omega^2
\end{aligned}
\right.
\qquad \text{at } \omega = c\dot{\gamma}
\tag{3.66a}
$$

Thus, the Spriggs theory predicts that the steady-state and dynamic characteristics are equivalent in form but are shifted relative to each other along the axis of log ω versus log $\dot{\gamma}$ by an amount log c which is the intrinsic parameter of the system.

The use of various methods of writing constitutive equations in the convected coordinate system with the subsequent transformation of these equations to the fixed coordinate system by means of various differential operators makes it possible to obtain various forms of correlations between the steady-state and dynamic characteristics of viscoelastic systems. With certain assumptions concerning the numerical values of the constants included in constitutive equations, relations (3.61), (3.63), and (3.66) are obtained.

3.4.3. Molecular Models

The molecular theories of polymeric systems are essentially based on the analysis of the behaviour of the same mechanical models and their combinations which are used to construct the phenomenological models considered above. Therefore, the results following from the two approaches practically coincide. The main difference between the molecular and phenomenological theories consists in that the constants of the constitutive equations that serve as empirical constants in a purely phenomenological approach, are tied up with the characteristics of polymeric chains—their length, rigidity, etc.

Molecular models and their dynamic characteristics have been considered in the preceding sections of this chapter. The results discussed in these sections can be used for comparing the dynamic and steady-state characteristics of polymer systems.

According to the bead model (the KSR theory), a polymeric chain that has a relaxation-time spectrum exhibits neither non-Newtonian behaviour nor normal stresses. Therefore, just as in the linear theory of viscoelasticity, in considering this model the problem of the correlation between the dynamic and steady-state characteristics of the material is solved negatively, except for the trivial case $\eta'(0) = \eta_0$ when $\omega \to 0$.

An essential generalization of the KSR model has been achieved for the case of finite strains. This required the introduction of differential operators dealt with in the analysis of the kinematics of a continuum and used for the construction of nonlinear theories of viscoelasticity. This method has furnished the same results as in the case of phenomenological models. Such an approach presupposes the solution of the problem of the correlation between the dynamic and steady-state characteristics of viscoelastic properties of polymers not within the framework of molecular models proper but by way of using the ideas of geometric nonlinearity as the cause of the effects observed. Therefore, it is natural that the use of the Jaumann derivative in the KSR model leads to the relation $\eta'(\omega) = \eta(\dot{\gamma})$ at $\omega = \dot{\gamma}$, and the use of the Green and Finger tensors for the description of finite strains leads to the relations following from the Pao theory.

The problem of the correlation between the dynamic and steady-state characteristics of polymer solutions is examined from a different standpoint by Bueche who proceeds from the model of a drained molecular coil rotating in the stream during shearing flow. As has been shown in the discussion of the viscoelastic properties of this model (see Sec. 3.1.3), the oscillations of segments are responsible for the effect of non-Newtonian flow behaviour and for the appearance of a relaxation spectrum. This leads to a simple form of the correlation between the functions $\eta(\dot{\gamma})$ and $\eta'(\omega)$, namely: $\eta(\dot{\gamma}) = \eta'(\omega)$, which conforms to relation (3.61).

Thus, as regards the correlation between the dynamic functions and stresses for steady-state flow the molecular models do not allow one to obtain results fundamentally different from those provided by phenomenological theories.

3.4.4. Experimental Results

The correlation between the dynamic and steady-state characteristics was detected in 1954 in the works of Philippoff, De Witt and coworkers and Ferry and coworkers, who conducted independent measurements of the functions $\eta(\dot{\gamma})$ and $\eta'(\omega)$ for a number of polymer solutions. The problem of the correlation between normal stresses and the modulus G' for polymer solutions was raised by Markovitz in 1957.

The general experimentally observed regularity concerning the correlation between the dependences characterizing the dynamic functions and stresses in steady-state shearing flow is the coincidence of the values of η_0 and η' and of ζ_0 and G'/ω^2 in the region of small values of the arguments when τ is proportional to $\dot{\gamma}$ and σ is proportional to $\dot{\gamma}^2$. When $\dot{\gamma}$ and ω increase, the functions $\tau\,(\dot{\gamma})$ and $G''\,(\omega)$, on the one hand, and the functions $\sigma\,(\dot{\gamma})/2$ and $G'\,(\omega)$, on the other, begin to diverge gradually, so that on the comparison condition $\dot{\gamma}=\omega$ there are observed the inequalities $\tau > G''$ and $\sigma/2 > G'$, although by the order of magnitude the quantities η and η' are close to each other. Therefore, the predictions of theories that made use of the Jaumann derivative provide nothing more than a qualitative estimation of the character of the correlation between $\eta\,(\dot{\gamma})$ and $\eta'\,(\omega)$. Though in some experiments there was observed a satisfactory agreement between $\eta\,(\dot{\gamma})$ and $\eta'\,(\omega)$, this should be regarded as an estimation of the comparable quantities rather than a method of their precise comparison.

The impossibility of establishing a direct correlation between $\eta\,(\dot{\gamma})$ and $\eta'\,(\omega)$ led to the search for more complex methods of finding the relation between these functions. The dependences $\eta\,(\dot{\gamma})$ and $\eta'\,(\omega)$ are, as a rule, similar in character, and therefore the most frequently used procedure of establishing a correlation between $\eta\,(\dot{\gamma})$ and $\eta'\,(\omega)$ was to shift the graphs of these dependences along the axis of log $\dot{\gamma}$ against log ω until a superposition is attained. This corresponds to the following form of the correlation between η and η': $\eta\,(\dot{\gamma})=\eta'\,(\omega)$ at $\dot{\gamma}=C\omega$ (where C is a constant).

Depending on the nature of the system under study and the range of shear rates and frequencies covered by measurements, the values of the coefficient C varied from unity, which corresponds to the simplest form of the use of the Jaumann derivative, to 5.3. In fact, the value of C is not a constant of the material but increases somewhat with increasing rate of shear and frequency, approximately within the range of 1 to 4.* Such a variation of C most exactly conforms to experimental data and thereby points to the complex character of the correlation between the components of the stress tensor in steady-state shear flow and the components of the complex shear modulus. Thus, the equalities $\eta\,(\dot{\gamma})=\eta'\,(\omega)$ and $\tau=CG''$ at $\dot{\gamma}=C\omega$ are only in qualitative agreement with experiment.

Among the diverse variants of the correlation between the dynamic and steady-state characteristics of polymers, of the greatest interest is the Cox-Merz empirical method**, according to which

$$\eta\,(\dot{\gamma})=|\,\eta^*\,|\,(\omega) \quad \text{at } \dot{\gamma}=\omega \tag{3.67}$$

where $|\,\eta^*\,|=[(G'/\omega)^2+(G''/\omega)^2]^{1/2}$ is the absolute value of the complex viscosity.

Relation (3.67) may be represented in another equivalent form:

$$\tau=|\,G^*\,|=G''/\sin\delta \tag{3.67a}$$

The relation between $\sigma\,(\dot{\gamma})$ and the dynamic functions can be obtained on the basis of Eq. (3.67a) and the relation fulfilled for some polymers and their solutions (Fig. 3.36):

$$\tan\delta=2\tau/\sigma \tag{3.68}$$

This gives

$$\sigma=2G'/\sin\delta \tag{3.69}$$

---

* Leonov, A. I. and A. Ya. Malkin, *Zhurn. Prikl. Mekh. i Tekh. Fiz.*, **4**, 107 (1965).
** Cox, W. P. and E. H. Merz, *J. Polymer Sci.*, **28**, 619 (1958).

Formulas (3.67) through (3.69) have no theoretical backing but they are in best agreement with experimental data.*

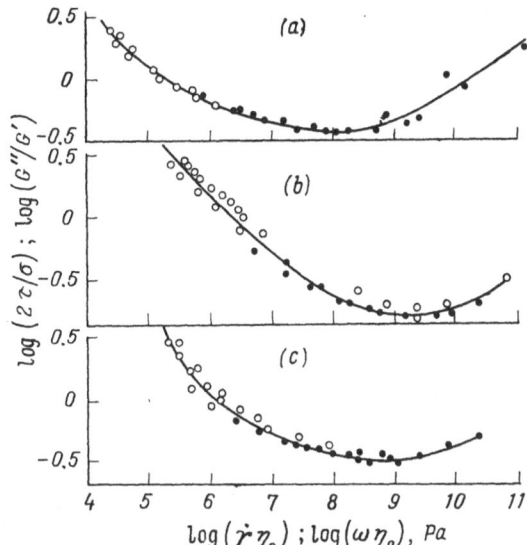

Fig. 3.36. Comparing the dependence of $(2\tau/\sigma)$ on γ (○) with that of (G''/G') on ω (●) for polyisobutylene (a), butyl rubber (b), and plasticized butyl rubber (c).
The abscissa has been normalized by the largest Newtonian viscosity.

In order to illustrate the comparative results of the use of various methods of reducing the dynamic to the apparent viscosity, Fig. 3.37 presents the

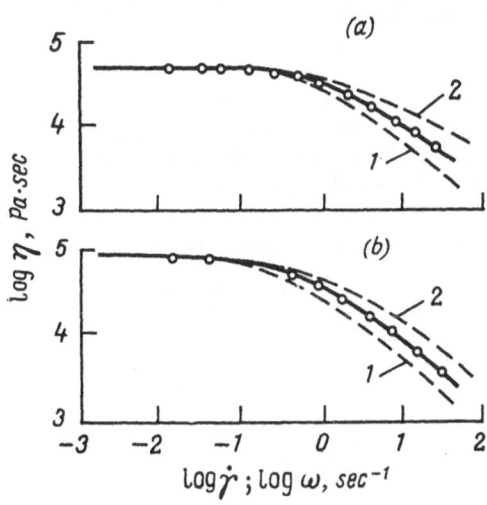

Fig. 3.37. Comparing the dependence $\eta\,(\dot\gamma)$ (solid lines) with frequency dependences of $|\eta^*|$ (unfilled circles), η' (1) and $\eta'/\sin^2\delta$ (2) for polymethyl methacrylate (a) and polyvinyl acetate (b). [Onogi, S. et al., J. Phys. Chem., 68, 1598 (1964); J. Polymer Sci., C, 15, 481 (1966); Koll.-Z., 222, 110 (1968).]

dependences $\eta'\,(\omega)$, $|\eta^*| = \eta'/\sin\delta$, and $\eta''/\sin^2\delta$, which are compared with the experimentally determined dependence $\eta\,(\dot\gamma)$. In the region of low

* Vinogradov, G. V., A. Ya. Malkin, Yu. G. Yanovsky, V. F. Shumsky, and E. A. Dzura, *Mekhanika Polimerov*, 1, 164 (1962); *Rheol. Acta*, 8, 490 (1969); Yanovsky, Yu. G. and A. Ya. Malkin, in: *Advances in Polymer Rheology*, edited by G. V. Vinogradov, Khimiya Publishers, Moscow, 1970, pages 52-78 (in Russian).

values of ω and $\dot{\gamma}$ when $\sin \delta \approx 1$, all methods of reduction give concordant results. Associated with the fulfillment of the same condition ($\delta \approx \pi/2$) is the simplicity of using the method of correlating η with η' for monodisperse polymers, for which the nonlinear effects are very weakly pronounced in a wide region of shear stresses.

A comparison of $\eta (\dot{\gamma})$ with $| \eta^* | (\omega)$ and other dynamic characteristics at high values of $\dot{\gamma} = \omega$ is also made in Fig. 3.38 which gives the dependences $\eta (\dot{\gamma})$, $| \eta^* | (\omega)$, and $\eta' (\omega)$ for polyisobutylene and plasticized butyl rubber.

Thus, the comparison of $\eta (\dot{\gamma})$ with $| \eta^* | (\omega)$ at $\dot{\gamma} = \omega$ is still the most reliable empirical method of correlating the dynamic and steady-state char-

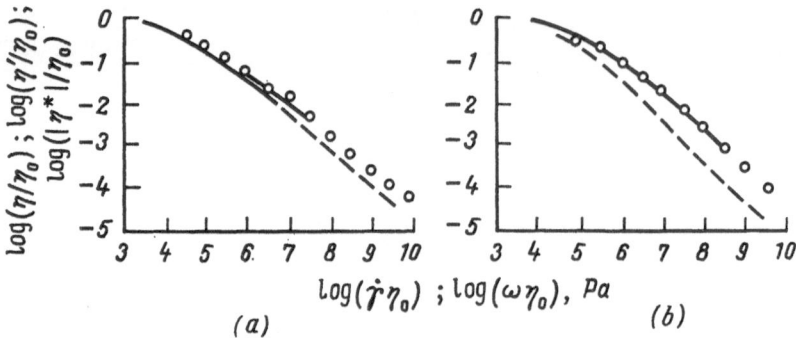

Fig. 3.38. Comparing the dependences $\eta (\dot{\gamma})$ (solid lines) with the dependences $| \eta^* | (\omega)$ (open circles) for polyisobutylene (a) and plasticized butyl rubber (b). Dashed lines represent the frequency dependences of the dynamic viscosity η'.

acteristics of polymers. This important experimental fact, which is valid in an enormous number of cases, must serve as the criterion for estimating the adequacy of the new phenomenological and molecular models proposed and also for checking up the correctness of the choice of the functions characterizing the viscoelastic properties of individual materials.

Below are given various forms of correlation between the dynamic and steady-state characteristics of the properties of viscoelastic materials:

Forms of correlation between stresses and the components of the complex shear modulus		Condition for comparison
$\tau (\dot{\gamma}) = G'' (\omega);$	$\sigma (\dot{\gamma}) = 2G' (\omega)$	$\dot{\gamma} = \omega$
$\tau (\dot{\gamma}) = \sqrt{3/2}\, G'' (\omega);$	$\sigma (\dot{\gamma}) = 3G' (\omega)$	$\dot{\gamma} = \sqrt{3/2}\, \omega$
$\tau (\dot{\gamma}) = G'' (\omega)/\sin^2 \delta (\omega);$	$\sigma (\dot{\gamma}) = 2G' (\omega)/\sin^2 \delta (\omega)$	$\dot{\gamma} = \omega$
$\tau (\dot{\gamma}) = C^{-1}G'' (\omega);$	$\sigma (\dot{\gamma}) = 2C^{-2}G' (\omega)$	$\dot{\gamma} = \omega/C$
$\tau (\dot{\gamma}) = G'' (\omega)/\sin \delta (\omega);$	$\sigma (\dot{\gamma}) = 2G' (\omega)/\sin \delta (\omega)$	$\dot{\gamma} = \omega$

The existence of a correlation between the components of the complex shear modulus and the components of the stress tensor is an effect specific to polymers exhibiting viscoelastic properties and capable of flowing. This effect is not observed with other non-Newtonian systems, say with plastic dispersed systems based on low-molecular-mass binders thickened with a solid filler.

3.5. Relaxation Properties of Polymers in Flow and Finite Deformations

3.5.1. Variation of Viscoelastic Properties in Steady-State Shear

Many of the specific properties of polymers are due to the effect of deformation on their viscoelastic characteristics. The essence of this effect consists in that during flow the material is subjected to continuous external loading, the characteristic time of which is determined by the quantity $\dot{\gamma}^{-1}$. (or the characteristic frequency—by the quantity $\dot{\gamma}$). The relaxation spectrum of the polymer consists of a set of relaxation times distributed as a rule over a wide range of values. Therefore, there exist relaxation times much longer than $\dot{\gamma}^{-1}$. Relaxation has no chance to be completed with respect to such relaxation times, and when the polymer is subjected to loading, the nonrelaxing part of the entanglement network is broken down. Conversely, for fast relaxation processes, for which $\theta_p \ll \dot{\gamma}^{-1}$, the effect of strains is insignificant and therefore the corresponding regions of the relaxation spectrum remain intact upon deformation. Finally, in the region of the relaxation spectrum close to $\dot{\gamma}^{-1}$ the application of strain causes a change in relaxation processes, which is the stronger the longer is θ_p.

Such a qualitative picture of the variation of the relaxation properties of polymers in steady-state shear flow becomes quantitatively spectacular if the dynamic functions [the dependences $G'(\omega)$ and $G''(\omega)$] are measured upon simultaneous superposition, say, of steady-state flow and harmonic oscillations.

There are two possible schemes of measurement of the dynamic properties of the material upon superposition of steady-state flow: harmonic small-amplitude deformation, which takes place parallel or orthogonal to the direction of flow.

If parallel small-amplitude harmonic oscillations are superimposed on steady-state flow, the rate of shear $\dot{\gamma}$ varies by the law

$$\dot{\gamma} = \dot{\gamma}_s + \xi \omega \cos \omega t$$

where $\dot{\gamma}_s$ is the rate of shear under steady-state shear flow; ξ is the amplitude of small deformations; ω is the frequency.

In this case, there arise shear stresses which vary in time $\tau(t)$ by the following law:

$$\tau = \tau_s + \tau_d \sin(\omega t + \delta)$$

Assuming, just as in the absence of flow, that the amplitude of the alternating stress τ_d varies in proportion to the amplitude of deformation, ξ, the complex shear modulus $\bar{G}^*$ and the loss angle δ may be introduced in the usual manner. But both these quantities, $\bar{G}^*$ and δ, depend not only on the frequency ω but also on the rate of shear $\dot{\gamma}_s$ of steady-state flow, i.e., $\bar{G}^* = \bar{G}^*(\omega, \dot{\gamma}_s)$ and $\delta = \delta(\omega, \dot{\gamma}_s)$. The dependences $\bar{G}^*(\dot{\gamma}_s)$ and $\delta(\dot{\gamma}_s)$ reflect

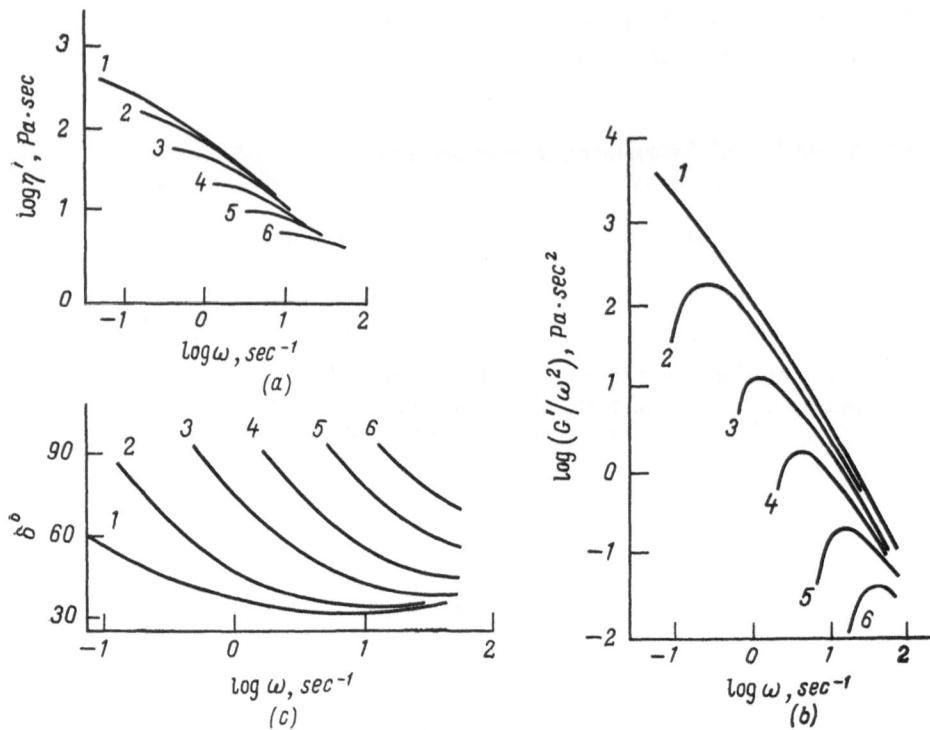

Fig. 3.39. Frequency dependences of the dynamic viscosity (a), the parameter (G'/ω^2) (b) and the loss tangent (c) for a 5-percent solution of the ethylene-propylene copolymer in decalin at various shear rates. The curves *1-6* refer to the following values of $\log \dot{\gamma}_0$: $-\infty$; -0.7; 0; 0.5; 1.0; 1.4, respectively. [Booij, H., *Rheol. Acta*, 5, 215 (1966).]

the influence of flow on the dynamic properties of the system, which are characterized by the function $\bar{G}^*$ (ω) and δ (ω) measured at different values of the parameter $\dot{\gamma}_s$. In an analogous way, one may introduce the concept of the dynamic viscosity under the conditions of steady-state flow, η', which depends on $\dot{\gamma}_s$ and ω.

On the basis of the results of measurements of the constant stress component under harmonic oscillations it is possible to formulate the criterion of the smallness of the amplitudes of harmonic oscillations: they must be so small as to prevent the influence of harmonic oscillations on the ratio $(\tau_s/\dot{\gamma}_s)$, i.e., so as to render the steady-state characteristics of the material independent of the superposition of harmonic oscillations. In this sense, small harmonic oscillations present the adestructive method of investigation of the material, which can be used to follow the change of the dynamic properties during flow, the steady-state flow conditions being left unaffected.

Typical results of investigations of the dynamic properties of a polymer measured when the oscillations are superimposed along the direction of steady-state flow are given in Fig. 3.39 which shows the frequency dependences of the dynamic viscosity η', the parameter $A_G = G'/\omega^2$, and the loss angle δ at different velocity gradients of steady-state shear flow. In inspecting this

figure, attention is immediately drawn to the peculiar character of the variation of the loss angle, depending on the deformation regime: when $\dot{\gamma}_s$ increases
in the region of low frequencies the phase angle δ increases and in this case
there exists such a frequency at which $\delta = 90°$. The frequency ω_0 at which
$\delta = 90°$ depends on $\dot{\gamma}_s$. The general relationship between ω_0 and $\dot{\gamma}_s$ has the
form*

$$\omega_0 = a\dot{\gamma}_s^b \qquad\qquad (3.70)$$

where the values of the constants a and b depend on the nature of a particular
material (in earlier works there was repeatedly given a simpler relation between
ω_0 and $\dot{\gamma}_s$: $\omega_0 = \dot{\gamma}_s/2$, which is, however, a special case).

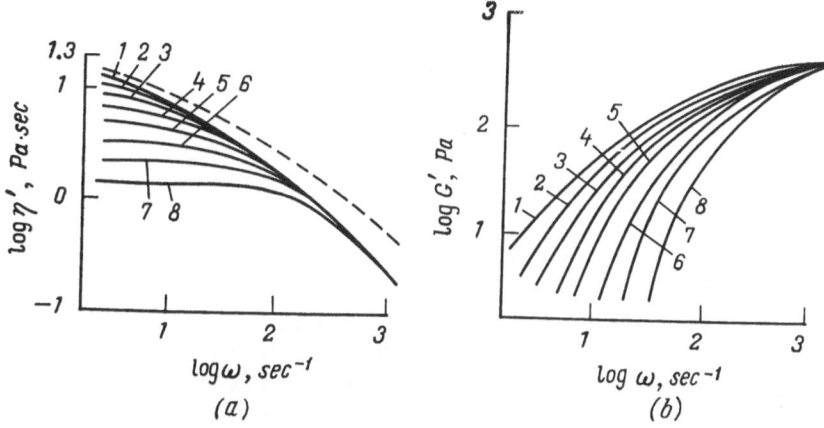

Fig. 3.40. Effect of the superposition of orthogonal shearing flow on the dependences
$\eta'(\omega)$ (a) and $G'(\omega)$ (b) for a 8.54-percent solution of polyisobutylene in cetane. The
curves 1-8 correspond to the following values of $\dot{\gamma}_0$ (in sec^{-1}): 0.2, 0.36, 12.7, 25.4, 50.9,
102.7, 204.8, 407. The dashed line represents the dependence $\eta(\dot{\gamma})$ measured for steady-
state flow at $\dot{\gamma} = \omega$. [Simmons, J. M., Rheol. Acta, 7, 182 (1968).]

Characteristic experimental data referring to the superposition of orthogonal shear oscillations upon steady-state flow are given in Fig. 3.40. These
data are qualitatively quite analogous to the corresponding results obtained
for oscillations coinciding in direction with steady-state flow (see Fig. 3.39).
Indeed, in the region of low frequencies ($\omega \ll \dot{\gamma}_s$) the dynamic viscosity is
no longer dependent on frequency, and the storage modulus G' tends to zero.
At very high frequencies ($\omega \gg \dot{\gamma}_s$) the dependence $\eta'(\omega)$ remains insensitive
to the superimposition of steady-state flow. All changes of η' and G' occur in
the region of frequencies close to $\dot{\gamma}_s$ in order of magnitude.
Thus, the relaxation properties of the material undergo change during
flow, and the relaxation times at which these changes occur are the shorter
the higher is the rate of deformation under steady-state shearing flow.
A quantitative interpretation of experimental data on the effect of the
superimposed shearing flow on the dynamic properties of polymer systems

* Macdonald, I. F., Trans. Soc. Rheol., 17, 537 (1973).

is based on the assumption that at the shear rate $\dot{\gamma}_s$ the relaxation spectrum of the material is changed.

The dependences η' (ω) and G' (ω) obtained by measuring the dynamic properties of a polymer under steady-state shearing flow conditions can be understood and quantitatively described if it is assumed that as the rate of flow increases the slow relaxation processes are suppressed the more strongly the higher is the rate of deformation. This conclusion follows spectacularly from the experimental data given in Figs. 3.39 and 3.40 and can formally be described by theories that take into account the effect of deformation on the rate of relaxation processes in the material. In a crude model, in order to obtain a qualitative picture of the specific features of the manifestation of the viscoelastic properties of the medium during flow it may be assumed that the change of the relaxation spectrum occurs in a stepwise manner at θ_0 (the value of θ_0 is of the order of $\dot{\gamma}^{-1}$). In a more precise model it should be taken into account that in fact the region of the change of the relaxation spectrum may be somewhat blurred.

3.5.2. Variation of the Relaxation Spectrum in Flow

The decrease of the apparent viscosity $\eta = \tau l \dot{\gamma}$ and the normal stress coefficient $\zeta = \sigma/2\dot{\gamma}^2$ with increasing shear rate may be tied up with the change of the relaxation properties of a polymer, which is spectacularly confirmed by the effect of superposition of shear flow on the dynamic characteristics of the material. Here it may be assumed that for each rate of shear in steady-state flow there exists a relaxation spectrum F (θ, $\dot{\gamma}$) which depends on $\dot{\gamma}$ and which changes, at $\dot{\gamma} \to 0$, to the initial relaxation spectrum of the system, F_0 (θ). The dependences η ($\dot{\gamma}$) and ζ ($\dot{\gamma}$) may then be represented in the form

$$
\begin{cases}
\eta(\dot{\gamma}) = \int_0^\infty \theta F(\theta, \dot{\gamma})\, d\theta \\[4mm]
\zeta(\dot{\gamma}) = \int_0^\infty \theta^2 F(\theta, \dot{\gamma})\, d\theta
\end{cases}
\tag{3.71}
$$

The two integral expressions given above are not, in principle, enough for the functions F (θ, $\dot{\gamma}$) to be calculated, but these formulas can be used as testing relations showing how well the supposed nature of the function F (θ, $\dot{\gamma}$) reflects the observed experimental facts. As regards the form of the function F (θ, $\dot{\gamma}$), various hypotheses and suppositions may be made.

For instance, a satisfactory agreement between calculations and experimental data is provided if account is taken of the possibility of a smooth variation of the relaxation spectrum in the region of values of relaxation times of the order of $\dot{\gamma}^{-1}$ and if it is assumed that the spectrum F (θ, $\dot{\gamma}$) is formed by way of the parallel shifting of the long-time end of the initial spectrum F_0 (θ), the amount of shifting being the larger the higher is the rate of deformation.* This character of the variation of the relaxation spectrum is shown schematically in Fig. 3.41.

The consideration of the effect of non-Newtonian behaviour as a consequence of the change of the relaxation properties of the material may be based on various theoretical representations of the function $\eta\ (\dot\gamma)$. For instance, as has already been said (see Chapter 2), it is very tempting to make use for this purpose of the Graessley theory. Using this theory, one can derive expressions** describing various nonlinear effects, say, the functions $G'\ (\omega,\ \dot\gamma)$, $G''\ (\omega,\ \dot\gamma)$, etc., which are written by way of replacement of $F_0\ (\theta)$ in the formulas of the linear theory of viscoelasticity by $F\ (\theta,\ \dot\gamma)$.

3.5.3. Dependence of Dynamic Properties on the Amplitude of Deformation

An important method of investigation of changes of the relaxation properties of polymer systems with increasing deformation intensity is provided by measurements of dynamic properties at large amplitudes.

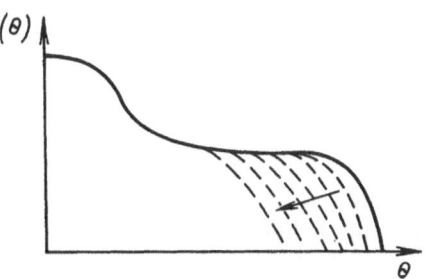

Fig. 3.41. Variation of the relaxation spectrum $F_0\ (\theta)$ with increasing rate of deformation. The direction of shift of the long-time part of the spectrum with increase of the rate of deformation is marked by an arrow. Dashed lines represent the spectral functions $\dot F\ (\theta,\ \dot\gamma)$.

The concept of dynamic functions is introduced into the theory of viscoelasticity on the basis of the analysis of the response of the system to the stress or deformation which varies with time by the harmonic law. Here it is assumed that there exist such small amplitudes of deformation γ_0 and stresses τ_0 that their ratio τ_0/γ_0 does not depend on τ_0 and γ_0 and is determined by the frequency alone. A graphical representation of this concept is the elliptical form of the figures obtained in recording the experimental results in the coordinates τ and γ after the elimination of time t which plays the role of a dummy parameter in the expressions for $\tau\ (t)$ and $\gamma\ (t)$.

If the amplitude of deformation γ_0 increases, the response of the system to the stress gradually ceases to be linear. To this there corresponds a gradual distortion of the form of the figure obtained in the coordinates τ and γ, as shown in Fig. 3.42. One can introduce the cycle-averaged characteristics of the dynamic properties of the material at large deformation amplitudes, which are determined by the ratio of the amplitude values of the stress to deformation and by the area of the figure in Fig. 3.42, which has the physical meaning of mechanical losses during a cycle of deformation. The parameters characterizing the response of the system to deformation are the abso-

---

* Malkin, A. Ya., in: *Advances in Polymer Rheology*, edited by G. V. Vinogradov, Khimiya Publishers, Moscow, 1970, pages 171-180 (in Russian).
** Shroff, R. N. and M. Shida, *J. Polymer Sci.*, C, **35**, 153 (1971); *Trans. Soc. Rheol.*, **21**, 327 (1977); Cote, J. A. and M. Shida, *Trans. Soc. Rheol.*, **17**, 401 (1973).

lute value of the modulus

$$| G^* | = \tau_0/\gamma_0 \tag{3.72}$$

and the loss angle

$$\delta = \arcsin{(A\pi/\gamma_0\tau_0)} \tag{3.73}$$

where A is the work dissipated as heat per cycle of deformation, which is equal to the area outlined by the curve in Fig. 3.42.

The extent of nonlinearity of the response of the system to external loading depends both on the quantity γ_0 and on the properties of the polymeric system itself. This can be seen by way of expanding the time dependence of stresses

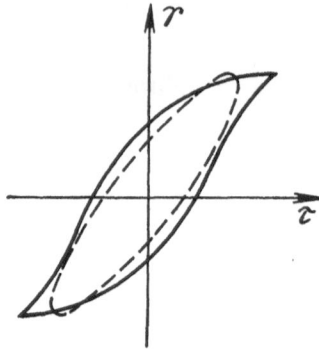

Fig. 3.42. Variation of the dependence of stress-strain relationship on cyclic deformation with large amplitudes (in the nonlinear region). The dashed line represents an ellipse observed when investigating a material with linear viscoelastic properties.

into a harmonic series and comparing the amplitudes of the first and higher harmonics.

Philippoff has shown* that for many polymer solutions, up to very large amplitudes of the order of several units (hundreds of per cent of deformation), there is only detected the third harmonic, apart from the first, the ratio of the amplitudes of the third and the first harmonic not exceeding 5 per cent. This means that the response of the system may be considered linear up to very large deformations and the dynamic properties can be characterized by the quantities G' and G'' determined in the same manner as was done in the region of small amplitudes. As the amplitude increases both these quantities, G' and G'', become dependent on γ_0 with a slight (often negligibly small) deviation of the character of variation of the stress from the specified sinusoidal deformation regime. This feature of the action of large amplitudes is typical of polymer systems capable of exhibiting very large recoverable deformations.

The character of the variation of the components of the dynamic modulus, G' and G'', with increasing amplitude of deformation is shown in Fig. 3.43 for various frequencies. From Fig. 3.43 it is seen that when the amplitude exceeds a certain critical value, the dynamic characteristics begin decreasing the more strongly the higher is the value of γ_0. This is a typical manifestation of the nonlinearity of the mechanical properties of a polymer system during intensive deformation. Therefore, the specification of vibrations with large amplitudes when G' and η' depend on γ_0 is no longer an adestructive method of investigation of the relaxational properties of the material and becomes a factor that alters its viscoelastic properties. In this sense, large-amplitude vibrations are similar in the character of influence on the

* Philippoff, W., *Trans. Soc. Rheol.*, **10**, 317 (1966).

relaxational properties of the system to the steady-state shearing flow; in both cases, the viscosity of the polymer (accordingly, the apparent viscosity η and the dynamic viscosity η') and its rigidity (the normal stress coefficient and the storage modulus) decrease.

The curves of G' and G'' versus γ_0 shown in Fig. 3.43 are quite typical of one-component polymer systems. When a filler is introduced into the system, the nonlinearity of the deformation regime is observed at substantially smaller amplitudes than with unfilled polymers.

The values of the dynamic characteristics η' and G' of polymers decrease with increasing amplitude of deformation when vibrations and steady-state

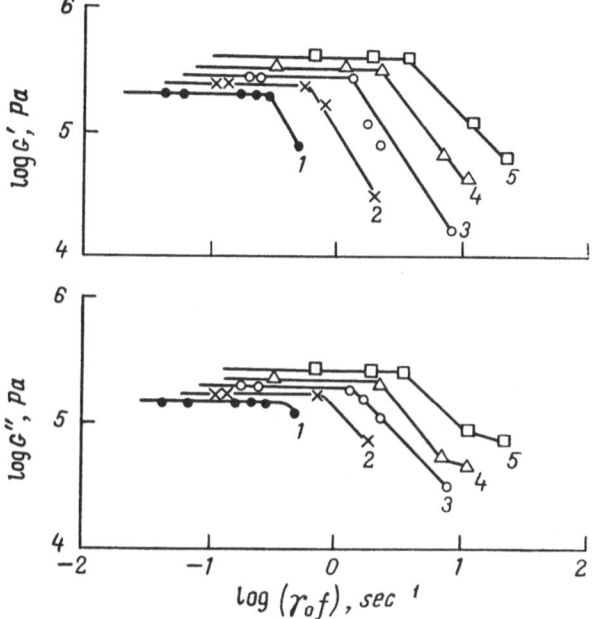

Fig. 3.43. Variation of the components of the dynamic modulus of polyisobutylene on harmonic deformations with different frequencies, Hz:
1—6; *2*—12; *3*—20; *4*—32; *5*—64. [Vinogradov, G. V. *et al.*, in: *Advances in Polymer Rheology*, Moscow, Khimiya, page 79, 1970.]

flow act simultaneously. At the same amplitude the action of orthogonal vibrations on the dynamic viscosity is the weaker the greater is the rate of deformation (Fig. 3.44). This is caused by the fact that the flow itself leads to the decrease of the dynamic viscosity, and therefore the increase of the amplitude proves to be only an additional factor affecting the viscoelastic properties of the system as compared to the stronger effect of shear flow.

Just as in steady-state shear flow the variation of the viscoelastic properties of a polymer is governed by the rate of deformation, so in harmonic oscillations with large amplitudes the determining factor is the amplitude of the deformation rate $\dot{\gamma}_0 = \gamma_0 \omega$. Typical results of measurements of the dependences of the absolute value of viscosity ($|\eta^*|$) on $\dot{\gamma}_0$ at different (constant in each case) frequencies ω are presented in Fig. 3.45 for polyisobutylene. The horizontal portions of the graphs correspond to the linear region, in which the dynamic viscosity is independent of the amplitude of deformation.

At a certain critical value of deformation $\gamma_{0,c}$ and the corresponding value
(at a given frequency) of the amplitude of the deformation rate $\dot{\gamma}_{0,c} = \gamma_{0,c}\omega$
the value of the complex viscosity begins to decrease. The downward shift
from one horizontal straight line to another corresponds to the increase of

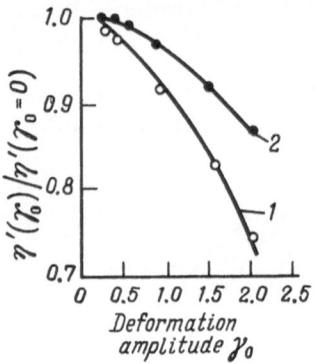

Fig. 3.44. The decrease of the dynamic viscosity η' of a solution of polyisobutylene in decalin with increase
of the amplitude of deformation γ_0 at
constant frequency of oscillations $\omega=$
$= 314$ sec^{-1} superimposed in the orthogonal direction upon steady-state
shearing   flow   at a shear rate of
51 sec^{-1} (1) and 203.5 sec^{-1} (2). [Sim-
mons, J. M., Rheol. Acta, 7, 184 (1968).]

the frequency, which leads to the decrease of $|\eta^*|$ in the linear region. Fig-
ure 3.45 also shows the dependence of the apparent viscosity on the rate
of shear, $\eta\,(\dot{\gamma})$, which is similar in form to the envelope curve. Therefore,

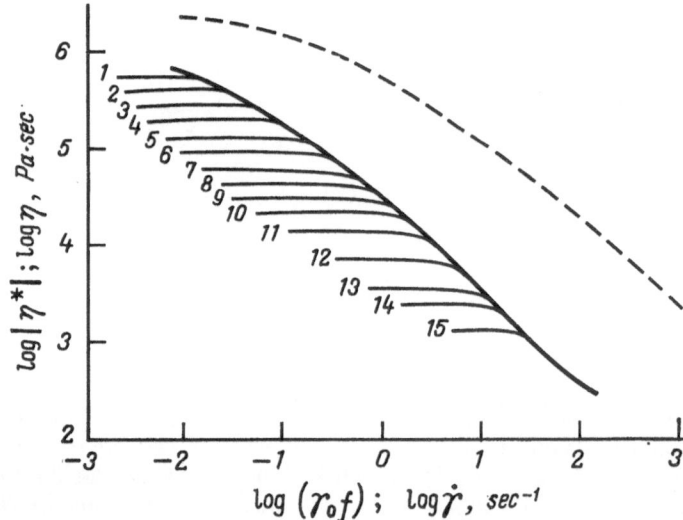

Fig. 3.45. The dependence of the $|\eta^*|$ of polyisobutylene on the amplitude of the defor-
mation rate at various frequencies. The curves 1-15 refer to the following values of
log ω (ω, sec^{-1}):
—1.2, —0.9, —0.7, —0.5, —0.2, 0, 0.3, 0.5, 0.7, 0.9, 1.2, 1.6, 1.9, 2.1, and 2.6, respectively. The
dashed line represents the dependence $\eta\,(\dot{\gamma})$. The amplitude of the rate of deformation has been calculated
as $(\gamma_0 f)$. [Vinogradov, G. V., et al., J. Polymer Sci., A-2, 8, 1239 (1970).]

we may write $|\eta^*|\,(\dot{\gamma}_0) = \eta\,(\dot{\gamma})$, assuming that $\log\dot{\gamma} = \log\dot{\gamma}_0 + a$. The
constant a is the distance between the curves in Fig. 3.45—the envelope
$|\eta^*|\,(\dot{\gamma}_0)$ and the function $\eta\,(\dot{\gamma})$—and depends on the individual properties

of the polymeric system. It may be said that the increase of the amplitude of the deformation rate $\dot{\gamma}_0$ leads to the alteration of the relaxation spectrum that corresponds to high values of relaxation times and that the region of fast relaxation processes is left intact. The situation here is the same as in the case of the change of the relaxation properties of the material under the influence of steady-state flow. This indicates the possibility of a qualitative treatment of the action of large-amplitude vibrations on the relaxation spectrum of the polymer as the analogue of the effect of steady-state flow on its viscoelastic properties.

While inspecting Fig. 3.45 it is important to provide an answer to the question of how far one can move down the envelope after leaving the linear region of constant values of $| \eta^* |$. For steady-state flow it is known that the rate of shear cannot be increased indefinitely because at a high rate of shear the polymer passes to the forced rubbery state, loses its ability to flow, and tears off from the solid wall of the measuring device. Quite an analogous situation arises during cyclic deformation with large amplitudes and high rates of deformation amplitude $\dot{\gamma}_0$. In this case too, there exist critical values of $\dot{\gamma}_0$ at which the vibration regime is upset because the polymer tears off from the solid wall of the measuring device. This limits the possible movement along the envelope in Fig. 3.45, which could have been expected if $\dot{\gamma}_0$ could have been increased with harmonic oscillations being maintained.

Periodic large-amplitude vibrations (just as steady-state flow) exert a destructive effect on the structure existing in the material.* The two components of the modulus, G' and G'', decrease as a result of large-amplitude vibrations and are gradually recovered at rest. This means that the effect of such vibrations on the state of the polymer is thixotropic. The rate of the change of the modulus during rest characterizes the kinetics of the recovery of the initial structure of the polymeric material, which has been changed by vibrations.

---

* Faitelson, L. A. and M. G. Tsiprin, *Mekhanika Polimerov*, 5, 927 (1968); Briedis, I. P. and Yu. P. Yakovlev, *Mekhanika Polimerov*, 3, 528 and 561 (1969).

CHAPTER 4

Normal Stresses in Shear
(the Weissenberg Effect)

4.1. The General State of Stress in Simple Shear

Suppose that flow takes place in the direction of the x_1 axis, the velocity gradient in the x_2 direction is $\dot{\gamma}$ and the rate-of-strain tensor has the form

$$\{\dot{\gamma}\} = \frac{1}{2} \begin{vmatrix} 0 & \dot{\gamma} & 0 \\ \dot{\gamma} & 0 & 0 \\ 0 & 0 & 0 \end{vmatrix}$$

Since we cannot say in advance which of the stress components are equal to zero, in order to characterize the state of stress we should write the complete stress tensor:

$$\{\sigma\} = \begin{vmatrix} \sigma_{11} & \sigma_{12} & \sigma_{13} \\ \sigma_{21} & \sigma_{22} & \sigma_{23} \\ \sigma_{31} & \sigma_{32} & \sigma_{33} \end{vmatrix}$$

A new coordinate system x_i' may be introduced at any point in the medium, which is determined in such a way that the new coordinates x_1', x_2', and x_3' are related to the old coordinates x_1, x_2, and x_3 by the following conditions:

$$x_1' = x_1; \quad x_2' = x_2; \quad x_3' = -x_3$$

This means that the coordinate system x_i' is the mirror image of the old coordinate system x_i relative to the plane formed by the axes x_1 and x_2. Then, use may be made of the general formulas for the transformation of the tensor components from some coordinates to others [see formula (1.34)] in application to the tensors $\{\dot{\gamma}\}$ and $\{\sigma\}$, their components being transformed from the coordinate system x_i to x_i'. This leads to the following expressions for the components of the tensors $\{\dot{\gamma}\}$ and $\{\sigma\}$, which are, respectively, designated as $\dot{\gamma}_{ij}'$ and σ_{ij}' in the new coordinate system:

$$\{\dot{\gamma}\}' = \begin{vmatrix} \dot{\gamma}_{11}' & \dot{\gamma}_{12}' & \dot{\gamma}_{13}' \\ \dot{\gamma}_{21}' & \dot{\gamma}_{22}' & \dot{\gamma}_{23}' \\ \dot{\gamma}_{31}' & \dot{\gamma}_{32}' & \dot{\gamma}_{33}' \end{vmatrix} = \frac{1}{2} \begin{vmatrix} 0 & \dot{\gamma} & 0 \\ \dot{\gamma} & 0 & 0 \\ 0 & 0 & 0 \end{vmatrix}$$

and

$$\{\sigma\}' = \begin{vmatrix} \sigma_{11}' & \sigma_{12}' & \sigma_{13}' \\ \sigma_{21}' & \sigma_{22}' & \sigma_{23}' \\ \sigma_{31}' & \sigma_{32}' & \sigma_{33}' \end{vmatrix} = \begin{vmatrix} \sigma_{11} & \sigma_{12} & -\sigma_{13} \\ \sigma_{21} & \sigma_{22} & -\sigma_{23} \\ -\sigma_{31} & -\sigma_{32} & \sigma_{33} \end{vmatrix}$$

It is obvious that with any choice of the subscripts, $\dot{\gamma}_{ij} = \dot{\gamma}'_{ij}$ and therefore $\sigma'_{ij} = \sigma_{ij}$. The latter can only be realized if $\sigma_{13} = -\sigma_{31}$ and $\sigma_{23} = -\sigma_{32}$; no restrictions are imposed on the other components. It follows that $\sigma_{13} = \sigma_{31} = 0$; $\sigma_{23} = \sigma_{32} = 0$ and, in the general case, in unidimensional shear flow the complete stress tensor may be represented in the following form:

$$\{\sigma\} = \begin{vmatrix} \sigma_{11} & \sigma_{12} & 0 \\ \sigma_{21} & \sigma_{22} & 0 \\ 0 & 0 & \sigma_{33} \end{vmatrix}$$

There is no ground for stating that all the components of the tensor $\{\sigma\}$, except σ_{12}, are equal to zero. Thus, in simple shear flow, apart from tangential stresses, there may arise normal stresses σ_{ii}. This phenomenon is called the Weissenberg effect or the normal stress effect.

The existence of non-zero diagonal components of the stress tensor in shear flow is a specific feature of the rheological properties of many polymers. The results obtained show that in steady-state shear flow the behaviour of a liquid may be considered to have been determined not only by the dependence τ $(\dot{\gamma})$, which has been considered in detail in Chapter 2, but also by the dependences of normal stresses on the rate of deformation, i.e., if the functions σ_{11} $(\dot{\gamma})$, σ_{22} $(\dot{\gamma})$, and σ_{33} $(\dot{\gamma})$ are known. For an incompressible liquid, all these three functions are independent. Indeed, if we subtract the hydrostatic pressure, then σ_{11}, σ_{22}, and σ_{33} will represent the components of the stress tensor deviator. Then the following equality must be fulfilled:

$$\sigma_{11} + \sigma_{22} + \sigma_{33} = 0$$

Therefore, of the three functions σ_{11} $(\dot{\gamma})$, σ_{22} $(\dot{\gamma})$, and σ_{33} $(\dot{\gamma})$ only two are independent. Since the choice of the functions that should be taken as the determining properties of the medium is immaterial, the normal stress differences are commonly used as the characteristics of the liquid:

$$\sigma = \sigma_{11} - \sigma_{22}$$

$$\sigma' = \sigma_{22} - \sigma_{33}$$

The quantities σ and σ' are called the first and second normal stress differences, respectively. The use of the normal stress differences is the more expedient that in this case the hydrostatic pressure is automatically excluded since the difference of the diagonal components of the stress tensor deviator is equal to that of the diagonal components of the stress tensor itself.

Thus, for the rheological flow properties of a liquid in simple shear flow to be described completely, it is necessary to determine the three functions τ $(\dot{\gamma})$, σ $(\dot{\gamma})$, and σ' $(\dot{\gamma})$. The same functions may be expressed in terms of the corresponding coefficients:

$$\eta\,(\dot{\gamma}) = \sigma_{12}/2\dot{\gamma}_{12} = \tau/\dot{\gamma}$$

$$\zeta\,(\dot{\gamma}) = \sigma/2\dot{\gamma}^2 = (\sigma_{11} - \sigma_{22})/2\dot{\gamma}^2$$

$$\beta\,(\dot{\gamma}) = \sigma'/2\dot{\gamma}^2 = (\sigma_{22} - \sigma_{33})/2\dot{\gamma}^2$$

The first of these is the viscosity coefficient, which has been discussed in detail in Chapter 2. The two others, ξ $(\dot{\gamma})$ and β $(\dot{\gamma})$, are the normal stress coefficients.

The existence of non-zero diagonal components of the stress tensor in the simple shear flow of a liquid leads to a number of spectacular manifestations of the specific properties of the material. These specific features of the behaviour of elastic liquids are usually collectively termed the Weissenberg effect. A number of manifestations of the Weissenberg effect are quoted by Bird, Armstrong and Hassager*. There is no doubt that new manifestations of this effect will be detected in the nearest future. Some examples of such mani-

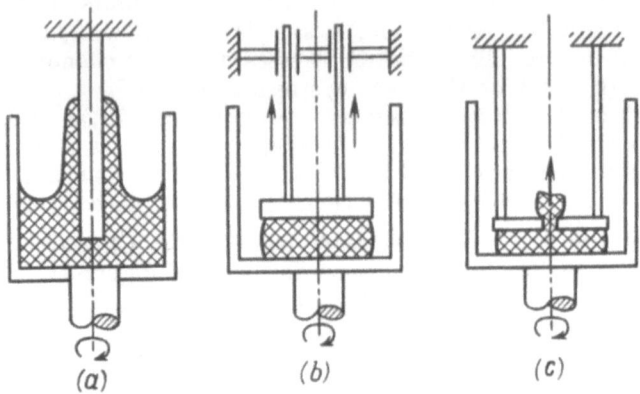

Fig. 4.1. Some manifestations of the Weissenberg effect in shearing flow:
a—the liquid climbs up the rod with the rotating beaker; b—movement of the rod (in the direction indicated by arrows) under the action of a vertical force arising upon relative rotation of the disks; c—squeezing of the liquid through the opening (in the direction of the arrow) upon relative rotation of the disks.

festations of the Weissenberg effect are shown in Fig. 4.1. For example, if a fixed rod-like stator is placed in a rotating cylindrical beaker containing such a liquid, the liquid will "climb up" the rod instead of being thrown off to the beaker walls, as is the case in an analogous experiment conducted with low-molecular-mass liquids. If the liquid is placed between two parallel disks, one of which is rotating relative to the common axis, there will develop a force normal to the surface of the disks. If the disks are not secured and can be moved along the axis, they will move apart under the influence of this force. And if a hole is made in the centre of one of the disks, the liquid being deformed will be forced through it. Other schemes of experiments are also possible, which show the effect of normal stresses developing in shear flow on the specific features of the flow of the liquid.

The possibility of development of the Weissenberg effect has been proved above for simple shear flow. In this scheme of deformation it is possible to elucidate most clearly the specificity of the state of stress for liquids, for which σ and σ' are different from zero. In more complex conditions of deformation the normal stresses may be different from zero because of the peculiarities of the geometric scheme of deformation, and therefore the manifestation of the Weissenberg effect is not so obvious as in simple shear. But materials that

---

* Bird, R. B., R. C. Armstrong, and O. Hassager, *Dynamics of Polymeric Fluids*, vol. 1, John Wiley and Sons, New York, 1977.

exhibit the Weissenberg effect in simple shear behave, in complex deformation schemes, differently from a liquid incapable of displaying this effect.
In order to define the Weissenberg effect in complex deformation schemes,
one should make recourse to the constitutive equation for liquids with arbitrary rheological flow properties and consider the differences in the nature of
developing normal (diagonal) stresses as compared with the stresses arising
during the flow of a Newtonian liquid.

The essence of the Weissenberg effect consists in that in liquids capable of
exhibiting this effect a strictly one-dimensional shear deformation is impossible: one-dimensional flow always leads to a three-dimensional pattern of
the state of stress.

4.2. Normal Stresses in Elastic Bodies

4.2.1. Geometrical Interpretation

Although the Weissenberg effect is specific to the shear flow of a liquid, the
physical causes of this phenomenon are, as a rule, tied up with the viscoelasticity of the material, the appearance of normal stresses being attributed to

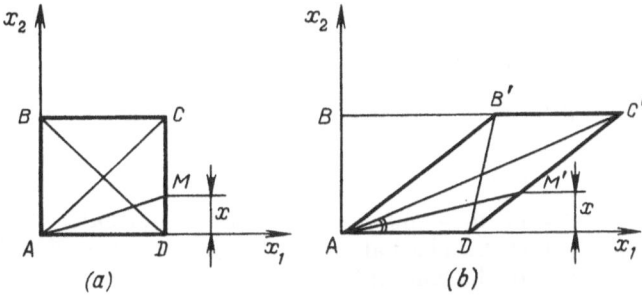

Fig. 4.2. Diagrams illustrating simple shear in an elastic body.

the development of finite elastic strains in the body. The supposition that the
observed outward manifestations of normal stresses are due to the elasticity
of the liquid was advanced by Weissenberg who was the first to describe the
effects under discussion. It is expedient to examine, in pure form, the consequences brought about by finite elastic deformations in solids incapable of
flowing, i.e., to consider the model of the physical phenomenon which is
usually referred to in order to explain the Weissenberg effect in liquids.

Normal stresses always arise on finite deformations of an elastic body.
This is proved by inspecting the scheme of simple shear of an elastic body
shown in Fig. 4.2. Let the distance between the parallel plates along the
vertical be equal to unity. We assume that it does not change in the course
of deformation. The element $ABCD$ in the initial state has the form of a
square and after being shifted (this shift being characterized by the quantity γ
equal to BB' as $AB = 1$) it occupies the position $AB'C'D$, in which case
$AB' = C'D$ and $AD = B'C' = 1$.

As a result of deformation, the diagonal line AC becomes elongated up to AC' and a certain portion AM drawn so that the point M is at a distance x from the x_1 axis is elongated up to the amount AM'. The amount of displacement γ corresponds to the portion BB'. The relative strain ε of the arbitrarily chosen portion AM is

$$\varepsilon = \frac{AM'}{AM} = \left[\frac{(1+\gamma x)^2 + x^2}{1+x^2} \right]^{0.5}$$

It is not difficult to show that when x varies from 0 to 1, the relative deformation ε increases with increasing γ, and therefore the maximum value of ε corresponds to the elongation of the diagonal AC up to AC', i.e., the principal extension in the scheme under consideration is equal to AC'/AC. Accordingly, the angle $C'AD$ corresponds to the angle between the directions of the maximum extension and shear. To simplify the further analysis, it may be assumed that in the direction of the main elongation AC and in the perpendicular direction there act the principal stresses σ_1 and σ_2. This supposition is strictly fulfilled only for infinitesimal deformations, but it is useful for obtaining spectacular results for large values of shear as well. Then χ represents the angle between the line of action of the principal stress σ_1 and the direction of shear. Formulas (1.1) and (1.4), considering that the tangential stresses in the direction in which the stress σ_1 acts are equal to zero, allow one to calculate all the components of acting stresses:

$$\sigma_{11} = \sigma_1 \cos^2 \chi + \sigma_2 \sin^2 \chi$$

$$\sigma_{22} = \sigma_1 \sin^2 \chi + \sigma_2 \cos^2 \chi \qquad\qquad (4.1)$$

$$\tau = \sigma_{12} = -\frac{1}{2}(\sigma_1 - \sigma_2) \sin 2\chi$$

The normal stress difference $\sigma = \sigma_{11} - \sigma_{22}$ is expressed in terms of the principal stresses:

$$\sigma = \sigma_{11} - \sigma_{22} = (\sigma_1 - \sigma_2) \cos 2\chi$$

Thus, at all times when $\sigma_1 - \sigma_2 \neq 0$, there exist normal stresses, i.e., the Weissenberg effect appears. The values of σ may be expressed in terms of the shearing stresses τ and the amount of shear.

From the relations written above it is seen that

$$\frac{\sigma}{2\tau} = \cot 2\chi$$

and from Fig. 4.2 it follows that with the principal stresses being oriented along and at right angles to the diagonal AC'

$$\tan \chi = (1+\gamma)^{-1}$$

After simple trigonometric transformations we find that

$$\cot 2\chi = \frac{(2+\gamma)\,\gamma}{2\,(1+\gamma)}$$

Then, the formula that relates the first normal stress difference to the tangential stresses and the amount of shear is written in the form

$$\frac{\sigma}{2\tau} = \frac{(2+\gamma)\,\gamma}{2\,(1+\gamma)} \qquad\qquad (4.2)$$

It follows that the normal stress difference is approximately proportional to the amount of shear γ since the factor $(2+\gamma)/2\,(1+\gamma)$ varies only within the range from 1 to 0.5. Therefore, at small γ the normal stresses are negligibly small as compared with the tangential stresses but they are commensurate with τ and exceed them with increasing recoverable shear γ. Therefore, in accordance with the results obtained σ is determined by the amount of elastic deformations developing in the material.

From formula (4.2) it follows that at infinitesimal deformations the following relation is fulfilled:

$$\sigma/2\tau = \gamma \qquad\qquad (4.3)$$

which is usually called the Lodge formula, and at finite deformations formula (4.2) becomes

$$\sigma/\tau = \gamma \qquad\qquad (4.4)$$

which is known as the Weissenberg-Mooney-Rivlin formula (see below). Formulas (4.3) and (4.4) are the limiting cases of the more general relation (4.2) which results from an approximate geometric interpretation of finite deformations if the direction of the diagonal of the deformed element in Fig. 4.2 is identified with the line of action of the principal tensile stress (generally speaking, this assumption is not quite strict).

The results obtained show that at finite elastic deformations, normal stresses will necessarily develop in the body by virtue of a purely geometric picture of extensions and rotations of the elements of the material during shear, irrespective of the physical mechanism allowing for finite deformations.

4.2.2. Normal Stresses and Strain-Energy Functions

The representation of strain energy (stored energy) as a function of the invariants of the strain tensor is the most general method of describing the stress-strain relationships for elastic bodies. Therefore it may be expected that this method of writing the constitutive equation of a material will allow us to predict the existence of normal stresses and will show how the normal stresses that arise in simple shear depend on the amount of shear γ. One of the most frequently used methods of representing the properties of an elastic body is the application of the Mooney-Rivlin strain-energy function written as (1.62b), i.e., in the form of the dependence of the strain energy function on the linear combination of the first invariants of the Green and Finger finite strain tensors.

The normal stresses dealt with in the present section are stresses which act at right angles to the areas oriented along the coordinate axes and which do not coincide with the principal stresses. Let the principal stresses in elastic shear strain be σ_1, σ_2, and σ_3 and the directions of their action coincide with those of the principal extensions $\varkappa_1$, $\varkappa_2$, and $\varkappa_3$, respectively. Since we are here dealing with the simple shear strain, the principal extensions in this type of strain are given by [see formula (1.25)]:

$$\varkappa_1 = \cot \chi; \quad \varkappa_2 = \tan \chi; \quad \varkappa_3 = 1$$

and the angle χ between the direction of shear and the direction of the principal stresses is expressed in terms of γ as follows (see page 37):

$$\chi = \frac{1}{2} \operatorname{arccot} \left(\frac{\gamma}{2} \right)$$

We have given above [see formulas (4.1)] expressions which allow us to find the normal stresses σ_{11} and σ_{22} through the principal stesses σ_1 and σ_2. The third principal stress σ_3 coincides with the normal stress σ_{33}, which, in its turn, is inversely proportional to the hydrostatic pressure p:

$$\sigma_3 = \sigma_{33} = -p$$

The hydrostatic pressure acts on a "free" area oriented perpendicularly to the axes x_1 and x_2 (see Fig. 4.2).

Considering that in the general case $\sigma_i = \lambda_i \, (\partial W/\partial \lambda_i)$ [see formula (1.54)], the principal stresses σ_1, σ_2, and σ_3 are expressed in terms of the Mooney-Rivlin elastic strain-energy function for simple shear strain in the following manner:

$$\begin{cases} \sigma_1 = 2C_1\lambda_1^2 - 2C_2\lambda_1^{-2} = 2\,(C_1 \cot^2 \chi - C_2 \tan^2 \chi) \\ \sigma_2 = 2C_1\lambda_2^2 - 2C_2\lambda_2^{-2} = 2\,(C_1 \tan^2 \chi - C_2 \cot^2 \chi) \\ \sigma_3 = 2C_1\lambda_3^2 - 2C_2\lambda_3^{-2} = 2\,(C_1 - C_2) \end{cases}$$

Substitution of the expressions for σ_1 and σ_2 into formulas (4.1) gives, after certain trigonometric rearrangements, the following expressions for the normal stress differences:

$$\begin{cases} \sigma_{11} - \sigma_{22} = (\sigma_2 - \sigma_1) \cos 2\chi = -2\,(C_1 + C_2)\,\gamma^2 \\ \sigma_{22} - \sigma_{33} = 2C_1\gamma^2 \\ \sigma_{11} - \sigma_{33} = -2C_2\gamma^2 \end{cases}$$

Thus, the normal stress differences are found to be proportional to the square of the amount of shear, and the proportionality factors are the constants figuring in the expression for the Mooney-Rivlin strain-energy function. If, instead of the two-term function, use is made of a one-term function as the initial constitutive equation, then at $C_1 = 0$

$$\begin{cases} \sigma_{11} - \sigma_{22} = \sigma_{11} - \sigma_{33} = -2C_2\gamma^2 \\ \sigma_{22} - \sigma_{33} = 0 \end{cases}$$

and at $C_2 = 0$

$$\begin{cases} \sigma_{11} - \sigma_{22} = \sigma_{33} - \sigma_{22} = -2C_1\gamma^2 \\ \sigma_{11} - \sigma_{33} = 0 \end{cases}$$

This corresponds to various possible relationships between the normal stresses.

The absolute values of normal stresses are determined if the hydrostatic pressure or the amount of one of the diagonal components of the stress tensor is known. This can be done by using relation (4.5), which enables us to relate the constants C_1 and C_2 to the pressure:

$$2\,(C_1 - C_2) = -p$$

Hence, we obtain the following expressions for the components of normal stresses:

$$\begin{cases} \sigma_{11} = -2C_2\gamma^2 - p \\ \sigma_{22} = 2C_1\gamma^2 - p \\ \sigma_{33} = -p \end{cases}$$

Thus, normal stresses, just like their differences, are proportional to the square of shear, but the absolute values of stresses are determined only within the hydrostatic pressure. The above discussion only refers to an incompressible liquid, for which the superposition of an arbitrary hydrostatic pressure does not affect the stress-strain relations or, more exactly, the values of stresses developing due to the assignment of a certain deformation.

The relation between the normal stress differences depends on the ratio of the constants C_1 and C_2. An additional characteristic of the rheological properties of the material is the ratio $\varepsilon = \sigma'/\sigma$, which is given in the following manner for an elastic liquid, whose properties are described by the Mooney-Rivlin function:

$$\varepsilon = \frac{\sigma_{22} - \sigma_{33}}{\sigma_{11} - \sigma_{22}} = - \frac{C_1}{C_1 + C_2} = - \frac{(C_1/C_2)}{1 + (C_1/C_2)} \tag{4.5}$$

According to the Weissenberg hypothesis, which was used by Weissenberg himself in formulating the constitutive equation for an elastic liquid, $\varepsilon = 0$, which is possible only if $C_1 = 0$. Thus, the Weissenberg hypothesis is obeyed by a material whose strain-energy function is proportional to the first invariant of the Finger finite strain tensor. As will be shown below, for real materials the value of ε is small and therefore the ratio of the constants (C_1/C_2) is also small.

The results obtained can be generalized for a strain-energy function of any form, which refers to an incompressible elastic body; the latter condition means that only the two invariants of the finite strain tensor are independent. Let the strain-energy function be a certain function of the invariants E_1 and E_2, which may be represented (see Chapter 1, Sec. 1.6) as a function of the first invariants of the Green or Finger strain tensors:

$$W = W(E_1^G, E_1^F) \tag{4.6}$$

Then, the following expression is valid for the derivatives of the strain-energy function with respect to the principal extension ratios:

$$\frac{\partial W}{\partial \varkappa_i} = \frac{\partial W}{\partial E_1^G} \frac{\partial E_1^G}{\partial \varkappa_i} + \frac{\partial W}{\partial E_1^F} \frac{\partial E_1^F}{\partial \varkappa_i} \quad (i = 1, 2, 3) \tag{4.7}$$

The derivatives $(\partial E_1^G/\partial \varkappa_i)$ and $(\partial E_1^F/\partial \varkappa_i)$ are equal to:

$$\frac{\partial E_1^G}{\partial \varkappa_i} = 2\varkappa_i; \qquad \frac{\partial E_1^F}{\partial \varkappa_i} = -2\varkappa_i^{-3}$$

Formulas (1.54) allow us to write the following expressions for the principal stresses:

$$\begin{cases} \sigma_1 = 2\left(\varkappa_1^2 \dfrac{\partial W}{\partial E_1^G} - \varkappa_1^{-2} \dfrac{\partial W}{\partial E_1^F}\right) = 2\left(\cot^2\chi \dfrac{\partial W}{\partial E_1^G} - \tan^2\chi \dfrac{\partial W}{\partial E_1^F}\right) \\[2mm] \sigma_2 = 2\left(\varkappa_2^2 \dfrac{\partial W}{\partial E_1^G} - \varkappa_2^{-2} \dfrac{\partial W}{\partial E_1^F}\right) = 2\left(\tan^2\chi \dfrac{\partial W}{\partial E_1^G} - \cot^2\chi \dfrac{\partial W}{\partial E_1^F}\right) \\[2mm] \sigma_3 = 2\left(\varkappa_3^2 \dfrac{\partial W}{\partial E_1^G} - \varkappa_3^{-2} \dfrac{\partial W}{\partial E_1^F}\right) = 2\left(\dfrac{\partial W}{\partial E_1^G} - \dfrac{\partial W}{\partial E_1^F}\right) \end{cases}$$

Since the stress σ_{33} is equal to the principal stress σ_3 and only the hydrostatic pressure acts on the free surface, it follows that

$$\sigma_{33} = -p = 2\left(\frac{\partial W}{\partial E_1^G} - \frac{\partial W}{\partial E_1^F}\right) \tag{4.8}$$

The normal stresses σ_{11} and σ_{22} are expressed, in accordance with formula (4.1), in terms of the derivatives of the strain-energy function with respect to the invariants $(\partial W/\partial E_1^G)$ and $(\partial W/\partial E_1^F)$:

$$\begin{cases} \sigma_{11} = 2\left[\dfrac{\partial W}{\partial E_1^G} - \dfrac{\partial W}{\partial E_1^F}\left(\dfrac{\sin^6\chi + \cos^6\chi}{\sin^2\chi\cos^2\chi}\right)\right] \\[2ex] \sigma_{22} = 2\left[\dfrac{\partial W}{\partial E_1^F}\left(\dfrac{\sin^6\chi + \cos^6\chi}{\sin^2\chi\cos^2\chi}\right) - \dfrac{\partial W}{\partial E_1^F}\right] \end{cases}$$

Formula (4.8) permits the elimination of one of the derivatives from the expressions for the normal stresses:

$$\begin{cases} \sigma_{11} = -2\left(\dfrac{\partial W}{\partial E_1^F}\right)\gamma^2 - p \\[2ex] \sigma_{22} = 2\left(\dfrac{\partial W}{\partial E_1^G}\right)\gamma^2 - p \\[2ex] \sigma_{33} = -p \end{cases}$$

Hence, the normal stress differences are written in the following manner:

$$\begin{cases} \sigma \equiv \sigma_{11} - \sigma_{22} = -2\left(\dfrac{\partial W}{\partial E_1^F} + \dfrac{\partial W}{\partial E_1^G}\right)\gamma^2 \\[2ex] \sigma' \equiv \sigma_{22} - \sigma_{33} = 2\dfrac{\partial W}{\partial E_1^G}\gamma^2 \end{cases} \qquad (4.9)$$

The expression for the tangential stress in simple shear is derived with the aid of formula (1.2) if the form of the dependence of the strain-energy function on E_1^G and E_1^F is arbitrary:

$$\tau \equiv \sigma_{12} = -\frac{1}{2}(\sigma_1 - \sigma_2)\sin 2\chi = -2\left(\dfrac{\partial W}{\partial E_1^G} + \dfrac{\partial W}{\partial E_1^F}\right)\gamma \qquad (4.10)$$

The formulas obtained show the structure of the dependences of the normal stresses on the properties of the material, which are expressed by the function $W(E_1^G, E_1^F)$ and by the elastic shear strain γ. Indeed, the normal stress effect is quadratic with respect to the amount of shear, whereas the tangential stresses depend on γ in a linear fashion. The quantity $-2\,(\partial W/\partial E_1^F + \partial W/\partial E_1^G)$ represents the "effective shear modulus", although, in a general case, of course this sum is not necessarily constant and may vary in a complicated way with increasing amount of shear, depending on the form of the function $W(E_1^G, E_1^F)$. Therefore, the dependences of τ, σ, and σ' on strain may, in fact, be nonlinear or non-quadratic; it is only essential that the expressions for the normal and tangential stresses contain γ^2 and γ, respectively.

A comparison of formulas (4.9) and (4.10) allows one to derive a simple expression that relates the first normal stress difference to the shear stress:
$$\sigma = \sigma_{11} - \sigma_{22} = \tau\gamma.$$
This formula has already been derived above from the approximate geometrical considerations referring to the case of large values of γ [see formula (4.4)]. It was first obtained by Weissenberg working from the assumption that $\sigma' = 0$, and later by Mooney, who proceeded from the function (1.62b) and, finally, by Rivlin for an arbitrary form of the strain-energy function in the same way as was done above. Therefore, as has already been said, this formula is called the Weissenberg-Mooney-Rivlin formula.

Relation (4.9) shows that the second normal stress difference is equal to zero only when the strain-energy function is independent of E_1^G and is only determined by the quantity E_1^F. Since for real materials σ_{22} is really equal to σ_{33} (or, at least, $\sigma' \ll \sigma$), it may be presumed that the strain-energy function is indeed a function only of E_1^F. Therefore, the following expressions for the normal stresses at the hydrostatic pressure p are valid:

$$\begin{cases} \sigma_{11} = -2\left(\dfrac{\partial W}{\partial E_1^F}\right)\gamma^2 - p \\ \sigma_{22} = \sigma_{33} = -p \end{cases}$$

And if $p = 0$, then there exists only one non-zero diagonal component of the stress tensor:

$$\sigma_{11} = -2\frac{\partial W}{\partial E_1^F}\gamma^2$$

The results obtained provide the most general predictions about the normal stresses arising during the shear of elastic bodies in the case of arbitrarily large deformations. Concrete calculations require a knowledge of the dependence $W(E_1^G, E_1^F)$ or of the dependence $W(E_1^F)$ if it is assumed in advance that $\sigma' = 0$. The latter dependence can be found for various materials from any scheme of deformation; it is most convenient to measure the dependence $\tau(\dot{\gamma})$ in simple shear or the dependence of the force on deformation in uniaxial extension. Any of these experiments allows us to find $W(E_1^F)$ and, hence, to calculate the normal stresses.

4.3. Normal Stresses in Flow*

4.3.1. Normal Stresses in Various Constitutive Equations

In one-dimensional shearing flow of a Newtonian liquid there exist no normal stresses different from the hydrostatic pressure. This follows immediately from the constitutive equation of a Newtonian liquid since the stresses that develop during its flow, σ_{ij}, depend only on the components of the rate-of-strain tensor with the same subscripts. Therefore, if $\dot{\gamma}_{ii} = 0$, then σ_{ii} is also equal to zero. In viscous liquids, whose rheological properties are described by more complex constitutive equations than those used for a Newtonian liquid, the appearance of normal stresses is possible in shearing flow.

For the Weissenberg effect arising in liquids to be explained and described quantitatively, use is commonly made of an approach based on the definition of a liquid as a viscoelastic body, whose properties are characterized by a certain relaxation spectrum (see Chapter 3 for more detail) and which is capable of developing large elastic deformations. The latter circumstance presupposes the necessity of using the kinematic relations formulated in Chapter 1 (Sec. 1.4), i.e., the constitutive equation is written for the neighbourhood of a certain point of a moving body (in the convected coordinate system) and is then transformed to the space-fixed coordinate system with the aid of the various differential operators.

---

* See also a review article: Malkin, A. Ya., *Mekhanika Polimerov*, 1, 173 (1975).

The procedure of using differential operators of various structure for generalization of constitutive equations with a discrete distribution of relaxation times was described in Chapter 2 (Sec. 2.5.10), where we also outlined methods of calculating the normal stresses through the constants of certain rheological models. This has enabled the normal stresses to be represented in the form of a function of the rate of shear. The form of this function depends, first, on the form of the differential operator used to transform the convected to the fixed coordinate system and, second, on the number of terms retained in the constitutive equation (1.104). We shall give here only the results of calculations based on the use of the most important differential operators as applied to a model with an arbitrary number of terms.

The generalization of a constitutive equation with the aid of the Oldroyd operator D_O leads to the following expressions for the normal stress differences (compare with the formulas given in Chapter 2, Sec. 2.5.10):

$$\begin{cases} \sigma_{22} = (A_2 - 2B_1 A_1)\,\dot{\gamma}^2 \\ \sigma_{11} = \sigma_{33} = 0 \end{cases}$$

The quantity A_1 has the meaning of the initial Newtonian viscosity. The Oldroyd model, which does not predict the non-Newtonian behaviour [see formula (2.48)], describes the appearance of normal stresses as an effect that is quadratic with respect to the rate of deformation.

More complicated results are obtained on the basis of the use of the Jaumann operators D_J and the Oldroyd nonlinear operator $D_{O,\,n}$ in the constitutive equation (1.104). In the first case [see formulas (2.52)]:

$$\sigma_{11} - \sigma_{22} = 2\eta_0 \dot{\gamma}\, \frac{BC - AD}{A^2 + B^2}\,; \qquad \sigma_{33} = 0$$

and in the second case [(see formulas (2.54)]:

$$\sigma_{11} - \sigma_{22} = \sqrt{6}\,\dot{\gamma}\, \frac{PS - RQ}{S^2 + R^2}\,; \qquad \sigma_{22} - \sigma_{33} = 0$$

Here the capital letters designate the functions of the rate of shear, which are the power series of $\dot{\gamma}$ with the coefficients included in the constitutive equations (1.104). The shape of these functions has been considered in detail in Sec. 2.5.10. Depending on the choice of the number of the members in the expansion series and the concrete values of the constants, the formulas derived for the normal stress differences can practically describe any experimentally observed results.

As seen from the results obtained, the use of the Oldroyd operator $D_{O,\,n}$ for any viscoelastic liquid predicts that the second normal stress difference is equal to zero. Therefore, if really $\sigma' \neq 0$, then it is necessary to use operators of a different structure. An example of such a model is the constitutive equation (1.127).

4.3.2. Normal Stresses in a Viscoelastic Liquid

The appearance of normal stresses during the shearing flow of a viscoelastic liquid is due to the fact that finite elastic strains are developing in this liquid. For this reason, it is necessary to write a constitutive equation for the volume element moving in space and to take into account the kinematics of the motion of the material. This fact has been reflected above by introducing differential operators of different structures into the constitutive equations of a material with a discrete distribution of relaxation times. The effect of the motion of the body in space on the stresses that arise is very spectacularly traced out in the analysis of the motion of a viscoelastic liquid with a continuous distribution of relaxation times, which is described by the Boltzmann superposition principle.

Let the constitutive equation of a linear viscoelastic liquid (1.79) be valid for a small neighbourhood of each point in the body being deformed and, hence, this equation is written in the convected coordinate system. The change to the fixed coordinate system is accomplished by means of the ordinary method of transformation of coordinates. Then, the components of the stress tensor in the space-fixed coordinate system* are written as follows:

$$\sigma_{ij}(x_i, t) = \int\limits_{-\infty}^{t} 2\varphi(t-t') \frac{\partial X_m}{\partial x_i} \frac{\partial X_n}{\partial x_j} \dot{\gamma}_{mn} \, dt' \qquad (4.11)$$

where X_i are the values of the coordinates of a point at each time in the convected coordinate system; x_i are the space-fixed coordinates of the point.

The relations between x_i and X_i were established in a general form in Chapter 1 (Sec. 1.4), and in simple shearing flow these relations assume the form [see Eqs. (1.32)]:

$$\begin{cases} X_1 = x_1 - \int\limits_{t'}^{t} V(x_2, t'') \, dt'' \\ X_2 = x_2 \\ X_3 = x_3 \end{cases}$$

where V is the velocity in the direction x_1, which depends, in a general case, on the coordinate x_2 and on time.

For steady-state flow $\dot{\gamma}$ is time-independent and $V = \dot{\gamma} x_2$, where $\dot{\gamma}$ is the specified velocity gradient. Then, for steady shearing flow the stress components may be expressed in the following way:

$$\begin{cases} \sigma_{12} = \sigma_{21} = \dot{\gamma} \int\limits_{0}^{t} \varphi(t-t') \, dt' \\ \sigma_{11} - \sigma_{22} = 2\dot{\gamma}^2 \int\limits_{0}^{t} \int\limits_{t'}^{t} \varphi(t-t') \, dt'' \, dt' = 2\dot{\gamma}^2 \int\limits_{0}^{t} (t-t') \varphi(t-t') \, dt' \\ \sigma_{22} - \sigma_{33} = 0 \end{cases}$$

---

* There are various methods of writing the stresses in the space-fixed coordinate system, which are based on a constitutive equation derived for a moving material element. Here, use is made of the method based on the Oldroyd transformation. For greater detail, see Malkin, A. Ya., *Rheol. Acta*, **7**, 335 (1968).

or

$$
\left\{
\begin{aligned}
& \tau = \dot{\gamma} \int_0^t \varphi(x)\, dx \\[2ex]
& \sigma = 2\dot{\gamma}^2 \int_0^t x\varphi(x)\, dx \\[2ex]
& \sigma' = 0
\end{aligned}
\right.
$$

Replacement of the relaxation function $\varphi(x)$ by the relaxation spectrum in accordance with formula (1.86) leads to the following expressions:

$$
\tau(t) = \dot{\gamma} \int_0^\infty \theta F(\theta)\, (1 - e^{-t/\theta})\, d\theta = \dot{\gamma} \int_0^\infty s^{-1} N(s)\, (1 - e^{-ts})\, ds
$$

$$
\sigma(t) = 2\dot{\gamma}^2 \int_0^\infty \theta^2 F(\theta) \left[1 - e^{-t/\theta}\left(1 + \frac{t}{\theta}\right) \right] d\theta = 2\dot{\gamma}^2 \int_0^\infty s^{-2} N(s)\, [1 - e^{-ts}(1 + ts)]\, ds
$$

The first of these formulas has already been obtained in considering the relations of the linear theory of viscoelasticity [see formula (1.97)]. The second describes the change of the first normal stress difference with time. For steady flow when $t \to \infty$

$$
\eta = \frac{\tau(\infty)}{\dot{\gamma}} = \int_0^\infty \theta F(\theta)\, d\theta = \int_0^\infty \frac{N(s)}{s}\, ds \tag{4.12}
$$

$$
\zeta = \frac{\sigma(\infty)}{2\dot{\gamma}^2} = \int_0^\infty \theta^2 F(\theta)\, d\theta = \int_0^\infty \frac{N(s)}{s^2}\, ds \tag{4.13}
$$

Formula (4.12) has already been discussed as one of the corollaries of the linear theory of viscoelasticity (see Chapter 1, Sec. 1.8). Therefore, the method used to generalize a constitutive equation does not predict the effect of non-Newtonian behaviour because it is identical with the use of the Oldroyd differential operator D_O in the constitutive equation of a viscoelastic body with a discrete distribution of relaxation times. As has been shown in Chapter 2 (Sec. 2.5.10), this method of formulating a constitutive equation fails to describe the dependences of the apparent viscosity on the rate of shear. This cannot be done by using the integral constitutive equation (4.11) either.

Formula (4.13) is a new result which does not follow directly from the theory of the mechanical properties of a linear viscoelastic body since in this case normal stresses develop only as a result of the movement of an element of the body deformed in space. This causes the appearance of the diagonal components of the stress tensor in simple shearing flow. According to formula (4.13), the normal stresses are proportional to the square of the rate of shear, as was also the case in the application of the Oldroyd operator to a constitutive equation with a discrete relaxation-time distribution. Therefore, the normal stress effect in a viscoelastic liquid is found to be quadratic (or a second-order effect) with respect to the rate of deformation.

According to formula (4.13), the normal stress coefficient is the second moment of the relaxation spectrum of the system. It has already been shown (see Sec. 1.8) that the second moment of the relaxation spectrum is equal to the product of the square of the viscosity η by the equilibrium compliance J_∞. Therefore, from the results obtained it follows that

$$\zeta = \eta^2 J_\infty \tag{4.14}$$

or, replacing the coefficients by the stress and elastic strain,

$$\sigma = 2\tau\gamma_e$$

This formula (the Lodge formula), as has been shown, is a geometric consequence of large deformations of an elastic body. Here this formula is derived as a corollary of the theory of finite deformations of viscoelastic liquids. It may therefore be assumed that it is valid when there is fulfilled the linear relationship between the shear stresses and the shear rate and the quadratic relation between the normal stresses and the rate of shear. The first of these relations holds for a liquid that does not exhibit non-Newtonian behaviour and for any liquid in the region of low rates of deformation, at least as a limiting case when $\dot\gamma \to 0$. It may be presumed that the second of these relations, namely, $\sigma \propto \dot\gamma^2$, is realized as a limiting case when (at $\dot\gamma \to 0$) the normal stresses prove to be of second-order. The Lodge formula must therefore be fulfilled also as a limiting case at $\dot\gamma \to 0$.

The relationship between the viscosity coefficient, the normal stress coefficient, and the modulus of rubber elasticity at high rates of deformation does not follow, in a general case, from the relations given because for large values of $\dot\gamma$ the formulas, from which there are derived the expressions for stresses and the equilibrium compliance in terms of the relaxation spectrum of a viscoelastic liquid, are not fulfilled.

At high rates of shear the viscoelastic properties of a liquid are changed and the relaxation spectrum depends on the rate of deformation. This question has been discussed in detail in Chapters 2 and 3.

A consequence of the dependence of the relaxation properties of the system on the rate of strain is the change of ζ as compared with its initial value expressed by formula (4.13) with increase of $\dot\gamma$. The dependence $\sigma(\gamma)$ is particularly sensitive to the character of the change of the relaxation spectrum since the deformation exerts an influence, in the first place, on the long-time part of the function $F(\theta)$, and it is exactly this region that has the strongest effect on ζ. This is associated with the structure of the expression for ζ [see formula (4.13)], which includes θ^2, whereas the expression for viscosity contains the relaxation time raised to the first power. However, the suppression of the role of slow relaxation processes when the rate of deformation increases is, however, the general factor responsible for the change of both η and ζ, depending on $\dot\gamma$, so that the two functions $\eta(\dot\gamma)$ and $\zeta(\dot\gamma)$, appear to be associated with the same mechanism of the change of the relaxation spectrum of the system under the influence of increasing rates of deformation.[*]

The results obtained allow one to depict the general course of the dependence of the normal stress coefficient on the velocity gradient in simple shearing flow of polymers. At small values of $\dot\gamma$ the quantity σ_{ii} is proportional

---

[*] Malkin, A. Ya., in: *Advances in Polymer Rheology*, edited by G. V. Vinogradov Khimiya Publishers, Moscow, 1970, pages 171-180 (in Russian).

to $\dot{\gamma}^2$, and therefore there exists a limited value of the function $\zeta\,(\dot{\gamma})$ at $\dot{\gamma} \to 0$. This limiting value of the function $\zeta\,(\dot{\gamma})$ may be called the initial normal stress coefficient ζ_0 by analogy with the initial viscosity coefficient. The quantity ζ_0 is expressed in terms of the relaxation spectrum of the system as the second moment of the spectrum, and therefore the integral (4.13) must be converging. With increasing $\dot{\gamma}$ the normal stress coefficient diminishes as compared with ζ_0, and to this there corresponds a slower (than a quadratic) growth of normal stresses with increasing rate of shear.

Theory predicts quite different relations between the components of normal stresses, depending on the choice of the constitutive equation of the body. Therefore, the question of the ratio of the coefficients ζ and β cannot be solved irrespective of the choice of the constitutive equation.

4.3.3. Normal Stresses and Transient Deformations

Investigations of the transient regimes of deformation of polymers, especially the prestationary regimes, and of the process of stress relaxation after abrupt cessation of steady flow are sensitive methods of determination of the manner in which the shear and normal stresses depend on various parts of the relaxation spectrum of polymers.

For the prestationary stage of shear the following formula has been obtained above:

$$\tau\,(t) = \dot{\gamma} \int\limits_0^\infty s^{-1} N\,(s)\,(1 - e^{-ts})\,ds = \dot{\gamma} \int\limits_0^\infty \theta F\,(\theta)\,(1 - e^{-t/\theta})\,d\theta$$

Then, the area S of the region formed by the straight line $\tau_s = \dot{\gamma}\eta_0$ and the $\tau\,(t)$ curve is calculated as

$$S = \int\limits_0^\infty [\tau_s - \tau\,(t)]\,dt = \dot{\gamma} \int\limits_0^\infty \int\limits_0^\infty \theta F\,(\theta)\,e^{-t/\theta}\,d\theta\,dt = \dot{\gamma} \int\limits_0^\infty \theta^2 F\,(\theta)\,d\theta \qquad (4.15)$$

The area under the relaxation curve is calculated in quite an analogous manner with the same result since in this case

$$\tau_R\,(t) = \int\limits_0^\infty \theta F\,(\theta)\,e^{-t/\theta}\,d\theta \quad \text{and} \quad S = \int\limits_0^\infty \tau_R\,(t)\,dt$$

A direct comparison of the formulas obtained with Eq. (4.13) shows that

$$S = \dot{\gamma}\zeta_0$$

and hence

$$\zeta = S/\dot{\gamma} \quad \text{and} \quad \sigma = 2\dot{\gamma}S$$

These formulas establish the relation between the time dependence of the shear stress at the pre-steady-state stage of shearing flow (or upon relaxation) and the normal stresses.

The physical significance of this result can be understood if we take into account that S corresponds to the elastic energy, W, stored in unit volume during the development of steady-state flow, the relation between S and W being found from the formula

$$W = \frac{1}{2} \dot\gamma S$$

A comparison of the formulas obtained shows that $\sigma = 4W$, i.e., the normal stresses directly characterize the elastic energy stored in the material in steady flow.

A further interesting relation can be obtained if the formula for ζ_0 is represented in the following manner:

$$\zeta_0 = \frac{S}{\dot\gamma} = \eta \int_0^\infty \frac{\tau(t)}{\tau_s} \, dt$$

where the integral might be treated as a certain characteristic relaxation time (according to its dimensions).

The variation of $\sigma(t)$ in transient deformation processes takes place much slower than that of the shear stresses. The corresponding formulas for the prestationary stage of shear flow have been given above. Analogously, the relaxation of normal stresses after the cessation of a rather slow shearing flow occurs by the law

$$\sigma(t) = 2\dot\gamma^2 \int_0^\infty \frac{N(s)}{s^2} e^{-ts} (1 + ts) \, ds = 2\dot\gamma^2 \int_0^\infty \theta^2 F(\theta) e^{-t/\theta} \left(1 + \frac{t}{\theta}\right) d\theta \qquad (4.16)$$

With any form of the relaxation spectrum $F(\theta)$ the variation of $\sigma(t)$ occurs more slowly than that of $\tau(t)$. This is ascribed to the above-mentioned high sensitivity of σ, as compared with τ, to slow relaxation processes and can be proved formally in a general case by a comparison of the expressions:

$$\tau/\tau_s = \int_0^\infty \theta F(\theta) e^{-t/\theta} d\theta \Big/ \int_0^\infty \theta F(\theta) \, d\theta$$

and $\sigma/\sigma_s = \int_0^\infty \theta^2 F(\theta) e^{-t/\theta} [1 + (t/\theta)] \, d\theta \Big/ \int_0^\infty \theta^2 F(\theta) \, d\theta$

It is easy to show that at every instant $\tau(t)/\tau_s \leqslant \sigma(t)/\sigma_s$, the equality being valid only at $t = 0$ and $t \to \infty$. This means that τ always relaxes more rapidly than σ does and reaches steady-state values under the $\dot\gamma = \text{const}$ regime upon completion of the prestationary stage of shear also more rapidly than σ. From this it also follows that the measurement of $\sigma(t)$ in transient deformation processes may be used for the determination of the relaxation properties of the material in that region of large values of relaxation times in which the measurement of $\tau(t)$ for this purpose becomes insensitive.*

---

* This problem has been examined in detail in Malkin, A. Ya. and M. P. Zabugina, *Mekhanika Polimerov*, 2, 335 (1975). This work also gives formulas for calculation of the relaxation spectrum from the measured dependence $\sigma(t)$. It should, however, be borne in mind that the correct measurement of normal stresses in transient deformation conditions requires the use of highly rigid transducers and is associated with rather serious methodological difficulties [see Meissner, J., *J. Appl. Polymer Sci.*, 16, 2877 (1972); *Rheol. Acta*, 14, 201 (1975); Crawley, R. L. and W. W. Graessley, *Trans. Soc. Rheol.*, 21, 19 (1977)].

The relation between the rates of relaxation of tangential and normal stresses, which derives from the above-considered formulas referring to the linear region, remains valid for the nonlinear region at high rates of shear. An illustration of this is Fig. 4.3 which compares the dependences of θ_τ and θ_σ on the rate of shear $\dot\gamma$. The quantities θ_τ and θ_σ signify the time intervals after a sudden cessation of steady-state shearing flow, during which the initial

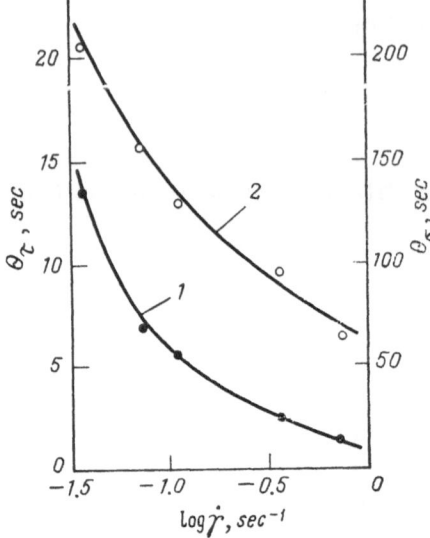

Fig. 4.3. Dependences of the relaxation times of the shear, θ_τ (1), and normal, θ_σ (2), stresses on the rate of shear (butyl rubber; 25°C).

values of the corresponding quantities diminish by e times. It is obvious that in all cases $\theta_\sigma > \theta_\tau$.

4.3.4. Dynamic Normal Stresses and the Viscoelasticity of the Polymeric Chain

The theoretical conceptions and experimental facts which have been considered in Chapter 3 refer to the behaviour of polymers caused by the viscoelastic properties of the macromolecular chain in one-dimensional deformations. Analysis of the spatial picture of deformation of the polymeric chain leads to the prediction of new facts and features of the manifestation of its properties, just as the discussion of the three-dimensional picture of the state of stress in one-dimensional shear leads to the prediction of the completely new effect of appearance of normal stresses. In an analogous way, in one-dimensional shearing vibrations of polymeric systems, whose viscoilasticity is due to the relaxation properties of the chain, there appear dynamec normal stresses. This effect can be predicted on the basis of the generalization of any of the models described in Chapter 3 if, instead of the projections of stresses onto the chosen axis, we consider the complete stress tensor under the conditions when the deformation of the macromolecular chain, whose properties are characterized by the relaxation-time spectrum (in shear), occurs by way of harmonic shear displacements.

The theory of this problem as applied to the Kargin-Slonimsky-Rouse (KSR) and Kirkwood-Riseman-Zimm (KRZ) models (see Chapter 3) and also to the intermediate model of the partially draining coil has been examined by Williams.* The fundamentally new result obtained by Williams consists in predicting that in shear taking place by the specified law $\gamma = \gamma_0 e^{i\omega t}$, apart from shear stresses, there also develop normal stresses. The first normal stress difference σ (ω, t) is expressed thus:

$$\sigma = \sigma_c(\omega) + \sigma_{osc}(\omega)\, e^{2i(\omega t - \delta)}$$

This means that at a specified frequency ω the quantity σ is composed of two components: the constant term σ_c (ω) and the harmonically varying stresses σ_{osc} (ω) $e^{2i(\omega t - \delta)}$ with an amplitude σ_{osc} (ω) and a frequency doubled with respect to the frequency of oscillations of shear stresses. The quantities σ_c and σ_{osc} depend on frequency. The second normal stress difference in the models of random coils is equal to zero at any frequency.

Just as is done for the description of normal stresses under the conditions of steady-state shearing flow, so for the dynamic normal stresses to be characterized there are introduced the normal stress coefficients determined in the following way:

$$\zeta(\omega) = \frac{\sigma}{\dot{\gamma}_0^2} = \frac{\sigma_c}{\dot{\gamma}_0^2} + \frac{\sigma_{osc}}{\dot{\gamma}_0^2}\, e^{2i(\omega t - \delta)}$$

or

$$\zeta(\omega) = \zeta_c + \zeta^* = \zeta_c + \zeta' - i\zeta''$$

The quantity $\dot{\gamma}_0 = \gamma_0 \omega$ represents the amplitude of the rate of harmonic oscillations. An important characteristic of normal stresses is the initial value of the normal stress coefficient ζ_0, which is given by

$$\zeta_0 = \lim_{\omega \to 0} \zeta = \zeta_c(0) + \zeta'(0) \qquad\qquad (4.17)$$

which has the meaning of the limiting point analogous to the initial normal stress coefficient in shearing flow, which is determined as $\lim \zeta(\dot{\gamma})$ at $\dot{\gamma} \to 0$.

If the material has a set of relaxation times determined by the KSR or KRZ theories, then the frequency dependences of the components of the dynamic viscosity $\eta^* = \eta' - i\eta''$ and of the normal stress coefficient $\zeta = \zeta_c + \zeta' - i\zeta''$ may be expressed in terms of the relaxation-time spectrum by means of the following formulas:

$$\begin{cases} \eta'(\omega) = \eta_s + NkT \sum_{p=1}^{n} \dfrac{\theta_p}{1 + (\omega\theta_p)^2} \\[3mm] \eta''(\omega) = NkT \sum_{p=1}^{n} \dfrac{\omega\theta_p^2}{1 + (\omega\theta_p)^2} \end{cases} \qquad (4.18)$$

* Williams, M. C., *J. Chem. Phys.*, **42**, 2988; **43**, 4542 (1965); Akers, L. C. and M. C. Williams, *J. Chem. Phys.*, **51**, 3834 (1969).

and

$$\left\{ \begin{array}{l} \zeta_c\,(\omega) = NkT \sum_{p=1}^{n} \dfrac{\theta_p^2}{1+(\omega\theta_p)^2} \\[2.5em] \zeta'\,(\omega) = NkT \sum_{p=1}^{n} \dfrac{(1-2\omega\theta_p)\,\theta_p^2}{[1+(\omega\theta_p)^2]\,[1+(2\omega\theta_p)^2]} \\[2.5em] \zeta''\,(\omega) = NkT \sum_{p=1}^{n} \dfrac{3\omega\theta_p^3}{[1+(\omega\theta_p)^2]\,[1+(2\omega\theta_p)^2]} \end{array} \right. \qquad (4.19)$$

The pair of relations (4.18) is quite analogous to the system (3.18) obtained earlier in Chapter 3, and formulas (4.19) are a new result of the theory, which expresses the frequency dependences of the components of dynamic normal stresses.

Quantitative calculations of the functions $\zeta_c\,(\omega)$, $\zeta'\,(\omega)$, and $\zeta''\,(\omega)$ require a knowledge of the distribution of relaxation times. If it conforms to

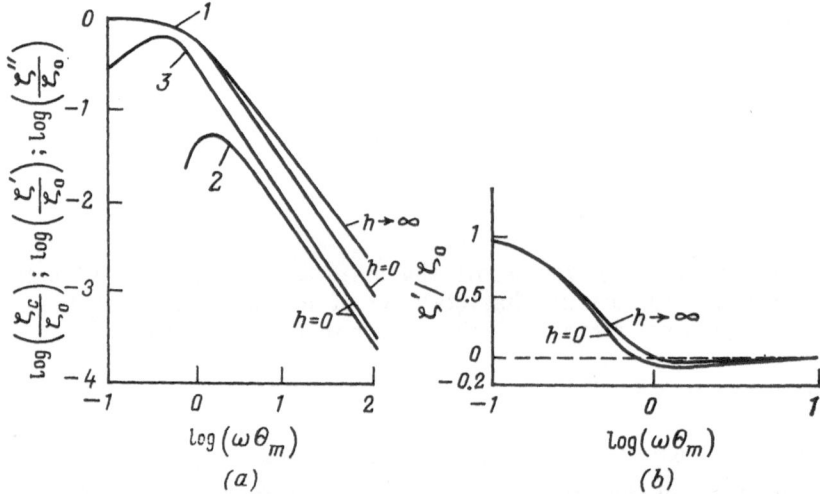

Fig. 4.4 Frequency dependences of the components of the dynamic normal stress coefficient ζ (1), ζ' (2), and ζ'' (3) calculated by M. Williams for the KSR model ($h = 0$), and ζ_c for the KRZ model ($h \to \infty$) (a); shown on the right is the effect of inversion of the sign of the function $\zeta'\,(\omega)$ at small values of the argument (b).

the predictions of the KSR theory (i.e., to the value of $h = 0$ in the theory of the partially draining coil), then the frequency dependences of the normal stress coefficients calculated for such a spectrum, and normalized by the initial value of the normal stress coefficient ζ_0, may be represented as shown in Fig. 4.4 as a function of the dimensionless argument $(\omega\theta_m)$. For comparison, Fig. 4.4 also gives the dependence of the ratio (ζ_c/ζ_0) on frequency for the other extreme case: the KRZ theory when $h \to \infty$. The effect of the parameter h (i.e., of the specific features of the polymer-solvent interaction) on the course of the frequency dependences of dynamic normal stresses is, on the whole, insignificant.

Of special interest is the theoretically predicted inversion of the sign of ζ' upon approach to $\omega\theta_m = 1$. This is shown in Fig. 4.4b for two extreme cases: $h = 0$ and $h \to \infty$. The inversion of the sign of ζ' cannot be shown in Fig. 4.4a because a logarithmic scale is chosen for describing theoretical results, and therefore only the negative values of ζ' are indicated for the region corresponding to large values of the dimensionless argument $(\omega\theta_m)$.

The dynamic normal stresses treated in the generalized kinetic models of a polymer as well as the dynamic functions discussed for these models in Chapter 3 refer to the region of small amplitudes where the normal stress coefficients, just like the moduli, are independent of the amplitude of deformation. Therefore, the test of theoretical results must be carried out during the measurements of dynamic normal stresses that arise at small amplitudes of deformation. This turns out to be a rather formidable experimental task since the normal stresses themselves at infinitesimal deformations constitute an effect of second-order with respect to shear stresses. Therefore, measurements of dynamic normal stresses may involve considerably larger experimental errors and a greater uncertainty of results than measurements of the elastic modulus. Nonetheless, experiments show that dynamic normal stresses that arise in shear small-amplitude oscillations are described qualitatively satisfactorily by the formulas derived for models of random coils.

An independent method of testing theoretical predictions may be proposed for the simultaneous measurement of the dynamic viscosity and dynamic normal stresses. This method relies on the fact that the values of the hydrodynamic interaction parameter h and the longest relaxation time θ_m are found from the results of measurements of $\eta'(\omega)$ and $\eta''(\omega)$ and the resulting values of these parameters are used to calculate the coefficients ζ_c and ζ^* without resort to additional assumptions and new constants [see formulas (4.18) and (4.19)]. The quantity ζ_c calculated theoretically is really in very good agreement with results of direct measurements. Williams also succeeded in observing the inversion of the sign of $\zeta'(\omega)$ predicted by theory in the region of $\omega = \theta_m^{-1}$.

The relationship among the coefficients η', η'', ζ_c, ζ', and ζ'', which follows from formulas (4.18) and (4.19), holds also in a general case for an arbitrary relaxation spectrum. For instance, it can be shown that the following relations are valid in the general case:

$$\zeta_c(\omega) = \omega^{-1}\eta''(\omega) = G'(\omega)/\omega^2$$

$$\zeta'(\omega) = -\omega^{-1}\eta''(\omega) + \omega^{-1}\eta''(2\omega) = [-G'(\omega) + 0.5G'(2\omega)]/\omega^2 \qquad (4.20)$$

$$\zeta''(\omega) = \omega^{-1}\eta'(\omega) - \omega^{-1}\eta'(2\omega) = [G''(\omega) - 0.5G''(2\omega)]/\omega^2$$

Thus, knowing the frequency dependences of the components of the complex viscosity, one can calculate the components of the dynamic normal stress coefficient. This is a very important achievement of the theory, especially as the functions $\eta'(\omega)$ and $\eta''(\omega)$ themselves are determined by a single general characteristic of the viscoelastic properties of the material—its relaxation spectrum. It turns out that the dynamic normal stresses are also determined by the relaxation spectrum.

A thorough experimental test of relations (4.20) has demonstrated that they are really fulfilled provided that the methodological requirements for such a fine experiment are strictly adhered to.*

---

* Kajiura, H., H. Endo, and M. Nagasawa, *J. Polymer Sci., Polymer Phys. Ed.*, **11**, 2371 (1973).

4.3.5. Oscillating Normal Stresses with Harmonic Deformations Superimposed on Shear Flow

The change of the relaxation properties of polymers in steady-state shear-ing flow has a bearing not only on the dynamic characteristics that deter-mine shear stresses (see Chapter 3) in shear oscillations but also on the oscil-lating normal stresses involved. If straining is effected by the law $\gamma =$ $= \dot{\gamma}_0 t + \xi \sin \omega t$, then not only shear stresses arise but also normal stresses which may be represented, in a general case, in the form of three terms: the constant component, harmonic oscillations with a frequency equal to 2ω, and harmonic oscillations with a frequency coinciding with the frequency of strain assignment.* At small harmonic oscillations without flow ($\dot{\gamma}_0 = 0$) the deformation occurring by the law $\gamma = \xi \sin \omega t$ causes itself the appear-ance of a constant component of normal stresses and oscillations with dou-

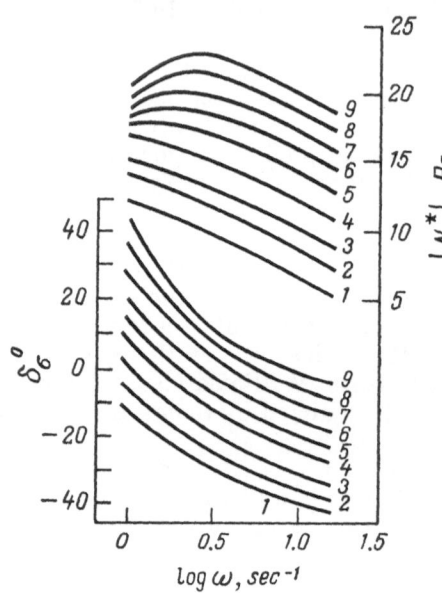

Fig. 4.5. Frequency dependences of the absolute value of the "modulus" of nor-mal stresses $|N^*|$ and the angle δ_σ for a 5-percent solution of the ethylene-propylene copolymer in decalin at different values of the velocity gra-dient of the superimposed flow, $\dot{\gamma}_0$: the curves 1-9 correspond to the following values of log $\dot{\gamma}_0$ (sec⁻¹): —0.3, —0.1, 0.1, 0.4, 0.6, 0.9, 1.1, 1.4, 1.6 (after H. C. Booij).

bled frequency. The superimposed shearing flow leads to a sharp increase of the constant normal-stress component, the suppression of the harmonic with doubled frequency and to the appearance of the harmonic with frequency ω. Therefore, in shearing flow with superimposed small oscillations taking place in the same direction in which the flow occurs the first normal stress differ-ence for the experimentally studied range of velocity gradients may be represented in the form

$$\sigma = \sigma_c + \sigma^* = \sigma_c + \xi \left(N' \sin \omega t + i N'' \cos \omega t \right)$$

where ξ is the amplitude of harmonic deformations, and N' and N'' are the quantities which have the meaning of the real and imaginary parts of the complex "modulus" of normal stresses since it turns out that in the studied range of deformation conditions the amplitude of oscillating normal stresses

---

* Booij, H. C., *Rheol. Acta*, **7**, 202 (1968).

$| \sigma^* |$ is proportional to ξ. Therefore, in the same way as is done for shear stresses it is expedient here to introduce the complex "modulus" $N^* = N' + iN''$ with the components N' and N'' and the angle of lag of normal stresses, δ_σ, so that these quantities appear to be connected by relations analogous to the relations between the components of the complex shear modulus and the phase angle, i.e.,

$$(| N^* |)^2 = (N')^2 + (N'')^2$$

$$N' = | N^* | \cos \delta_\sigma; \quad N'' = | N^* | \sin \delta_\sigma$$

The experimentally determined frequency dependences of $| N^* |$ and δ_σ for a 5-percent solution of an ethylene-propylene copolymer in decalin are shown in Fig. 4.5. The main feature of these experimental results is the existence of a maximum of $| N^* |$ at a certain frequency, whose value depends on the superimposed velocity gradient $\dot\gamma_0$, and also the change of the sign of the angle δ_σ, which occurs at frequencies which are the greater the higher is the value of $\dot\gamma_0$; here the value of N' is always greater than zero, and $N'' > 0$ if $\omega \ll \dot\gamma_0$ and $N'' < 0$ if $\omega \gg \dot\gamma_0$.

Thus, the superposition of shearing flow on harmonic oscillations taking place at small amplitudes alters very sharply the entire picture of frequency dependences of the stress components.

4.4. Experimental Relations Between Normal and Shear Stresses and Rates of Shear

4.4.1. Effect of Shear Rate on the Relation Between Shear and Normal Stresses

Considerable experience has accumulated at present in measurements of the dependences $\tau (\dot\gamma)$ and $\sigma (\dot\gamma)$ for various polymer systems. This allows one to draw a general picture of the effect of the rate of deformation in steady-state shear flow on the values of η and ζ and also on the relation between various components of the stress tensor.

Most of the available experimental data obtained in investigations of normal stresses refer to the quantity $\zeta = \sigma/2\dot\gamma^2$ since, first, its measurement is carried out much simpler than the measurement of β, and, second, because $\sigma' \ll \sigma$. Therefore, the main part in the characterization of the stress state of a body, apart from the shear stress, is played by the first normal stress difference ζ and not by β.

Typical dependences $\tau (\dot\gamma)$ and $\sigma (\dot\gamma)$ are given in Fig. 4.6 for polyisobutylene at different temperatures and in Fig. 4.7 for solutions of polystyrene in decalin at various concentrations. In all the examples presented there has been experimentally attained such a region of low rates of shear where τ is proportional to $\dot\gamma$ and σ is proportional to $\dot\gamma^2$, and therefore the theoretical predictions referring to the region of infinitely low values of $\dot\gamma$ are fulfilled.

Here the normal stresses prove to be significantly lower than the shear stresses, in which case one may speak of normal stresses as a second-order effect. As the rate of shear increases the values of τ and σ increase, but the rate of growth of normal stresses is faster and therefore at certain values of $\dot{\gamma}$ the normal stresses become equal to and then greater than the shear stresses. As seen from Fig. 4.6, the condition $\tau = \sigma$ is fulfilled at the same values of stresses for all the temperatures studied. This is a rather typical phenomenon if no specific effects are observed in the polymer in the temperature range used for the measurements.

Examples of the dependences of all the stress components on the rate of shear are presented in Fig. 4.8 for two relatively dilute solutions.* Apart

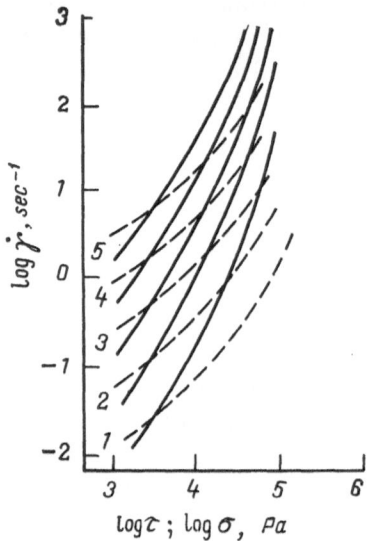

Fig. 4.6. Dependences of shear stresses (solid lines) and the first normal stress difference (dashed lines) on the rate of shear for polyisobutylene: 1—22°C; 2—40; 3—60; 4—80; 5—100°C. [Vinogradov, G. V., A. Ya. Malkin, and V. F. Shumski, *Rheol. Acta*, 9, 155 (1970).]

from the above-described relation between τ and σ, Fig. 4.8 shows the relationship between the various normal stress components, namely, it is seen that σ' is always smaller than σ.

At low rates of shear [as directly follows from formula (4.14)]

$$\sigma = 2J_\infty \tau^2 \tag{4.21}$$

Since $J_\infty = G_0^{-1}$, where G_0 is the initial elastic shear modulus, formula (4.21) may be written in the following equivalent ways:

$$\sigma = \frac{2}{G_0} \tau^2 = 2 \left(\frac{\zeta_0}{\eta_0^2} \right) \tau^2$$

or

$$\zeta = (\zeta_0/\eta_0^2) \, \eta^2$$

When the rate of shear increases, the relationship between σ and τ cannot be determined in advance because it depends on the nature of the effect of

---

* These data are taken from: Markovitz, H., in: *Rheology, Theory and Applications*, edited by F. Eirich, vol. 4, New York-London, Academic Press, 1967, pp. 347-410. In original works, the quantity $\sigma + \sigma' = \sigma_{11} - \sigma_{33}$ is often given instead of σ, but since $\sigma' \ll \sigma$, $\sigma + \sigma' \approx \sigma$. See also: Huppler, J. D., *Trans. Soc. Rheol.*, 9, 273 (1965); Spriggs, T. W., J. D. Huppler, and R. B. Bird, *Trans. Soc. Rheol.*, 10, 191 (1966).

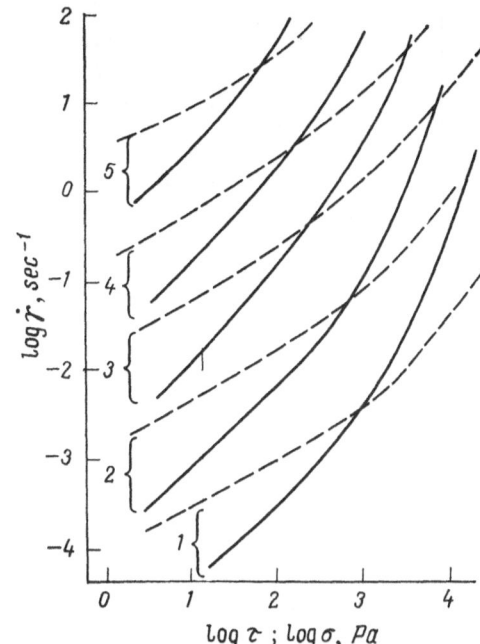

Fig. 4.7. Dependences of shear stresses (solid lines) and the first normal stress difference (dashed lines) on the rate of shear for various solutions of polystyrene in decalin:

1—0.573; 2—0.466; 3—0.380; 4—0.290; 5 —0.184 volume fractions. [Vinogradov, G. V., A. Ya. Malkin, and G. V. Berezhnaya, *Vysokomol. Soedin.*, A13, 2793 (1971).]

the rate of deformation on the relaxation spectrum of the system. Experiment, however, shows* that at rather high shear rates too, in the region of

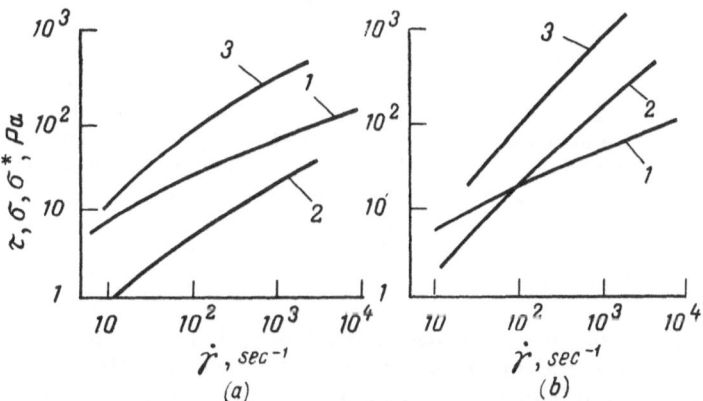

Fig. 4.8. Dependences of the stresses τ (*1*), σ (*2*), and σ' (*3*) on the rate of shear for approximately 1-percent aqueous solutions of hydroxyethylcellulose (*a*) and polyethylene oxide (*b*).

sharply pronounced non-Newtonian behaviour and decrease of the normal stress coefficient as compared with ζ_0 there is still fulfilled, in many cases, a quadratic relation between normal and shear stresses.

---

    * Vinogradov, G. V., A. Ya. Malkin, and V. F. Shumsky, *Vysokomol. Soedin.*, A10, 2672 (1968); A11, 663 (1969); *Rheol. Acta*, 9, 155 (1970); Vinogradov, G. V., A. Ya. Malkin, and G. V. Berezhnaya, *Vysokomol. Soedin.*, A13, 2793 (1971).

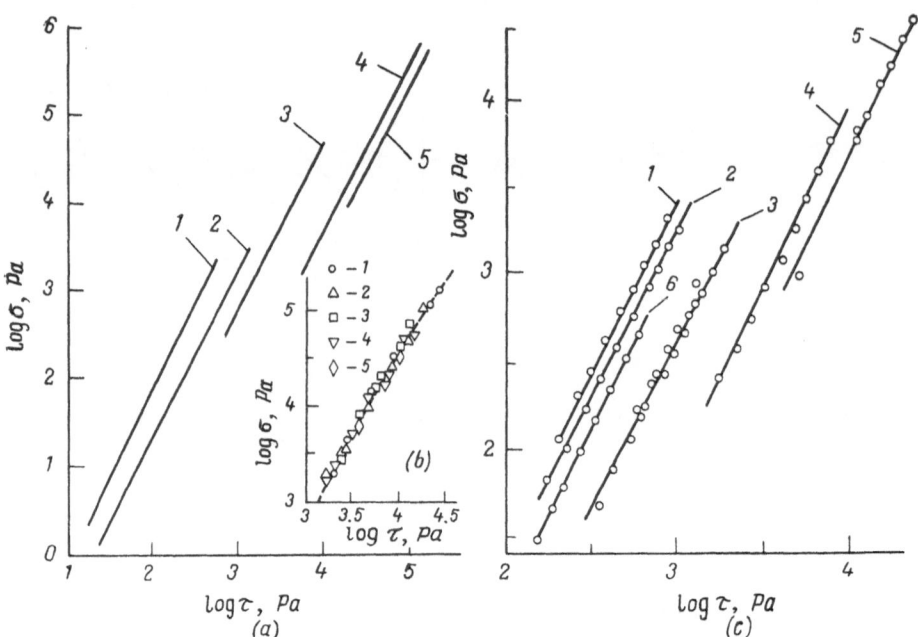

Fig. 4.9. Relation between normal and shear stresses for various polymers:
a—solutions of butyl rubber in cetane with the following contents of polymer in the system: 1—0.09; 2—0.20; 3—0.51; 4—0.85; 5—1.00 volume fractions [Vinogradov, G. V., A. Ya. Malkin, and V. F. Shumski, Vysokomol. Soedin., A10, 2672 (1968); A11, 663 (1969).]; b—polyisobutylene (the symbols are the same as in Fig. 4.6); c—solutions of monodisperse polystyrene with molecular mass 4.11×10^{5} at different polymer contents and different temperatures: 1—2.5%, 23°C; 2—5%, 23°C; 3—10%, 23°C; 4—20%, from 22 to 60°C; 5—50%, 120°C; 6—100%, 200°C (after H.J.M.A. Mieras).

Examples that illustrate the character of the dependence of σ on shear stresses for some systems are given in Fig. 4.9. Over a rather wide range of the variables the dependence σ (τ) is practically quadratic, but at large stresses there appear departures of the experimental points from the strictly quadratic relation, and the rate of growth of normal stresses is somewhat slowed down. This is seen from the results obtained for polyisobutylene (Fig. 4.9b). It should, however, be borne in mind that in the case of high-molecular-mass polymers of narrow MMD the dependence of σ on τ may prove to be stronger than quadratic near the region of their transition to the forced rubbery state.*

The existence of the quadratic dependence of σ on τ predetermines the character of the deviation of the viscosity and normal-stress coefficients, η and ζ, from their initial values, η_0 and ζ_0, respectively. An example of the dependences (η/η_0) and (ζ/ζ_0) on $\dot{\gamma}$ is given in Fig. 4.10; the graph of the function $\zeta (\dot{\gamma})$ (the dashed line) corresponds almost exactly to the square of the function $\eta (\dot{\gamma})$. By virtue of the existing relation between the functions $\zeta (\dot{\gamma})$ and $\eta (\dot{\gamma})$, the deviations of the normal-stress coefficient from the initial value ζ_0 becomes noticeable at lower rates and are more sharply pronounced than the deviation of the apparent viscosity from the initial (Newtonian viscosity) values. That σ is proportional to τ^2 is believed to be an impor-

* Brizitsky, V. I. et al., J. Appl. Polymer Sci., 20, 25 (1976).

tant experimental fact. But unfortunately, there are no criteria indicating the upper stress limit of this rule.

Indeed, the analysis of a very large body of experimental evidence for polystyrene has shown* that over a rather wide range of rates σ is proportional to τ^a. For polydisperse samples, on an average, $a = 1.66$, and it is only for monodisperse polystyrenes that $a = 2.0$. An analogous generalization of

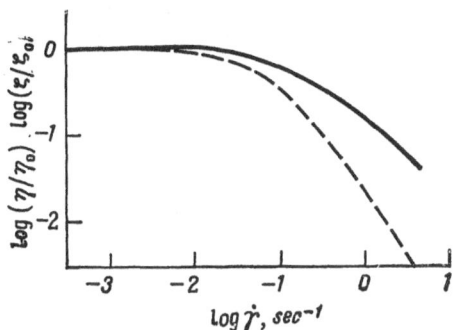

Fig. 4.10. Deviation of viscosity (th solid line) and of the normal stres coefficient (the dashed line) from their initial values with increasing rate of shear for a 46.6-percent solution of polystyrene in decalin (see the reference to Fig. 4.7).

experimental data has been made also for solutions of polyisobutylene**, in which case a was found to be close to 2.

As regards the analytical form of the dependence $\sigma(\dot{\gamma})$, it can be deduced from the relation $\sigma \propto \tau^2$ and the equation of the flow curve. Over a wide range of sufficiently high shear rates the latter may be represented in the form of a power function: $\tau \propto \dot{\gamma}^n$. From this it follows that σ is proportional to $\dot{\gamma}^{2n}$. Dependences of this kind have been repeatedly observed experimentally for various polymers and their solutions.

4.4.2. Temperature- and Concentration-Invariant Characteristics of Normal Stresses

The experimental dependences of normal stresses on rates of shear obtained for a single polymer at different temperatures or for one polymer-solvent pair with different contents of the polymer in the system can be generalized with the aid of the temperature or concentration superposition method.

The possibility of constructing temperature-invariant plots of normal stresses is seen from Fig. 4.9b, which presents data referring to different temperatures and forming a set of parallel dependences of σ on τ. In Chapter 2 we have discussed in detail the question of constructing a master curve for the dependences of shear stresses or viscosity on the shear rate. Theoretical considerations and experimental findings show that for temperature-invariant curves of shear stresses to be plotted the argument should be represented in dimensionless form as the product $(\dot{\gamma}\theta)$, where θ is the characteristic relaxation time of the system. Since as a function of temperature θ varies in proportion to η_0, the argument of the temperature-invariant plots of shear stresses is the product $(\dot{\gamma}\eta_0)$. On the basis of the data of Fig. 4.9b and other

* Oda, K., J. L. White, and E. S. Clark, *Polymer Eng. Sci.*, **18**, 25 (1978).
** Tanner, R. I., *Trans. Soc. Rheol.*, **17**, 365 (1973).

similar data it may be stated that experimental data on the dependences $\sigma\ (\dot\gamma)$ obtained at different temperatures can be generalized in an analogous manner. This is shown in Fig. 4.11 which presents master curves for both shear stresses and the first normal stress difference for polyisobutylene, the quantity $(\dot\gamma\eta_0)$ being used as the argument.

A typical example that illustrates the effect of concentration on the character of the build-up of normal stresses in polymer solutions is given in

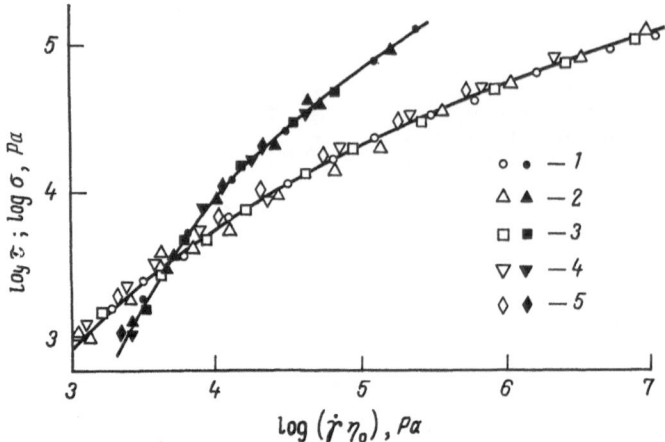

Fig. 4.11. Temperature-invariant dependences of shear stresses (unfilled symbols) and normal stresses (filled symbols) on the reduced shear rate for polyisobutylene (for the designations and the reference, see Fig. 4.6).

Fig. 4.12. From this figure it can be seen that the increase of the polymer content in the solution shifts the onset of a departure from the initial value of

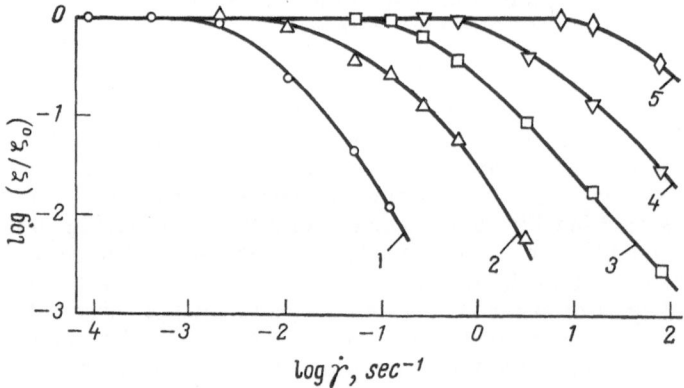

Fig. 4.12. Effect of concentration on the variation of the normal stress coefficient as a function of the rate of shear for solutions of polystyrene in decalin (for the designations and the reference, see Fig. 4.7).

the normal stress coefficient to the side of lower values of the rate of shear.

Generalization of experimental data on the dependences $\sigma\ (\dot\gamma)$ obtained for solutions of different concentrations (except dilute solutions) boils down to the correct choice of the concentration dependence of the shift factor or the characteristic relaxation time $\theta\ (c)$. Just as has been described in the construction of master curves of flow properties (see Chapter 2), the reduction of the dependences $\sigma\ (\dot\gamma)$ obtained for solutions of different concentrations to a single master curve can be done by shifting the $\sigma\ (\dot\gamma)$ curves along the log $\dot\gamma$ axis until the curves are superposed, which allows the concentration shift factor to be found empirically.

In works devoted to this topic, use has been made of various empirical forms of the dependence $\theta\ (c)$, which are usually reduced to an expression of the following type:

$$\theta\ (c) \propto \eta_0\ (c)\ c^{-a}$$

where the exponent a varies from 1 to 3.

A better grounded approach to the construction of concentration-invariant plots of normal stresses is to make use of two independent functions of con-

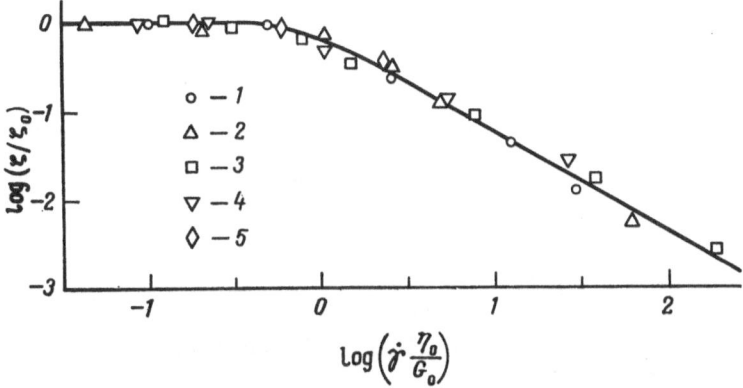

$$\log\left(\dot\gamma\ \frac{\eta_0}{G_0}\right)$$

Fig. 4.13. Concentration-invariant characteristic of normal stresses for solutions of polystyrene in decalin (for the designations and the reference, see Fig. 4.7).

centration: the initial Newtonian viscosity $\eta_0\ (c)$ and the initial modulus $G_0\ (c)$. The concentration dependence of the relaxation time, or of the shift factor, will then be expressed in the following way:

$$\theta_0\ (c) = \eta_0\ (c)/G_0\ (c)$$

This approach allows one to construct the concentration-invariant dependence of σ on $(\dot\gamma\theta_0)$ by deducing θ_0 from results of measurements of η_0 and G_0. A graph of this kind is shown in Fig. 4.13 for solutions of polystyrene in decalin, which confirms the validity of the calculation of $\theta_0\ (c)$ as the ratio (η_0/G_0) in plotting the concentration-invariant curves not only of viscosity but also of normal stresses.

The use of the rubbery elasticity modulus as the parameter which, along with viscosity, determines the concentration dependence of the characteristic relaxation time allows one to generalize experimental data on the ratio of normal and shear stresses measured for solutions of various concentration. To do this, one should consider the dependence of (σ/G_0) on (τ/G_0); here the concentration dependence of viscosity is taken into account by choosing the stresses (and not the rates of shear) as the argument and, besides, the stresses are normalized by the value of the modulus G_0, which depends on the content of the polymer in solution. In this way, there is obtained a master curve for stresses for a polymer-solvent pair with different ratios of the com-

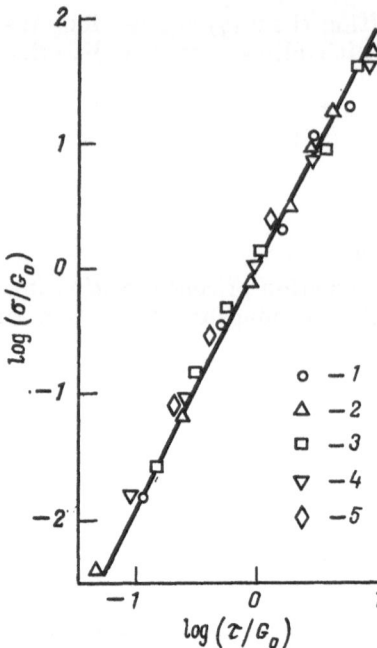

Fig. 4.14. Concentration-invariant relation between the first normal stress difference and the first shear stress difference for solutions of polystyrene in decalin (the designations and the reference are the same as in Fig. 4.7).

ponents (Fig. 4.14) with the variation of the viscoelastic properties of solutions over wide ranges.

4.5. Non-Newtonian Viscosity and Normal Stresses

4.5.1. General Relationships*

The possibility of establishing a relationship between the various components of the stress tensor is associated with their representation in terms of the relaxation spectrum of the system. In practice it is most readily done through the use of known experimental facts pertaining to the correlation between the components of the complex shear modulus $G'(\omega)$ and $G''(\omega)$ and the quantities τ and σ on the comparison condition $\omega = \dot{\gamma}$. This question has already been discussed in detail in the preceding chapter. Here we shall make use of the final results which allow us to derive the equation relating the dependences $\tau(\dot{\gamma})$ and $\sigma(\dot{\gamma})$.

---

* Malkin, A. Ya., *Mekhanika Polimerov*, 3, 506 (1971).

As has been shown in Sec. 3.4, for a large number of polymer systems there are fulfilled the following relations:

$$\tau^2(x) = [G'(x)]^2 + [G''(x)]^2$$

$$\sigma(x) = 2\tau(x)\frac{G'(x)}{G''(x)} \tag{4.22}$$

where $x = \omega = \dot{\gamma}$.

The subsequent treatment is based on these formulas and refers only to those cases where relations (4.22) are fulfilled.

The functions $G'(x)$ and $G''(x)$ are not the independent characteristic of the material and may be represented in the form of integral transforms of the relaxation spectrum of the material with the aid of formulas (1.89) and (1.90). In deriving the relation between the functions $\tau(x)$ and $\sigma(x)$ it is important that the relaxation spectrum be eliminated from equations (1.89) and (1.90) in order to define the function $G'(\omega)$ in terms of $G''(\omega)$ or vice versa.

Considering that

$$G''(x) = x\eta'(x)$$

we can express G' through $\eta'(x)$:

$$G'(x) = \frac{2}{\pi}\int_0^\infty \eta'(\alpha)\frac{x^2}{x^2-\alpha^2}\,d\alpha \tag{4.23}$$

The integral on the right-hand side of formula (4.23) is understood in the sense of its principal value according to Cauchy. Joint consideration of relations (4.22) and (4.23) leads to the following formula:

$$[\tau(x)]^2 = [x\eta'(x)]^2 + \left[\frac{2}{\pi}\int_0^\infty \eta'(\alpha)\frac{x^2}{x^2-\alpha^2}\,d\alpha\right]^2 \tag{4.24}$$

or

$$[\eta(x)]^2 = [\eta'(x)]^2 + \left[\frac{2}{\pi}\int_0^\infty \eta'(\alpha)\frac{x}{x^2-\alpha^2}\,d\alpha\right]^2 \tag{4.24a}$$

which may be regarded as the equation for $\eta'(x)$ with the dependence of the apparent viscosity on the rate of shear, $\eta(x)$, being known.

Formulas (4.22) and (4.23) yield the expression for the function $\sigma(x)$:

$$\sigma(x) = \tau(x)\frac{\sqrt{[\tau(x)]^2-[G''(x)]^2}}{G''(x)} = \tau(x)\,\frac{\dfrac{2}{\pi}\displaystyle\int_0^\infty \eta'(\alpha)\frac{x^2}{x-\alpha^2}\,d\alpha}{x\eta'(x)}$$

or the expression for the normal stress coefficient:

$$\zeta(x) = \frac{2\eta(x)}{\pi\eta'(x)}\int_0^\infty \eta'(\alpha)\frac{d\alpha}{x^2-\alpha^2} \tag{4.25}$$

Thus, the computation of the function $\sigma(x)$ from $\tau(x)$, or, what is the same thing, the calculation of $\zeta(x)$ from the known dependence of the apparent viscosity on the rate of shear, reduces to the solution of Eq. (4.24) for $\eta'(x)$ and to the calculation of $\zeta(x)$ with the aid of formula (4.25).

For the arbitrarily assigned function $\eta(x)$ the analytical solution of Eq. (4.24) for $\eta'(x)$ is not considered possible. It is therefore necessary to resort to numerical methods with the use of computers. The most straightforward method of solving Eq. (4.24) is the method of successive approximations when an $(n+1)$st approximation of the function $\eta'(x)$ is computed as

$$\eta'_{(n+1)}(x) = \left\{ [\eta(x)]^2 - \left[\frac{2x}{\pi} \int_0^\infty \eta'_{(n)} \frac{d\alpha}{x^2 - \alpha^2} \right]^2 \right\}^{0.5}$$

where $\eta'_{(n)}(x)$ is the nth approximation.

Any function satisfying the condition $\eta'_{(0)}(x) \leqslant \eta(x)$ can be used as the zero approximation. In particular, the function $\eta(x)$ may be taken as the zero approximation. Calculations have shown* that the method of successive approximations is but slightly sensitive to the choice of the zero approximation and very rapidly, beginning with the third or fourth approximation, boils down to the solution of Eq. (4.24), giving a result that deviates from the exact solution by not more than 10-15 per cent.

4.5.2. Calculation of the Initial Normal Stress Coefficient

Of special importance in the analysis of the dependence $\zeta(\dot{\gamma})$ is the initial normal stress coefficient which is defined through the second moment of the relaxation spectrum [see formula (4.13)] and which is connected with the relation between the flow and rubber-like properties of the system [see formula (4.14)]. Therefore, the evaluation of the quantity ζ_0 on the basis of the general equation (4.24) without using computer methods is of special interest.

The calculation of ζ_0 is based on the approximate relation

$$\eta^*(x) \approx [\eta'(x)]$$

which holds at small values of x (see Sec. 3.4). Then the ratio included in formula (4.25) is close to unity and in the region of small values of x the function $\zeta(x)$ is expressed as follows:

$$\zeta(x) = \frac{2}{\pi} \int_0^\infty \eta'(\alpha) \frac{d\alpha}{x^2 - \alpha^2} \approx \frac{2}{\pi} \int_0^\infty \eta(\alpha) \frac{d\alpha}{x^2 - \alpha^2}$$

This formula may be written in the form

$$\sigma(\dot{\gamma}) = \frac{4}{\pi} \dot{\gamma}^2 \int_0^\infty \eta(\alpha) \frac{d\alpha}{\dot{\gamma}^2 - \alpha^2}$$

which allows one to calculate the function $\sigma(\dot{\gamma})$ in the region of small values of $\dot{\gamma}$. The limiting relation has the form

$$\zeta_0 = \frac{2}{\pi} \lim_{x \to 0} I(x)$$

---

* Vinogradov, G. V., A. Ya. Malkin, G. V. Berezhnaya, S. A. Bostandjiyan, and A. M. Stolin, *Mekhanika Polimerov*, 4, 714 (1971).

where

$$I(x) = \int\limits_0^\infty \eta(\alpha) \frac{d\alpha}{x^2 - \alpha^2}$$

The formulas obtained enable one to establish the existence of the initial normal stress coefficient by proceeding from the measured dependence $\eta(\dot\gamma)$ and to calculate its value.

It is possible to find the general condition that must be satisfied by the function $\eta(\dot\gamma)$ at small values of $\dot\gamma$ in order to provide the existence of a limited and non-zero value of the initial normal stress coefficient. This condition is associated with the requirements that must be met by the function $\eta(\dot\gamma)$ near the region of Newtonian flow in order to ensure the existence of a finite value of ζ_0.

The condition of finiteness of the value of ζ_0 is determined by the existence of a limit:

$$I_0 = \lim_{x \to 0} I(x) = \lim_{x \to 0} \int\limits_0^\infty \eta(\alpha) \frac{dx}{x^2 - \alpha^2}$$

where the function $\eta(\alpha)$ by its physical meaning is the dependence of the apparent viscosity on the rate of shear.

It can be shown that

$$\lim_{x \to 0} \int\limits_0^\infty \eta(\alpha) \frac{d\alpha}{x^2 - \alpha^2} = - \int\limits_0^\infty \frac{1}{\alpha} \left[\frac{d\eta(\alpha)}{d\alpha} \right] d\alpha \qquad (4.26)$$

and hence

$$\zeta_0 = -\frac{2}{\pi} \int\limits_0^\infty \frac{1}{\alpha} \left[\frac{d\eta(\alpha)}{d\alpha} \right] d\alpha$$

Therefore, the necessary condition of the existence of a finite value of ζ_0 is the requirement of the equality to zero of the derivative of viscosity with respect to the rate of shear at $\dot\gamma = 0$. If at the zero point $d\eta(\alpha)/d\alpha < 0$, then the integral diverges and there is no limited initial value of the normal stress coefficient.

The result obtained means that not every function $\eta(\dot\gamma)$ specified analytically satisfies the requirement of the existence of the finite value of the normal stress coefficient. For example, this requirement is not satisfied by functions such as

$$\eta(\dot\gamma) = \eta_0 e^{-a\tau}; \quad \eta(\dot\gamma) = \frac{\eta_0}{1 + a\dot\gamma}$$

where η_0 is the initial Newtonian viscosity; a is an empirical constant.

For these functions the value of the derivative $d\eta(\dot\gamma)/d\dot\gamma$ at $\dot\gamma = 0$ is not equal to zero.

The resulting formula (4.26) does not imply that the requirement $d\eta/d\dot\gamma = 0$ must be fulfilled over any wide range of shear rates: this is only necessary for the point $\dot\gamma = 0$.

In principle, experiment cannot provide an answer to the question of the exact analytical form of the dependence η $(\dot\gamma)$ at small values of $\dot\gamma$ because the same set of experimental data can be described by any number of methods. However, for finite values of ζ_0 to exist it is necessary that the method chosen involve the zero shear-rate derivative of viscosity at $\dot\gamma = 0$. This is schematically illustrated in Fig. 4.15, where the "forbidden" form of the function η $(\dot\gamma)$ is shown on the left and the function satisfying the indicated requirement of the equality to zero of the derivative of viscosity at $\dot\gamma = 0$ is given

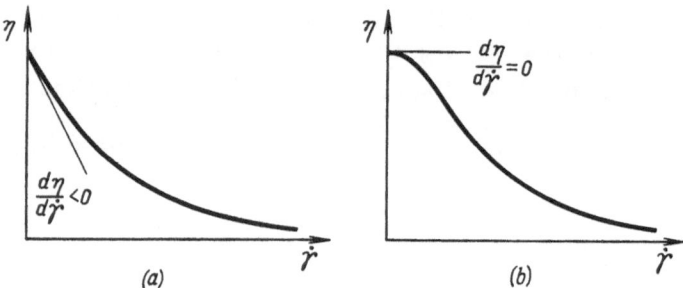

(a)   (b)

Fig. 4.15. Specific features of the dependence of apparent viscosity on the rate of shear at $\dot\gamma \to 0$:

a—$d\eta/d\dot\gamma < 0$, no limited value of ζ_0 is present; b—$d\eta/d\dot\gamma = 0$, there is a limited value of ζ_0.

on the right and, hence, only in the second case do there exist limited values of ζ_0.

A convenient form of representation of the function η $(\dot\gamma)$ near $\dot\gamma = 0$ is, for example, the following analytical dependence:

$$\eta = \frac{\eta_0}{1+(\theta\dot\gamma)^2} \tag{4.27}$$

where η_0 is the initial or Newtonian viscosity; θ is an empirical constant which has the meaning of a certain effective relaxation time.

This function satisfies the condition $d\eta$ $(\dot\gamma)/d\dot\gamma = 0$ at $\dot\gamma = 0$. Then, substitution of the function (4.27) into formula (4.26), after performing appropriate computations, gives a simple formula for the determination of ζ_0 from two constants of the material—the viscosity η_0 and the relaxation time θ:

$$\zeta_0 = \eta_0\theta \tag{4.28}$$

Other analytical methods of calculating ζ_0 are also possible. They are based on the empirically determined dependences η $(\dot\gamma)$ in the region of small values of $\dot\gamma$, which meet the condition of existence of the limit of integration I (x) at $x \to 0$.

The problem of the effect of non-Newtonian flow behaviour on the value of the initial normal stress coefficient is qualitatively illustrated by the following example. For the convenience of calculations, the entire flow curve is split into two regions—the region of shear rates, where the viscosity remains constant, η_0, and the region of non-Newtonian behaviour, where the flow curve is described by a power law of the form

$$\eta(\dot{\gamma}) = K\dot{\gamma}^{-n}$$

where K and n are certain empirical constants.

The boundary of these two regions corresponds to the value of shear rate $\dot{\gamma} = B$. At the point $\dot{\gamma} = B$ there must be satisfied the condition of fitting of the two branches of the flow curve:

$$\eta_0 = KB^{-n}$$

Thus, the function $\eta(\dot{\gamma})$ is represented in the form

$$\eta(\dot{\gamma}) = \begin{cases} \eta_0 & \text{at } \dot{\gamma} < B \\ K\dot{\gamma}^{-n} & \text{at } \dot{\gamma} > B \end{cases}$$

Calculations of ζ_0 with the aid of formula (4.26) yield the following expression:

$$\zeta_0 = \frac{2n}{\pi(n+1)} \frac{\eta_0}{B} \tag{4.29}$$

which allows one to analyse the effect of the main parameters of the flow curve on the normal stresses. First of all, the details of non-Newtonian behaviour, which are characterized by the value of n, have a slight effect on ζ_0 since in real cases the factor $n/(n+1)$ varies within narrow limits. The quantity ζ_0 is much more strongly influenced by the viscosity of the system, η_0, and by B which corresponds to the onset of non-Newtonian behaviour. At the same viscosity the normal stresses are the higher the lower are the shear rates at which the region of non-Newtonian behaviour sets in. On the contrary, if non-Newtonian behaviour does not manifest itself up to very high rates of shear, as is observed, for example, with monodisperse polymers (see Chapter 2), no high normal stresses will arise and the polymer will behave like an ordinary highly viscous Newtonian liquid over a wide range of shear rates.

Formula (4.29) serves for illustrative purposes because the resolution of the function $\eta(\dot{\gamma})$ into two parts is arbitrary. It clearly shows, however, the structure of the effect of various factors on the coefficient ζ_0.

Instead of formula (4.29), we may write the following relation:

$$\zeta_0 = m\eta_0/B$$

where the coefficient m is found empirically on the basis of the evaluation of the experimentally measured values of ζ_0, η_0, and B. Treatment of experimental data obtained as a result of the investigation of the rheological properties of many polymer melts and solutions has shown that, on an average, $m = 0.36$ and

$$\zeta_0 = 0.36\eta_0/B \tag{4.30}$$

This formula enables one to calculate ζ_0 with an error of up to ± 20 per cent, which corresponds to the possible experimental error of measurement of ζ_0.

Some results of calculations of the initial normal stress coefficient ζ_0 on the basis of the function $\eta(\dot{\gamma})$ are presented in Fig. 4.16 for a number of polymers and their solutions. In Fig. 4.16 the experimental values of ζ_0 are compared with the results of calculations based on formulas (4.28) and (4.30).

Thus, the experimental and theoretical results discussed in this section show the existence of an intimate relationship between the two effects spe-

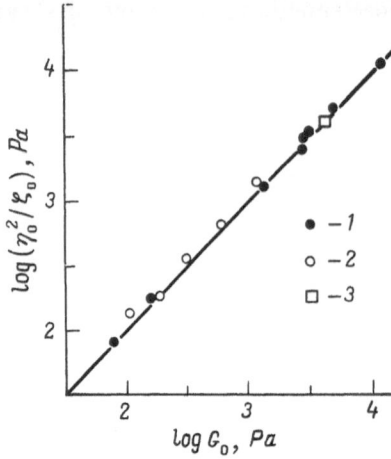

- 1
- 2
- 3

Fig. 4.16. Comparing calculated and experimental values of the initial normal stress coefficients ζ_0. The unfilled symbols represent data calculated by formula (4.28), the filled circles, data calculated by formula (4.30) for polyisobutylene (1), butyl rubber (2), and solutions of polystyrene in decalin (3).

cific to polymer melts and solutions—the appearance of non-Newtonian behaviour and the development of normal stresses.

4.6. Dependence of Normal Stresses on Temperature, Molecular Properties of Polymers and Solution Concentration

The basic regularities of the effect of temperature, solution concentration, and the molecular properties of polymers on normal stresses in steady-state shearing flow are quantitatively due to a common factor—the fact that normal stresses create an effect which is quadratic with respect to shear stresses. This can be explained with the aid of formulas (4.14) and (4.21).

4.6.1. Temperature Dependence of Normal Stresses

As a general rule, the equilibrium compliance is independent of temperature or, at least, depends on it much less strongly than does the viscosity. Therefore, the temperature dependence of normal stresses may be expressed in the following manner:

$$\frac{\partial \sigma}{\partial T} \propto 2 \frac{\partial \tau}{\partial T}$$

or

$$\frac{\partial \ln \sigma}{\partial (1/T)} = 2 \frac{\partial \ln \tau}{\partial (1/T)} \quad \text{and} \quad \frac{\partial \ln \zeta}{\partial (1/T)} = 2 \frac{\partial \ln \eta}{\partial (1/T)} \tag{4.31}$$

From this it follows that the temperature dependence of the normal stress coefficient must be stronger than the temperature dependence of viscosity. This is shown in Fig. 4.17 on the basis of the experimental data obtained for polyisobutylene. Formula (4.31) shows that the coefficient $\partial \ln \zeta / \partial (1/T)$

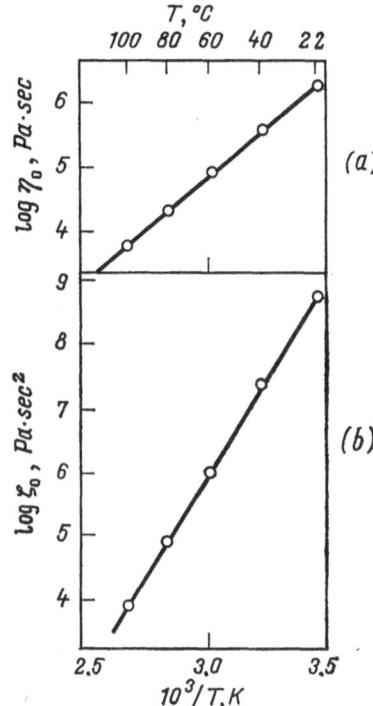

Fig. 4.17. Temperature dependences of the initial viscosity coefficients (a) and the initial normal stress coefficients (b) for polyisobutylene (see the reference to Fig. 4.6).

must be equal to the doubled activation energy of viscous flow. This inference is consistent with experimental facts.*

Thus, normal stresses are found to be sensitive to the change of temperature; by varying the temperature one can also considerably change the ratio of the normal stresses to the shear components of the stress tensor.

4.6.2. Dependence of Normal Stresses on Solution Concentration

The concentration dependence of normal stresses can also be analysed by using formula (4.14), on the basis of which one can assert that the form of the function $\zeta (c)$ is determined by two dependences: $\eta (c)$ and $J_\infty (c)$. The first of these has been described in detail in Chapter 2; the dependence of the rubbery modulus on concentration is discussed in this chapter. Therefore, at this point, without dwelling on the nature of the function $J_\infty (c)$, and proceeding from the formula

$$\zeta (c) = J_\infty (c) \cdot \eta^2 (c)$$

* Vinogradov, G. V., A. Ya. Malkin, and V. F. Shumsky, in: *Advances in Polymer Rheology*, edited by G. V. Vinogradov, Khimiya Publishers, Moscow, 1970, pp. 206-228 (in Russian).

one can state that the dependence of normal stresses on concentration must be very strong but somewhat weaker than η^2 (c) since J_∞ (c) is a decreasing function of concentration, though it is weaker than the viscosity. An example illustrating the character of the concentration dependence of viscosity and the initial normal stress coefficient is given in Fig. 4.18, which has been plotted for solutions of butyl rubber in cetane. Obviously, the normal stresses can be affected very strongly by varying the content of the polymer in solution; in the example given the normal stress coefficient varied by a factor of up to 10^{13}.

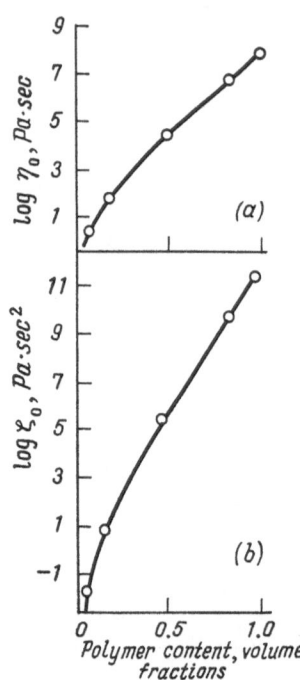

Measurements of normal stresses over the entire range of compositions have shown that the function ζ_0 (c), like the function η_0 (c) (Fig. 4.19), consists of two parts. In the region of high concentrations ζ_0 (c) is described by a power dependence. The exponent is determined by the nature of the functions η_0 (c) and J_∞ (c) and usually has a value close to 5-6. In the region of low concentrations the function ζ_0 (c) becomes weaker and if, as before, it is described by a power relation, its exponent will be substantially smaller, approaching 2. Here, the variation of the character of the dependence ζ_0 (c) takes place at the same critical concentration indicated in Fig. 4.19 in which case there is observed the variation of the concentration dependence of viscosity (see Chapter 2). For polymers of different molecular masses the dependences ζ_0 (c) are generalized into a single function* if the product of the concentration by the intrinsic viscosity c $[\eta]$ (Fig. 4.20) is used as the argument, just as in the case of the concentration dependences of the viscosity of polymers having different molecular masses.

Fig. 4.18. Concentration dependences of the viscosity coefficient (a) and the initial normal stress coefficient (b) for solutions of butyl rubber in cetane (see the reference to Fig. 4.9a).

Normal stresses in dilute polymer solutions arise as a result of the deformations and orientation of individual molecular coils in shear flow. The response of such a coil to the change of its shape and/or orientation is the appearance of not only shear but normal stresses as well.

A typical peculiarity of dilute solutions in which normal stresses can be detected is the absence of rubber-like elasticity. Here, it is characteristic that in the region of critical concentrations, where the rate of the dependences η_0 (c) and ζ_0 (c) undergoes change, the character of the concentration dependence of the "rigidity" of the solution represented by the parameter $(\eta_0 - \eta_s)^2/\zeta_0$, where η_s is the viscosity of the solvent, is changed. As seen from Fig. 4.21, the minimum of this quantity corresponds to the critical concentration (shown by a dashed line), at which there occurs a change from the local elasticity of individual non-linked chains to the elasticity of the network of macromolecules. The quantity $(\eta_0 - \eta_s)^2/\zeta_0$ has the meaning of the rubbery

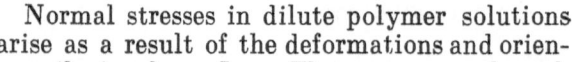

* Malkin, A. Ya., *Rheol. Acta*, **13**, 486 (1973)

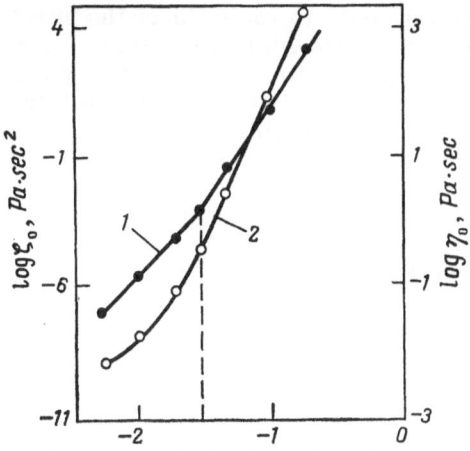

Fig. 4.19. Dependence of the initial viscosity coefficient and the initial normal stress coefficient for solutions of polybutadiene $(\overline{M}_v = 3.2 \times 10^5)$ in methyl naphthalene on the content of polymer in solution at 25°C. The dashed line indicates the critical concentration [Malkin, A. Ya., G. V. Berezhnaya, and G. V. Vinogradov, *Mekhanika Polimerov*, 5, 896 (1972).]: $1-\zeta_0$; $2-\eta_0$.

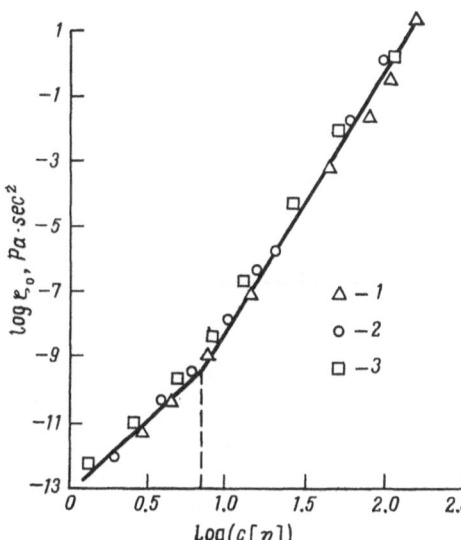

$\triangle - 1$
$\circ - 2$
$\square - 3$

Fig. 4.20. Dependence of the initial normal stress coefficient on the dimensionless concentration for solutions of polybutadienes in methyl naphthalene at 25°C. The molecular masses of the polymers are as follows $(M \times 10^{-5})$: $1-1.52$; $2-2.4$; $3-3.2$. The dashed line indicates the critical concentration (see the reference to Fig. 4.19).

Fig. 4.21. Concentration dependence of the "rigidity" of solutions of polybutadienes $(\overline{M}_v = 3.2 \times 10^5)$ in methyl naphthalene at 25°C (see the reference to Fig. 4.19).

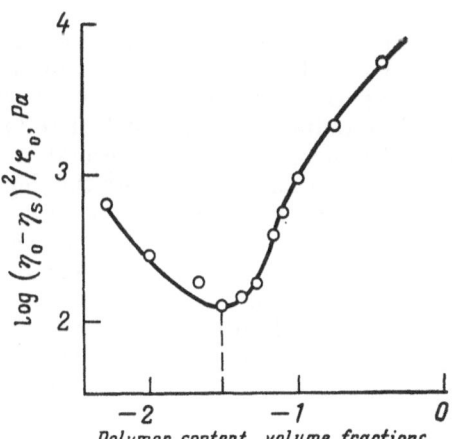

modulus only to the right of the minimum in Fig. 4.21 [in this region $\eta_0 \gg$ $\gg \eta_s$ and, according to formula (4.14), $\eta_0^2/\zeta_0 = G_0 = J_\infty^{-1}$], but it does not represent the resistance to rubbery deformations to the left of the minimum of the curve in Fig. 4.21 since dilute solutions are incapable of developing rubbery deformations.*

4.6.3. Effect of Molecular Mass on Normal Stresses

The general approach to the analysis of the problem of the effect of molecular mass on normal stresses developing in shearing flow must also be based on formula (4.14). Assuming that the rubbery modulus (or the equilibrium

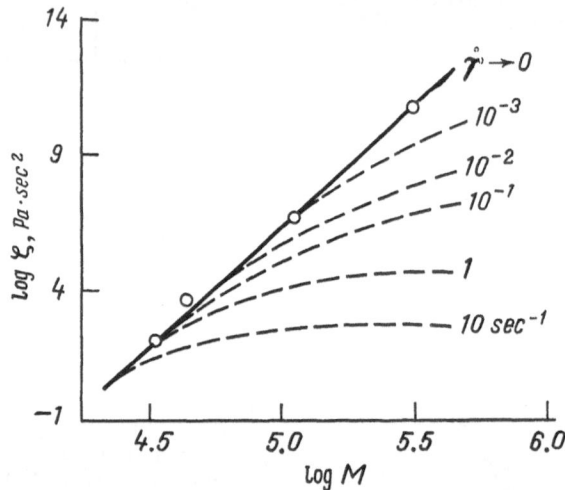

Fig. 4.22. The normal stress coefficient versus the molecular mass of *cis*-polybutadienes at various rates of shear (indicated at the curves) at 25°C. [Malkin, A. Ya. *et al.*, *Vysokomol. Soedin.*, **A12**, 120 (1970).]

compliance) is independent of molecular mass M in the region of high values of molecular masses (see Chapter 5), it may be presumed that the dependence $\zeta(M)$ must be quadratic with respect to $\eta(M)$. If the dependence $\eta_0(M)$ is expressed by the power law $\eta_0 \propto M^\alpha$ with the exponent α being close to 3.5 (see Chapter 2), then the dependence $\zeta_0(M)$ must also be expressed by a power law but with the exponent approaching 7. Figure 4.22 shows the results of an experimental measurement of the dependences $\zeta_0(M)$ and $\zeta(M)$ for *cis*-polybutadienes, from which it follows that the predicted character of the effect of molecular mass on normal stresses is really fulfilled. The solid line in this figure, which represents the dependence $\zeta_0(M)$, is drawn with a slope equal to 8. With increasing shear rate the dependence $\zeta(M)$ becomes much weaker and at high rates of shear the normal stress coefficient practically ceases to be dependent on molecular mass. The situation here is, in principle, the same as in the case of the effect of the shear rate on the dependence of viscosity on molecular mass. The difference is that usually normal stresses constitute a second-order effect as compared with shear stresses.

---

* Malkin, A. Ya., G. V. Berezhnaya, and G. V. Vinogradov, *Mekhanika Polimerov*, **5**, 896 (1972); *J. Polymer Sci., Symposium*, **42**, 1111 (1973).

4.6.4. Molecular-Mass Distribution and Normal Stresses

The problem of the effect of molecular-mass distribution on normal stress-es arising in steady shearing flow of polymers, just as in the case of viscosi-ty, reduces to the choice of such an averaged value of molecular mass $\overline{M}$ for which the dependence $\zeta_0\,(\overline{M})$ must coincide with the dependence $\zeta_0\,(M)$ measured for monodisperse polymers. A very limited number of experimental data are avail-able on the effect of molecular-mass distribu-tion on normal stresses. The existing experi-mental evidence points to the stronger (than in the case of viscosity) effect of the higher moments of molecular-mass distribution on the quantity ζ_0. At any rate, the use, as the argument, of the dependence of the mass-aver-age molecular mass, with the aid of which it is possible to describe well the experimen-tal data on the flow properties of polymers having arbitrary molecular-mass distributions, is found to be unsatisfactory for normal stress-es. For example, for polydimethyl siloxanes with different molecular-mass distributions the initial normal stress coefficient is found to be a single-valued function of the product of two average molecular masses—the mass-average and the z-average molecular masses (Fig. 4.23). But it is not known whether this argument will be suitable for the description of the effect of molecular mass and molecular-mass distri-bution on the normal stresses arising during the shearing flow of other polymers.

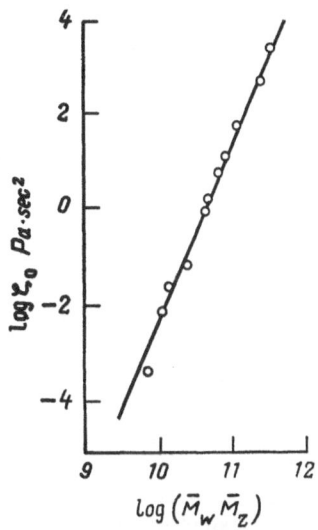

Fig. 4.23. The initial normal stress coefficient of polydimeth-yl siloxanes versus the average molecular masses. [Mills, N J., Europ. Polymer J., 5, 675 (1969).]

Referring to the general formula (4.14), we can assert that the experimentally observed effect of molecular-mass distribution on normal stresses is associated with the role played by this factor[i] in the man-

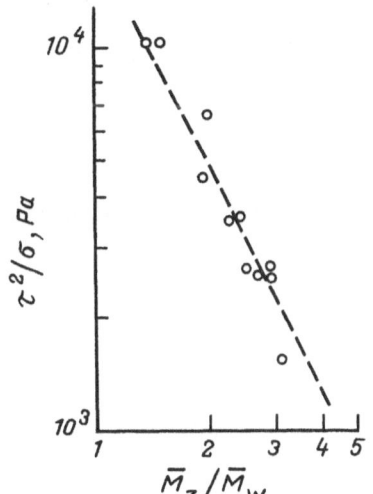

Fig. 4.24. Dependence of the ratio of the stresses (τ^2/σ) on the MMD cha-racteristic for melts of polypropylene. [Mieras, H. J. M. A. The paper pre-sented to the conference "Advances in Rheology", Glasgo, 1969; Mieras, H. J. M. A. and C. F. H. van Rijn, Nature, 218, 865 (1968); J. Appl. Polymer Sci., 13, 309 (1969).

ifestation of the rubber-like properties of the polymer, i.e., with the existence of the effect of molecular-mass distribution on the equilibrium compliance of polymer systems. This effect will be discussed in greater detail in Chapter 5. At this point it is worthwhile to focus attention on the effect of molecular-mass distribution on the relation between the normal and shear stresses. Figure 4.24 presents the dependence of the parameter (τ^2/σ) on the ratio of two "average" molecular masses. These data show that even at not too broad a molecular-mass distribution (up to $\overline{M}_z/\overline{M}_w \approx 3$) the quantity (τ^2/σ) may diminish by a decimal order. Therefore, by varying the molecular-mass distribution of the polymer one can vary the normal stresses at the same shear stress by tens of times.

Thus, normal stresses are highly sensitive to the variation of both the average molecular mass of the polymer and its molecular-mass distribution.

4.6.5. Some Special Cases of Development of Normal Stresses

One of the most interesting cases of the Weissenberg effect is the appearance of normal stresses during the flow of solutions of aluminium naphthenates and similar systems. In such solutions, enormous recoverable deformations are readily developed. Apparently, these systems are similar, in their structure, to solutions of flexible-chain polymers, which is what accounts for the occurrence of the same mechanical phenomena in the cases being compared.

More specific is the case of development of normal stresses in dispersions of anisotropic particles. In such systems, to steady-state flow there correspond equilibrium distributions of the orientations of particles, which are different from the uniform distribution. Owing to the tendency towards the return to a uniform distribution of orientations of particles, which is due to the Brownian motion and which is dependent on the viscosity of the material, there arise stresses normal to the shear plane. The magnitude of these stresses and their dependence on the rate of shear are determined by the size and shape of the dispersed particles and also by the intensity of molecular motions. Effects of this kind are extensively discussed in various kinetic models proposed in the literature for the explanation of the basic regularities of the behaviour of polymeric and colloid systems.

4.7. Birefringence and the State of Stress in Polymeric Systems

4.7.1. General Remarks

The most important optical characteristic of a substance is the refractive index. It is determined by the polarizability of atomic groups and bonds in molecules. Most molecules are optically anisotropic. However, if the molecules are randomly oriented in space, they form optically isotropic materials. If the molecules are oriented under the deformation of the body, the

material becomes anisotropic. The anisotropy of the material is characterized by its birefringence—the refractive indices appear to be different in different directions.

In a general case, the birefringence of the material is determined by the
refractive index tensor $\{n\}$. It means that in any anisotropic material there
always exist such three mutually perpendicular directions in which the
diagonal components of the tensor $\{n\}$ assume the extremal values n_I, n_{II},
and n_{III}. They are the principal values of the tensor $\{n\}$. In the directions
corresponding to the principal components the non-diagonal components of
the tensor $\{n\}$ are equal to zero.

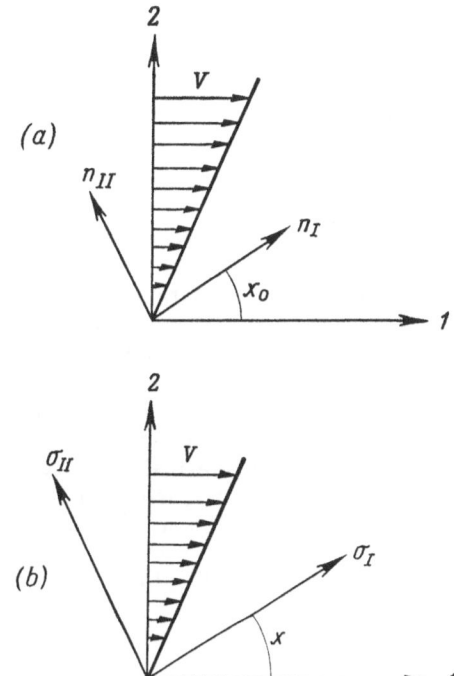

Fig. 4.25. Designations of axes,
stresses, and refractive indices in
simple shear.

The most extensive information concerning the measurement of the birefringence of fluid polymeric systems is available for simple shear.

Let us examine Fig. 4.25. Here 1 designates the direction of shear, 2 is
the direction of the velocity gradient. Hence, shear occurs in the plane 1-2.
The third, so-called neutral, direction in which the light propagates is normal to the plane of the drawing. For simple shear, the principal directions of
the birefringence tensor are defined by the vectors n_I and n_{II}. They characterize the orientation of the birefringence ellipsoid in space. Experimental
determination of this orientation is carried out as follows. Use is made of a
system of crossed polarizing devices—the planes of polarization of light in
the polarizer and analyser are arranged at an angle of 90°; in this case the
light does not reach the observer situated after the analyser. The object
under study is placed between the analyser and the polarizer. The plane-
polarized light of wavelength λ when passing through a doubly refracting
material of thickness d is split into two beams which propagate along the
principal directions of the refractive index ellipsoid. Having passed

through the analyser, these two beams interfere with a phase shift of $\delta =$ $= 2\pi d\,(\Delta n/\lambda)$. Here $\Delta n = n_{\bar{I}} - n_{II}$. It is this quantity that is a measure of birefringence; it is measured in brewsters; 1 Br $= 10^{-12}$ Pa^{-1}. The sign of the value of birefringence is determined by the ratio of the principal refractive indices n_I and n_{II}.

Upon synchronous rotation of the crossed polarizer and analyser the light will be extinguished. The angle of complete light extinction is taken to be the smaller of the two angles between the plane of polarization and the direction of shear. Then, by definition, the angle χ is smaller than or equal to 45°. For low-molecular-mass liquids the angle χ is 45° within ordinary values of shear stresses and rates. For polymer melts and solutions the angle χ is strictly equal to 45° only within the limits of zero shear stress. As the stress increases the refractive index ellipsoid rotates relative to the neutral axis and the extinction angle diminishes. In the limiting case, the refractive index ellipsoid is found to be oriented in the direction of the axis 1 and the angle is equal to zero.

4.7.2. The Optical Stress Rule. Polymer Melts

The method of quantitative measurement of stresses in fluid polymeric systems is based on the use of the stress optical rule, which is sometimes called the optical-stress law. It holds true for a large group of industrially important polymers, such as polyethylenes, polystyrenes, and many other thermoplast melts and solutions, randomly cross-linked (cured) elastomers. This law is obeyed by polymeric systems consisting of linear flexible-chain molecules. The stress optical rule establishes an unambiguous linear relationship between the stress tensor and the birefringence tensor, which means that the principal axes of the stress and refractive index ellipsoids coincide. This relationship is expressed through the proportionality factor C, called the stress optical coefficient, which is a fundamental characteristic of a polymer:

$$C = \frac{2\pi}{45} \frac{(n^2+2)^2(\alpha_1 - \alpha_2)}{NkT} \tag{4.32}$$

where n is the mean value of the refractive index, which is determined from the relation $3n = n_I + n_{II} + n_{III}$; α_1 is the polarizability of the monomeric unit in the direction of the axis of macromolecules; α_2 is the polarizability in the normal direction; k is Boltzmann's constant; T is the absolute temperature; N is the number of segments in unit volume. The above expression for C and the stress optical rule follow from the theory of ideal networks in rubber-like polymers.*

The use of the stress optical rule for the estimation of the state of stress in fluid polymeric systems is based on the idea** of the existence of a fluctuating entanglement network, which manifests itself like a network formed by covalent bonds in cross-linked (cured) elastomers. Indeed, a typical distinctive feature of polymeric systems possessing a well-developed entanglement network is that the macromolecular chains between the entanglement points undergo numerous conformational transformations during the time required for small displacements of the centres of gravity of the macromolecules. The high rate of conformational changes between entanglement network junctions

---

* Treloar, L., *The Physics of Rubbery Elasticity*, 2nd edition, Oxford, 1958.
** Lodge, A., *Elastic Liquids*, Academic Press, New York, 1964.

and between nodes formed by covalent bonds in cross-linked elastomers is responsible for the fact that both types of polymer systems have many common properties and, in particular, they obey the stress optical rule.

From the foregoing we can conclude that the most important criterion of the applicability of the stress optical rule to uncured polymers must be the existence of a random entanglement network. That the sizes of segments obey the Gaussian distribution must serve as a criterion of such a state.

There is an indication* that the condition $C = $ const has already been observed to be fulfilled for by polystyrene sample for which $M = 5 \times 10^4$, which is close to the value of M_c.

A remarkable feature of the stress optical rule is that over a wide range of stresses the quantity C does not depend on the deformation and the rate of strain. At the same time, it is well known that in the case of simple shear the rate of shear depends strongly on the shear stress. The constancy of C under simple shear for high-molecular-mass polymeric systems may be easily explained by proceeding from the earlier discussed conceptions; it was shown above that the nonlinear relation between the shear rate and the shear stress for polydisperse polymers is due to the successive transition of the highest-molecular-mass fractions to the rubbery state. This reduces hydrodynamic losses, thereby changing the viscosity of polymers, but the transition of the polymer from the fluid to the rubbery state is not accompanied by a change in the quantity C. Therefore, for high-molecular-mass polydisperse polymers containing no low-molecular-mass fractions (at any rate when $M > M_c$) the stress optical rule is satisfied over a wide range of shear stresses.

An important limitation on the use of the stress optical rule and, hence, on the constancy of the stress optical coefficient C is associated with the approach of polymeric systems to the glassy state, irrespective of whether it is attained under the influence of the fall of temperature or as a result of the increase of the rate of deformation (forced vitrification).

The stress optical rule is not met by filled polymers even in those cases when they exhibit high optical transparency and the depolarization of light is insignificant. This also applies to some copolymers with a heterogeneous microstructure.

Turning to the quantitative aspect of the stress optical rule, it is useful to refer once more to Fig. 4.25 which, apart from the geometrical characteristic of the relationships between simple shear and birefringence, shows a graph that gives the idea of the system of forces acting in this case. Assuming a linear relation between the birefringence and stress tensors, we obtain the following system of equations:

$$n \sin 2\chi = 2n_{12} = 2C\sigma_{12}$$

$$n \cos 2\chi = n_{11} - n_{22} = C\,(\sigma_{11} - \sigma_{22}) = \sigma$$

$$n_{22} - n_{33} = C\,(\sigma_{22} - \sigma_{33}) = \sigma' \qquad\qquad (4.33)$$

$$n_{11} - n_{33} = C\,(\sigma_{11} - \sigma_{33})$$

$$\cot 2\chi = \frac{n_{11} - n_{22}}{2n_{12}} = \frac{\sigma_{11} - \sigma_{22}}{2\sigma_{12}} = \cot 2\chi_m$$

---

* Wales, J., *The Application of Flow Birefringence to Rheological Studies of Polymer Melts*, Delft University Press, 1976.

23*

The determination of the stress optical coefficient C and the test of its constancy is usually carried out by using the relation

$$\Delta n = 2C \frac{\tau}{\sin 2\chi} \qquad\qquad (4.34)$$

or

$$C = (\Delta n \sin 2\chi)/2\sigma_{12}$$

Since the middle of the fifties when Lodge* extended the concept of the network structure of cross-linked (cured) polymers to elastic liquids, there have been carried out a large number of measurements of stresses in melts, concentrated solutions, and uncured elastomers. Special mention should be made here of the fundamental works of the schools of Janeschitz-Kriegl** and Philippoff***.

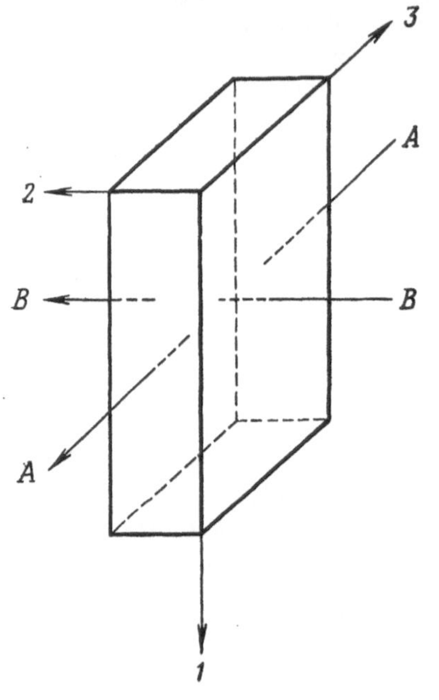

Though measurements of stresses by the optical method are not direct, they are of greatest interest for the following reasons. The method makes it possible to conduct measurements in the stream of polymers without flow perturbations that could be caused by the various measuring devices. The optical method is of special importance for the estimation of the first and second normal stress differences. Such measurements can be made in practically absolutely rigid measuring devices. In the measurement of these values by means of mechanical devices, considerable difficulties arise, which are due to the deformability of various dynamometers, this being essential for transient (in particular, prestationary) deformation regimes.

Fig. 4.26. Scheme of birefringence measurements with a slit die.

In rheology, use is commonly made of rotational instruments at relatively low shear stresses (as a rule, lower than 10^4 Pa because at higher stresses polymers are squeezed out from the gaps between the measuring surfaces) and different capillary viscometers at high stresses.

Let us consider briefly the latter method of measurement. Figure 4.26 shows the scheme of measurements of the normal stress differences in a slit

* Lodge, A., *Nature*, 4487, 838 (1955); *Trans. Farad. Soc.*, 52, 120 (1956).

** Janeschitz-Kriegl, H., "Flow Birefringence of Elastico-Viscous Systems", *Adv. Polymer Sci.*, 6, 170 (1969); Noordermeer, J., *A Flow Birefringence Study of Polymer Conformation*, S-Gravenhage, 1974; Gortemaker, F., *A Flow Birefringence Study of Stresses in Sheared Polymer Melts*, Kips Repro, Meppel, 1976; Wales, J., *The Application of Flow Birefringence to Rheological Studies of Polymer Melts*, Delft University Press, 1976; a series of papers by Janeschitz-Kriegl and coworkers have been published since the early seventies in *Rheologica Acta*.

*** Brodnyan, J. G., F. H. Gaskins, and W. Philippoff, *Trans. Soc. Rheol.*, 1, 95 (1957); Philippoff, W., *ibid.*, 4, 159 (1960); 5, 149 and 163 (1961); Dexler, F. D., Miller, J. C., and W. Philippoff, *Trans. Soc. Rheol.*, B, 193 (1961); Philippoff, W. and S. J. Gill,

capillary instrument. Wales* has succeeded in carrying out measurements, of the quantities $n_{11} - n_{22}$ and $n_{11} - n_{33}$ in the AA and BB directions, respectively. In this way one can find the second normal stress difference σ' which is proportional to the value of $(n_{11} - n_{33}) - (n_{11} - n_{22})$. According to these measurements, $\sigma' < 0$ and is about 10 per cent of σ. This is in good agreement with the data obtained by direct mechanical measurements of σ and σ'**.

We may write the following system of equations:

$$\frac{\sigma_{11} - \sigma_{22}}{2\sigma_{22}} = J_\infty \sigma_{12} = \cot 2\chi_m = \cot 2\chi = \cot \delta = \gamma$$

where J_∞ is the quasi-equilibrium value (corresponding to steady-state flow) of the compliance; δ is the phase angle; γ is the recoverable deformation.

Based on these formulas one can determine the compliance and recoverable deformation corresponding to steady-state flow conditions using polarization-optical measurements. It must, however, be emphasized that the determination of the recoverable deformation based on polarization-optical measurements can yield satisfactory results only at relatively low values of σ_{12}.

4.7.3. The Concentration Dependence of the Stress Optical Coefficient in Polymer Solutions

Tsvetkov*** and collaborators showed that the quantity C remains constant over the entire range of polymer concentrations, from infinitely dilute solutions up to molten polymers. In the early forties W. Kuhn and H. Kuhn, using Gaussian statistics for a polymer coil showed that the principal stresses must be proportional to the square of deformations of the unperturbed sphere. W. Kuhn and F. Gruen established the same relationship for the difference between the polarizabilities of the originally optically isotropic coil. The ratio of these quantities determines C and must not vary at higher concentrations when the coils are involved in a hydrodynamic interaction, which has the same effect on the stress and birefringence. What has been said emphasizes once more the determining role of stresses (and not of rates of deformations).

With a low content of the polymer in the solution the properties of the solvent begin to play an important role, and the estimation of the optical properties of the solution must in this case be carried out with account taken of the contribution of the effect of stresses on the appearance of optical anisotropy of the solvent to the stresses and optical anisotropy. For a comparative evaluation it may be pointed out that whereas the stress optical coefficient of polyisobutylene has a value close to 1500-1600 Br, solvents, such as cetane and methylnaphthalene, have the values of C equal to 1100 and 1900 Br, respectively. Therefore, depending on the ratio of the optical

ibid., 7, 33 (1963); Philippoff, W., *ibid.*, 7, 45 (1963); Philippoff, W. and R. A. Stratton, *ibid.*, 10, 467 (1966); Philippoff, W., *Proc. of the Fifth Intern. Congress on Rheology*, 4, 3 (1968); Philippoff, W. and R. A. Stratton, *ibid.*, 13.
    * loc. cit.
   ** Walters, K., *Rheometry*, Chapman and Hall, London, 1975 (Table 4.2).
  *** Tsvetkov, V. N., in: *Newer Methods of Polymer Characterization*, Chapter 14, Interscience, New York, 1965.

properties of polymer and the solvent, the stress optical coefficient may
vary with change of the concentration of the polymer solution.

The general method of determining normal stresses in solutions with the
aid of optical measurements consists in separating the contributions of the polymer and the solvent to the total observed birefringence into the components
which are added up as vectors (in the case of the biaxial state of stress) since
the deformation of the solvent is accompanied by the appearance of only
shear stresses and that of the polymer, by the development of shear and
normal stresses. Therefore, the difference of the refractive indices of the
system, Δn, is expressed in terms of the difference between the refractive
indices of the solvent, Δn_0, and the polymer, Δn_1, with the aid of the following vector equality:

$$\Delta n_1 = \Delta n - \Delta n_0$$

This is shown in Fig. 4.27a which also gives the observed angle χ and the
angle χ_1 associated with stresses that arise because of the presence of a polymer in the solution. The angle χ due to the deformation of the solvent is equal

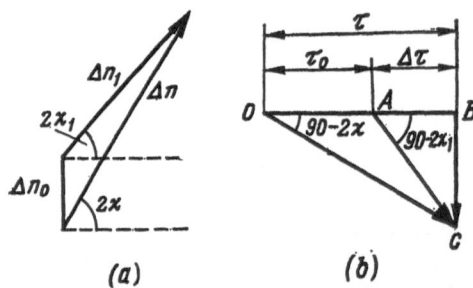

(a)   (b)

Fig. 4.27. Vector diagram of the plane
stress state for a polymer solution:
the ratio of the refractive indices (a)
and of the stresses (b) for the solution and the polymer.

to 45° since no normal stresses arise during its flow. The diagram of stresses
developing in the solution is shown in Fig. 4.27b, where τ is the total shear
stress acting in the system and τ_0 is the component of the total stress, which
is due to the flow of the pure solvent. The difference $\Delta\tau = \tau - \tau_0$ is the contribution introduced by the presence of a polymer in solution to the shear
stress (and, hence, to the viscosity of the system). The vector BC corresponds
to the normal stress difference $(\sigma_{11} - \sigma_{22})$, and the vector OC is the difference between the principal stresses operating in solution. It is important to
note that the angle χ between the direction of shear and the direction of
principal stresses in solution is not equal to χ_1. Of special interest in the
case of dilute solutions is the question of the relation between the angle χ_m
and the stress-optical properties of the polymer. Experimental investigation
of this problem has shown that the following equality is fulfilled:

$$\chi_m = \chi_1 \tag{4.35}$$

but $\chi_m \neq \chi$. And the stress optical law for the polymer is written in the form

$$\Delta n_1 = 2C \frac{\Delta\tau}{\sin 2\chi_1} \tag{4.36}$$

where C is the stress optical coefficient of the polymer, which is independent
of the concentration of the solution and the nature of the solvent.

The obvious differences between formula (4.36) and (4.34) consist in that,
instead of the total difference between the refractive indices, Δn, measured

experimentally, in formula (4.36) use is made of the quantity Δn_1, and the total shear stress τ is replaced by the quantity $\Delta \tau$ which refers to shear stresses that arise additionally because of the presence of a polymer in the system.

Concrete calculations with the aid of the above relations require a knowledge of two parameters of the solvent: its viscosity η_s and its stress optical coefficient C_0. Then

$$\tau_0 = \dot{\gamma}\eta_s$$

$$\Delta n_0 = 2C_0\tau_0 = 2C_0\dot{\gamma}\eta_s$$

The quantities Δn and χ are determined experimentally and the method of determining Δn_1 and χ_1 with the aid of the vector triangle is quite clear from Fig. 4.26a. The difference of normal stresses arising during the shearing flow of the solution is then calculated by the formula

$$\sigma = 2\Delta \tau \cdot \cot 2\chi_1 \qquad\qquad\qquad (4.37)$$

For concentrated solutions the solvent makes a negligibly small contribution to the value of stress and birefringence, and therefore $\chi_m = \chi = \chi_1$ and the decisive role is played by the stress-optical properties of the polymer.

If the stress optical coefficient of the polymer is unknown, then, according to formula (4.36), it can be found from the results of measurements carried out for dilute solutions.

4.7.4. Some Applications of the Polarization-Optical Method of Measuring Stresses

As an example of the use of the polarization-optical method, we consider the measurement of the normal stresses arising when the polymer flows from the reservoir into a rectangular duct.*

Figure 4.28 shows the dependence of the extensional stress σ_{11} on the dimensionless length Z/H, i.e., the distance along the flow axis referred to the duct width. To each curve in Fig. 4.28 there corresponds a constant value of shear stress τ_w on the duct wall in the region of the fully developed velocity profile. The positive values of Z/H refer to the preentrance region, and their negative values refer directly to the duct. The value of $Z/H = 0$ corresponds to the edge of the duct. Inspection of Fig. 4.28 shows that in the preentrance region of the duct the extensional stresses increase, reaching a maximum; then relaxation begins, which terminates inside the duct.

It has been found that the maximum of extensional stresses is always located over the duct entrance at a distance of (0.2-0.3) H from the duct edge and that the position of the maximum depends neither on the molecular mass and MMD of the polymers investigated (polybutadienes and polyisoprenes have been studied) nor on the value of shear stresses acting at the duct wall in the region of developed flow.

The highest extensional stresses as well as the dimensionless quantity Z_0/H (where Z_0 is the distance along the flow axis over which the extensional stress relaxes to zero) are unambiguously determined by the value of τ_w.

The dimensionless parameter Z_0/H is strongly dependent on MMD; one may assume that it is determined by the ratio of the rate of deformation of

---

* Brizitsky, V. I. et al., J. Appl. Polymer Sci., 21, 755 (1977).

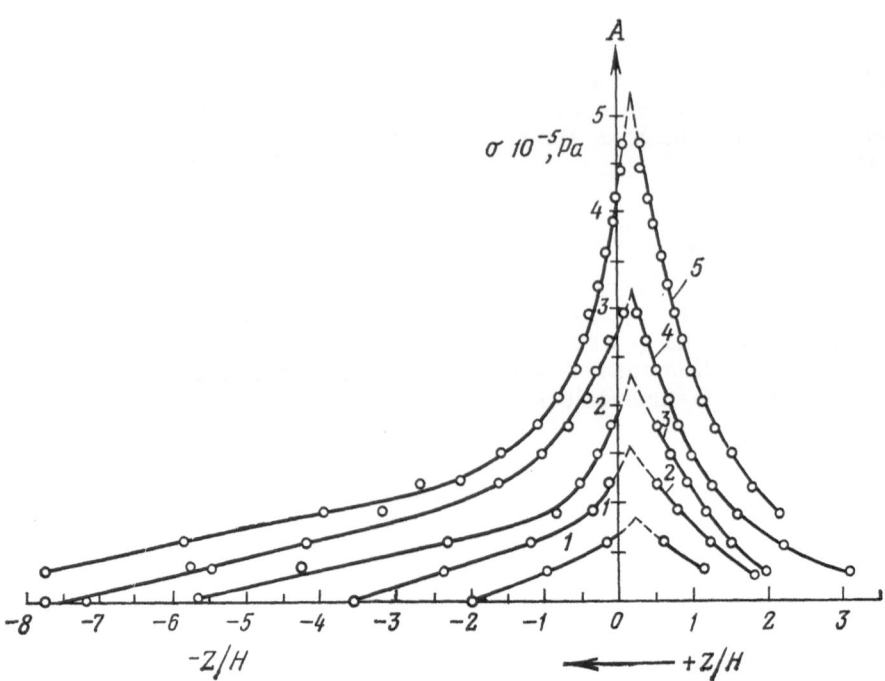

Fig. 4.28. Tensile stresses along the flow axis at the entrance into the slit die for polybutadiene (molecular mass 2.4×10^5; $M_w/M_n = 3.0$; 25°C). The arrow indicates the direction of flow. The curves *1-5* correspond to stresses at the duct wall in the zone of developed flow, MPa: 0.051; 0.098; 0.141; 0.21; 0.25.

the polymer in the duct to its characteristic relaxation time θ_0, which in its turn is determined by the initial (Newtonian) viscosity η_0 and the rubbery modulus. For high-molecular-mass narrow-distribution polymers the modulus is independent of molecular mass and θ_0 is proportional to M^α, where α is an exponent which characterizes the dependence $\eta_0 = f(M)$ for high values of M. For narrow-distribution polymers, to each given value of τ_w there corresponds a value of $\dot{\gamma}$ which is inversely proportional to M^α. Hence, the product of the characteristic relaxation time and the deformation rate at a given value of τ_w is a constant. This explains why Z_0/H is independent of molecular masses for each given constant value of τ_w. For broad-distribution polymers the value of $\dot{\gamma}\theta_0$ is always greater than for narrow-distribution polymers. Therefore, the dependence of Z_0/H on τ_w for broad-distribution polymers is always higher than for narrow-distribution ones.

The quantity Z_0/H is associated with the conformational characteristic of the polymer chain. This characteristic is determined by the ratio of the second moment of the chain end-to-end distance in the molecule, $\langle r^2 \rangle$, and the number of independent segments, N. For a free-jointed chain this ratio is proportional to l^2, where l is the segment length.

The attention of investigators has recently been drawn to the optical method of measuring stresses under the conditions of sinusoidal oscillating shear.*

* Grischenko, A. E. *et al.*, *Vysokomol. Soedin.*, B19 504 (1977).

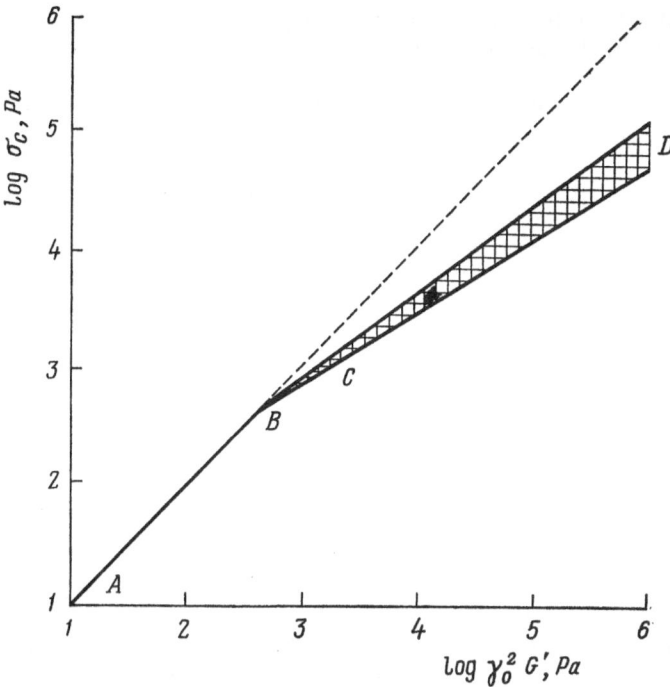

Fig. 4.29. Master curve showing the dependence of the constant component of the first normal stress difference on the amplitude of deformation. The portion of the curve ABC refers to polystyrene solutions according to data from Endo, H. and M. Nagasawa, *J. Polymer Sci.*, A-2, 8, 371 (1970); the band CD refers to a series of polybutadienes and polyisoprenes according to data from Vinogradov, G. V. *et al.*, *J. Appl. Polymer Sci.*, 22, 665 (1978).

It has been shown above (see Sec. 4.4) that normal stresses measured by the dynamic method consist of two components: the constant term σ_c and the oscillating stress with doubled frequency as compared with the frequence of the specified shear stresses. The constant component characterizes the long-time end of the relaxation spectrum since the stress has no time to relax during each deformation cycle.

The quantity σ_c under the linear deformation regime is equal to $\gamma_0^2 \cdot G'$, where γ_0 is the amplitude of deformation and G' is the storage modulus. Figure 4.29 shows this relation for a series of polybutadienes and polyisoprenes having molecular masses equal to $(0.76\text{-}3.8)\ 10^5$ and narrow MMD. This figure also presents data for solutions of polystyrenes. Note that the dependence obtained is invariant to the oscillation frequency. At low values of σ_c it is proportional to $\gamma_0^2 \cdot G'$. The portion AB refers to polystyrene solutions and the portion BC describes data for the above-mentioned polybutadienes and polyisoprenes. At high amplitudes the proportionality between these quantities is upset, but a master curve is obtained, which is shown in the form of a narrow band in Fig. 4.29.

4.7.5. Visualization of the Motion of Polymer System in Polarized Light

In Sections 2.6.6 and 2.6.7 we discussed critical deformation regimes for flexible-chain linear polymers and the associated spurt of their stream in ducts.

However, it is impossible to obtain from theories any information as to what happens to polymers during spurt and under above-critical regimes when elastic turbulence is observed. Answers to these questions are supplied by the visualization of a polymer stream. Visualization of deformation processes is particularly useful in investigations of polymer flows through channels of complicated geometrical forms.*

What is achieved by the method of visualization of the flow of polymers can be illustrated by data obtained in experiments on the flow of narrow-distribution polybutadiene of molecular mass 1.5×10^5 through a flat slit. The stream lines were recorded by the movement of 10-20μ glass beads in the polymer. The stress distribution in the flow was determined by using circular-polarized monochromatic light. Figure 4.30 shows that low shear rates are associated with the regular nature of stream lines in the duct and at its entrance.

Let us now see what can be achieved by observation of a polybutadiene flow in circular-polarized light (Fig. 4.31).

Photographs a and b depict, respectively, the entrance and exit zones in the duct and also a strictly regular distribution of interference bands inside the duct. The linear relationship between the shear stresses and the order of the band is fulfilled at stresses of up to about the critical value.

---

* Vinogradov, G. V., *Rheol. Acta*, **12**, 357 (1973).

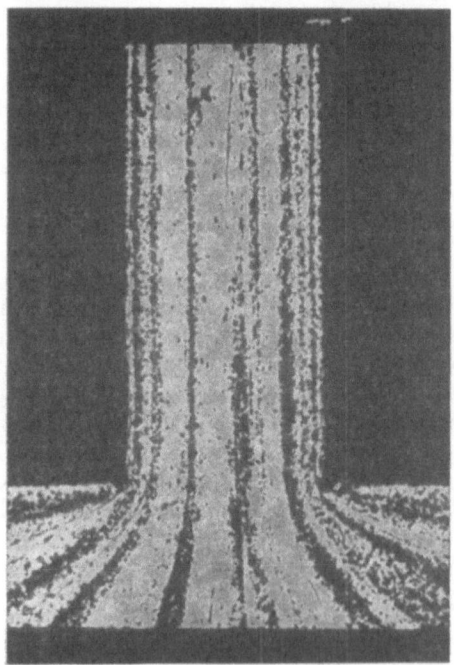

Fig. 4.30. Flow lines of polybutadiene at shear stresses of 6.31×10^5 dyn/cm² at the walls of a rectangular duct.

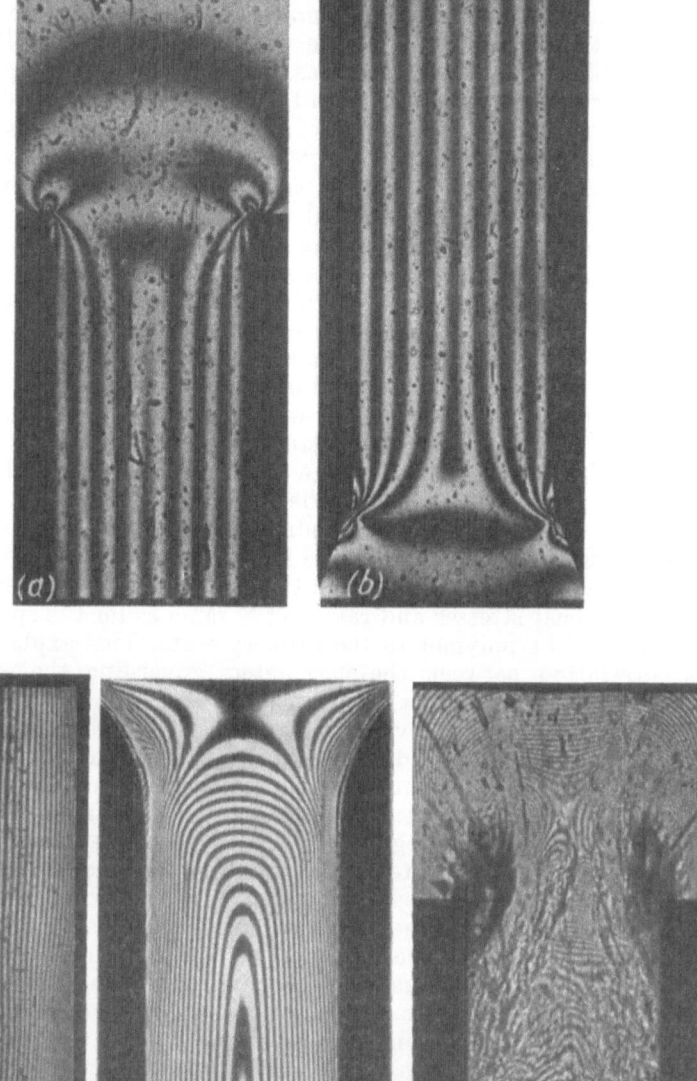

Fig. 4.31. Interference band patterns in a rectangular duct during the movement of poly-butadiene:

(a) duct entrance at a stress of 6.31×10^5 dyn/cm²; (b) same at duct exit; (c) middle part of the duct at a stress of 3.16×10^6 dyn/cm²; (d) smooth entrance at a stress of 3.98×10^6 dyn/cm²; (e) duct entrance at a stress of 5.62×10^6 dyn/cm² (all shear stresses are at duct walls).

It should be pointed out that at the entrance into the duct and at its exit (in the zones adjacent to the edges) stress concentration is observed near the walls. The shear stress in these zones may achieve the critical value, whereas inside the duct the shear stress at the wall is still substantially lower than critical. If at the duct entrance a critical stress is reached, the polymer passes into the rubbery state, its adhesion to the wall diminishes, and a local spurt occurs. As a result of the detachment of the polymer from the walls the stress acting on it falls off, the polymer relaxes and then begins to flow in the layer adjacent to the wall. This continues as long as the shear stress inside the duct at the walls is below the critical value.

When the polymer emerges from the duct, the stress concentration up to critical values also leads to a local tear-off of the polymer from the duct wall. As a result, a small portion of more or less relaxed polymer is ejected from the duct. It is followed by a portion of the polymer under a higher stress. Accordingly, sections of larger and smaller diameter appear in the polymer extrudate leaving the duct. An estimate of the stress concentration near the duct edge in the exit zone shows that when the stress at the walls inside the duct reaches a value at which the distortion of the extrudate begins, it exceeds the critical value in the edge zone. This explains why small-scale periodic distortion in the extrudate shape may be observed at the duct exit during the Newtonian flow of a monodisperse polymer in the duct. Another important conclusion is that the cause for the appearance of small-scale distortions of the extrudate shape and the enhancement of these distortions with increased shear stresses and rates is the same as for the spurt effect, viz. the transition of the polymer to the rubbery state. This explains the unambiguous correlation between the parameters governing the spurt and the visual manifestation of elastic turbulence.

Let us now consider the movement of a polymer in the duct when critical values of pressure gradients and flow rates are achieved inside it. When observing the movement of solid particles in the wall zone, a stick-slip process is clearly registered. This process is very sharply reflected in the polarization-optical pattern shown in photograph c. One can see the alternating narrowing and broadening of the interference bands along the duct walls. The narrow zones correspond to an increase in stress up to the critical value and the transition of the polymer to the rubbery state. This causes a reduction in the adhesion of the polymer to the wall and also results in slippage. Then, the stress relaxes, the polymer passes to the fluid state and sticks to the wall, the stress again increases, etc. Local spurts occurring along the length of the duct cause longitudinal fluctuations in the optical pattern. At the duct entrance where the stress is above the critical value, no stress relaxation occurs and the polymer does not exhibit a steady-state flow.

When the shear stress at the duct walls becomes critical, stress concentration occurs at the sharp edge at the duct entrance, which leads to a rupture of the polymer—the deformed material loses its continuity. This very important process will be discussed in more detail at a later time. But since it masks continuous slippage to some extent, it is convenient to use a smooth-entrance duct for observing slippage in its pure form. This is shown in photograph d. The figure clearly demonstrates the continuous slippage of a polymer and gradual relaxation of stresses along the duct walls. This is testified by the fact that the order of the bands decreases along the length of the duct from its entrance. Besides, in smooth-entrance ducts the disturbing influence of various factors affecting the asymmetry of the flow at the en-

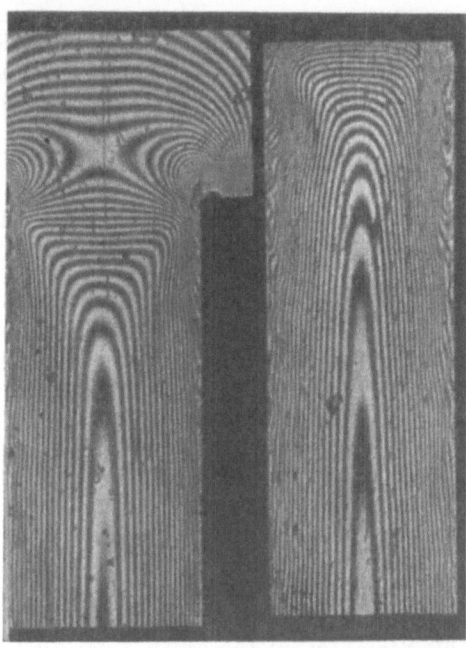

Fig. 4.32. Interference band patterns during the flow of polybutadiene in a rectangular duct under spurt conditions.

trance diminishes, and the polarization-optical pattern of the polymer flow proves to be strictly symmetrical.

Inspection of the interference band pattern in photographs *c* and *d* shows that at above-critical shear stresses and rates the routine methods of calculating the shear rate of the polymer at duct walls are meaningless. As for the calculation of the shear stress from the pressure drop in duct, the stress thus determined is actually a certain averaged value, and the nature of this averaging essentially depends on the velocity of the polymer in the duct.

It is important to ascertain what happens when the stress at the duct walls reaches or exceeds the critical value. It is obvious that in the zones of stress concentration near the edges at the duct entrance the stresses are still higher. Under the effect of high stresses the polymer breaks into pieces of irregular shape, and the flow in the duct becomes chaotic. This is shown in photograph *e*. The dark spots at the sharp edges appear as a result of light scattering in the non-uniform medium.

Of great interest is the observation of flow regimes corresponding to a pressure gradient a little larger than the critical value. This is depicted in Fig. 4.32 where the duct entrance is shown in the left part, and the duct proper in the right part; the lower edge of the photograph corresponds to the middle part of the duct along its length. With pressure gradients that create the flow pattern presented here the ruptures in the flowing polymer extend over a limited zone adjoining the edges, and the random movement of the polymer takes place only in layers adjacent to the wall. As polymer lumps advance along the duct walls the stress in them relaxes, and a distribution of interference bands of ever improving regularity appears near the walls.

The change in the slippage regime of the polymer—the movement of lumps of the polymer which are randomly displaced relative to one another—explains the increase in the pressure gradient with increasing rate of flow under above-critical regimes of polymer movement in the duct.

It should be emphasized that with sufficiently high pressure gradients and flow rates a break in the continuity of the polymer occurs with a smooth entrance as well. The sharp edges at the duct entrance only facilitate this process but its origin is determined by the fact that in the rubbery state the polymer undergoes a rupture when the deformation rates and stresses exceed certain critical values.

The photographs presented above show that at high pressure gradients (and the volume rates of flow) the movement of the material has nothing to do with the usual laminar flow of the liquid. Thus, the visualization of polymer movement in the ducts under spurt conditions, and all the more so under the above-critical regimes, shows that for such deformation regimes the concepts of viscosity and flow curves lose their meaning.

The visualization of the deformation process in ducts makes it possible to explain the small (in absolute scale) but regular effect of the duct size on the critical parameters associated with the transition of polymers from the fluid to the rubbery state.

If one estimates the deformation process by a definite shear stress, then the pressure gradient should be directly proportional to the length and inversely proportional to the diameter of the cylindrical duct. In such a case the increase of the duct length-to-diameter ratio corresponds to an increase in the pressure gradient. Under the deformation regime directly preceding the spurt the resistance of the duct decreases as a result of the development of wall slippage. This is registered formally as a reduction in the critical values of shear stresses and rates (up to 35 and 40% respectively).

A reduction in the duct diameter at a constant length-to-diameter ratio entails the shortening of the duct. In this case the role of wall slippage in the duct diminishes, this leading to an overstatement of the critical parameters in the range of 15-25%.

Bagley[*] and coworkers were the first to describe the appearance of loops of self-oscillation on the curves showing the dependence of the flow rate on the pressure gradient. The nature of this effect is reminescent of the mechanism of spurt.

---

[*] Bagley, E. B., I. M. Cabbot, and D. C. West, *J. Appl. Phys.*, 29, 109 (1958).

CHAPTER 5

The Rubber-Like Behaviour
of Polymers in Flow

5.1. Introduction

The ability to develop large recoverable (rubbery) deformations is the most important specific feature of the mechanical properties of polymeric materials. The most extensive investigations of the rubbery elasticity have been accomplished for cross-linked elastomers since this property is their main characteristic. But large recoverable deformations are also typical for polymers in the fluid state. Rubber-like elasticity is also inherent in many polymer solutions.

For polymers in the fluid state, the total deformation is composed (the instantaneous component being neglected) of the irrecoverable deformation of viscous flow and the rubber-like deformation which recovers after the cessation of flow. The ratio of the plastic (irrecoverable) and elastic components of deformation at a specified temperature depends on the loading conditions and the duration of deformation. For the steady-state flow, to each shear rate and stress there corresponds a definite quasiequilibrium rubbery deformation which is retained in the system for any straining period. After the load is removed there takes place a prolonged change of the shape, this change occurring in different ways, depending on whether the sample is allowed to change its shape freely, while being deformed in any direction (free recovery) or its elastic recovery takes place strictly only in a direction opposite to the direction of the preceding shearing flow (constrained recovery). The first case occurs, for example, when the polymer extrudate emerges from the duct and it can freely change its dimensions (swelling) due to the release of the elastic energy stored during the flow in the capillary. The second case is observed, as a rule, in a quantitative investigation of the effect of elastic recovery when the polymer is in the working gap in rotational devices and the sample is deformed under the specified conditions of simple shear. After the cessation of the forced rotation the sample is allowed to return to its original shape by turning one of the confining surfaces, which has one degree of freedom, in a direction opposite to the direction of the preceding shear. In both cases the process of elastic recovery takes place for a more or less prolonged period of time.

The elastic recovery of the shape of elastic liquids is analogous in many respects to the elastic recovery of rubbers after removal of the external load. But, in contrast to rubbers, the elastic deformations stored during the flow of polymeric systems are capable of relaxing. This means that if there is a certain time period between the moment when the forced deformation is discontinued and the onset of elastic recovery, the rubbery deformation measured is found to be the lower the longer is the duration of relaxation of the stored rubbery deformation. In contrast to this, in rubbers the equilibrium elastic deformation (except for special cases) is independent of the

time during which the sample is being stressed. This difference in the behaviour of cured rubbers and polymeric liquids is the same as the difference in
the equilibrium stress: in rubbers it is retained (theoretically) for an infinitely long period of time and in polymeric liquids the stresses relax down to
zero.

In what follows we shall only be concerned with the "equilibrium" values
of rubbery deformations corresponding to steady-state flow conditions, i.e.,
independent of the time of deformation before the onset of elastic recovery.

5.2. Viscoelasticity and Rubbery Deformations

If during steady-state flow rubbery deformations are accumulated, which
are characterized by the amount of elastic recovery under the conditions of
simple shear, γ_e (in the case of constrained recovery), then the modulus of
rubber elasticity G_e is taken as a characteristic of the resistance of the material to elastic strains; the modulus of rubber elasticity is defined as

$$G_e \quad \tau/\gamma_e$$

where τ is the shear stress during the flow preceding the elastic recovery.

The rubber-like properties of polymer systems are also characterized by the
inverse quantity—the equilibrium compliance:

$$J_e = G_e^{-1} = \gamma_e/\tau$$

The values of the constants G_e and J_e are determined by a set of relaxation
properties of the polymer system. The relationship between the relaxation
properties of the material and its ability to exhibit rubbery deformations is
established, in the most simplest way, for viscoelastic materials whose properties are described by the relation of theory of linear viscoelasticity. The
corresponding values of G and J are designated as G_0 and J_∞. As has been
shown in Sec. 1.8, in this case the equilibrium compliance may be expressed
in terms of the relaxation spectrum of the system in the following way:

$$J_\infty = G_0^- = \frac{\int\limits_0^\infty \theta^2 F(\theta)\, d\theta}{[\int\limits_0^\infty \theta F(\theta)\, d\theta]^2} = \frac{1}{\eta_0^2} \int\limits_0^\infty \theta^2 F(\theta)\, d\theta \qquad (5.1)$$

Therefore, if from some theoretical considerations the relaxation spectrum
of the system is known or has been experimentally determined, then it is
easy to find the modulus of rubber elasticity or the equilibrium compliance.
With relation (5.1) being fulfilled the quantities G_0 and J_∞ are the constants
of the material which are independent of the deformation regime and therefore the rubbery deformations must be proportional to the shear stresses.
This corresponds to the fulfillment of Hooke's law in shear.

Based on the general equations of the theory of linear viscoelasticity, one
can express the quasiequilibrium compliance directly in terms of the experimentally measured characteristic of the systems: the relaxation function
$\varphi(t)$ or the components of the complex shear modulus. Thus, the following
formula, with the aid of which the quasiequilibrium compliance is expressed

in terms of the relaxation spectrum, is valid:

$$J_\infty = \frac{1}{\eta_0^2} \int_0^\infty t\varphi(t)\, dt = \frac{1}{\eta_0^2} \int_0^\infty t^2\varphi(t)\, d\ln t \qquad (5.2)$$

Here η_0 is the initial viscosity.

If the components of the complex modulus, G' and G'', (or of the compliance, J' and J'') are determined experimentally, the following series of relations can be obtained for γ_e:

$$\gamma_e = J_\infty \tau = \lim_{\omega \to 0}(J'G'') = \lim_{\omega \to 0}\left[\frac{G'G''}{(G')^2+(G'')^2}\right] = \lim_{\omega \to 0}\left(\frac{G'}{G''}\right) = \lim_{\omega \to 0}\cot\delta \qquad (5.3)$$

With G' and G'' being replaced by their expressions in terms of the spectral functions of relaxation time distribution formula (5.1) obtains.

The quantity J_∞ can also be calculated if we know the law of relaxation of shear stresses, $\tau_R(t)$ after cessation of steady shearing flow. Since $\int_0^\infty \tau_R(t)\, dt$ represents $\dot\gamma \int_0^\infty \theta^2 F(\theta)\, d\theta$ (see page 314), it is not difficult to derive the following expression with the aid of formula (5.1):

$$J_\infty = \frac{\dot\gamma}{\tau_s^2} \int_0^\infty \tau_R(t)\, dt = \frac{1}{\eta_0} \int_0^\infty \frac{\tau_R(t)}{\tau_s}\, dt \qquad (5.4)$$

where τ_s is the stress corresponding to steady-state flow conditions.

Thus, the determination of rubbery deformations during the flow of a viscoelastic body described by the relations of linear viscoelasticity theory, just as in the case of any characteristic of such a body, is carried out with the aid of the concept of the relaxation spectrum and can be quantitatively accomplished either directly by finding the limit of the creep function on very prolonged loading or by means of the written theoretical relationships.

A quantitative description of the dependence of the modulus of rubber elasticity on the shear rate or stress in the nonlinear region requires the introduction of new assumptions about the fundamental features of the properties of the polymer to the same extent as in the case of the description of the dependence of apparent viscosity on deformation conditions. Such a description may be based on the idea that the development of rubbery deformations in the flowing polymer, just as in rubber, corresponds to the accumulation of potential energy, which depends on the properties of the material and recoverable deformation. Therefore, it may be presumed that a quantitative substantiation of the observed dependences $\tau(\gamma_e)$ must rely on the consideration of the possible forms of the strain-energy function (the stored-energy function), just as is done in the analysis of the rubber-like properties of elastic bodies (see Sec. 1.6).

As has been discussed in the treatment of simple shear of elastic bodies, the frequently used Kuhn-Mark and Mooney-Rivlin strain-energy functions predict that in shear the dependence of rubbery deformations on stress must be linear. This contradicts, on the whole, the experimental facts available for a wide range of stresses and shows the inapplicability of these strain-energy functions for a quantitative treatment of the rubber elasticity of linear polymers above the glass-transition temperature. This is probably associated with the general inadequacy of the simplest Gaussian statistics for the discussion of the behaviour of polymer melts and solutions at finite deformations.* Therefore, for the rubbery deformations of such materials to be discussed, use must be made of other strain-energy functions of more complicated structure. However, this point has not been studied in any systematic way and the problem of the quantitative description of rubbery properties during the flow of polymers on the basis of general phenomenological considerations still remains open. One of the possible ways for the solution of this problem is the utilization of the strain-energy functions of a "fractional type".** Some of the calculations carried out and a comparison of the results obtained with experiment show that, with the aid of a strain-energy function of this type, it is possible to correctly describe elastic deformations in polymer melts both in shear and in uniaxial extension, i.e., for different types of the state of stress. This is especially important for critical experimental tests of strain-energy functions (elastic-energy functions) of different structure.

5.3. Dependence of Rubber-Like Behaviour on Stress, Shear Rate and Temperature

Rubbery deformations that correspond to the steady-state flow conditions are determined by the rate of deformation and temperature. An example illustrating the general character of the change of rubbery deformations as a function of these parameters is shown in Fig. 5.1. The monotonic increase of

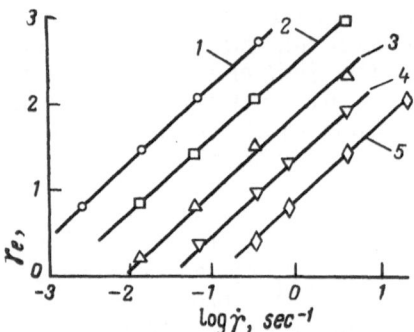

Fig. 5.1. Rubbery deformations in steady-state flow of polyisobutylene at different temperatures [Vinogradov, G. V., A. Ya. Malkin, and V. F. Shumski, *Rheol. Acta*, 9, 155 (1970).]: *1*—22; *2*—40; *3*—60; *4*—80; *5*—100°C.

rubbery deformations with increasing rate of shear is characteristic of all the cases investigated, with the result that no tendency towards the saturation of the values of rubbery deformations has been detected. The extent of non-Newtonian behaviour increases simultaneously with increase of γ_e (Fig. 5.2), but it is characteristic that rather considerable rubbery deformations can develop even in the region of a slight departure from non-Newtonian flow. This is typical of linear flexible-chain polymers of narrow MMD, for which the region of a slight departure from non-Newtonian flow is extended up to very high shear stresses and in this case recoverable deformations of the order of 100 per cent are developing.

---

  * Worth, R. A., *Trans. Soc. Rheol.*, 21, 493 (1977).
 ** Bloch, R., W. V. Chang, and N. W. Tschoegl, *Trans. Soc. Rheol.*, 22, 1 (1978)

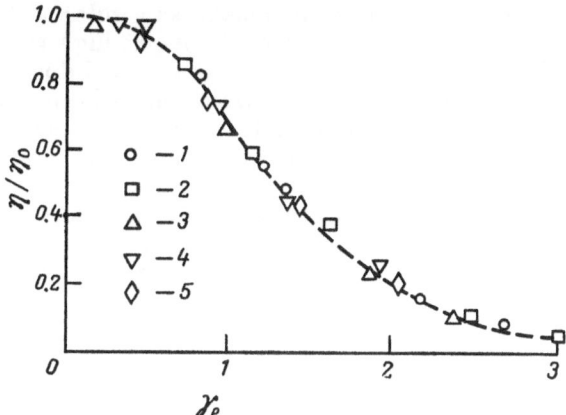

Fig. 5.2. Correlation between the degree of non-Newtonian flow behaviour and the amount of rubbery deformations in steady-state flow of polyisobutylene (see Fig. 5.1 for the designations and the reference).

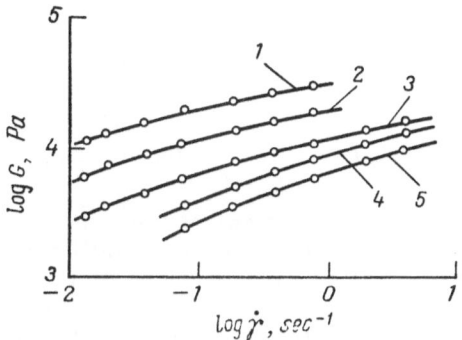

Fig. 5.3. The rubbery modulus of butyl rubber versus the rate of shear at different temperatures [Vinogradov, G. V., M. P. Zabugina, and V. F. Shumski, *Vysokomol. Soedin.*, A11, 1221 (1968).]:
1—25; *2*—60; *3*—80; *4*—100; *5*—120°C.

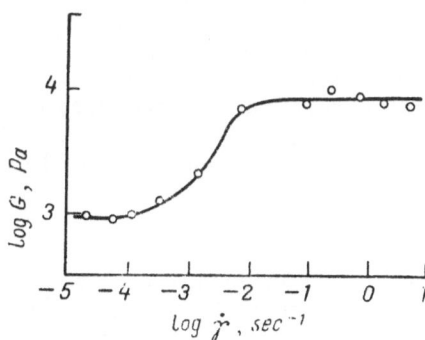

Fig. 5.4. The rubbery modulus of a 60-percent solution of polystyrene in diethyl phtthalae versus the rate of shear. [Malkin, A. Ya. *et al.*, *Rheol. Acta*, **10**, 336 (1971).]

24*

The modulus of rubber elasticity increases, as a rule, with increasing rate of shear (Fig. 5.3), but it may be expected that at high rates of shear there will take place the saturation of the values of the modulus. This is illustrated by experimental results (Fig. 5.4), which have been obtained for solutions of polystyrene in diethyl phthalate. Evidently, such a mode of variation of the modulus with the rate of shear (Fig. 5.4) is rather common to polymer systems.

The generalized representation of the dependence of the modulus of rubber elasticity on the rate of shear and temperature is attained by using the method of reduced variables, which has been described in discussing the problem of

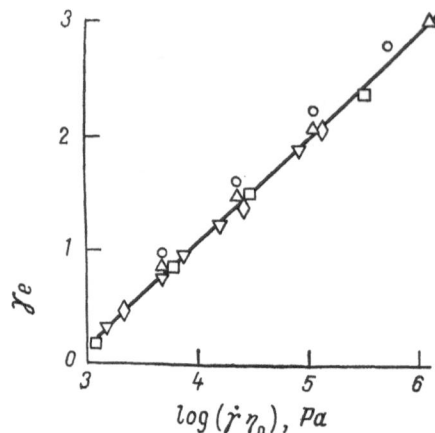

Fig. 5.5. Temperature-invariant dependence of rubbery deformations of polyisobutylene on the rate of shear (see Fig. 5.1 for the designations and the reference).

constructing the temperature-invariant characteristics of shear stresses and the first normal stress coefficient. The applicability of this method to the generalization of the results of investigations of the rubber-like properties of polymers is shown in Fig. 5.5 which is based on the use of the same experimental data as those of Fig. 5.1. The possibility of reducing the dependences $\gamma_e (\dot{\gamma})$ to a single dependence on the argument $(\dot{\gamma}\eta_0)$ is an indication that the factor determining rubbery deformations is, in the first place, stresses since there exists a relationship between the argument of the temperature-invariant characteristics—the product $(\dot{\gamma}\eta_0)$ and the shear stress. Similar results have been obtained for solutions of poly-α-methylstyrene.[*]

At the same stress the temperature exerts a rather slight effect on rubbery deformations. The situation here is basically the same as with rubbers for which the effect of temperature on their rubber-like properties (within the limits of the rubbery state) proves to be a secondary factor as compared with stresses. For rubbers, according to the entropy theory of elasticity, the stresses corresponding to the condition $\gamma_e = $ const depend linearly on the absolute temperature. As regards fluid polymers, for them the dependence of stresses on temperature at $\gamma_e = $ const is somewhat more complicated. Figure 5.6 gives examples of experimental data illustrating various possible situations. In the simplest case, for polyisobutylene, the stresses increase with temperature in the same way as is observed for rubbers. The dependence $\tau (T)$ at $\gamma_e = $ const turns out to be weakly nonlinear. In a more complicated case,

---

[*] Berry, G. C., B. L. Hager, and C.-P. Wong, *Macromolecules*, **10**, 361 (1977).

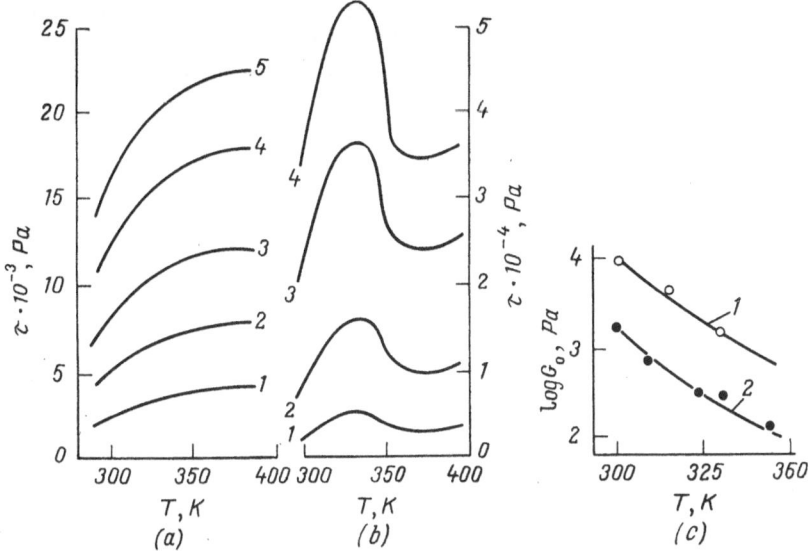

Fig. 5.6. Temperature dependences of rubber-like properties for a number of polymeric systems in the form of the dependences $\tau(T)$ at $\gamma_e = $ const for polyisobutylene (a), butyl rubber (b) and of the dependence $G_0(T)$ for solutions of polymethyl methacrylate ($\varphi_2 = 0.70$) in toluene (1) and acetyl cellulose ($\varphi_2 = 0.25$) in cyclohexanone (2) (c):

The values of γ for the curves 1-5 in (a): 0.8, 1.2, 1.5, 1.8, 2.0; for the curves 1-4 in (b): 1.0, 2.0, 3.0, 3.5. [Vinogradov, G. V., A. Ya. Malkin et al., Vysokomol. Soedin., A12, 1044 (1970); Dreval, V. E., A. Ya. Malkin, G. V. Vinogradov, and A. A. Tager, Europ. Polymer J., 9, 85 (1973).]

for butyl rubber, the course of the dependence of τ on temperature may be extremal, which is due to the secondary transition that occurs in this polymer in the region of temperatures 60-80°C. Finally, for certain solutions the decrease of the modulus with rise of temperature is due to the progressing breakdown of the structure of the system under the influence of thermal motion. It follows that the dependence of the modulus (or the compliance) on temperature may be different in different temperature ranges. Indeed, for a number of polymers, for which the dependence $J_\infty(T)$ has been measured over a very wide temperature range, it has been found* that J_∞ is proportional to T^{-1} in the region of $2 < T/T_g < 3$ (where T_g is the glass-transition temperature). At the same time, upon approach to T_g (in the region of $T/T_g < 2$) the compliance of polydimethyl siloxane remains to be temperature-independent and in the direct vicinity of T_g (at $T/T_g < 1.2$) the values of J_∞ for polystyrene and polyisobutylene sharply decrease. Thus, when considering the temperature dependence of J_∞ (or G_0) one should take into account the closeness or remoteness from the glass-transition temperature and the possibility of occurrence of secondary relaxation transitions at $T > T_g$.

According to the various phenomenological theories discussed in Chapter 4, there exists a direct relation between the components of the stress tensor and deformations. Therefore, a comparison of the quantities τ, σ, and γ_e under various steady-state flow conditions, characterized by certain rates of deformation, is a very interesting and important method of general evaluation of the rheological properties of polymeric systems and of the validity

Plazek, D. J. and A. J. Chelko, Jr., Polymer, 18, 15 (1977).

of various phenomenological theories. According to the theoretical results considered in Chapter 4, the relation between the three parameters in question (τ, σ, and γ_e) is described by the Lodge formula:

$$\gamma_e = \sigma/2\tau \tag{5.5}$$

or by the Weissenberg-Mooney-Rivlin formula

$$\gamma_e = \sigma/\tau \tag{5.6}$$

The same formulas may be rewritten for the coefficients characterizing the quantities under discussion:

$$J_\infty = G_0^{-1} = a\zeta_0/\eta_0^2 \tag{5.7}$$

where a is equal to 1 according to formula (5.5) and to 2 according to formula (5.6).

A large body of contradicting data are available in the literature on the value of the coefficient a. The validity of theoretical considerations and es-

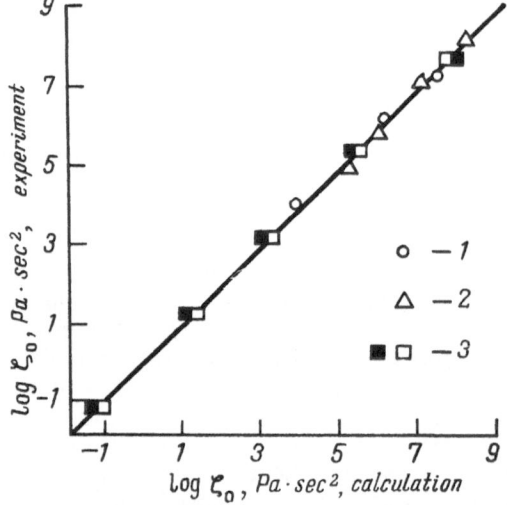

Fig. 5.7. Comparing the experimentally determined initial values of the coefficients G_0, η_0, and ζ_0 (the solid line corresponds to the equation $\eta_0^2/\zeta_0 = G_0$):
1—solutions of butyl rubber in cetane [Vinogradov, G. V., A. Ya. Malkin, V. F. Shumski, and M. P. Zabugina, *Vysokomol. Soedin.*, A11, 2002 (1969).]; 2—solutions of polystyrene in decalin [Malkin A. Ya., G. V. Berezhnaya, and G. V. Vinogradov, *Mekhanika Polimerov*, 5, 896 (1972).]; 3—polyisobutylene at different temperatures. [Vinogradov G. V., A. Ya. Malkin, and V. F. Shumski, *Rheol. Acta*, 9, 155 (1970).]

pecially the applicability of conflicting formulas must be tested on the basis of independent and reliable measurements of all the three quantities concerned (τ, σ, and γ_e). In doing so, it is very important that the experimental results under discussion refer to a strictly linear region of the mechanical behaviour of the material, when τ is proportional to $\dot{\gamma}$, σ to $\dot{\gamma}^2$ and γ_e to τ. A thorough test of theoretical predictions has convincingly (Fig. 5.7) confirmed the applicability of the Lodge formula to the linear region of the mechanical behaviour of viscoelastic liquids.* Formula (5.6) holds for nonfluid elastic

---

* Malkin, A. Ya., G. V. Berezhnaya, and G. V. Vinogradov, *Mekhanika Polimerov*, 5, 896 (1972); Stratton, R. A. and A. F. Butcher, *J. Polymer Sci.*, A-2, 9, 1703 (1971).

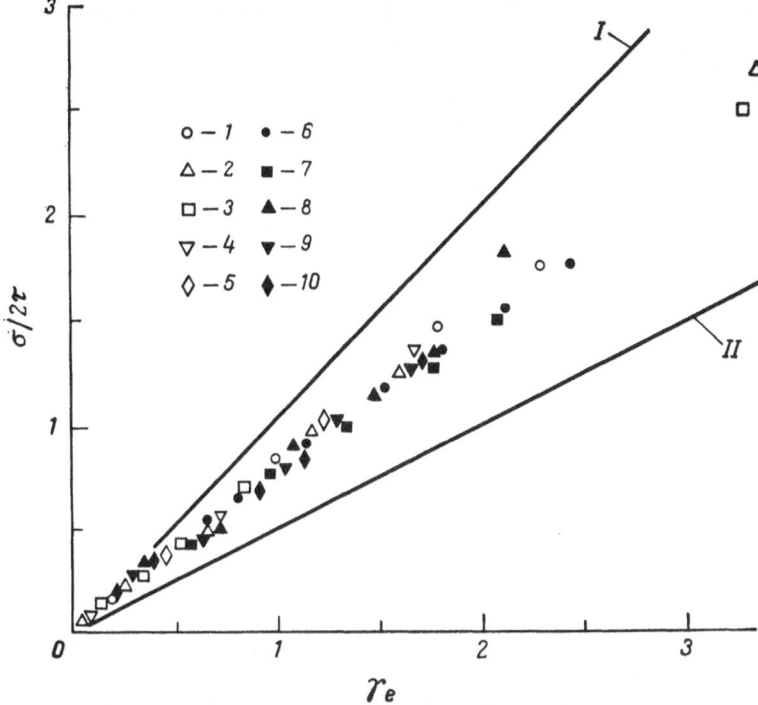

Fig. 5.8. Correlation between the ratio of stresses to rubbery deformation in steady-state shearing flow of solutions of polystyrene in decalin with the following contents of the polymer in solution, φ_2: 1—0.573; 2—0.466; 3—0.380; 4—0.290; 5—0.184; and of poly-isobutylene at the following temperatures, °C: 6—25; 7—40; 8—60; 9—80; 10—100. The line I corresponds to the equation $\sigma = 2\tau\gamma_e$, and the line II to $\sigma = \tau\gamma_e$.

bodies. Since during the flow of liquids there always takes place the dissipation of the work of external forces, for them the stresses cannot be uniquely related to elastic strains, in contrast to cross-linked elastomers, for which all the stress tensor components under equilibrium conditions are determined by the form of the elastic energy function which fully characterizes the type of the rheological behaviour of the material.

The proof of the validity of formula (5.5) refers to the linear region of the mechanical behaviour of polymeric systems. This formula should not be expected to be obeyed in the general case at high rates of deformation. Indeed, as the rate of shear increases γ_e increases more rapidly than the ratio (σ/τ), which leads to the progressively increasing deviations from formula (5.5). This is illustrated by the experimental data presented in Fig. 5.8.

5.4. Rubbery Deformations in Polymer Solutions

The ability to exhibit rubbery deformations is most prominently manifested during the flow of solutions of flexible-chain polymers, such as, for example, rubber glue. In simple shearing flow the rubbery deformations in solutions may reach enormous values: of the order of tens or even hundreds of thousands of per cent, which is usually unattainable for polymer melts. The observed rubbery deformations measured under the conditions of constrained recovery depend significantly on the total deformation of the material, which has been produced prior to elastic recovery. Rubbery deformations are also very strongly dependent on the previous history of the solution—the preliminary straining and rest conditions. The situation here is, on the whole, analogous to what is known for stresses which are also dependent on the duration of straining and the history of the system, but all these factors tell, in the most strongly way, on the rubber-like properties of polymer solutions.

This fact may be qualitatively accounted for on the basis of a comparison of the contribution of relaxation processes with long relaxation times to the flow and rubber-like properties of the system. Thus, for the region of low stresses [see formula (5.1)] the expression for the compliance includes the square of the relaxation time and the expression for shear stresses contains the relaxation times raised to the first power. Therefore, it is exactly the long-time region of the relaxation spectrum of the system that has a decisive effect on the rubber elasticity of the system.

5.4.1. Dependence of the Initial Modulus on Concentration

The effect of concentration on the rubber-like properties of polymer solutions in various solvents in the region of low stresses is shown in Fig. 5.9. As seen from this figure, the nature of the solvent does not influence the

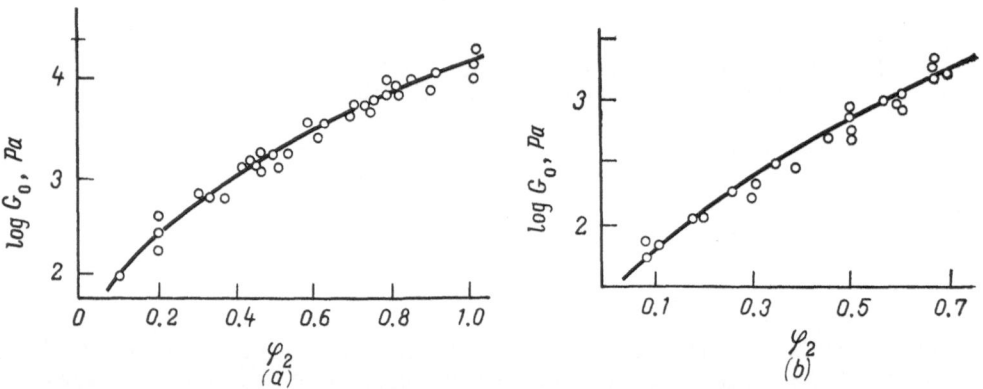

Fig. 5.9. Concentration dependences of the initial rubbery modulus for solutions of polyisobutylene at 20°C in six different solvents (a) and of polystyrene in decalin and toluene (b). [Dreval, V. E., A. Ya. Malkin, G. V. Vinogradov, and A. A. Tager, *Europ. Polymer J.*, 9, 85 (1973).]

modulus of rubber elasticity G_0 of equally concentrated solutions of flexible chain polymers in various solvents. As the polymer is diluted with a low-molecular solvent, the modulus decreases rather strongly and, hence, the compliance increases. The general character of the concentration dependence of the modulus $G_0 (\varphi_2)$ is similar for different polymer solutions (φ_2 is the mass content of the polymer in solution).

The experimental data presented in Fig. 5.9 and other known experimental facts point to a rather strong effect of the content of the polymer in the system on its rubber elasticity. The observed character of the dependence of the initial modulus of rubber elasticity on concentration is described by a formula of the type

$$G_0 = K e^{mc}$$

where K is a constant and c is the volume content of the polymer in solution; the constant m depends on the nature of the polymer (for example, for polyisobutylene $m = 4.6$).

Separate regions of the concentration dependence of the modulus can also be described by other empirical formulas, for example, by a power law of the type

$$G_0 = K' c^n$$

where K' is a constant and n is an exponent close to 2-3.

Thus, the concentration dependence of the initial modulus of rubber elasticity is rather strong, though it is weaker than the concentration depen-

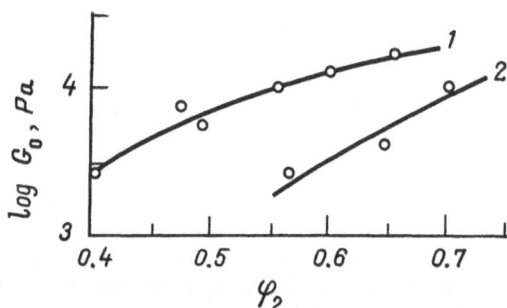

Fig. 5.10. Effect of the nature of the solvent on the concentration dependences of the initial rubbery modulus for solutions of polymethyl methacrylate in dihexyl oxalate (*1*) and toluene (*2*) (for the reference, see Fig. 5.9).

dence of viscosity: as was pointed out in Chapter 2, over a wide range of compositions the viscosity is proportional to the concentration raised to the fifth power.

That the modulus of rubber elasticity of equiconcentrated solutions of high-molecular-mass flexible-chain polymers is independent of the nature of the solvent is associated with almost the same junction density of the fluctuating entanglement network. If, however, the polymer chain stiffness is high, the quantity G_0 should then be expected to depend on the character of the interaction of macromolecules with the solvent. Effects of this kind have been experimentally observed (Fig. 5.10) for solutions which are also found to exhibit the temperature dependence of the modulus, which is not typical of flexible-chain polymers.

Since the rubber elasticity of polymer solutions is associated with the existence of a three-dimensional fluctuating network, it may be presumed that in dilute solutions at concentrations lower than a certain critical value $(c < c_{cr})$ the rubber elasticity will disappear. The quantity c_{cr} (see Chapter 2) is taken to mean such a concentration region where the character of the concentration dependence of viscosity varies because of the formation of a three-dimensional network. This inference is verified by experiment.* However, for

* Malkin, A. Ya., G. V. Berezhnaya, and G. V. Vinogradov, *J. Polymer Sci.,* Symposium, **42** 1111 (1973).

all the ranges of compositions, including dilute solutions, it is possible to find the effective measure of the rigidity of the solution, which is defined as $(\eta_0 - \eta_s)^2/\zeta_0$, where η_s is the viscosity of the solvent, and ζ_0 is the initial normal stress coefficient. Instead of this composite quantity, an equivalent quantity, $(G'' - \omega\eta_s)^2/G'$, can also be measured. For concentrated solutions $\eta_0 \gg \eta_s$, $G'' \gg \omega\eta_s$ and the following equalities are fulfilled: $\eta_0^2/\zeta_0 = = (G'')^2/G' = G_0$, i.e., these composite quantities are virtually equal to the modulus of rubber elasticity [see formulas (5.3) and (4.14)]. This is, however, valid only in the region of concentrations $c > c_{cr}$. At $c < c_{cr}$ the determination of G_0 has no physical meaning because no rubbery deformations can be developed in the solution. But the measurement of ζ_0 and G' is still possible at $c < c_{cr}$. In this range of concentrations the elastic effects are caused by the deformations of individual chains that are not united into a single three-dimensional entanglement network (fluctuating contacts). The difference in the mechanism of elasticity of solutions at $c > c_{cr}$ and $c < c_{cr}$ leads to different courses of the concentration dependence of the rigidity of the system: at $c < c_{cr}$ the "rigidity" falls off with increasing concentration and at $c > c_{cr}$ it increases, so that to the region of $c \approx c_{cr}$ there correspond the minima of the quantities $(\eta_0 - \eta_s)^2/\zeta_0$ and $(G'' - \omega\eta_s)^2/G'$. For the first of these quantities this effect was shown in Fig. 4.21, and for the second it has been described by Ferry and collaborators*.

The rubber-like behaviour of polymer solution is also influenced by the branching of macromolecules. As has been shown for solutions of linear and star-shaped poly-α-methylstyrenes**, in the region of relatively low concentrations the rubbery modulus of solutions of branched polymers is higher (or the compliance is lower) than that of linear polymers having the same molecular mass. But in the range of high concentrations branching practically has no effect on the values of the rubbery modulus of solutions of equal concentration. Admittedly, the difference in the effect of branching for different concentration ranges is associated with its influence on the conditions of attainment of the critical concentration.

5.4.2. Certain Theoretical Relations

In order to ascertain the character of the concentration dependence of the modulus of rubber elasticity, one should resort to the theoretical concepts of the viscoelasticity of polymer solutions. The problem here reduces to the elucidation of the form of the concentration dependence of the relaxation spectrum, whose integral characteristics define the rubbery modulus [see formula (5.1)]. The relevant calculations have already been considered in detail in Chapter 3, which is specially devoted to the discussion of this problem. Here we shall only be concerned with certain final results of theoretical calculations referring to the dependence of the rubbery modulus on the solution concentration and the molecular mass of the polymer.

The model of the viscoelastic polymeric chain yields the following expression for the dependence of the modulus on the concentration c and the molecular mass M of monodisperse polymers in the region o. sufficiently high

---

* Holmes, L. A., K. Ninomiya, and J. D. Ferry, *J. Phys. Chem.*, **70**, 2714 (1966); Holmes, L. A. and J. D. Ferry, *J. Polymer Sci.*, C, **23**, 291 (1968).

** Kajiura, H., Y. Ushiyama, T. Fujimoto, and M. Nagasawa, *Macromolecules*, **11**, 894 (1978).

concentrations where the contribution of the solvent to the flow properties of the solution may be neglected:

$$G_0 = a \frac{cRT}{M} \tag{5.8}$$

The constant a depends on the model chosen for computations; for the most frequently considered KSR model (see Sec. 3.1) $a = 5/2$.

Formula (5.8) predicts the linear dependence of the modulus on the concentration in the region of high values of c. As has been pointed out above, the dependence of G_0 on c is ordinarily stronger. But here we are only interested in the general tendency of the effect of concentration on the modulus, and formula (5.8) qualitatively correctly predicts the growth of the modulus with increasing polymer content in the system. In the region of low concentrations, where the effect of the solvent on the viscoelastic properties of the system cannot be ignored, formula (5.8) becomes incorrect and a more exact relation must be written:

$$G_0 = a \frac{cRT}{M} \left(\frac{\eta}{\eta - v_1 \eta_s} \right)^2 \tag{5.9}$$

where η is the viscosity of the solution; η_s is the viscosity of the solvent; v_1 is the volume content of the solvent.

At very low concentrations of the polymer $\eta \approx \eta_s$ and formula (5.9) assumes the form:

$$G_0 = a \frac{RT}{cM} \tag{5.10}$$

Hence, in the region of small concentrations, as the content of the polymer in the solution diminishes the modulus must increase. This leads to the prediction of the existence of a minimum on the concentration dependence of the modulus.

The position of the region of the minimum on the curve of G_0 versus c is determined by the molecular mass of the polymer. Quantitatively, this may be expressed in dimensionless form if we consider the dependence of the dimensionless (reduced) modulus $(G_0 M/cRT)$ on $(c\,[\eta])$, where $[\eta]$ is the intrinsic viscosity of the polymer in the solvent chosen. From formula (5.9) it follows that this dependence must have the form:

$$\frac{G_0 M}{cRT} = \left(\frac{1 + c\,[\eta]}{c\,[\eta]} \right)^2 \tag{5.11}$$

Then, at large values of the argument $(c\,[\eta])$ the right-hand side of the formula tends to unity and equality (5.8) is fulfilled, and at small values of $(c\,[\eta])$ formula (5.11) turns into relation (5.10), so that at $c \to 0$ the modulus must infinitely increase.

The concentration dependence of the modulus as predicted by theory is of qualitative nature since, first, at large c values the dependence $G_0(c)$ is non-linear and, second, rubbery deformations should not be expected to be possible in the region of strongly dilute solutions. As was discussed in Sec. 5.4.1, the appearance of rubbery deformations in solutions with a concentration lower than the critical value is impossible, though the solution may exhibit viscoelastic properties.

Nonetheless, theory correctly describes the general character of the concentration dependence of the "rigidity" of polymer solutions—the existence of a minimum at a certain concentration. Here the effect of the concentration on the properties of the solution is governed by the dimensionless parameter

$(c\ [\eta])$, and at some critical value of this parameter, $(c\ [\eta])_{cr}$ there takes place a transition from the region of dilute solutions to concentrated solutions. According to the available experimental data $(c\ [\eta])_{cr} \approx 5\text{-}6$.

5.4.3. Rubber-Like Behaviour of Solutions at High Shear Rates and Stresses

A typical dependence of rubbery deformations on the rate of shear is shown in Fig. 5.11 for solutions of polystyrene in decalin, taken as an example. Obviously, with increasing concentration there are developing, at the same rate of shear, large recoverable deformations, which are the larger the

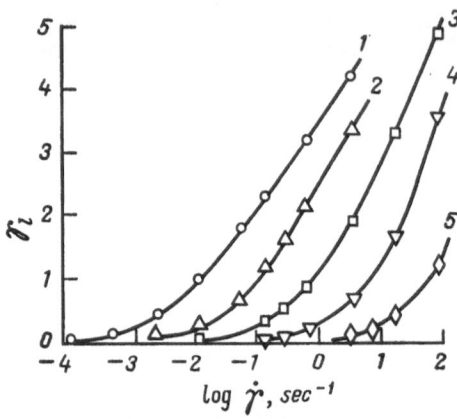

Fig. 5.11. Dependence of rubbery deformation on the concentration of solutions of polystyrene in decalin under the conditions of steady-state flow at various rates of shear. The content of the polymer in solution is as follows (in volume fractions): 1—0.573; 2—0.466; 3—0.380; 4—0.290; 5—0.184. [Vinogradov, G. V., A. Ya. Malkin, and G. V. Berezhnaya, Vysokomol· Soedin., A13, 2793 (1971).]

higher is the polymer content in the solution. In this sense concentrated solutions and melts should be regarded as more "elastic" than dilute solutions. With dilution of the polymer with a low-molecular solvent the rubbery modulus decreases. This has been shown above for the initial modulus G_0. A similar regularity holds for the moduli measured at increased stresses. Therefore, at the same stress the rubbery deformations are the larger the lower is the modulus and, consequently, dilute solutions prove to be, from this standpoint, more "elastic" than concentrated solutions.

Thus, the results of the estimation of the effect of concentration on the rubbery elastic behaviour of solutions depend on the comparison conditions, i.e., on whether the rubbery deformations under the $\dot{\gamma} = \text{const}$ or $\tau = \text{const}$ conditions are compared. This result is general for polymer solutions. It is associated with the fact that the viscosity which determines the stresses acting in steady-state flow varies, depending on the composition to a larger extent than does the rubbery modulus.

The concentration dependence of viscosity may be excluded from the discussion of the rubber elasticity of solutions if shear stresses are used as the argument that defines the state of the system. Taking into account an additional change of the initial rubbery modulus as a function of concentration, one can plot a concentration-invariant curve of rubbery deformations for polymer solutions. To do this, the quantity $(\gamma_e G_0)$ should be plotted as a function of shear stresses (Fig. 5.12). In the region of low τ and $\dot{\gamma}$ values $\gamma_e = \tau/G_0$; this is shown by a dashed line in Fig. 5.12. Thus, up to certain

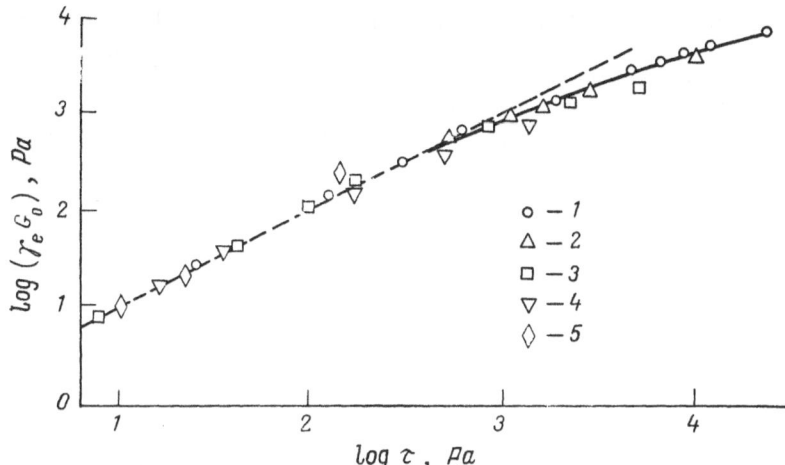

Fig. 5.12. Concentration-invariant dependence of rubbery deformations on the shear stress for solutions of polystyrene in decalin (the symbols are the same as in Fig. 5.11). The dashed line corresponds to the equation $\gamma_e = \tau/G_0$.

stresses (equal to about $10^{2 \cdot 5}$ Pa in the case of solutions of polystyrene in decalin), Hooke's law in shear is fulfilled. But at higher stresses there are observed systematic departures of experimental data from the dashed line. This is accounted for by a gradual increase of the rubbery modulus, as compared with G_0, with increasing shear stresses.

The dependence of rubbery deformations on the concentration and the stress may be represented in the form of the product of two independent functions, each with a single argument. One of them, $G_0 (c)$, which is the initial rubbery modulus, depends on the concentration only, and the other, $f (\tau)$, shows the effect of shear stresses on the variation of the viscoelastic properties of the solution. The dependence $\gamma_e (\tau, c)$ is thus given in the form

$$\gamma_e = f (\tau)/G_0 (c)$$

The function $f (\tau)$ is equal to τ at low shear stresses and increases more weakly than τ when the stresses continue to increase.

5.5. Dependence of the Rubber-Like Properties of Polymers on Molecular Mass and Molecular-Mass Distribution

5.5.1. Effect of Molecular Mass on Rubber-Like Behaviour

All the experimental data available indicate that in the region of sufficiently high molecular masses the modulus G_0 is independent of molecular mass, though at relatively low molecular masses there is observed the tendency for the modulus to diminish with increasing molecular mass. The fact that the modulus G_0 is independent of

the molecular mass of high-molecular-mass samples is illustrated by the experimental data presented in Fig. 5.13 for polybutadienes and polyisoprenes
with narrow molecular-mass distributions. The constant value of G_0 equal
to 1.4×10^5 Pa has also been observed for polystyrenes produced by the
method of anionic polymerization*, which are of narrow molecular-mass

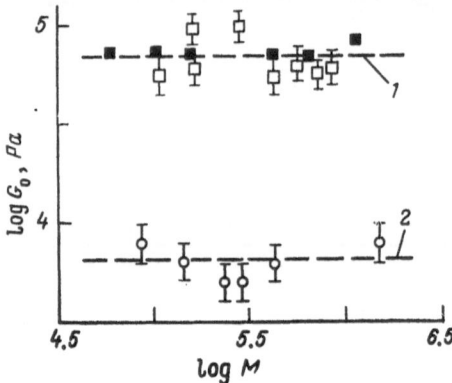

Fig. 5.13. Values of the initial rubbery modulus of polyisoprenes (*1*) and
polybutadienes (*2*) of narrow MMD.
The unfilled symbols are the data from
Malkin, A. Ya., G. Zh. Zhangereeva,
and M. P. Zabugina, *Vysokomol. Soedin.*, A18, 572 (1976); the filled symbols represent the data from Nemoto,
N. *et al.*, *Macromolecules*, 4, 2, 215
(1971).

distribution. The condition $G_0 = $ const is probably general for high-molecular-mass samples. It is due to the fact that for polymers having very high

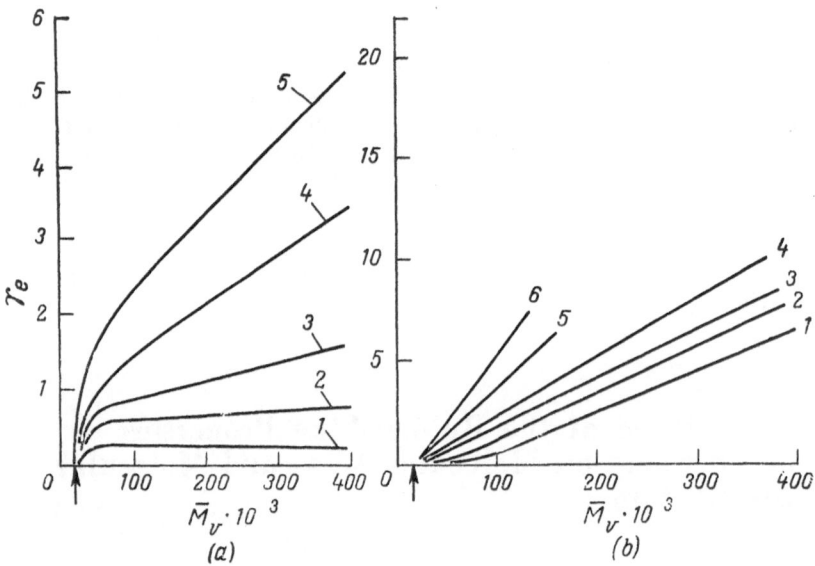

Fig. 5.14. Dependences of rubbery deformations on the molecular mass of *cis*-polybuta-
dienes at $\tau = $ const (*a*) and at $\dot{\gamma} = $ const (*b*):
The values of τ (Pa): *1*—1×10^2; *2*—5×10^2; *3*—1×10^3; *4*—3×10^3; *5*—1×10^4. The values of
γ (sec^{-1}) are: *1*—0.01; *2*—0.04; *3*—0.1; *4*—0.4; *5*—2.5; *6*—10. The arrow indicates the critical values
of molecular mass. [Malkin, A. Ya., V. G. Kulichikhin, M. P. Zabugina, and G. V. Vinogradov, *Vysokomol. Soedin.*, A12, 120 (1970).]

---

* Oda, K., J. L. White, and E. S. Clark, *Polymer Eng. Sci.*, 18, 25 (1978).

molecular masses the decisive role is played by the length of the chain between neighbouring junctions, which does not depend on the molecular mass
of the entire chain.

What has just been said is illustrated by the experimentally determined
dependences of rubbery deformations on the molecular mass at different shear
rates and stresses (Fig. 5.14). Evidently, there exists a distinctly pronounced
tendency towards the decrease of rubbery deformations down to zero when
the molecular mass approaches the critical value (marked by arrows on the
abscissas in Fig. 5.14). Here the critical molecular mass M_c has been deter-

Fig. 5.15. Dependence of rubbery deformations developing during flow at
a shear rate $\dot{\gamma} = 1$ sec^{-1} at 50°C on the
relative length of the molecular chain
for polybutadienes containing:
1—60; *2*—80; and *3*—88% units in the *cis*-
position (see the reference to Fig. 5.14).

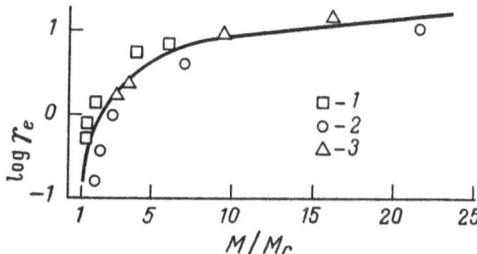

mined from the point at which the character of the dependence of viscosity
on molecular mass is changed, as discussed in Chapter 2.

The form of the functions $\gamma_e (M)$ shown in Fig. 5.14 remains valid for a
series of polymers with different values of the critical molecular mass. This
is shown in Fig. 5.15. The ratio (M/M_c) is chosen as the argument in the
dependence for rubbery deformations for three series of polybutadienes with
different microtacticity and different values of the critical molecular mass.
The resulting master curve of γ_e versus (M/M_c) with the sharply prominent
tendency of $\gamma_e \rightarrow 0$ at $(M/M_c) \rightarrow 1$ shows that the attainment of the critical
molecular mass is a necessary and sufficient condition which provides the
possibility of developing rubbery deformations in the shearing flow of polymers.

5.5.2. Effect of Molecular-Mass Distribution on Rubber-Like Behaviour

Since the very first works devoted to the study of the rubber-like properties
of polymer melts and solutions—all authors have emphasized the important
role played by the molecular-mass distribution of a polymer on its rubber
elasticity. In the early works it was asserted that the characteristic of the
distribution that determines the elasticity of the polymer is the composite
quantity $(\overline{M}_w/\overline{M}_z\overline{M}_{z+1})$, where $\overline{M}_w$, $\overline{M}_z$, and $\overline{M}_{z+1}$ are, respectively, the
mass-average molecular mass and the higher averages of the molecular-mass
distribution. This places a special emphasis on the role of high-molecular-mass
fractions which have a decisive effect on the rubbery modulus of the system.
If we apply the relation

$$G_1 \propto (\overline{M}_w/\overline{M}_z\overline{M}_{z+1})$$

to a monodisperse polymer, in which case $\overline{M}_w = \overline{M}_z = \overline{M}_{z+1}$, the modulus should be expected to be inversely proportional to the molecular mass. It has already been pointed out above that this theoretical prediction contradicts the known experimental facts, according to which $G_0 = $ const. It may therefore be asserted that the composite quantity $(\overline{M}_w/\overline{M}_z\overline{M}_{z+1})$ is not a parameter that determines the rubber-like properties of polymer systems.

The effect of the molecular-mass distribution on the rubber elasticity of a polymer does exist and manifests itself very strongly. This is illustrated in Fig. 5.16 by experimental data obtained when measuring the modulus G_0 of blends of monodisperse polyisoprenes and a series of polydimethyl siloxanes with different values of the mass-average molecular mass.

The modulus of polymer blends is lower than the moduli of the polymers of which they are composed, in spite of the fact that all the averages of molecular mass, $\overline{M}_w$, $\overline{M}_z$, and $\overline{M}_{z+1}$ and, hence, the values of the composite quantity $(\overline{M}_w/\overline{M}_z\overline{M}_{z+1})$ lie between the molecular masses of the original

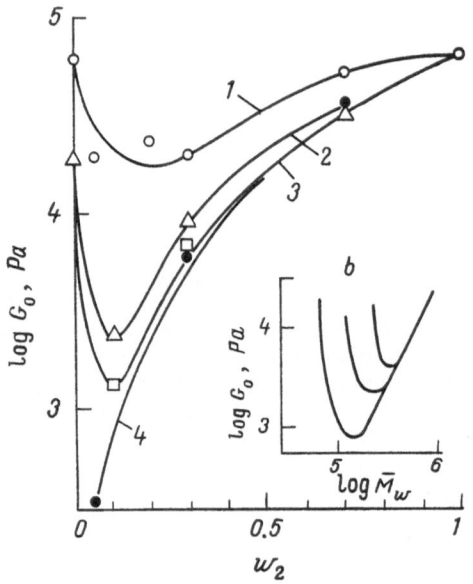

Fig. 5.16. Dependence of the initial rubbery modulus on the composition of a binary mixture for various polymers:

a—polyisoprenes (Zhangereeva, G. Zh., M. P. Zabugina, and A. Ya. Malkin, in: *Rheology of Polymers and Disperse Systems and Rheophysics.* Edited by G. V. Vinogradov and Z. P. Shulman, Part I, Minsk, Institute of Heat and Mass Transfer, of the BSSR Academy of Sciences, 1975, in Russian; b—polydimethyl siloxanes. [Prest, W. M., Jr., *Am. Chem. Soc. Polymer Preprint,* 10, 137 (1969); J. Polymer Sci., A-2, 8, 1897 (1970).] For polyisoprenes, there is shown the dependence of G_0 on the content in the mixture of the high-molecular component, w_2, for pairs with the following molecular masses: *1*—1.6 × 10⁵/5.75 × 10⁵; *2*—6.5 × 10⁴/5.75 × 10⁵; *3*—6.5 × 10⁴/8.3 × 10⁵; *4*—7.3 × 10⁵/5.75 × 10⁵. For polydimethyl siloxanes, the argument is the mass-average molecular mass of the mixture.

components of the blend. It is characteristic that addition of a low-molecular-mass polymer to a high-molecular-mass polymer sharply increases the rubber elasticity (the modulus decreases), i.e., the role of the factor contributing to the increase of the compliance of the system is played not only by the high-molecular-mass but also by the low-molecular-mass fractions or, in other words, the elasticity of the polymer depends on the width of the molecular-mass distribution and not on some "averages" of molecular mass.

The character of the dependence of the modulus of polydisperse polymers on the composition of the system, as shown in Fig. 5.16, is common to various polymers and their solutions. Therefore, by controlling the molecular-mass distribution one can considerably vary the rubber-like properties of polymers by changing G_0 by tens of times.

Since the rubber-like properties of a polymer are determined by the width of the molecular-mass distribution, a quantitative comparison of values of the modulus with the characteristics of the distribution requires the use of the ratio of the various moments, so that in going to monodisperse polymers

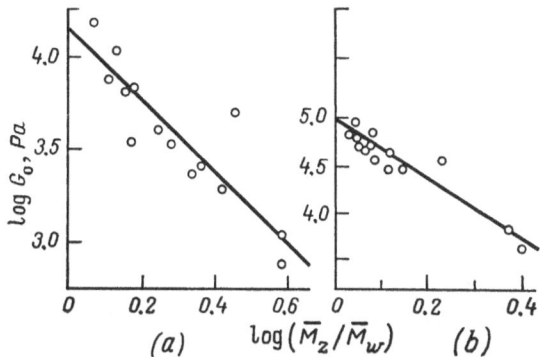

Fig. 5.17. Dependences of the rubbery modulus of polydimethyl siloxanes (a) and polystyrene melts (b) on the ratio of the average molecular masses, $\bar{M}_z/\bar{M}_w$. The experimental data have been taken from various literature sources.

the effect of the molecular mass is eliminated. It has been suggested* that the ratio $(\bar{M}_z/\bar{M}_w)$ be used as such a characteristic when considering the dependence of the modulus G_0 on this parameter. Figure 5.17 presents the results of treatment of published experimental data on polystyrenes and polydimethyl siloxanes [in the coordinates $\log G_0$ vs. $\log (\bar{M}_z/\bar{M}_w)$].

As seen from Fig. 5.17, in both examples given the dependence of the modulus on the polydispersity of polymers may be expressed by the following empirical equation:

$$G_0 = G_{0,m} (\bar{M}_z/\bar{M}_w)^\beta \qquad (5.12)$$

where $G_{0,m}$ is the rubbery modulus of monodisperse polymers, and β is an exponent characterizing the effect of polydispersity within a given polymer-homologous series on the rubber elasticity of the members of the series.

According to available experimental data, for polystyrenes $\beta = 3$ and $G_{0,m} = 9 \times 10^4$ Pa (though other values of $G_{0,m}$ have also been reported, for example, it was pointed out on page 370 that for anionic polystyrenes $G_{0,m} = 1.4 \times 10^5$ Pa).

Formula (5.12) is, however, of a special case. It is inapplicable, for example, to blends of polymers whose components have different $G_{0,m}$ values. Of more general importance** is the empirical formula for the compliance of

* Mieras, H. J. M. A. and C. F. N. van Rijn, *Nature*, 218, 865 (1968); Mills, N. J., *Nature*, 219, 1249 (1968); *Europ. Polymer J.*, 5, 675 (1969).
** Prest, W. M., Jr., *J. Polymer Sci.*, A-2, 8, 1897 (1970); Malkin, A. Ya., G. Zh. Zhangereeva, and M. P. Zabugina, *Vysokomol. Soedin.*, A18, 572 (1976).

the blend, J_∞, which is written as follows:

$$J_\infty = \overline{M}_w^{-n} \sum_i w_i J_i M_i^n$$

where M_i are the molecul masses of the components blended; J_i are their compliances; w_i are the mass fractions of the components in the blend; $\overline{M}_w$ is the mass-average molecular mass of the polydisperse polymer; n is an arbitrarily chosen quantity.

The value of the factor n may be varied. For most of the cases investigated it has been found to be equal to $(\alpha - 1)$, where α is an exponent in the dependence: $\eta_0 \propto M^\alpha$.

An important experimental fact is the existence of such a composition for which the modulus is minimal (see Fig. 5.15). As the molecular mass of the low-molecular component in the blend decreases the concentration of the high-molecular polymer, corresponding to the minimum of G_0, decreases and the minimum value of G_0 falls off.

5.6. Free Elastic Recovery

5.6.1. Qualitative Observations

In the preceding sections we have dealt with the constrained recovery of a deformed polymer after the cessation of steady-state flow taking place in a direction opposite to the direction of flow. These observations have been made under the conditions of simple shear with the field of stresses being homogeneous.

The free recovery of the shape after cessation of flow is effected in directions that may not coincide with the direction of the forced preliminary deformation. Therefore, the transition from the undeformed state to the state in which the sample is at the onset of elastic recovery and the reverse transition to the undeformed state can be accomplished in different ways.

The free elastic recovery of shape of the sample subjected (prior to the onset of recovery) to the conditions of simple shearing flow has not practically been studied quantitatively. The elastic recovery of the extrudate emerging from the capillary has been investigated rather extensively. This special case differs from the cases considered above in that the sample in the capillary is not subjected to the action of homogeneous stresses and, hence, different elastic strains are accumulated in different circular cross-sectional areas of the sample. The observed effect of elastic recovery is therefore integral, being associated with the elastic recovery of the material which is different along the radius of the extrudate. Nonetheless, examination of this case is of interest since some general regularities of this phenomenon may be deduced by studying the free elastic recovery of the extrudate. It must, however, be kept in mind that the results of measurements of this kind may be distorted for low-viscosity polymer solutions because of the rapid stress relaxation, and for polymer melts because of the vitrification or crystallization of the polymer unless special measures are taken for the annealing of samples.

The effect under discussion consists in that the extrudate leaving the capillary of circular cross-section (or, generally speaking, of any other cross-section) undergoes dimensional changes as compared with the dimensions of the outlet cross-section of the capillary, so that the diameter of the extrudate D becomes, as a rule, larger than that of the capillary, d. This phenomenon is known as the Barus effect. A quantitative characteristic of this effect is the ratio $\alpha = D/d$ called the swelling ratio (or the elastic recovery ratio).

An example of the dependence of the swelling ratio on the shear stress τ on the capillary wall* is shown in Fig. 5.18 for a number of polymers. This

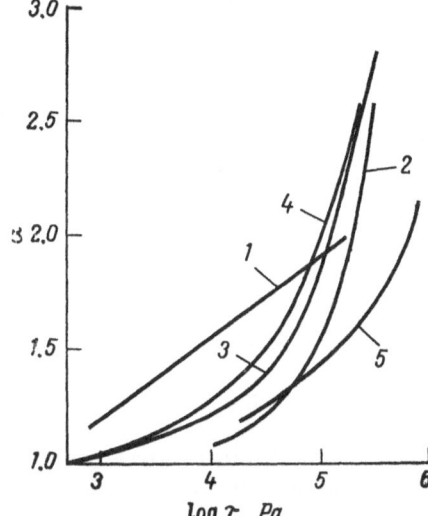

Fig. 5.18. Dependence of the swelling ratio on the shear stress at the capillary wall for various polymers:

1—high-pressure polyethylene; 2—polypropylene; 3—polyisobutylene; 4—polystyrene; 5—styrene-butadiene rubber (according to the measurements carried out by N. V. Prozorovskaya).

variation of α as a function of the stress is typical and the monotonic increase of α occurs prior to the onset of unsteady-state flow conditions. At higher stresses the measurements of the swelling ratio become ambiguous and unreliable.

The effect of extrudate swell is observed with any rubber-like polymers and solutions. Of particular interest are the dependences of the swelling ratio on the shear stress for such polymer solutions for which it is possible to measure the complete flow curve since it becomes possible to evaluate the complete dependence of the swelling ratio α on the shear stress. The data presented in Fig. 5.19 for solutions of polystyrene in diethyl phthalate show that the dependences of $\log \dot{\gamma}$ and α on τ are similar in many respects. At low stresses corresponding to the region of Newtonian flow the swelling ratio is small. The increase of the swelling ratio begins at somewhat lower stresses than those at which there are observed substantial departures from the Newtonian behaviour. This is associated with the fact that, as has been pointed out in Sec. 5.3, appreciable rubbery deformations may develop even in the region of Newtonian flow. As the non-Newtonian viscosity increases and the apparent viscosity decreases the swelling ratio α increases and reaches a maximal value equal to about 4 in the region of the lowest Newtonian viscosity.

---

* The shear stresses in the capillary with the developed velocity profile and steady-state flow conditions vary linearly along the radius, so that the shear stress is equal to zero along the capillary axis and is maximal at the capillary wall. Therefore, for the stress state of the material in the capillary to be characterized, it will suffice to know the shear stress at the capillary wall. But upon approach to the outlet section of the capillary and also at the duct edge the stress profile may undergo change (see Section 4.7.6).

The dependence of the swelling ratio α on the shear stress is invariant to the dimensions of the capillary (for long capillaries) and temperature. This is a consequence of the weak effect of temperature on the rubbery deformations

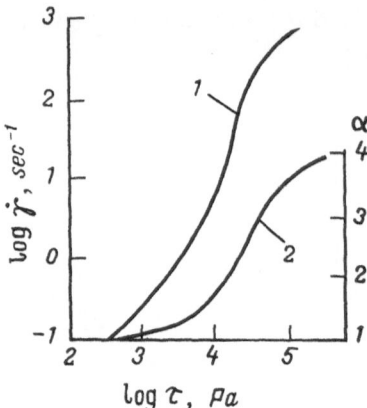

Fig. 5.19. Comparing the flow curve (*1*) and the dependence of the swelling ratio (*2*) on the shear stress for a 40-percent solution of polystyrene in diethyl phthalate. [Shishido, S. and I. Ito, *Nippon kagaku Zasshi*, 85, 13 (1964).]

which are compared at constant values of the shear stress. However, if measurements of the swelling ratio are carried out on short capillaries, there is observed a monotonic decrease of α with increasing capillary length. This effect corresponds to the change of the rubbery deformation stored in the

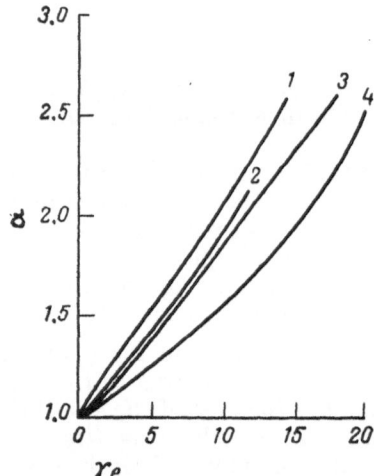

Fig. 5.20. Relation between the swelling ratio and rubbery deformations at equal values of shear stress at the capillary wall for various polymers: *1*—polystyrene; *2*—polypropylene; *3*—linear polyethylene; *4*—high-pressure polyethylene. [Kishi, N., and H. Iizuka, *J. Polymer Sci.*, 2, 399 (1964).]

polymer at the pre-stationary stage of straining, depending on the total deformation in simple shear. In what follows we shall be only concerned with the "steady-state" values of the swelling ratio measured on long capillaries when α becomes independent of the duration of straining of the polymer in the capillary. When the round capillary is replaced by a slit capillary, the basic regularities of the manifestation of the Barus effect remain unchanged, in particular to the same shear stresses on the capillary wall there correspond the same swelling ratios.

The experimental data given in Fig. 5.20 show, first, the existence of a relation between α and γ_e and, second, the ambiguity of this relation, depending on the nature of the polymer. Therefore, though there is no doubt that the change of the size of the extrudate after leaving the capillary is an effect caused by the elasticity of the polymer, the constrained recovery and the free recovery occur differently and the relation between the parameters that characterize both these effects is influenced by the nature of the polymer.

The undoubtedly existing relation between the effect of extrudate swell and the elasticity of the melt allows one to seek a correlation between the swelling ratio α and the molecular parameters of the polymer, which affect the elacticity of its melt. For example, it has been repeatedly noted that the swelling ratio α is very small for monodisperse polymers, which is associated, in an obvious way, with the low values of the compliance characteristic of them. As the molecular-mass distribution is broadened the swelling ratio α increases. This enables one to suggest a method of estimating the molecular-mass distribution on the basis of the results of the measurement of the swelling ratio. However, though the existence of such a correlation is obvious from the qualitative standpoint, its quantitative form remains to be indefinite.*

5.6.2. Concerning the Theory of the Free Elastic Recovery of the Extrudate

The essence of the elastic recovery of the extrudate after it has emerged from the capillary consists in that the elastic energy stored in the material from the onset of deformation to the beginning of steady-state flow is released. If the velocity distribution along the capillary radius is $V(r)$ and the dependence of the shear stress on the rubbery deformation is $\tau(\gamma_e)$, then the elastic energy w stored upon deformation under steady-state flow conditions is given by

$$w = \int_0^R 2\pi r V(r) \left[\int_0^r \tau \, d\gamma_e \right] dr \tag{5.13}$$

where R is the capillary radius.

Concrete computations of w require a knowledge of the properties of the material, which determine the form of the functions $V(r)$ and $\tau(\gamma_e)$. To illustrate the results obtained in such computations, let us consider the simplest case of a liquid, whose flow properties are described by Newton's law and whose elastic properties are defined by Hooke's law with the modulus G. Then

$$V(r) = \frac{2Q}{\pi R^2} \left[1 - \left(\frac{r}{R} \right)^2 \right]$$

where Q is the volume rate of flow in unit time.

Since the distribution of stresses along the capillary radius is linear, with the maximum value at the capillary wall being equal to τ_m and the corresponding recoverable deformation corresponding to this stress, $\gamma_m = \tau_m/G$, the inner integral in formula (5.13) is calculated in the following manner:

$$\int_0^r \tau \, d\gamma_e = \frac{\tau^2}{2G} = \frac{\tau_m}{2G} \left(\frac{r}{R} \right)^2 = \frac{1}{2} G \gamma_m^2 \left(\frac{r}{R} \right)^2$$

---

* Mendelson, R. A. and F. L. Finger, *J. Appl. Polymer Sci.*, **19**, 1061 (1975).

Substitution of this expression into formula (5.13) gives

$$w = \frac{1}{6} QG\gamma_m^2 = \frac{Q\tau_m^2}{6G} = \frac{Q}{6} \gamma_m \tau_m$$

or the elastic energy stored in unit volume of the material during flow through the capillary is found to be equal to

$$W = \frac{w}{Q} = \frac{\tau_m^2}{6G} = \frac{G\gamma_m^2}{6} = \frac{1}{6} \gamma_m \tau_m \qquad (5.14)$$

This formula shows the general structure of the dependence of W on the parameters that determine the properties of the material. For materials whose viscoelastic properties are more complicated than those of the above-considered liquid with constant viscosity and constant elastic shear modulus, the expression for W will have, of course, a different form. Thus, if it is assumed that the dependences $\gamma_e(\dot{\gamma})$ and $\tau(\dot{\gamma})$ are described by power laws, it appears* that the formula for W will have a form similar to (5.14), namely:

$$W = k\gamma_m \tau_m \qquad (5.15)$$

where the coefficient k is expressed in terms of the exponents in the dependences $\gamma_e(\dot{\gamma})$ and $\tau(\dot{\gamma})$. The value of this coefficient reduces to 1/6 for the models of a Newtonian liquid and a Hookean elastic body.

For the case of flow in a flat duct the structure of the dependence of W on γ_m and τ_m remains the same as before, and only the numerical value of the coefficient k is somewhat changed.

Thus, it can be shown that for various rheological models of viscoelastic liquids the quantity W must be proportional to the product $\gamma_m \tau_m$.

After the extrudate has emerged from the capillary it behaves like a rubber-like body, being changed in the diameter by α times and in height by α^2 times, accordingly. This process may be regarded as a compression (or stretching) of a rubbery cylinder without change of its volume with the principal extensions

$$\lambda_1 = \alpha^2; \quad \lambda_2 = \lambda_3 = \alpha^{-1}$$

Hence, for a rubber-like body [see formula (1.53)], the following expression obtains for the stored energy function:

$$W = \frac{G}{2} (\alpha^4 + 2\alpha^{-2} - 3)$$

Now this expression for W should be equated to the W value from Eq. (5.14) since it is the elastic energy stored during the flow of the material through the capillary that is consumed for the rubbery recovery of the extrudate.

Therefore, for the model leading to formula (5.14) γ_m can be expressed as:

$$\gamma_m = \sqrt{3(\alpha^4 + 2\alpha^{-2} - 3)} \qquad (5.16)$$

Formula (5.16) gives the form of the dependence of the elastic recovery ratio (swelling ratio) on the rubbery deformation corresponding to the shear stress τ_m at the capillary wall.

The result obtained should be regarded, of course, as a qualitative estimate illustrating the method of calculating the swelling ratio of the free extrudate for a viscoelastic material with the simplest properties.

The establishment of the relation between the rubbery deformations measured during constrained recovery and free expansion of the extrudate is pos-

---

* Malkin, A. Ya., V. V. Goncharenko, and V. V. Malinovsky, *Mekhanika Polimerov*, **3**, 487 (1976).

sible in the case under consideration because the direction of elastic recovery is known in advance. In more complicated cases the possibility of computation of the swelling ratio α for free rubbery deformations of the material remains to be problematic.

In a more rigorous treatment of the problem of elastic recovery of the extrudate emerging from the capillary one should also take into account the rearrangement of the velocity profile, which takes place from the moment the extrudate leaves the capillary to the attainment of the steady state. From the requirement of the law of momentum conservation it follows that the extrudate emerging from the capillary must contract. This effect is well known for low-molecular-mass liquids which do not exhibit rubber-like properties; in this case, the change of the diameter is 0.87. For polymers, the extrudate swell produced must be somewhat lower than that calculated from the value of the stored elastic energy, due to the rearrangement of the velocity profile after the extrudate leaves the capillary.

Attempts have been repeatedly made in the literature to relate the ratio α to the normal stresses that develop during the shearing flow in the capillary.* This problem is associated with the consideration of the hydrodynamic problem of the flow of an elastic liquid near the exit before it emerges from the capillary and its behaviour under free (unconstrained) conditions. Such a consideration requires the use of the constitutive equation of the material (which makes the results dependent on this choice) and is based on the analysis of the balance of the forces acting on the free stream. Of course, this problem is outside the scope of polymer rheology and is the concern of hydrodynamics.

Because of the complexity of an exact solution of the problem of relating the quantities α and γ_e and the attractiveness of a simple method of measuring α, many investigators have tried to find semi-theoretical simplified relations that would permit calculation of γ_e from the values of α measured at different shear rates and stresses in the capillary. A large variety of formulas have been obtained. An example is Eq. (5.16). A comparison of the various formulas reported in the literature has, however, shown** that, in spite of the difference in their form, the predicted character of the dependence $\gamma_e(\alpha)$ was found to be similar in many cases. Therefore, for practical purposes, use may be made of a very simple relationship:

$$\gamma_e = m\alpha^2$$

where the coefficient m is dependent, to a certain extent, on the nature of a particular polymer (as is clearly seen, for example, from Fig. 5.20 which shows the absence of a "universal" correlation between γ_e and α), but, on an average, the coefficient m is close to 1.4-1.5.

---

* White, J. L., *Trans. Soc. Rheol.*, 19, 271 (1975); Nakamura, R. and N. Yoshioka, *J. Chem. Eng. Japan*, 9, 287 (1976); Boger, D. V. and R. R. Huigol, *Trans. Soc. Rheol.* 21, 447 (1977); Vlachopoulos, J., *Rubber Chem. and Techn.*, 51, 133 (1978).
** Malkin, A. Ya., V. V. Goncharenko, and V. V. Malinovsky, *Mekhanika Polimerov*, 3, 487 (1976); Vinogradov, G. V., A. I. Isayev, V. I. Brizitsky, Yu. Ya. Podolsky, A. Ya. Malkin, and M. P. Zabugina, *Mekhanika Polimerov*, 1, 116 (1977).

CHAPTER 6

Rheological Properties of Filled Polymers

6.1. Introduction

Filled polymers are multi-component and multi-phase systems. Such systems display a rather complex set of properties owing to the peculiarities of their structure. This makes it difficult to draw any general conclusions as to their behaviour. Nevertheless, there exist simple model situations, for which the effect of the filler concentration on the properties of the composition can be disclosed quite distinctly and unambiguously. On this basis one can pass over to the analysis of the behaviour of more complex multi-component polymeric systems.

The class of filled (loaded) polymers may formally include various multi-component systems containing a polymer; such are polymer foams (a polymer with a gaseous filler), filled polymers proper (with a solid filler of various geometric form, from spherical to fibrous), and polymer blends (i.e., systems in which one polymer is a matrix and the other is a filler dispersed in it). For all the compositions indicated there exists the general problem of the effect of the composition on the viscosity, elasticity, and relaxation properties of the material, i.e., on the same characteristics of the behaviour which are studied in the case of homogeneous polymers. Just as in the preceding chapters, the discussion of the rheological properties of compositions will be restricted mainly to the region in which flow is possible.

6.2. Polymers with a Solid Filler

There exist a large variety of compositions, in which the polymer forms a continuous matrix (plays the role of a binder), and the solid filler forms a dispersed phase. As the content of the filler increases there become possible, and then even inevitable, contacts between filler particles, and when a certain concentration is attained, the filler also forms a continuous phase (or a "network").

The simplest model of the filled composition is a diluted suspension of rigid spherical particles. Further complications arise in the following directions: as the concentration of particles increases it becomes necessary to take into account the hydrodynamic effects of the interaction of dispersed particles. Then, one should consider the role of departures from the spherical shape of particles, these departures extending to deformed filler particles, which can exhibit flexibility and change their form in the stream; this has an effect on the behaviour of the dispersion as a whole. Finally, one should take into account the interaction between filler particles, as a result of which

they form a continuous three-dimensional network, which is partly destroyed
upon deformation, and also the filler-polymer interaction. All the phenomena
mentioned may develop in different ways, depending on the properties of
the matrix, i.e., whether the dispersion medium is a viscous fluid or a viscoelastic liquid.

6.2.1. Dilute Suspensions*

Such systems present the simplest model that makes it possible to reveal
the role of the introduction of a solid filler into the polymeric matrix. A
classical solution of the problem of the viscosity of an infinitely dilute suspension was given by A. Einstein (1906; 1911). The statistical approach developed by Einstein and the results obtained remain to be the cornerstone of
any theoretical treatment of analogous problems. The formula derived by
Einstein for the specific viscosity η_{sp} has the form:

$$\eta_{sp} = 2.5\varphi \tag{6.1}$$

where $\eta_{sp} = (\eta - \eta_s)/\eta_s$; η is the viscosity of the suspension; η_s is the viscosity of the dispersion medium (solvent); φ is the volume content of the
solid phase.

From Eq. (6.1) it follows that for a suspension of spherical particles the
inherent viscosity $\eta_{inh} = (\eta - \eta_s)/\eta_s\varphi = 2.5 = \text{const}$, and the intrinsic viscosity $[\eta] = \lim_{\varphi \to 0} \eta_{inh}$ always must be equal to 2.5.

It would not be superfluous to stress the basic conditions under which formula (6.1) should be expected to be fulfilled. First, this is the complete absence of any hydrodynamic interaction between the filler particles (this is
possible if $\varphi \ll 1$). Second, no physico-chemical interaction must occur between the filler particles and the dispersion medium. Third, the shape of the
particles must be strictly spherical and they must not undergo deformations
during flow. Fourth, the flow must remain laminar. It is clear that there is
no point in discussing the various experimental facts in terms of the Einstein theory [i.e., formula (6.1)] if the basic conditions under which it has
been obtained are not fulfilled, though it is sometimes done.

Formula (6.1) does not include the particle size, i.e., it is assumed that it
must be fulfilled for any suspensions of mono- or polydisperse particles of any
size provided they are spherical in shape. Formula (6.1) does dot apply to a
suspension of non-spherical particles.

The simplest example of non-spherical particles is ellipsoids. A theoretical analysis of the behaviour of a suspension of ellipsoidal particles was
first undertaken by Jeffery (1922), and the results obtained later were generalized by Goldsmith and Mason** and subsequently by Brenner.*** The principal and new (as compared with a suspension of spherical particles) circumstance in the model of a suspension of ellipsoidal particles is the possibility
of appearance of the predominant orientation of the axes in the flow. As a
result of this, there| arise| new mechanical phenomena. The phenomena that

---

* See also: Pokrovsky, V. N., *Statistical Mechanics of Dilute Suspensions*, Nauka,
Moscow, 1978 (in Russian).
** Goldsmith, H. L. and S. G. Mason, in: *Rheology*, edited by F. Eirich, Academic
Press, New York, vol. 4, 1967.
*** Brenner, H., in: *Progr. in Heat and Mass Transfer*, edited by W. R. Schowalter,
Pergamon Press, New York, vol. 5, 79, 1972; *Intern. J. Multiphase Flow*, 1, 195 (1974).

are of primary importance to polymer rheology are the following: (1) the appearance of the dependence of apparent viscosity on shear rate; (2) the development of normal stresses in shearing flow; (3) the dependence of the viscosity (as a measure of the resistance to flow) on the flow geometry, i.e., on whether the flow takes place, say, under shear or extension.

The general mechanism of all these phenomena consists in the rotation of flowing particles in the stream, which is determined by the superposition of two factors: the orientation under the action of mechanical forces and the disorientation caused by Brownian motion. It is evident that the latter factor plays an essential part only if the particle size is sufficiently small. A quantitative measure of the ratio of the factors indicated, which governs the rheological properties of the suspension, is the dimensionless ratio $\dot{\gamma}/D_r$, where D_r is the rotational diffusion coefficient. This quantity has the dimensions of the inverse time and therefore the inverse quantity may be regarded as the characteristic relaxation time. The corresponding relaxation process is, not, however, associated with the viscoelasticity of the suspension, though it is of molecular-kinetic nature. The quantity D_r is related to the rotational friction coefficient W which is a measure of the energy dissipated during the motion of particles in the liquid. This relationship is expressed by the classical Einstein-Debye equation:

$$D_r = kT/W$$

where k is Boltsmann's constant and T is the absolute temperature. The quantity W, and, hence, D_r, depends on the shape of the particles.

To stress the fact that infinitely dilute suspensions are meant here, the intrinsic viscosity $[\eta]$ is used as a measure of their properties. The intrinsic viscosity is defined in an ordinary way as the limit of the ratio (η_{sp}/φ) at $\varphi \to 0$. For a suspension of rigid spherical particles $[\eta] = 2.5$, but in a general case, for a suspension of particles having an arbitrary geometric form $[\eta]$ depends on their shape and size.

For particles with ellipsoidal symmetry $[\eta]$ is determined by the ratio of the lengths of the axes of the ellipsoid, (L/l). The dependence of $[\eta]$ on (L/l) has been calculated by many authors and can be given in the form of a graph shown in Fig. 6.1, the left part of which refers to oblate ellipsoids and the right part to prolate ellipsoids. The central point at which $L/l = 1$ corresponds to the value of $[\eta]$ equal to 2.5. These results refer, however, to the region of low shear rates since at high rates of shear $[\eta]$ becomes dependent on $\dot{\gamma}/D_r$.

For the present discussion, not so much the particular form of the dependence of viscosity on the ratio $\dot{\gamma}/D_r$ for various types of the stress state is important as the very fact of non-Newtonian behaviour and the existence of the dependence of apparent viscosity on the geometry of the flow. To this category of phenomena also belongs the appearance of normal stresses in the shearing flow of a suspension of ellipsoids*, which is predicted as a second-order effect with respect to the rate of shear, which is generally typical for normal stresses (see Chapter 4). In simple shear the normal stress coefficient, like the viscosity, depends on the factor $(\dot{\gamma}/D_r)$. Here it is essential that the development of normal stresses, like non-Newtonian flow, is not associated with the viscoelasticity of the dispersion medium or suspended particles and is

---

* Giesekus, H., *Rheol. Acta*, **2**, 50 (1962).

found to be a consequence of purely geometrical effects. The factor that determines the appearance of normal stresses is the dimensionless criterion $(\dot{\gamma}/D_r)$ associated with the rotational diffusion of the particles. If $D_r \to \infty$, which corresponds to the statistically random distribution of the orientations of the ellipsoid axes in space, no normal stresses will develop.

The effect of appearance of normal stresses in dispersions of rigid ellipsoids may be looked upon as a consequence of a certain relaxation process of transition from the predominant orientation of particles to the statistically random distribution of the orientations of the axes, which occurs with a characteristic relaxation time D_r^{-1}.

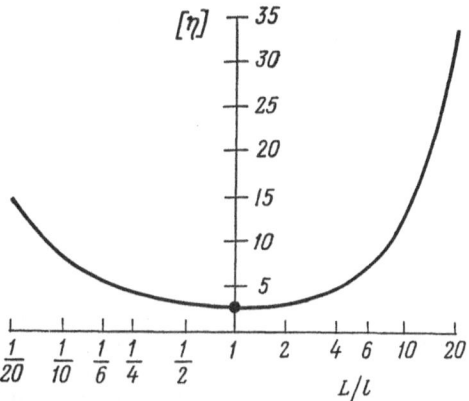

Fig. 6.1. Dependence of the intrinsic viscosity on the ratio of the lengths of the ellipsoid axes (after W. Kuhn and H. Kuhn).

Nevertheless, this mechanism is different from the mechanism of development of normal stresses in the shearing flow of polymer melts and solutions, the latter being associated, first of all, with the change of the conformations of macromolecular chains.

Thus, identical mechanical phenomena (non-Newtonian behaviour, normal stresses in shearing flow, the difference between the shear and extensional viscosity coefficients) can be caused by various factors and may be connected with different molecular mechanisms. Therefore, from the rheological effects observed one cannot unambiguously judge about the factors that cause them.

Most theoretical calculations are based on an analysis of the behaviour of particles of idealized shape in flow, in particular ellipsoids. It is important, however, that a sufficiently simple relationship can be established between the basic geometrical characteristic of ellipsoids—the ratio of the lengths of the semiaxis, $r_e = L/l$, and the basic geometrical characteristic of cylindrical particles—the ratio of their length to their radius, r_p, the fulfillment of which leads to equivalent hydrodynamic consequences:[*]

$$r_e = 1.14 r_p^{0.844}$$

Then, for the viscosity of the suspension of cylindrical particles (say, short fibres) to be determined, use may be made of the results of calculations sum-

---

[*] Harris, J. B. and J. F. T. Pittman, *J. Colloid and Interface Sci.*, 50, 280 (1975).

marized in Fig. 6.1 if from the known value of r_p for cylindrical particles we find the equivalent value of r_e and make use of Fig. 6.1.

The dependences of the basic characteristics of the rheological properties of a suspension on the characteristic sizes of dispersed particles (which determine the value of D_r) and on the dimensionless factor $\dot{\gamma}/D_r$ have been calculated by many authors. Such calculations have been, however, tied up not with the analysis of the behaviour of filled polymers but with the modelling of the rheological behaviour of polymer solutions. In this respect, the approach in question consists in constructing kinetic models and predicting the relaxation properties of polymers. An analysis of the physical basis for such calculations and of the inferences that follow from them has been given in Chapter 3. Nonetheless, though the development of this approach refers to the theory of polymer solutions, the conclusions based on the simplest model considerations for suspensions of non-interacting particles of various shapes are of fundamental importance to the understanding of the behaviour of filled polymeric systems.

6.2.2. Concentrated Suspensions of Non-Interacting Particles

When the content of suspended particles in a composition is increased, there appear new hydrodynamic effects even if there is no physico-chemical interaction between the filler and the dispersion medium. These effects are caused by the fact that the conditions of streamlining of each particle begin to be dependent on the presence of other neighbouring particles. This means that in concentrated suspensions the particles are involved in a hydrodynamic interaction which affects their rheological properties. The corresponding problem of calculating the viscosity (as a measure of the rate of dissipation of the work of viscous friction) has been repeatedly studied by many authors. The results obtained are varied, but on the whole they are approximations of higher orders (in powers of φ) with respect to formula (6.1). Therefore, in the most general case, the dependence η (φ) may be given as a power series:

$$\eta_{sp} = \sum_{n=1} A_n \varphi^n \qquad (6.2)$$

where $A_1 = 2.5$, and the values of the coefficients A_n (for $n > 1$) are given by a corresponding hydrodynamic theory. The values of A_n obtained by different authors are somewhat different. For instance, the values cited for A_2 lie in the range from 4.4 to 14.1. It should be remembered, however, that an essential part in the estimation of A_2 is played by the choice of the number of terms in the sum (6.2) and, in particular, the coefficient before the cubic term.

Formula (6.2) establishes that η_{sp} must be a unique function of the degree of loading φ, irrespective of the average sizes of particles and the particle size distribution. Numerous model experiments have shown that this is not always the case. Therefore, in a general case, formula (6.2) should be regarded as a model formula. In such an approach, the values of the coefficients A_n are chosen empirically and they will depend on the number of terms retained in the sum. One can choose such a number of terms that will be sufficient for describing the experimental dependence η (φ) with any desired degree of accuracy. But with such an empirical approach the description of the concentration dependence of viscosity is devoid of any theoretical generality and no physical significance should be sought in the experimental values of the coefficients A_n obtained by treatment of data.

An enormous number of empirical formulas have been proposed in the literature for describing the dependence η (φ). A review and comparison of these formulas with various experimental data constitutes an independent task which has no general solution. Though the empirical approach is of some value and has its own fields of practical application, nevertheless numerous attempts to choose "convenient" formulas for the dependence η (φ) have no general physical significance for polymer rheology. Nonetheless, some of these formulas describe rather well experimental data on the concentration dependence of the viscosity of various filled systems.

A general method for improving the agreement between a mathematical approximation and experimental data is to take into account the idea of the maximum possible loading of a polymer. This is achieved by using the following quantity as the argument of the concentration dependence of viscosity:

$$\Phi = \frac{2.5\varphi}{1-(\varphi/\varphi^*)}$$

where φ^* is a constant.

At low concentrations ($\varphi \to 0$) this quantity coincides with the argument in the Einstein formula: $\Phi = 2.5\varphi$, but at $\varphi \to \varphi^*$ the "effective loading" $\Phi \to \infty$. Therefore, φ^* has the meaning of the concentration corresponding to the limiting high degree of loading of the volume when the polymer is no longer capable of flowing.

A typical example of formulas that make use of the argument Φ is the Eilers equation (1941)*:

$$\eta_r = \left(1+\frac{\Phi}{2}\right)^2 = \left[1+\frac{1.25\varphi}{1-(\varphi/\varphi^*)}\right]^2 \tag{6.3}$$

where η_r is the relative viscosity: $\eta_r = \eta/\eta_s$, and η_s is the viscosity of the dispersion medium.

At $\varphi \to 0$ the Eilers equation naturally turns into the Einstein formula (6.1).

The most widely used formula in this group of semi-empirical formulas is the Mooney equation:

$$\eta_r = \exp{(1+\Phi)} = \exp\left[1+\frac{2.5\varphi}{1-(\varphi/\varphi^*)}\right] \tag{6.4}$$

from which, just as from formula (6.3), it follows that at $\varphi \to 0$ the Einstein equation must be fulfilled and at $\varphi \to \varphi^*$ the viscosity of the dispersion must increase infinitely.

Attention should be paid to the analogy between the structure of formula (6.4) and the Tobolsky-Lyons equation for the concentration dependence of the viscosity of polymer solutions, which was discussed in Chapter 2. This analogy is not accidental and is associated with the model representation of a polymer solution as a suspension of dispersed macromolecular particles. Therefore, in the analysis of the rheological properties of polymer solutions

---

* The applicability of this formula to the description of the concentration dependence of the viscosity of various dispersions has been demonstrated, for example, in: Chong, T. S., *J. Appl. Polymer Sci.*, **15**, 2007 (1971); Fedors, R. F., *Polymer*, **16**, 305 (1975).

and filled polymers analogies are of general importance for all theoretical investigations in this field, beginning from dilute solutions (or suspensions), as was discussed in the previous section, up to highly concentrated (filled) systems.

For the simplest cases of the ideal packing φ^* can be found on the basis of geometrical considerations. For example, if the filler is composed of spherical particles then for the hexagonal or cubic-centred packing $\varphi^* = 0.74$; for the statistically close packing $\varphi^* = 0.637$. In the case of fibrous fillers with a parallel hexagonal packing $\varphi^* = 0.907$, and with a random packing $\varphi^* = 0.52$. Usually, φ^* is of the order of 0.5-0.7, which is close to theoretical estimates. For real filler particles the value of φ^* is difficult to calculate with required accuracy, though it is possible to choose such a value for φ^* in formula (6.5) that this formula will satisfactorily describe the observed concentration dependence of viscosity over a wide range of compositions.

The estimation of φ^* and the concentration dependence becomes highly complicated when one is dealing with a polymer containing a polydisperse filler. The use of particles of different sizes allows one to considerably increase the maximum permissible degree of loading. The role of the fractional composition of the filler and its total concentration has the strongest effect on the viscosity of the dispersion in the region of high concentrations close to φ^*. At the same value of φ^* the value of the ratio φ/φ^* depends on the fractional composition, and at high values of φ/φ^* even slight changes of this parameter have a very strong effect on the viscosity of the dispersion.

A possible approach to the calculation of the viscosity of compositions with a polydisperse filler is based on the method of additivity.* According to this method, the relative viscosity, η_r, of a system containing, say, two types of fractions can be found as the product of the relative viscosities η_r' and η_r'' of the compositions, each of these quantities corresponding to the presence of only one component in the dispersion medium, independently of the other component. Then

$$\eta_r = \eta_r' \cdot \eta_r''$$

where η_r' and η_r'' are calculated from one of the formulas discused above.

A rather simple (though devoid of any physical significance) method of describing the concentration dependence of the viscosity of compositions that contain a polydisperse filler is the use of correlation equations with empirically determined coefficients.**

The physical meaning of the dependence $\eta\,(\varphi)$ is quite obvious so far as one is speaking of simple shearing flow. But it is not evident whether it is possible to extend this formula to the case of extensional flow. Indeed, Brenner points out*** that the application of the dependence $\eta(\varphi)$ to extensional flow is incorrect and that the viscosity of concentrated suspensions depends on the flow geometry. Therefore, the concept of the viscosity of a suspension of spherical particles is no longer unambiguous for the region of not infinitely dilute systems and, hence, the constants A_n in formula (6.2) are not the intrinsic characteristics of the composition but depend on the type of flow. The situation here is basically quite the same as in the case of a suspension of anisodiametric particles.

---

     * Farris, R. T., *Trans. Soc. Rheol.*, **12**, 281 (1968).
    ** Hsieh, H. P., *Polymer Eng. Sci.*, **18**, 928 (1978).
   *** See the references to reviews by Brenner on page 381.

This does not cover all the new effects caused by the increase of the concentration of a suspension. The possibility of coupled interactions between filler particles in the presence of Brownian motion leads to the appearance of the dependence of the apparent viscosity in shearing flow on the rate of deformation. In order to describe this phenomenon quantitatively, one should construct a dimensionless criterion that would characterize the relation between the rate of shear and the rate of relaxation (which is again associated with geometrical phenomena but not with the deformation or orientation of filler particles).

This criterion P has the form*

$$P = \eta_s R^3 \dot{\gamma}/kT \tag{6.5}$$

where R is the radius of the filler particles; η_s is the viscosity of the dispersion medium; kT is the energy factor.

Evidently, the quantity $(\eta_s R^3 \dot{\gamma}/kT)$ has the dimensions of time and may be regarded as the characteristic time of a certain structural process of interac-

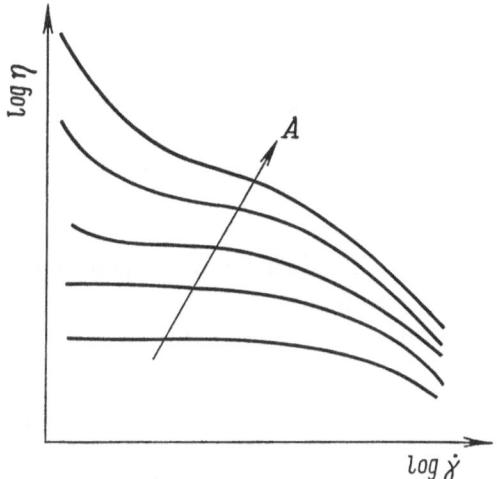

Fig. 6.2. Variation of the course of the dependence of viscosity on rate of shear with increase of the filler content. The arrow A corresponds to the increase of the filler concentration.

tion between suspended particles, which determines the development of non-Newtonian behaviour.

Thus, on flow of concentrated suspensions there must be observed the dependence of the apparent viscosity on the dimensionless parameter P even if the dispersion medium has the properties of a Newtonian liquid. Such an approach has been proved to be valid for polyvinyl chloride plastisols.**

In the range of sufficiently high concentrations the rigid filler particles suspended in the polymer form a structure (a structural framework) which possesses a certain strength. If there is no specific interaction between the particles, a pseudo-structure may be formed as a result of their mechanical contacts. Therefore, at high rates of shear such a composition behaves as a liquid, but below a certain stress limit the composition behaves as a solid-like body and a high-filled system and as a whole is a viscoplastic material with a characteristic yield value. A typical example of such behaviour is shown

---

* Krieger, T. M. and T. J. Dougherty, *Trans. Soc. Rheol.*, 3, 137 (1959).
** Willey, S. J. and C. W. Macosco, *J. Rheol.*, 22, 525 (1978).

schematically in Fig. 6.2 (according to data given in works* devoted to the investigation of model compositions which are dispersions of glass beads in polymer solutions or melts). It is seen that at low concentrations of the filler the flow curves of the compositions differ in no way from those of the dispersion medium—a polymer solution. An increase of φ only leads to the increase of the viscosity, which is stronger in the region of Newtonian flow than at high shear rates. As the content of the filler increases on the flow curve in the region of low $\dot{\gamma}$ values there appears an inflection to the side of high viscosity values, which is the stronger the higher is the concentration of the filler. This phenomenon is associated with the attainment of the yield limit, i.e., with the appearance of the plastic behaviour of the high-filled composition.

The effect of the unlimited increase of viscosity at low shear stresses is associated with the formation of a filler structure having a certain strength and its own (as distinct from the polymer) relaxation properties. The strength of such a structure is not high because in the case under consideration it is only produced by the mechanical contacts between the particles. As regards the specific relaxation properties of the system, which are associated with the presence of a filler, they manifest themselves most distinctly if a master flow curve is constructed for filled compositions. This is achieved by shifting the flow curves along the log η and log $\dot{\gamma}$ axes until they are superposed, as described in Chapter 2 (see Sec. 2.10). It has been found* that in the region of high shear rates, where the filler structure is broken down, the polymer behaves like any solution or melt and the experimental points lie on a single curve of η_r versus $\dot{\gamma}_r$, where η_r is the reduced viscosity and $\dot{\gamma}_r$ is the reduced shear rate. The shift factors depend on φ. But in the region of low rates of shear the dependences η_r $(\dot{\gamma}_r)$ diverge for different φ values and no master flow curves can be obtained. This is due to the appearance of a new relaxation mechanism associated with specific interactions. This effect gives rise to new relaxation times, and their concentration dependence is fundamentally different from the concentration dependence of relaxation times that are responsible for non-Newtonian flow in the region of high shear rates.

The loading of a polymer dispersion medium with rigid spherical particles, just as in the case of other liquids, leads to the growth of the viscosity if one measures it in shear. More interesting is the reverse effect—the fall of extensional viscosity as the polymer is filled with rigid spherical particles.** This rather unexpected experimental result once more shows that the role of the filler may be different, depending on the deformation scheme and the predominant molecular mechanism responsible for the resistance to flow. Upon extension (as will be discussed in detail in Chapter 7), the decisive role is played by the elasticity of the specimen being extended. Inclusion of a filler suppresses the rubbery elasticity of the polymer. This may cause a decrease in the extensional viscosity when the polymer is loaded with a filler. In contrast to this, in shearing flow the main part is played by hydrodynamic effects associated with the streamlining of rigid filler particles with the liquid, which leads to the growth of dissipation losses, a measure of which is the shear viscosity of the composition.

---

* Nicodemo, L. and L. Nicolais, *J. Appl. Polymer Sci.*, 18, 2809 (1974); Kataoka, T., T. Kitano, M. Sasahara, and K. Nishijima, *Rheol. Acta*, 17, 149 (1978).
** Nicodemo, L., B. De Cindio, and L. Nicolais, *Polymer Eng. Sci.*, 15, 679 (1975).

6.2.3. Polymers with a Fibrous Filler

The formation of the filler structure and its effect on the flow behaviour of the composition are traced out well in measurements of the properties of the polymer, to which is added a small amount of a fibrous filler. Such a system experiences no chemical or physico-chemical interactions, but a structure may be formed due to the entanglement and mechanical sticking of separate fibres. Since fibres are relatively very long (their length-to-diameter ratio is high) and are flexible, the filler particles are brought into contact with one another even at very low content of fibres. A polymer loaded with a fibrous filler may be regarded, on the one hand, as a model of a polymer solution

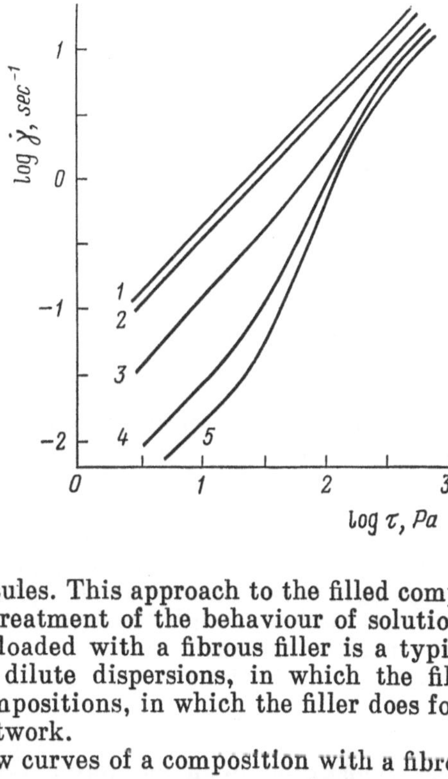

Fig. 6.3. Effect of the relative length of the fibre in the composition oligo-mer-fibrous filler on the flow proper-ties of the system. The original oligo-mer is a Newtonian liquid (1); oligo-mer with 1 per cent of powder filler (2); oligomer with 1 per cent of fibre 15.3 mm in diameter and with the length, mm: 2 (3); 5 (4); 10 (5). [A. Ya. Malkin, G. V. Epplé, and A. I. Gritsuk, Kolloid. Zhurn., 34, 550 (1972).]

containing flexible-chain macromolecules. This approach to the filled compositions is typical for the molecular treatment of the behaviour of solutions. But, on the other hand, a polymer loaded with a fibrous filler is a typical filled system intermediate between dilute dispersions, in which the filler particles do not interact, and the compositions, in which the filler does form a three-dimensional structural network.

A characteristic example of the flow curves of a composition with a fibrous filler is shown in Fig. 6.3, which compares the dependences $\tau(\dot{\gamma})$ for compositions containing only 1 per cent of filler.* It is seen that the inclusion of

---

* Attention should be focused here on serious methodological difficulties that arise in investigations of the rheological properties of polymeric compositions loaded with a fibrous filler. When rotational viscometry is used, the fibres are wound on the rotor of the instrument so that the filler content of the composition is reduced. When capillary viscometry is employed, there is observed the filtration effect, which consists in that the fibres form a filter-packet at the entrance into the capillary through which the dispersion medium (binder), containing a substantially lower amount of fibres than in the original homogeneous dispersion, is forced. Many of the methodological problems of this kind have been discussed in: Mashmeyer, R.O. and R.T. Hill, Trans. Soc. Rheol., 21, 183 (1977). Even if the formation of a packet of fibres at the capillary entrance and the changes of the average filler concentration in the extrudate do not take place, separate fibres stuck

1 per cent of a powder filler (of the same material of which the fibres are composed) increases the viscosity but very slightly, the form of the flow curve
being unaffected. As the relative length of the fibre (the length-to-diameter
ratio, which is, respectively, 130, 327, and 654 for the compositions being
compared) increases, its total content in the composition being maintained
constant, the viscosity of the composition sharply increases (by tens of times)
at low rates of shear; in the region of high shear rates the viscosity is changed
relatively slightly.* Analogous flow curves have been observed for other compositions with fibrous fillers.**

The effect of very long fibres on the viscosity may be regarded as the result
of the manifestation of a pseudo-structural skeleton which is gradually
"destroyed" in a wide range of shear rates. The low strength of structural
links of the "entanglement" is attributed to weak mechanical contacts between
the fibres. In this respect, compositions with fibrous fillers simulate melts
of polydisperse polymers in which the effect of non-Newtonian flow is also
"spread" over a wide range of shear rates (compare with Figs. 2.17 and 2.50).
As the length of fibres increases the slope of the straight-line portions of the
flow curves, which characterizes the extent of non-Newtonian behaviour of
the composition, also increases.

A further factor that should be taken into account in considering the viscosity of compositions with a fibrous filler is the rigidity of the material of
which the fibres are made. The increase of the flexibility of fibres leads to a
considerable enhancement of the effects shown in Fig. 6.3, i.e., it intensifies
the non-Newtonian behaviour of the composition and increases the viscosity
in the region of low shear rates.***

Here it is expedient to compare the flow curves of polymers loaded with a
fibrous filler, liquid-crystalline polymer solutions**** and high-filled systems
(see below). For all these materials, in distinction to homogeneous polymer
melts, there is observed an increase of viscosity at low shear stresses. In
the latter two cases this is associated with the existence of the yield point.
This term is here somewhat arbitrary since below the yield point there is
possible creeping flow with a very high viscosity. The case shown in Fig. 6.3
may be treated as intermediate between the limiting case of the brittle-frac-
tured structure of the filler with a very sharp decrease of viscosity (by thousands of times) and the flow of a non-Newtonian liquid, for which the region
of Newtonian behaviour is attained at low shear rates.

6.2.4. Compositions with an Active Filler

The term filler "activity" reflects qualitatively the possibility of strong
specific interactions taking place in filled polymers. Such interactions may
occur between the filler particles and the polymer dispersion medium and/or
between the filler particles and the polymer macromolecules.

---

at the entrance into the capillary may give a distorted picture of the manifestation of the
rheological properties of the composition, for example, a seemingly monotonic increase
of the apparent viscosity; see Malkin, A. Ya., S. G. Kulichikhin, V. M. Martynyuk,
and E. Z. Bokareva, *Plasticheskie Massy*, 3, 58 (1980).

   * Malkin, A. Ya., G. V. Epplé, and A. I. Gritsuk, *Kolloid. Zhurn.*, 34, 550 (1972).
  ** Faitelson, L. A. and V. P. Kovtun, *Mekhanika Polimerov*, 2, 326 (1975).
 *** Malkin, A. Ya., G. V. Epplé, and A. I. Gritsuk, *Kolloid. Zhurn.*, 34, 550 (1972).
**** Papkov, S. P., V. G. Kulichikhin, A. Ya. Malkin, V. D. Kalmykova, A. V. Vo-
lokhina, and L. I. Gudim, *Vysokomol. Soedin.*, 14B, 244 (1972).

If the filler is capable of interacting with the polymer, the hydrodynamic effects described above are supplemented with specific surface phenomena which may be treated as the formation of a transient layer between the matrix and the filler. If the polymer interacts chemically with the filler, this transient layer is found to be grafted onto rigid particles. If the interaction is weaker, there appear adsorbed layers or develops a predominant orientation at the rigid surface. In any case, the interaction of the dispersion medium with the filler may be treated as the appearance of a layer with a certain effective thickness δ on the surface of filler particles, a layer which is more or less firmly bound to the surface and which increases the apparent volume of the dispersed filler phase.* For the range of dilute suspensions this effect may be interpreted in terms of the Einstein treatment of the viscosity of dispersions. Indeed, suppose that Eq. (6.1) is fulfilled, as before, i.e., let η_{sp} be equal to 2.5 φ_{eff}, where φ_{eff} is taken to mean not the intrinsic volume of the filler but the volume of the particles increased by the thickness δ of the adsorbed layer. If $\delta \ll d$ (where d is the diameter of the filler particles), then

$$\varphi_{eff} = \frac{\pi}{6} (d+2\delta)^3 \approx \frac{\pi d^2}{6} \left(1+\frac{6\delta}{d}\right) = \varphi\left(1+\frac{6\delta}{d}\right)$$

that is, formula (6.1) may be given in the form:

$$\eta_{sp} = 2.5 \left(1+\frac{6\delta}{d}\right) \varphi = K\varphi \qquad (6.6)$$

Then, having determined experimentally the coefficient K in the dependence of η_{sp} on φ in the region of low concentrations of the filler, where the dependence of η_{sp} on φ is still linear, and having found that $K > 2.5$, we may regard this effect as the formation of an adsorbed layer on the surface of filler particles. The thickness of this layer can be computed from the measured values of K on the basis of Eq. (6.6) as follows:

$$\delta = \frac{K-2.5}{15} d \qquad (6.7)$$

This approach may also be applied to the region of moderately concentrated suspensions, for which the concentration dependence of viscosity is described by formula (6.4). In such cases, the formation of an adsorbed layer results in a change of the numerical value of the coefficient K. A comparison of the experimentally measured value of this coefficient with a theoretical case where the polymer is filled with spherical particles with the effective diameter being increased due to the formation of an adsorbed layer, enables one to find the thickness of this layer, just as was done in the above-considered case.**

The rheological effects caused by the interaction of the filler particles with the polymer dispersion medium are not restricted to the variation of the coefficient K in the formula for the concentration dependence of the viscosity of suspensions. Here, diverse and sometimes unexpected phenomena are encountered. For instance, at a low filler content the viscosity may not only increase with respect to the viscosity of the dispersion medium but it may also diminish, which is ascribed to the formation of an additional free volume at the polymer-filler interface.***

* Malinsky, Yu. M., I. V. Epelbaum, L. I. Ivanova, and G. V. Vinogradov, *Vysokomol. Soedin.*, 8, 1886 (1966).
** Faitelson, L. A. and A. I. Alekseenko, *Mekhanika Polimerov*, 3, 478 (1976).
*** Prokopenko, V. V., O. K. Petkevich, Yu. M. Malinsky, and N. F. Bakeev, *Doklady Akad. Nauk SSSR*, 214, 389 (1974); Prokopenko, V. V., O. K. Titova, N. S. Fesik, Yu. M. Malinsky, and N. F. Bakeev, *Vysokomol. Soedin.*, A19, 95 (1977).

The specific polymer-filler interactions play, generally, the decisive role in the manifestation of the flow properties of a composition. Of special interest in this respect are the results of measurements of the viscosity of a composition to which is added a modifier capable of forming a bridge of chemical bonds between the inorganic filler and the polyolefin matrix.* It has been found that the introduction of such an agent brings about a decrease in the viscosity of the filled composition. This is possibly associated with the change of the hydrodynamic situation developing around the filler particles since their binding with the matrix alters the conditions of enclosing of rigid particles and, hence, the total dissipation losses.

6.2.5. Filled Polymers (at Temperatures above T_g) as Plastic Disperse Systems

The interaction between the filler particles may result in the formation of a structural skeleton extending throughout the entire volume of the polymeric system. The strength of the skeleton is governed by the strength of the filler particles, their specific interaction, and the number of contacts between the particles. The consequences of the formation of such a skeleton are best traced out on typical plastic disperse systems (PDS) in which the dispersion medium is a relatively low-viscosity liquid.

The rheological behaviour of a PDS are determined, first of all, by the strength characteristics of the structural skeleton. They are manifested when the system at rest is deformed. Relevant results are presented in the generalized schematic form in Fig. 6.4a. This figure shows the dependence of the shear stress on the deformation measured with a rotational elastoviscometer fitted with a very rigid dynamometer by the method of assignment of a constant rate of deformation. In the region of infra-low values of the shear rates (usually, $\dot{\gamma} < 10^{-6}$ sec^{-1}) below a certain critical value of $\dot{\gamma}_A$ the dependences $\dot{\gamma}(\tau)$ are monotonic (curves 1-3). Under such regimes, in the initial period of straining the system is subjected to work hardening, which is accompanied by a decrease in the rate of accumulation of irrecoverable deformation. This work-hardening effect can be eliminated by the shearing of the system in the] opposite direction. To these regimes there corresponds the linear dependence $\dot{\gamma}(\tau)$ and viscosities of the order of 10^9 to 10^{11} Pa·sec. They are described by the right-hand horizontal branches of the curves in Fig. 6.4a. At $\dot{\gamma}$ equal to about 0.8 of the critical value $(\dot{\gamma}_A)$, it is found that the reinforced (packed) structure of the system begins to break down. It is reflected in the nonlinear portion of the curve of $\dot{\gamma}$ vs. τ.

To the attainment of $\dot{\gamma}_A$ there corresponds a characteristic value of the shear stress τ_y—the yield stress. The transition through the yield stress is accompanied by a catastrophic breakdown of the structural skeleton, as a result of which the resistance of the PDS to deformation is sharply reduced. It may decrease by a factor of hundreds of thousands to tens of millions. This is shown by curves 4 and 5 with a sharply pronounced maximum in Fig. 6.4a. With further increase of $\dot{\gamma}$ the role of viscous resistance is enhanced. At very high shear rates the viscous resistance masks the structural breakdown

---

* Han, C. D., C. Sanford, and H. J. Yoo, *Polymer Eng. Sci.*, 18, 849 (1978).

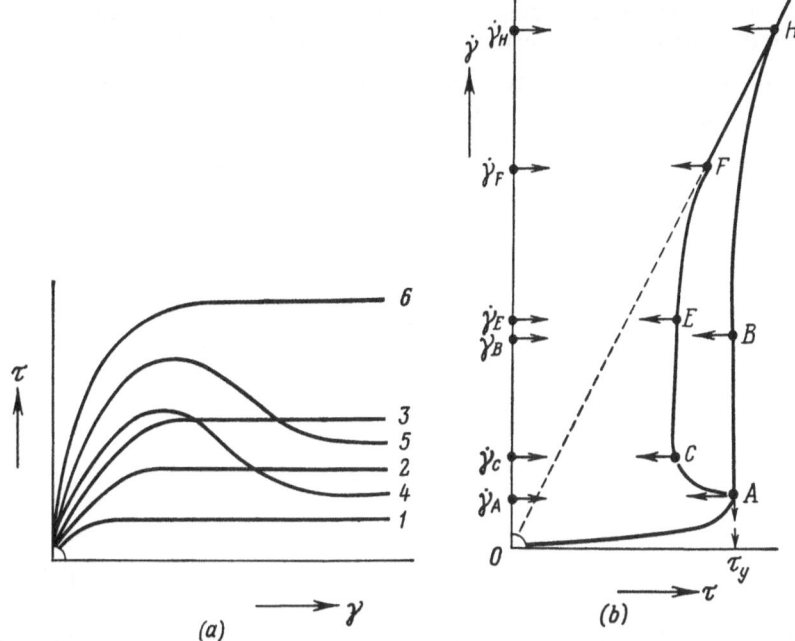

Fig. 6.4. Schematic representation of the deformation properties of plastic disperse systems over a very wide range of rates of deformation:

(a) Dependence of the rate of shear on the amount of deformation at $\dot\gamma$ = const. The rate of deformation increases from *1* to *6* (curves). (b) The ultimate and steady-state value of stress at different rates of deformation.

so strongly that the maximum on the curve of τ vs. $\dot\gamma$ disappears. Hence, in a general case, the ultimate shear strength of PDS is determined not only by the strength of the structural skeleton but also by the viscous resistance offered by the structural skeleton being destroyed. It is for this reason that the lowest shear stress at which the structural skeleton is destroyed, i.e., the lowest value of the ultimate strength, is taken as the yield stress.

Typical dependences of the shear stress on the rate of deformation (which was changed by more than 10^{10} times) for a PDS are presented in Fig. 6.4b in a generalized form. The portion OA on the curve corresponds to the above-considered creep process. At shear rates slightly exceeding the critical value $\dot\gamma_A$, to each value of $\dot\gamma$ in Fig. 6.4a there correspond two values of shear stress. One of these, τ_y, corresponds to the ultimate shear strength, and the other, τ_{st}, reflects the resistance in the steady flow of a PDS. Accordingly, two curves start from the point A in Fig. 6.4b; one of them (the curve ABH) represents the dependence of τ_y, and the other (the curve $ACEH$) describes the dependence τ_{st} ($\dot\gamma$).

In a qualitative consideration of the problem it is useful to represent τ_y and τ_{st} as the sum of the structural and viscous components. The first includes the resistance to the breakdown of the brittle skeleton and the resistance to the motion of rigid particles relative to one another. The viscous drag is caused by the relative displacement of the layers of the dispersion medium and by the fragments of the broken-down skeleton streamlined by the dispersion medium.

The form of the dependences τ_y $(\dot{\gamma})$ and τ_{st} $(\dot{\gamma})$ is determined by the relative contribution of the structural and viscous components. The viscous components of τ_y and τ_{st} are strongly changed with variation of $\dot{\gamma}$. In contrast to this, the structural components either depend weakly on $\dot{\gamma}$ (for τ_{st}) or do not depend on $\dot{\gamma}$ (for τ_y) at all. It is clear that because of the weak effect of the viscous component on τ_y at low $\dot{\gamma}$ the values of τ_y are independent of $\dot{\gamma}$. Accordingly, the portion AB in Fig. 6.4b appears to be a vertical straight line. When $\dot{\gamma}$ increases considerably (when $\dot{\gamma} > \dot{\gamma}_B$), there is observed an increasing contribution of the viscous component to τ_y and the curve ABH begins to depart to the right.

The portion AC in Fig. 6.4b characterizes a phenomenon called the superanomalous viscosity. By anomalous viscosity* is understood the non-Newtonian flow characterized by the weaker rather than linear change of τ with increasing $\dot{\gamma}$. The superanomalous viscosity is determined by the decrease of τ with increasing $\dot{\gamma}$.

Above the point C on the curve CEH there is a portion CE which is close to a vertical. It reflects a very weak dependence of τ_{st} on $\dot{\gamma}$. The cause of this phenomenon is that the total resistance to deformation on the portion CE is due mainly to the rupture of the bonds between the structural units of the skeleton and, as has been pointed out above, is almost independent of the viscous drag.

As the resistance of the liquid layer between the particles of the structural skeleton increases the portion of the curve above the point E changes into a curve typical of non-Newtonian liquids. With increase of $\dot{\gamma}$ this resistance rapidly increases. As a result of this, at sufficiently large $\dot{\gamma}$ values the viscous component of the resistance becomes predominant.

Beginning from certain values of $\dot{\gamma}$ (designated as γ_F in Fig. 6.4b) a PDS practically becomes a Newtonian liquid; the linear dependence of $\dot{\gamma}$ on τ is easily attained for systems with a very viscous dispersion medium.

If at a certain value of $\dot{\gamma}$ the resistance to the rupture of the links in the structural skeleton becomes negligibly small as compared with the viscous drag, the values of τ_y and τ_{st} become practically identical. In other words, the curves of τ_y and τ_{st} vs. $\dot{\gamma}$ converge (Fig. 6.4b).

Concluding the consideration of the rheological behaviour of PDS with a brittle structural skeleton, it is necessary to note that in the case of low-viscosity dispersion media the wall effect is of great importance.** It resembles outwardly the wall slippage, though its nature is quite different. The wall effect is caused by the increased content of the dispersion medium at the interface with the solid surface, relative to which there occurs a shear deformation of a PDS. The neglect of this effect may give rise to erroneous conclusions. Owing to the high viscosity of polymeric dispersion media, the wall effect is, as a rule, unimportant.

---

* Pawlow, W. P., G. W. Winogradow, W. W. Sinizyn, and Y. E. Deinega, *Rheol. Acta*, I, 470 (1961).
** Vinogradov, G. V. and G. B. Froishteter, *Rheol. Acta*, 17, 620 (1978).

The transition through the yield point may be non-brittle if the interactions between the filler particles are weak or the filler itself is not rigid, i.e., has viscoelastic properties. Then, the breakdown of the structure formed by it is extended in time and occurs during a prolonged period of deformation. The long-term strength (i.e., the time to break) of such a non-rigid structure depends on the applied stress.* In this respect, the long-term strength of the structure formed by the filler resembles that of solids; the time to break for them depends on the level of the applied stress. A typical case is shown in

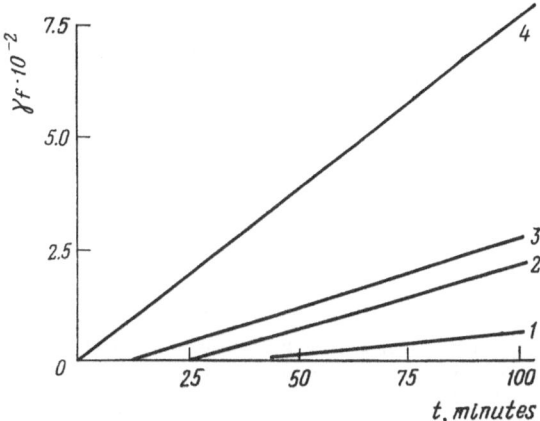

Fig. 6.5. The development of irrecoverable deformations γ_f with time upon application of shear stresses τ to bitumen with a high content of asphaltenes. The values of $\tau \times 10^{-2}$, Pa: 2 (*1*); 5 (*2*); 10 (*3*); 20 (*4*).

Fig. 6.5. The portions cut off along the time axis by the straight lines of γ_f (where γ_f is the plastic deformation) vs. t are the values of the long-term durability t^* of the filler structure, i.e., the loading time, during which there is no viscous flow and only elastic deformations are possible. Upon attainment of t^* the structure is broken down and the flow of the composition becomes possible. It is obvious that t^* depends on τ and this dependence is rather weak. It is logarithmic for the data presented in Fig. 6.5: t^* is proportional to $\log \tau$. The long-term effects of structure breakdown upon transition through the yield point are characteristic of a viscoelastic filler.

Of great interest is the case of the so-called "active" fillers. This term is, strictly speaking, ambiguous. By "activity" is understood the effect of the filler on the relative change of the various properties of the composition. For different polymeric systems the "activity" of the same filler may be different. But, as a general rule, the "activity" of the filler increases with increasing surface of the dispersed particles. In this respect, a typical "active" filler is carbon black.

The flow curves of a polymer loaded with carbon black are shown in Figs. 6.6 and 6.7, which also present, for comparison, the flow curves of a low-molecular-mass analogue of the same polymer, which is loaded with the same carbon black in the same proportions.** Analogous experimental data

* Malkin, A. Ya., O. Yu. Sabsai, E. A. Verebskaya, V. A. Zolotarev, and G. V. Vinogradov, *Kollotd. Zhurn.*, **38**, 181 (1976).
** Vinogradov, G. V., A. Ya. Malkin, E. P. Plotnikova, O. Yu. Sabsai, and N. E. Nikolaeva, *Intern. J. Polymeric Mater.*, **2**, 1 (1972).

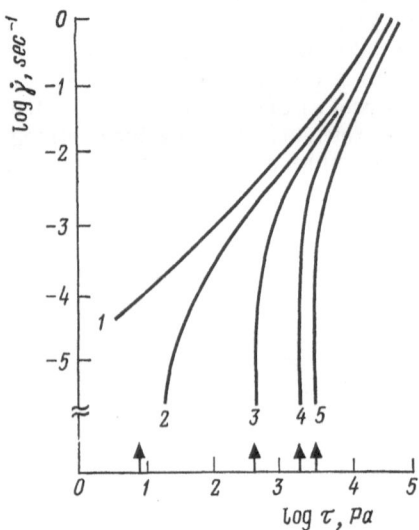
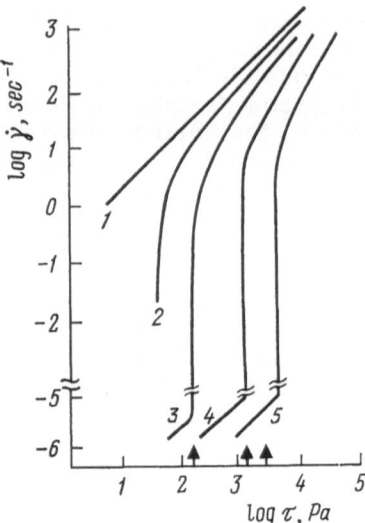

Fig. 6.6. The flow characteristic of polyisobutylene loaded with carbon black.
The filler content (per cent by volume):
0 (*1*); 2.5(*2*); 5 (*3*); 9 (*4*); 13 (*5*). The arrows
indicate the values of the yield point. [Vi-
nogradov, G. V., A. Ya. Malkin, *et al.*,
Intern. J. Polymeric Materials, 2, 1 (1972).]

Fig. 6.7. The flow curves of a low-
molecular-mass (oligomeric) polyisobutylene loaded with carbon black.
The designations and references are
the same as in Fig. 6.6.

have been obtained for dispersions of glass powder in oligomers (epoxide
resins and their solutions).* The form of the flow curves shown in these figures
are of general importance for filled polymeric systems. A polymeric binder exhibits all the most important properties of concentrated polymer solutions and molten polymers: a very high viscosity, the existence of a wide region of non-Newtonian flow, and rubber-like elasticity.

It is easy to see that the form of the flow curves of the polymer filled with
carbon black in Fig. 6.6 practically coincides with those in Fig. 6.7 and with
the scheme given in Fig. 6.4*b*. In all these cases the transition through the
yield point and the region of flow with the structure destroyed are sharply
pronounced at high shear rates and stresses. For these compositions there
also exists a region of very slow creep with a viscosity not lower than 10^9
Pa·sec.

It is important to note that such a distinct manifestation of the three regions
of the complete flow curve is observed only if one considers a very wide region of shear rates and stresses. Otherwise, the experimentator only deals
with an ordinary flow curve with a strongly pronounced non-Newtonian
region but, in distinction to the typical flow curves for polymer solutions and
melts, no transition to the limiting region of Newtonian flow is detected.
Therefore, one cannot limit oneself to the region of high shear rates since the
existence of the yield value and, especially, of the region of creep is only
detected in investigations carried out in the region of infralow shear rates
and stresses.

---

* Lipatov, Yu. S. *et al.*, *Vysokomol. Soedin.*, A15, 2106 and 2243 (1973).

The similarity of the flow curves in Figs. 6.6 and 6.7 is retained at low shear stresses until the yield point is passed. This similarity consists in the closeness of the values of the viscosity of compositions with undestroyed structure (they are of the order of 10^9 Pa·sec) and with the same yield point τ_y (Fig. 6.8), regardless of the enormous difference in viscosity between the dispersion media. An analogous absence of dependence of τ_y on the nature of the dispersion medium has been observed for polymer melts with glass beads of various sizes.*

The specific properties of the polymeric binder are only manifested after the yield point is passed since at $\tau > \tau_y$ the dispersion medium begins to

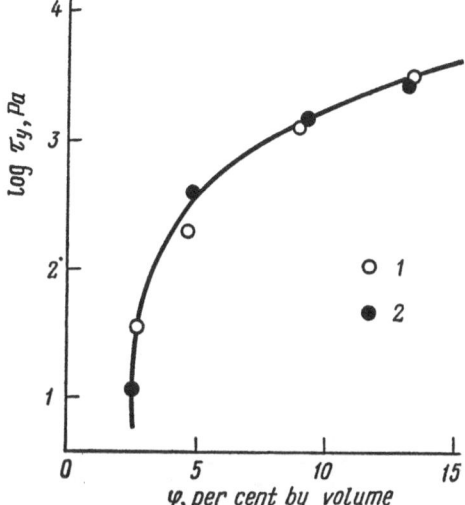

Fig. 6.8. Dependence of the yield point on the volume content of filler in low-molecular-mass (1) and high-molecular-mass (2) polyisobutylenes. The reference is the same as in Fig. 6.6.

flow together with the dispersed filler particles or clusters formed by the fragments of the destroyed structure. At practically equal values of viscosity in the region of very low shear rates, the jump of the viscosity, which is caused by the breakdown of the structure of the filler, is found to be sharply different; depending on the viscosity of these media, these jumps are as different as the viscosities of these liquids.

As regards the region of the transition through the yield point, the value of τ_y is determined, first of all, by the nature and concentration of the filler and, to a certain extent, by the nature of thin intermediate layers of the dispersion medium, which are retained between the filler particles. The viscosity of the dispersion medium plays here a secondary role (Fig. 6.8). But the beginning of the departure from the vertical branch on the flow curve depends on the viscosity of the dispersion medium. The change of the mechanism of deformation upon transition through the yield point is especially spectacular when comparing the flow curves corresponding to different temperatures (Fig. 6.9). The rise of temperature changes the viscosity of the dispersion medium but it practically does not affect the nature of the interaction between the filler particles. Therefore, τ_y does not change with rise of temperature, just as τ_y did not change upon replacement of the polymer by

---

* Kataoka, T., T. Kitano, M. Sasahara, and K. Nishijima, *Rheol. Acta*, **17**, 149 (1978).

a low-molecular-mass oligomer. But the viscosity at $\tau > \tau_y$ and, hence, the positions of the flow curves depend strongly on temperature.

A reflection of the change of the deformation mechanism with increasing stress is the dependence of the activation energy of viscous flow, E, on stress. As has already been said (see Chapter 2), the value of E for polymer melts does not depend on τ. This also applies to filled polymers at $\tau \gg \tau_y$ when the deformation takes the form of the flow of the polymer with fragments of the structural skeleton of the filler being dispersed in it. But upon approach to τ_y the resistance to deformation is no longer dependent on the viscosity of the dispersion medium since the decisive factor here is the contribution of

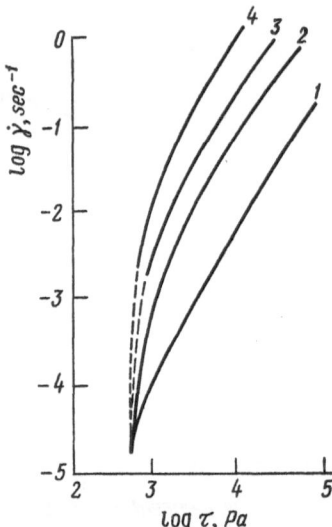

Fig. 6.9. Effect of temperature on the flow curves of high-molecular-mass polyisobutylene containing 5 per cent of active carbon black. The temperatures are, °C, 22 (*1*); 60 (*2*); 80 (*3*); and 100 (*4*).

the structural skeleton. At $\tau \approx \tau_y$ the concept of the activation energy of viscous flow becomes inapplicable ($E = 0$) because viscous flow itself disappears and the plastic deformation occurs by a deactivation mechanism. Obviously, there exists a transient region of stresses where the activation energy changes from zero to values characteristic of the flow of the dispersion medium (a polymer melt).

6.2.6. The Viscoelasticity of Filled Polymers

The inclusion of a solid filler in a polymer leads, as a rule, to the increase of the rubbery modulus and reduces the ability of the polymer to undergo rubbery deformations. This function of solid fillers was first demonstrated in experiments that have shown that the inclusion of glass beads in a styrene-acrylonitrile copolymer brings about a decrease in the swelling of the extrudate emerging from the capillary.* There was also observed** the decrease of the normal stresses in shearing flow of molten polypropylene to

---

* Newman, S. and Q. A. Trementozzi, *J. Appl. Polymer Sci.*, 9, 3071 (1965); analogous data were reported at a much later time in: Agarwal, P. K., E. B. Bagley, and C. T. Hill, *Polymer Eng. Sci.*, 18, 282 (1978).
** Han, C. D., *J. Appl. Polymer Sci.*, 18, 821 (1974).

which $CaCO_3$ was added and also the decrease of the entrance corrections which are associated with elastic deformations stored in elastic liquids during their capillary flow.*

A direct evidence of the increase of the rigidity of the polymer when it is loaded with a filler is the increase of the rubbery modulus G_e with increasing filler content and the attending decrease of recoverable deformations γ_e under comparable shear stresses (Fig. 6.10). These data show that the presence

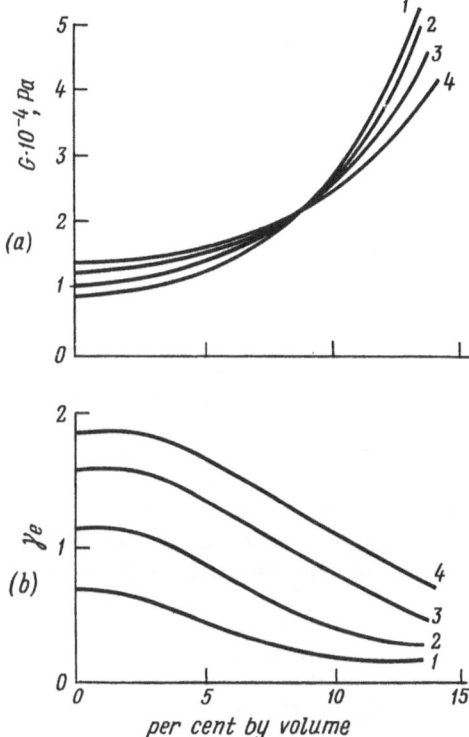

Fig. 6.10. Dependence of the elastic modulus (a) and rubber-like deformations on the filler content in a polymer at different shear stresses, $\tau \times 10^{-4}$: 1 (1); 1.5 (2); 2 (3); 2.5 (4). [Vinogradov, G. V., A. Ya. Malkin, E. P. Plotnikova, O. Yu. Sabsai, and N. E. Nikolaeva, *Intern. J. Polymeric Materials*, 2, 1 (1972).

of the structural skeleton of the filler or of its fragments impedes the manifestation of the viscoelasticity of the dispersion medium and requires an increase of the stresses necessary for the development of deformations that were produced in a pure polymer at lower stresses.

A consequence of the repression of the viscoelasticity of the polymer shown in Fig. 6.10 upon addition of a filler is the decrease of the degree of swelling of the extrudate leaving the capillary. This effect is observed upon addition of a solid filler to a polymer melt. In particular, it has been observed for the polymeric systems, whose viscoelasticity is presented in Fig. 6.10 as a function of the degree of loading.

The decrease of the degree of swelling caused by the loading of a polymer is easily understandable if one takes into account the fact that the degree of swelling is determined by the elastic energy stored during flow through the capillary. The elastic energy depends on γ_e and the quantity γ_e falls with

---

* White, J. L. and J. W. Crowder, *J. Appl. Polymer Sci.*, 18, 1013 (1974); Faulkner D. L., Schmidt L. R., *Polymer Eng. Sci.*, 17, 657 (1977).

increasing degree of loading. Besides, as the polymer is loaded with a filler the relative volume of the polymer itself, i.e., the volume of the material capable of storing elastic energy on shear, is reduced.

Investigations of normal stresses σ in the shearing flow of loaded compositions are scarce and the results obtained are rather contradicting.

The introduction of a filler into a polymer leads not only to the shift of the relaxation spectrum to the side of its long-time side but also to an additional contribution made by the formation of a structural skeleton to that part of the spectrum.* What has been said is accounted for by the increase of the initial normal stress coefficient. Under the conditions of steady-state flow the structural skeleton is broken down and the filler which is not physically bound with the matrix increases the viscosity of the composition but it has no effect on the normal stresses. If we assume the equality $\tau = \text{const}$ as the condition of comparison of the matrix with the filled composition,** then the apparent rubbery deformations determined [according to Lodge; see Chapter 4, formula (4.3)] as $\sigma/2\tau$ decrease and, accordingly, the effective modulus $2\tau^2/\sigma$ increases. If we compare the value of normal stresses that arise during the flow of the matrix and the filled system on the $\dot{\gamma} = \text{const}$ condition, then σ increases with increasing degree of filling.***

As regards the effect of a fibrous filler on the rubber-like properties of melts, this problem has not been studied in a systematic way. The few experimental data available in the literature show that in this case an important part may be played by the relative length of dispersed fibres, and that the normal stresses in shearing flow of a 20 per cent suspension of fibres in a melt may be higher than the normal stresses arising during the flow of the matrix at the same shear rate.****

The inclusion of a fibrous filler as well as of powdered filler has a strong influence on the relaxation low-frequency properties of the composition; here too the relative length of the fibre plays an important part. A theoretical analysis of this problem may be based on the same considerations that are used in the molecular theory of the viscoelastic properties of dilute polymer solutions, i.e., in this case too suspensions of long fibres are identified with polymer solutions.*****

6.2.7. Analytical Description of the Flow Curves of High-Filled Compositions

The practical importance of the hydrodynamic problems associated with the flow of high-filled polymeric (and non-polymeric) compositions requires the development of a rheological model of their flow behaviour. Such a model must reflect two most important features of the flow of compositions: the existence of the yield stress and the non-Newtonian flow curve at stresses exceeding the yield value.

* Han, C. D., *J. Appl. Polymer Sci.*, **18**, 821 (1974).
** Faitelson, L. A. and E. E. Yakobson, *Mekhanika Polimerov*, **6**, 1075 (1977); 1 172 (1978).
*** Faitelson, L. A. and A. I. Alekseenko, *Mekhanika Polimerov*, **2**, 332 (1978).
**** Chan, Y., J. L. White, and Y. Oyanagi, *J. Rheol.*, **22**, 507 (1978).
***** Attanasio, A. and U. Bernini, *J. Colloid. and Interface Sci.*, **29**, 81 (1969).

A rather general empirical equation that reflects the specific features of the flow curves of filled compositions is the formula*

$$\tau^{1/n} = \tau_y^{1/n} + (\mu\dot{\gamma})^{1/m} \tag{6.8}$$

where τ_y is the yield point; μ, n, and m are empirical constants. In the simplest case where $m = n = 1$ formula (6.6) reduces to the equation of the Bingham linear viscoplastic body. At $m = n = 2$ formula (6.8) turns into the well known Casson equation which satisfactorily describes the properties of many filled compositions over a rather wide range of shear rates.**

At $\tau < \tau_y$ the composition is not flowing; at $\tau = \tau_y$ there is attained the yield point (the ultimate strength of the structural skeleton), and at $\tau >$ $> \tau_y$ viscous flow becomes possible. According to Eq. (6.6) filled compositions behave as non-Newtonian liquids, and at $\tau \gg \tau_y$ their flow properties are described by a power law: $\tau = (\mu\dot{\gamma})^{n/m}$. This behaviour is characteristic of many polymer melts with a solid filler which forms a three-dimensional skeleton.

Determination of the constants in formula (6.6) is based on a graphical representation of the dependence of $\tau^{1/n}$ on $\dot{\gamma}^{1/m}$ and the extrapolation of the straight line obtained to $\dot{\gamma} = 0$, which gives the value of τ_y. Here the values of m and n must be chosen so as to provide the linearization of the dependence $\tau(\dot{\gamma})$ when using the coordinates indicated. Rather often (but not always) it is sufficient to assume $m = n = 2$, i.e., to use the Casson formula. The freedom of choice of the constants in formula (6.8) allows one to describe satisfactorily various experimental data on the dependence $\eta(\dot{\gamma})$ for filled compositions but it makes such a description ambiguous because various combinations of the constants m and n may prove to be permissible.

The values of τ_y determined by extrapolation may be compared with the filler concentration φ, just as was done above in Fig. 6.8. Here one finds the critical concentration φ_0, above which there appears the yield value τ_y. The concentration φ_0 is a quantitative characteristic of the ability of the filler to form a structure. The form of the dependence $\tau_y(\varphi)$ also reflects the intensity of structure formation in the filled polymer. Unfortunately, up to now this problem has not been studied in any systematic manner on various types of filler.

Another method of describing the flow properties of compositions with a yield value is the use of a rheological equation of state (a constitutive equation) of the form***:

$$\dot{\gamma} = \begin{cases} 0 & \text{at } \tau < \tau_y \\ k(\tau - \tau_0)^n & \text{at } \tau \geqslant \tau_y \end{cases} \tag{6.9}$$

where k, n, and τ_0 are constants, τ_0 being not equal to τ_y in a general case.

---

* Shulman, Z. P. in: *Heat and Mass Transfer*, Nauka i Tekhnika Publishers, Minsk, **10**, 3 (1968) (in Russian).
** Casson, N., in: *Rheology of Disperse Systems*, edited by C. C. Mill, Pergamon Press, Oxford, page 84, 1959; Matsumoto, T., A. Takashima, T. Masuda, and S. Onogi, *Trans. Soc. Rheol.*, **14**, 617 (1970); Kataoka, T., S. Ueda, R. Onooka, and K. Nishijima, Kobunshi Kagaku, **30**, 260 (1973).
*** Malkin, A. Ya., I. A. Glushkov, and V. A. Rozhkov, *Inzhenerno-Fizicheskii Zhurn.*, **28**, 356 (1975).

The character of the dependence $\dot{\gamma}$ (τ) described by formula (6.9) is shown in Fig. 6.11. The use of formula (6.9) makes it possible to describe the appearance of a curvilinear portion of the flow curve near τ_y, which is observed for high-filled compositions. At $\tau > \tau_y$ formula (6.7), just like formula (6.6), changes to an ordinary power law: $\dot{\gamma} = k\tau^n$. Formula (6.7) correctly reflects all the main features of the flow curves of polymeric (and not only polymeric) compositions with a high filler content. In contrast to ordinary equations for the flow curves of plastic materials, formula (6.7) reflects the existence of a break along the axis of shear rate at τ_y. In fact, this break exists at all

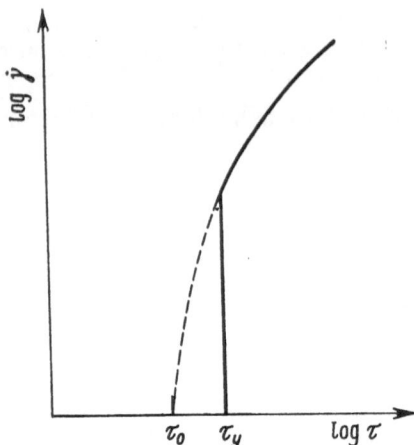

Fig. 6.11. The generalized form of the dependence $\dot{\gamma}$ (τ) for viscoplastic compositions according to Eq. (6.7).

times, while formulas of the type (6.8) predict that the flow curve starts with the "zero" shear rate (at $\tau = \tau_y$) and $\dot{\gamma}$ must increase smoothly with increasing shear stress. In the case of viscoplastic compositions there is always actually observed a jump of the shear rate upon attainment of τ_y. Therefore, in solving applied problems associated with the calculation of the flow of high-filled compositions it is expedient to rely on formula (6.9) or a similar formula but not on formulas of the type (6.8), in spite of the wide use of the latter.

6.3. Rheology of Polymer Blends

6.3.1. The Flow Properties of Polymer Blends

Polymers of various homologous series are, as a rule, incompatible. Their compatibility is usually observed in a very limited region of compositions. In those cases when polymers are compatible and form single-phase solutions, the change of their physical state (say, temperature) may be accompanied by the precipitation of one of the polymers as an individual phase. Owing to the high viscosity of the system it may prove to be in the metastable state.

To this state there may correspond the extremal changes of the viscosity. Sometimes anomalous situations are observed, such as the decrease of the viscosity of the polymer melt upon addition of small amounts of another, more viscous polymer.* Analogous phenomena which take place in the region of phase separation are known for solutions of incompatible polymers (Fig. 6.12). There have also been described cases with a minimum of the viscosity of blends with an intermediate range of compositions for various pairs of polymers, say, polypropylene and poly-1-butene.**

Two extreme cases of the structure of two-phase systems are possible: either one of the polymers forms a matrix (the dispersion medium) and the

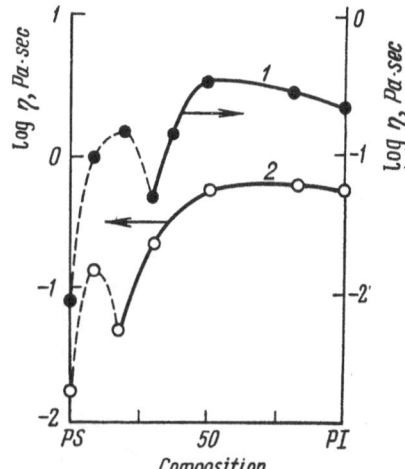

Fig. 6.12. Dependence of the solution viscosity on the composition of a blend consisting of polystyrene (PS) and polyisoprene (PI) in decalin at the total content of polymers: 2 per cent (1) and 2.25 per cent (2). The dashed lines correspond to the region of single-phase solutions. [Kuleznev, V. N., L. B. Kandyrin, and V. D. Klykova, Kolloid. Zhurn., 34, 231 (1972).]

other forms non-interconnected particles dispersed in the former; or both polymers form a continuous interpenetrating structure. The situation here is analogous to what is known for polymeric systems containing a solid dispersed phase.

The rheological behaviour of polymer blends depends on the structure which is primarily governed by the degree of dispersion of the components and by the mode of their distribution. The structure of polymer blends depends on the method of their preparation. Mechanical mixing leads, as a rule, to the formation of thermodynamically non-equilibrium (metastable) systems. A most uniform distribution of the components of the blend can be achieved by using a common solvent.***

For blends of the same composition it is often possible to produce a system in which one and the same component may be either a dispersion medium

---

* Vinogradov, G. V., A. Ya. Malkin, V. N. Kuleznev, V. F. Larionov, Kolloid. Zhurn., 28, 809 (1966).

** Deri, F., R. Genillon, and J.-F. May, Angem. Makromol., 57, 29 (1977). The viscosity minimum in the intermediate range of compositions, which has been described in this work, manifests itself in different ways, depending on the shear rate at which the viscosity is measured. This could possibly be associated with the effect of the rate of deformation on the phase separation occurring in a two-component system, similar to that described in: Malkin, A. Ya. and S. G. Kulichikhin, Vysokomol. Soedin., 19B, 701 (1977); 20B, 842 (1978); Kolloid. Zhurn., 41, 141 (1979).

*** Kuleznev, V. N., O. L. Mel'nikova, and V. D. Klykova, Europ. Polymer J., 14, 455 (1978).

or a dispersed phase. Therefore, for polymer blends it is difficult to obtain a unique composition-property relationship.

The general approach to the description of the dependence of the properties of two-phase systems on the composition hinges on the analysis of the behaviour of simple mechanical models.* Such an approach allows one to obtain, for example, the following equation for the viscosity of a blend in which the "drops" of one polymer are distributed in the continuous matrix of the other:

$$\eta = \eta_1 \frac{3\eta_1 + 2\eta_2 - 3(\eta_1 - \eta_2)\, \varphi_2}{3\eta_1 + 2\eta_2 + 2(\eta_1 - \eta_2)\, \varphi_2} \qquad (6.10)$$

where η_1 and η_2 are the values of the viscosity of the original components; φ_2 is the volume fraction of the particles of the dispersed phase.

The applicability of this equation is illustrated by Fig. 6.13 in which two cases are considered—an inhomogeneous coarsely dispersed blend and a

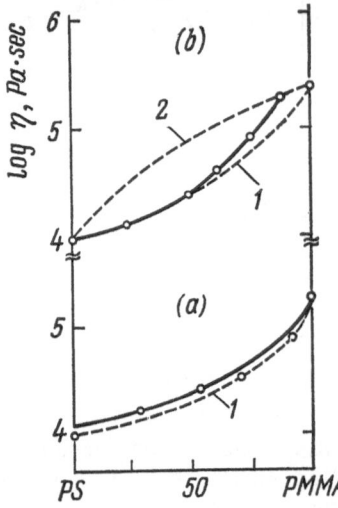

Fig. 6.13. Dependence of viscosity (at $\tau = 1.78 \times 10^5$ Pa) on the composition of a blend of polystyrene and polymethyl methacrylate for coarsely dispersed (a) and finely dispersed (b) compositions. The experimental data are represented by dots and solid lines. The dashed lines represent calculations according to formula (6.10) for the "drop" model of the more viscous component in the less viscous matrix (1) or a drop of the low viscous component in the more viscous matrix (2). [Kandyrin, L. B. and V. N. Kuleznev, *Kolloid. Zhurn.*, **36**, 473 (1974).]

homogenized finely dispersed blend of the same components. In the first case, the theoretical curve is calculated on the assumption that the more viscous component of the blend remains to be the dispersed phase (forms "drops"). In the second case, the less viscous component may be regarded as a dispersion medium only if its content exceeds 30 per cent. There is then observed a phase inversion, and the points "jump" from the lower curve corresponding to the system "the drop of the more viscous component in the matrix of the less viscous component" to the upper curve which refers to the system with an inverted phase distribution. From the data of Fig. 6.13 it follows that the viscosity of the system is most strongly affected by the inversion of a system with a high content of the more viscous component. Equation (6.10) describes the experimental data fairly well, but for this formula to be applicable in the region of the phase inversion the assignment of the indices 1 and 2 to the components of the system must be altered. Analogous results have been obtained in the work of Japanese authors cited above for blends of polyethylene and poly-1-butene melts. It should be noted that the theory does not specify the conditions of phase inversion, i.e., the critical value of φ_2 at which the assignment of the indices in the formula for $\eta\,(\varphi_2)$ should be

---

* Uemura, S. and M. Takayanagi, *J. Appl. Polymer Sci.*, **10**, 113 (1966).

changed since this point depends not only on the properties of the original components but also on the mode of their dispersion.

As regards the particular form of the formula for the dependence $\eta\,(\varphi_2)$, it may be varied. For instance, instead of Eq. (6.10), a simpler empirical equation has been proposed*, which is applicable to blends of various incompatible polymers:

$$\log \eta = \varphi_2^2 \log \eta_2 + (1 - \varphi_2^2) \log \eta_1 \qquad\qquad (6.11)$$

where φ_2 is the volume content of the dispersed phase of the blend. For the phase inversion the assignment of the indices 1 and 2 must also be changed.

The flow properties of a two-phase blend of incompatible polymers are first of all determined by the properties of the component which is the dispersion medium (a continuous phase). This is especially clearly seen from the example of addition of a low-viscosity component to a high-viscosity melt. So far as the latter forms a continuous phase the viscosity of the blend remains to be high. But as soon as the phase inversion is attained the viscosity of the blend falls very sharply even with a relatively low content of the low-viscosity component. Therefore, the S-shaped concentration dependence of the viscosity of a blend of incompatible polymers is, as a rule, an indication of the phase inversion, and the point of inversion corresponds to the point of inflection on the experimental curve.

The temperature dependence of the viscosity of blends is determined, as a rule, by the activation energy of the viscous flow of the dispersion medium. Nonetheless, the presence of a second component may exert a noticeable effect on it. For instance, for a blend of polyethylene and polyoxymethylene there has been observed a sharp change of the course of the temperature dependence of viscosity when the melting temperature of the second component is exceeded, though the dispersion medium is not altered.**

A peculiar case of two-phase polymer blends is provided by systems in which the dispersed phase is formed by a microgel—particles of a spatially cured polymer. The microgel may be a polymer of the same nature as the dispersion medium; for example, if molten polyethylene is mixed with particles of irradiated polyethylene.*** Even in such a case, however, there is formed a two-phase system. Such blends must be classified as a mixture consisting of a polymer and a viscoelastic filler. An interesting feature of molten polymers and rubbers containing microgel particles is the simultaneous increase of the viscosity and suppression of viscoelasticity.****

The increase of viscosity, depending on the content of the microgel, may occur non-monotonically and the addition of certain amounts of the microgel may contribute to the formation of compositions with improved technological properties.

The treatment of a polymer blend as a two-phase system opens up interesting possibilities for the modification of the properties of compositions by making use of the specific role of interface boundaries. Thus, when a third component is introduced into a blend it should be expected to be distributed at the interface between the phases. The addition of a low-viscosity plastis-

---

* Kandyrin, L. B. and V. N. Kuleznev, *Kolloid. Zhurn.*, **36**, 473 (1974); Kuleznev, V. N., O. L. Mel'nikova, V. D. Klykova, V. P. Skvortsov, and V. S. Glukhovskoy, *Kolloid. Zhurn.*, **37**, 273 (1975).
** Shumsky, V. F. *et al.*, *Kolloid. Zhurn.*, **38**, 949 (1976).
*** Vinogradov, G. V., V. N. Kuleznev, A. Ya. Malkin, A. V. Igumnova, and O. G. Polyakov, *Kolloid. Zhurn.*, **29**, 186 (1967).
**** Rosen, S. L. and F. Rodriguez, *J. Appl. Polymer Sci.*, **9**, 1601, 1615 (1965).

izer incompatible with each of the components of the blend enables the formation of a low-viscosity interphasial layer. As a result, the viscosity of
the composition is substantially reduced.* The decrease of the viscosity upon
addition of small quantities of plastisizer (of the order of 1 per cent) to the
blend appears to be much stronger in this case than upon addition of the
same plastisizer in an equivalent amount to each of the components of the
blend separately. This effect of the interphasial plasticization may evidently
be compared with the interstructural plasticization, which leads to a sharp
decrease of the viscosity of the melt upon addition of microquantities of oligomers (see Fig. 2.44 and the relevant discussion).

6.3.2. Specific Fibre-Formation in Blends of Polymer Melts

A large number of works ha e lately been devoted to the study of the flow
of two-component polymeric systems.** In this section we shall only be concerned with problems associated with specific fibre-formation during the flow
of incompatible polymers in capillary nozzles. The phenomenon of specific
fibre-formation consists in that during the flow of certain blends when certain
relationships between the rheological properties of the components are
fulfilled one of the polymers forms a continuous matrix while the other is
separated in the form of ultrathin fibres dispersed in it, which are oriented
in the flow direction. Under appropriate crystallization conditions these
fibres can be fixed and separated during the preparation of samples. The formation of such structures is of theoretical and technological interest. Here,
the main emphasis is placed on the rheological properties of blends in which
there arises the effect of specific fibre-formation.
   The possibility of specific fibre-formation is associated with the properties
of the original components of the blend and also with the flow conditions.
A highly dispersed system of fibres is formed when both components are in
the molten state. The viscosity of the matrix is important here; it must not
be too low to hinder the coalescence of the particles of the dispersed phase.
Geometrical factors also play a certain part. Thus, the deformation of the
stream during its flow from the reservoir to the die must be sufficiently great;
otherwise, the formation of microfibres is found to be impossible. The best
results are obtained when short dies are used. Therefore specific fibre-formation is associated not only with the rheological properties of the components
in the blend but also with the hydrodynamics of the flow, i.e., with the mode
of the development of shear and tensile stresses and with their interplay.

---

   * Kuleznev, V. N. and V. D. Klykova, Kolloid. Zhurn., 39, 969 (1977).
   ** Tsebrenko, M. V., M. Yakob, M. Yu. Kuchinka, A. V. Yudin, and G. V. Vinogradov, Intern. J. Polymeric Materials, 3, 99 (1974); Ablazova, T. I., M. V. Tsebrenko,
A. V. Yudin, G. V. Vinogradov, and B. V. Yarlykov, J. Appl. Polymer Sci., 19, 1781
(1975); Polymer, 16, 609 (1975); Krasnikova, N. P., E. V. Kotova, G. V. Vinogradov,
and Z. Pelzbauer, J. Polymer Sci., 22, 2081 (1978). The hydrodynamics of two-layer flow
of incompatible liquids has been examined in detail in: Malkin, A. Ya. et al., J. Appl.
Polymer Sci., 19, 375 (1975). Those who are interested in this problem are referred to the
following publications: White, J. A. et al., J. Appl. Polymer Sci., 16, 1313 (1972); Everage, A. E., Jr., Trans. Soc. Rheol., 17, 629 (1973); Han, C. D. et al., J. Appl. Polymer
Sci., 17, 1289 (1973); Trans. Soc. Rheol., 19, 245 (1975); Polymer Eng. Sci., 16, 697 (1976);
Southern, J. K. and R. L. Ballman, J. Polymer Sci., Polymer Phys. Ed., 13, 863 (1975);
Uhland, E., Polymer Eng. Sci., 17, 671 (1977).

In this connection, it will be recalled that it is the entrance into the capillary which is exactly the site where maximum tensile stresses develop (see Chapter 4, Sec. 4.7.5).

In the range of relatively low shear stresses the intensive fibre-formation is accompanied by a considerable increase of the resistance to flow. The increase of shear stresses up to the values exceeding 1 MPa leads to the change of the mode of flow and fibre-formation. This effect is evidently associated with deformation regimes approaching the flow spurt, and this entails a change in the fibre-formation—the composition acquires a lamellar-fibrillar structure. The flow becomes ordered, which brings about an appreciable decrease of the resistance and is recorded as a decrease of the shear stress (and viscosity).

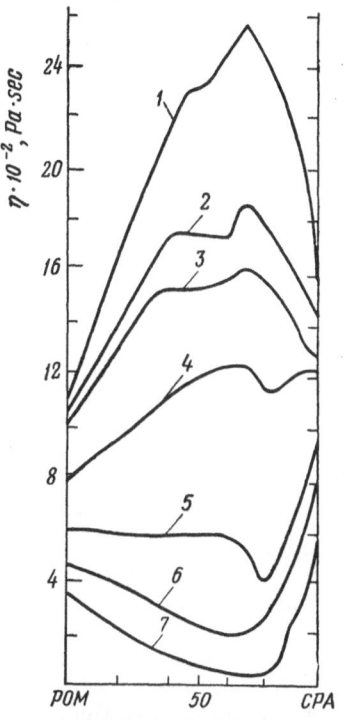

A typical pair of polymers that has been observed to exhibit specific fibre-formation effect consists of polyoxymethylene and copolyamide. The dependence of the viscosity on the composition of this polymer pair is shown in Fig. 6.14 for various shear stresses. It is clear that the non-monotonic course of the curves in Fig. 6.14 cannot be described by formulas such as (6.8) or (6.9). Non-monotonic dependences of viscosity on the composition of blends have also been observed for other pairs of polymers.*

A substantial variation of the course of the concentration dependence of viscosity with increasing shear stress attracts attention. At $\tau < 6.3 \times 10^4$ Pa there exist such blend compositions whose viscosity is greater than the viscosity of each of the components taken separately. At $\tau > 6.3 \times 10^4$ Pa a minimum appears on the concentration-viscosity curve, which lies near the 20-30 per cent cencentration of polyoxymethylene (POM). The curvature of the flow curve in this region of compositions is found to be much greater than for other blends. Therefore, while the viscosity of homogeneous samples decreases by a factor of about 3 as τ increases from 1.2×10^4 to 6.3×10^5 Pa, for blends containing 30 or 20 per cent of POM the viscosity falls by a factor of about 57 and 38, respectively.

Fig. 6.14. Concentration dependence of viscosity for a blend of polyoxymethylene (POM) with copolyamide (CPA) at different stresses: $\tau \cdot 10^{-4}$, Pa: 1.27 (*1*); 3.93 (*2*); 5.44 (*3*); 6.30 (*4*); 12.6 (*5*); 19.2 (*6*); 31.6 (*7*). Temperature, 190°C. [Ablazova, T. I. et al., J. Appl. Polymer Sci., 19, 1781 (1975).]

The formation of a system of ultrathin fibres leads to a sharp increase of the extrudate swell which may be 10-fold for a polyamide-polyoxymethylene composition containing 30 per cent of polyoxymethylene. The degree of extrudate swell characterizes the rubber-like properties of the composition (see Chapter 5, Sec. 5.6). Therefore, from the measured value of the swelling factor α one can find the stored rubbery deformations and normal stresses.

---

* Han, C. D., and T. C. Yu, *J. Appl. Polymer Sci.*, 15, 1163 (1971); *Polymer Eng. Sci.*, 12, 81 (1972); Han, C. D. and R. R. Lamonte, *Polymer Eng. Sci.*, 12, 77 (1972).

Relevant estimations* show that in the region of specific fibre-formation the calculated apparent recoverable deformations γ_e reach 100 units, which exceeds by tens of times the values of γ_e for homogeneous melts of plastic melts; the ratio of normal to shear stresses reaches 40, which also considerably exceeds the respective values for polymer melts. Here, a considerable difference is observed between the elastic behaviour of blends, which is measured during their flow through short and long capillaries. The role of tensile stresses acting in the entrance zone is especially important during the flow through short dies. All these effects are caused by the accumulation of large elastic deformations (with the corresponding elastic energy being stored) when an enormous number of fibres of one polymer are oriented in the matrix of the other under the action of tensile stresses in the entrance zone of the duct and of the partial relaxation that takes place during the flow of the composition through the capillary.

6.3.3. The Rheological Properties of Block Copolymers

Polymers whose linear chain is made up of blocks of different chemical nature form a special class of "mixtures" in which the incompatible components (blocks) are chemically united into a single chain. The incompatibility of blocks forming a single chain results in the tendency towards the aggregation of blocks of a single type and in the formation of a two-phase system. Its peculiarity consists in that, in spite of the obvious two-phase nature of the system, the phases cannot be separated from each other. Such systems exhibit many structural and mechanical properties characteristic of polymer blends.

A typical example is three-block copolymers produced by the anionic polymerization; two extreme blocks are built up of a relatively rigid polymer (say, polystyrene or poly-α-methylstyrene) and the central one is a flexible-chain polymer (say, polybutadiene or polyisoprene). The number of such blocks may exceed 3. The principal feature of the chemical structure of such copolymers is that blocks of the same structure have a similar length. Owing to the constancy of the lengths of the central blocks, the aggregates of extreme blocks are at the same distance from each other, which is why they are capable of forming a quasicrystalline lattice; at the lattice points there are domains formed by the aggregated blocks of the component with a rigid chain.

The specific features of the structural organization of block copolymers and its relationship to the past history of the material have been investigated and described in a large number of publications. At present the relevant problems have been so thoroughly studied** that they are expounded in textbooks and

---

* Tsebrenko, M. V., T. I. Ablazova, G. V. Vinogradov, and A. V. Yudin, *Vysokomol. Soedin.*, **18A**, 420 (1976); Tsebrenko, M. V., *Vysokomol. Soedin.*, **20A**, 93 (1978).
** Among the first works devoted to the study of this problem, mention should be made of the following: Kraus, G., C. W. Childers, and J. T. Gruver, *J. Appl. Polymer Sci.*, **11**, 1581 (1967); Holden, G. and E. T. Bishop, *J. Polymer Sci.*, **C26**, 37 (1969). For a review of the earlier works, see: Estes, G. M., S. L. Cooper, and A. V. Tobolsky. *J. Macromol. Sci.*, **C4**, 312 (1970). This topic is still of interest to investigators. For example, the phase state of block copolymers of the A—B—A structure as a function of the chemical nature of the components and the solvent used has been thoroughly examined in: Pico, E. R. and M. C. Williams, *J. Appl. Polymer Sci.*, **22**, 445 (1978).

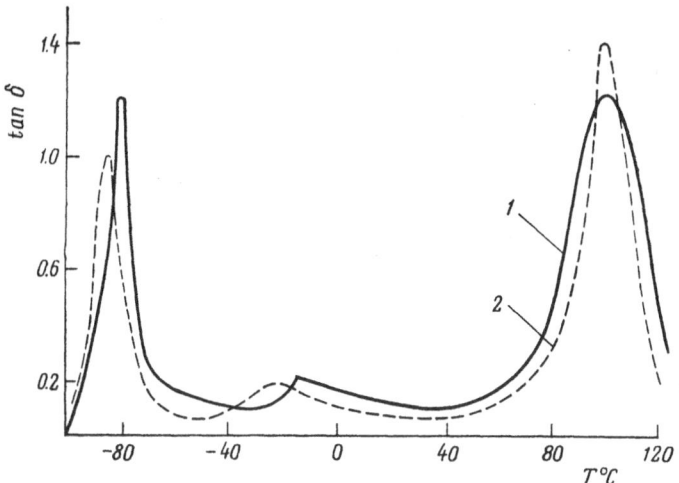

Fig. 6.15. Temperature dependences of tan δ for styrene-butadiene block copolymers of the S—B—S structure. Molecular masses of blocks: $B = 1.58 \times 10^5$; $S = 0.58 \times 10^5$ (1) and $B = 9.56 \times 10^4$; $S = 6.94 \times 10^2$ (2). [Oskin, V. N., Yu. G. Yanovsky, A. Ya. Malkin, V. N. Kuleznev, V. S. Altzitzer, and I. A. Tutorsky, *Vysokomol. Soedin.*, A14, 2020 (1972).]

need not be repeated here. The manifestations of the rheological properties of such systems have been studied much less thoroughly and they are therefore the subject of our discussion.

The concepts of the rheological properties of block copolymers are based on the separation of their relaxation spectrum into two components, each of which is determined by the molecular motions of blocks of a single type. Such a separation is possible because the points of the relaxation transitions of the blocks are at a considerable distance from one another on the temperature scale. Therefore, after the glass-transition temperature T'_g of the blocks of a flexible-chain polymer is exceeded the relaxation properties associated with these blocks are manifested. But the rigid-chain polymer remains to be in the glassy state and behaves as a solid filler, restricting the molecular motions of the flexible-chain blocks. Upon attainment of the glass-transition temperature T''_g of the rigid-chain blocks, they manifest their molecular mobility. Here the relaxation times associated with the molecular motions of the flexible-chain blocks appear to be very short since at a certain temperature $T_0 > > T''_g$ these blocks are in a state far away from their T'_g because $T''_g \gg T'_g$ and, accordingly, $T_0 \gg T'_g$.

The separation of the glass-transition temperatures T'_g and T''_g in a block copolymer is clearly seen in Fig. 6.15 in which to the transition temperatures of the blocks there correspond sharply prominent low- and high-temperature maxima of the loss tangent. The intermediate weak maximum possibly corresponds to the relaxation process taking place in the transient layer between the phases. It is essential that the temperature dependences of relaxation times, which correspond to both components of the relaxation spectrum, may differ considerably from each other since the activation energies of the molecular motions of blocks of different chemical structure are different. Therefore, it is impossible to achieve a simple temperature-time superposition of the viscoelastic characteristics of this type of block copolymers in any wide

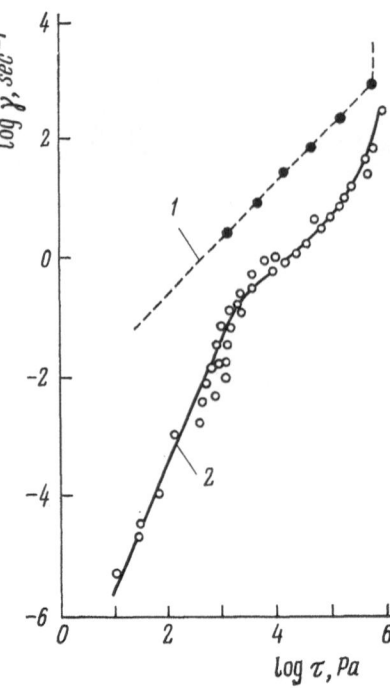

Fig. 6.16. Flow curves of monodisperse polybutadiene (*1*) and block copolymer S—B—S (*2*) with $M = 8 \times 10^5$ at 150°C. [Vinogradov, G. V., V. |E. Dreval, A. Ya. Malkin, *et |al., Rheol. Acta*, 17, 258 (1978).]

range of temperatures (times). In this sense, the block copolymers considered here should be classed as thermorheologically complex systems.*

It is usually presumed that the rheological properties of block copolymers above the glass-transition temperature T_g'' (i.e., the vitrification of a high-temperature block) are reminiscent, in all their manifestations, of the rheological properties of melts of any polymers. This impression is, however, misleading and is associated with a narrow range of shear rates and stresses studied. A more detailed investigation** in which more attention is devoted to the range of low shear stresses shows the existence of differences between the behaviour of block copolymers and the flow properties of melts of homopolymers and their similarity with filled polymers.

Typical in this respect are the data presented in Fig. 6.16 which compares the flow curves of polybutadiene and a three-block polystyrene-polybutadiene copolymer of the S—B—S structure. The molecular masses of homopolybutadiene and the polybutadiene block in the copolymer are the same. The inclusion of polystyrene blocks in the polymeric chain leads to an increase of the viscosity. This situation is quite analogous to that observed upon addition of a more viscous polymer to a less viscous one, this resulting in the formation of a blend consisting of incompatible components. At the same time, in block copolymers there may also be observed more complicated extremal dependences of viscosity; on composition, where the extremal points corres-

---

* Fesko, D. G. and N. W. Tschoegl, *J. Polymer Sci.*, C35, 51 (1971); Lim, K. K., R. E. Cohen, and N. W. Tschoegl, in: *Multicomponent Polymer Systems*, edited by R. F. Gould, *Amer. Chem. Soc.*, Washington, D.C., 1972.
** Vinogradov, G. V., V. E. Dreval, A. Ya. Malkin, Yu. G. Yanovsky, V. V. Barancheeva, E. K. Borisenkova, M. P. Zabugina, E. P. Plotnikova, and O. Yu. Sabsai, *Rheol. Acta*, 17, 258 (1978).

pond to the conditions of phase separation (this has been demonstrated*
for solutions of a block copolymer of polydimethyl siloxane and polyacrylate
of the A—B—A structure).

The most interesting difference between the rheological properties of a
homopolymer and a block copolymer appears to be the course of the flow
curves at low shear stresses. Whereas a monodisperse homopolymer typically
exhibits Newtonian flow over a very wide range of shear rates and stresses,
the viscosity of the block copolymer begins to increase sharply with decreasing shear stress, just as in the case of molten polymers with a solid active
filler illustrated in Fig. 6.6. Here the role of such a filler is played by rigid
(say, polystyrene) blocks which aggregate into domains.

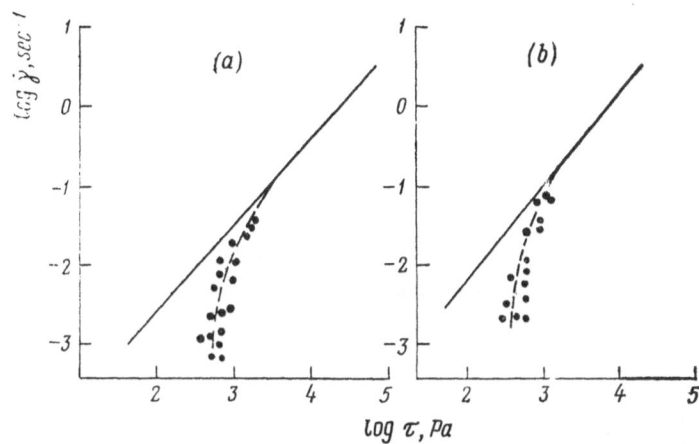

Fig. 6.17. Thixotropic behaviour of block copolymers S—B—S at 130° (a) and 150°C (b)
(see the reference to Fig. 6.16).

Like filled polymers, block copolymers exhibit thixotropy in their properties. This means that their viscosity depends on the past deformation and rest
history of the sample. A typical example is shown in Fig. 6.17 which compares
the flow curves of a sample which has preliminarily been subjected to deformation (solid lines) and fresh samples which after being charged into the
measuring device were allowed to "rest" for 4 hours (the points and the dashed
lines drawn through them). Thixotropy is observed only at sufficiently low
shear stresses.

Thixotropic changes in the structure of block copolymers can also be traced
out from the variation of their dynamic mechanical properties in the low-
frequency region.

The specific features of manifestations of the flow properties of block copolymers depend on temperature at which the measurements are carried out.
As T_g'' is exceeded the restraints imposed by the low rate of relaxation of more
rigid blocks are removed. This is reflected, for example, in the decrease of
the activation energy of viscous flow from values characteristic of more rigid blocks to much lower values corresponding to the viscous flow of flexible-
chain homopolymers. Structural changes in block copolymers, which lead
to the change of the temperature dependence of their viscosity, are not com-

---

* Nekhaenko, E. A., L. Z. Rogovina, Ya. V. Genin, G. L. Slonimsky, P. M. Valetsky,
S. B. Dolgoplosk, E. I. Levin, and L. B. Shirokova, *Vysokomol. Soedin.*, A20, 1736 (1978).

pleted at T_g''. Even at a higher temperature (140-150°C) there is observed a noticeable change of the activation energy which is supposedly associated with the morphological transitions taking place at this temperature.*

The role of a low-molecular-mass solvent introduced into a block copolymer depends on the character of its interaction with each of the components of which a copolymer is composed. If the solvent used is common to both components, then the behaviour of the single-phase solution formed is quite analogous to the behaviour of other polymer solutions. But if the solvent is good for one component and poor for the other, various anomalies of their properties may appear. For example, when a styrene-bytadiene block copolymer of star-like structure was dissolved in chlorohexadecane, in which the solubility of polystyrene is strongly dependent on temperature, no single master temperature-invariant curve of relaxation properties was found to exist,** this being typical of two-phase block copolymers (see above). This phenomenon is quite analogous to the existence of two different relaxation spectra separated with respect to temperature and exhibiting different temperature dependences of relaxation times, which has been described above for block copolymers.

---

* Chung, C. I. and J. C. Gale, *J. Polymer Sci., Polymer Phys. Ed.*, **14**, 1149 (1976); Chung, C. I. and M. I. Lin, *J. Polymer Sci., Polymer Phys. Ed.*, **16**, 545 (1978).
** Osaki, K., B.-S. Kim, and M. Kurata, *Polymer J.*, **10**, 353 (1978).

CHAPTER 7

Uniaxial Extension

7.1. Introduction

The uniaxial extension of polymer melts and solutions is one of the most important modes of their deformation. It is widely used in the fibre, spinning process, in the production of films and sheets, and is often combined with shearing flow in various technological processes. The extension conditions may be very complicated (in the sense of the dependence of stresses and rates of strain on time) and the specimens being stretched may be non-uniform along the length.

In this chapter we shall be concerned with the main rheological regularities of the extension of polymers, which are determined by the relationship among stresses, strains, and the longitudinal velocity gradient and also by the dependence of these parameters on time, temperature, and molecular properties of polymers. To establish the indicated interrelations one may limit oneself to the analysis of the simplest deformation regime—the isothermal uniform extension of samples.

In extension, just as in the case of shear, there can be realized steady-state flow conditions under which certain (sometimes, very considerable) rubbery deformations are developed. With increasing rate of straining the steady-state flow regime may appear to be unattainable. At high rates of deformation high-molecular-mass polymers and their concentrated solutions pass to a state similar in its behaviour to the state of cured (cross-linked) elastomers. This allows us to interpret this effect as a kind of transition to the forced rubbery state when the ability of the material to flow (to accumulate infinitely large irrecoverable deformations) is suppressed. The deformability of polymers in this state is limited, which predetermines the inevitability of their fracture at high rates of deformations.*

In order to choose rheological models describing the behaviour of polymeric systems, it is important to compare data obtained by studying simple shear and extension. Some models and the relevant constitutive equations (rheological equations of state), which describe well the properties of polymeric systems under the conditions of simple shear, prove to be unsuitable for the description of their behaviour in uniaxial extension, so that searches for general rheological models invariant to the deformation regime are required. The interest towards the processes of extension of polymeric systems has been lately increased, which is dictated by the above-mentioned theoretical considerations and by the very great practical importance of these processes in the technology of fibres and films. Special issues** and reviews devoted to the discussion of these problems have been published.***

---

* Vinogradov, G. V., *Polymer*, **18**, 1275 (1977).
** For example: *J. Non-Newton. Fluid Mechan.*, **4**, 1 and 2, Special Issue on Stretching Flow, 1978.
*** Hill, J. W. and J. A. Cucollo, *J. Macromolec. Sci.*, C14, 107 (1976); Malkin, A. Ya. and V. G. Kulichikhin, *Khim. Volokna*, 2, 4 (1975); C. J. Petrie, Elongational Flows, Pitman, 1979.

7.2. Kinematics of Extension

Let us consider kinematic relations that are fulfilled in the uniaxial extension of a polymeric sample which has a cylindrical form. It is assumed that
the cylinder is sufficiently long and therefore the end effects (the clamp effect
and the nonuniformity of distribution of stresses and strains near the edges
of the specimen) are negligibly small. Hence, the extension may be considered uniform along the entire length of the cylinder and the results of measurements of stresses under the specified kinematic conditions (or the rate of deformation at specified stresses) do not depend on the size of the specimen and
are completely governed by the rheological properties of the material being
stretched.

Suppose that the left end of the cylinder (Fig. 7.1) is rigidly gripped and a
force F is applied to the right end, and the sample is moving at a velocity V.

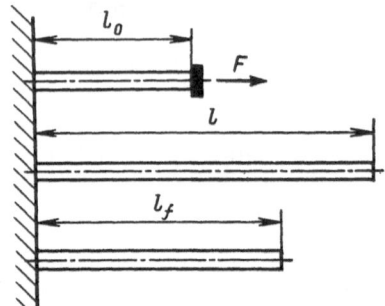

Fig. 7.1. Extension of a polymer
stream with initial length l_0.

The initial length of the sample is l_0 and the radius R_0. At a certain time t
the length of the sample becomes equal to l and the radius to R. Assuming the
absence of any volume change, we obtain:

$$l_0 R_0^2 = l R^2 \tag{7.1}$$

When the velocity changes with time, $V(t)$,

$$l(t) = l_0 + \int_0^t V(t')\, dt'$$

In particular, at $V = V_0 = \text{const}$

$$l(t) = l_0 + Vt \tag{7.2}$$

If at time t the load is removed from the right end of the sample and it is
freed, then owing to the rubbery deformations stored on extension there will
begin elastic recovery. Upon completion of this process the length of the
sample becomes equal to l_f, so that to the length difference $(l - l_f)$ there
corresponds the rubbery part of elongation and to the length difference
$(l_f - l_0)$ the elongation that results from viscous flow. The elastic recovery
of polymer samples may be accompanied by the change of the shape under
the influence of surface-tension forces.*

---

* The role of surface tension becomes especially important in the investigation of
the behaviour of dilute polymer solutions in uniaxial extension. The measurement of
their characteristics under strictly specified conditions of uniaxial extension is a very

The total Hencky strain* may be used as a quantitative measure of strain in uniaxial extension. This is necessary since the two strain components (recoverable and irrecoverable) are invariably superposed, and the result of summation must not depend on the mode and sequence of deformation development. As has been shown in Chapter 1, it is this property which is possessed by the Hencky measure of strain.

The total extension ε is expressed thr ough the extension ratio $\varkappa$:

$$\varepsilon = \ln \varkappa = \ln (l/l_0)$$

The relative deformation of viscous flow, ε_f, is expressed in an analogous manner:

$$\varepsilon_f = \ln (l_f/l_0)$$

The rubbery component of the total strain, ε_e, is defined as

$$\varepsilon_e = \ln (l/l_f)$$

It is essential that in the calculation of the rubbery deformation the quantity l refers not to the initial length of the sample, l_0, but to the quantity l_f, i.e., to the length which the sample acquires as a result of viscous flow which takes place simultaneously with the development of rubbery deformation. The indicated choice of the method of determination of rubbery deformation provides the fulfillment of the natural condition of the equality of the total relative strain to the sum of the flow and rubbery components of deformation:

$$\varepsilon = \varepsilon_f + \varepsilon_e$$

The above-given definitions of the components of the total deformation and of the total deformation itself correspond to the direction of extension and are the components of the strain tensor with the lower index 11. The other components are found from the condition of volume constancy:

$$\varepsilon_{11} + \varepsilon_{22} + \varepsilon_{33} = 0$$

Therefore, on extension of the cylinder the strain tensor $\{\varepsilon\}$ is expressed in the following way:

$$\{\varepsilon\} = \begin{vmatrix} 1 & 0 & 0 \\ 0 & -0.5 & 0 \\ 0 & 0 & -0.5 \end{vmatrix} \varepsilon$$

The rate-of-strain tensor $\{\dot{\varepsilon}\}$ has an analogous form since the rate of the relative change of volume is also equal to zero. Therefore,

$$\{\dot{\varepsilon}\} = \frac{d}{dt} \{\varepsilon\} = \begin{vmatrix} 1 & 0 & 0 \\ 0 & -0.5 & 0 \\ 0 & 0 & -0.5 \end{vmatrix} \dot{\varepsilon}$$

The rate of extension $\dot{\varepsilon}$ is expressed in the following manner:

$$\dot{\varepsilon} = \frac{d}{dt} \left[\ln \left(\frac{l}{l_0} \right) \right] = \frac{1}{l} \frac{dl}{dt} = \frac{V}{l}$$

---

complicated problem, for the solution of which numerous methodological difficulties must be overcome; see, for example, Baid, M. and A. B. Metzner, *Trans. Soc. Rheol.*, 21, 237 (1977).

* In this chapter (in contrast to Chapter 1) this measure of strain is denoted by the symbol ε without the upper index.

where V is the velocity, i.e., the rate of displacement of the free end of the sample.

If the extension occurs along the length of the specimen uniformly, then

$$\dot{\varepsilon} = \frac{dV}{dx} \qquad (7.3)$$

where the direction of the coordinate x coincides with the direction of the axis of the sample. The rate of extension is thus found to be equivalent to the longitudinal velocity gradient (rate of strain).

Suppose that extension occurs under constant velocity of one of the ends of the sample: $V = V_0 = $ const, and the second end remains fixed. This deformation regime is most easily effected in ordinary testing devices. The longitudinal velocity gradient appears then to be variable with time:

$$\dot{\varepsilon} = \frac{V_0}{l} = \frac{V_0}{l_0 + V_0 t} = \frac{\dot{\varepsilon}_0}{1 + \dot{\varepsilon}_0 t}$$

(where $\dot{\varepsilon}_0 = V_0/l_0$ is the initial velocity gradient). At $t \ll \dot{\varepsilon}_0^{-1}$ or $t \ll (l_0/V_0)$ the $V = $ const regime may be regarded as being equivalent to the $\dot{\varepsilon} = \dot{\varepsilon}_0 = $ const conditions. But, in general, the rate of deformation diminishes by a hyperbolic law, falling off at large values of t to zero. Therefore, the $V = $ const experiment, which is very often employed in laboratory practice, does not provide the constancy of the rate of strain. Thus, at different times or at different stages of deformation the sample is under non-equivalent kinematic conditions since the characteristic of strain kinematics is the rate of deformation rather than the rate of extension.

The equivalency of kinematic conditions at the various stages of extension is provided by the fulfillment of the $\dot{\varepsilon} = \dot{\varepsilon}_0 = $ const condition. Then

$$\ln (l/l_0) = \dot{\varepsilon}_0 t \qquad (7.4)$$

that is, the change of the length of the specimen with time may take place according to the law:

$$l = l_0 e^{\dot{\varepsilon}_0 t} \qquad (7.5)$$

The rate of elongation $V = dl/dt$ is expressed as follows:

$$V(t) = l_0 \dot{\varepsilon}_0 e^{\dot{\varepsilon}_0 t} = V_0 e^{\dot{\varepsilon}_0 t} \qquad (7.6)$$

Thus, the kinematic uniformity of straining in time is provided if the extension regime is experimentally realized in accordance with the law given by formula (7.6). This is fulfilled in practice if the operating conditions of the drive that provides the extension of the sample is determined by a programmed device* which operates by the law (7.6). Another possibility of effecting extension under the $\dot{\varepsilon} = $ const regime is the deformation of the stream between two pairs of rollers which rotate in different directions at different speeds.** The length of the specimen extended is in this case constant, which leads to the fulfillment of the $\dot{\varepsilon} = $ const condition.

The basic kinematic feature of the experiment carried out under extension conditions consists in the change of the length and cross-section of the speci-

---

* Vinogradov, G. V., B. V. Radushkevich, and V. D. Fikhman, *J. Polymer Sci.*, A-2, **8**, 1 (1970).

** Meissner, J., *Rheol. Acta*, **8**, 78 (1969); *Trans. Soc. Rheol.*, **16**, 405 (1972).

men. This complicates the measurements with the dynamic test conditions being specified. For example, if a constant force F_0 is applied to the specimen, the true tensile stress σ varies by the law:

$$\sigma = \frac{F_0}{\pi R_0^2} = \sigma_0 \frac{l}{l_0} = \sigma_0 \left[1 + l_0^{-1} \int_0^t V(t)\, dt \right] \tag{7.7}$$

where $\sigma_0 = F_0/\pi R_0^2$ is the initial stress corresponding to the condition $(l - l_0)/l_0 \ll 1$.

As the specimen becomes thinner and thinner the stresses increase considerably, which causes the rate of deformation to increase. Therefore, in order to ensure the regime of uniaxial extension at a constant true stress it is necessary to carry out measurements with the force varying in time. According to Eq. (7.7), for the $\sigma = \sigma_0 = $ const condition to be provided, it is necessary that the force as a function of time, $F(t)$, vary by the law:

$$F(t) = \frac{\sigma_0 \pi R_0^2}{1 + l_0^{-1} \int_0^t V(t)\, dt} \tag{7.8}$$

In practice, this can be done by means of programmed control of the force according to a specified law*; this control must be based on the variable length of the specimen or on its radius.

In realizing the above-considered extension regimes it is possible to find the total deformation of specimens. But for a quantitative estimation of their flow and rubber-like properties it is necessary to separate the total deformation into the irrecoverable and recoverable components. After the prestationary deformation regime is completed, when the rubbery deformation reaches an equilibrium value, all the further deformation is caused by viscous flow. The viscosity of the material can then be evaluated from the rate of development of total strain (which is equal to the rate of irrecoverable flow) without resort to the separation of deformation into the components. This is possible only for the extension under the $\sigma = $ const and $\dot{\varepsilon} = $ = const conditions since otherwise, because of the non-constancy of the deformation conditions, the rubbery deformation is continually changed and, hence, the total rate of deformation is not equal to the rate of deformation of viscous flow.

Various methods of investigation of polymers in extensional flow have been repeatedly proposed in the literature.** The most simple method consists in, for example, the measurement of the pressure gradient during the flow of polymers in converging ducts where the extensional velocity gradient is known to arise. This method has been proposed by Cogswell*** and has since been repeatedly used.**** Methods of this type have provided, in a number of cases, results that correlate well with data obtained by direct methods. Because of the absence of an adequate theoretical substantiation, however, such methods lack the required generality, and that is why the

---

* Vinogradov, G. V., V. D. Fikhman, and B. V. Radushkevich, *Rheol. Acta*, 11, 286 (1972); Dealy, J. et al., *Trans. Soc. Rheol.*, 20, 455 (1976).
** Kulichikhin, V. G. and A. Ya. Malkin, *Khim. Volokna*, 1, 16 (1975); Dealy, J. M., *J. Non-Newton. Fluid. Mech.*, 4, 9 (1978).
*** Cogswell, F. N., *Polymer Eng. Sci.*, 12, 64 (1972).
**** Shroff, R., L. V. Cancio, and M. Shida, *Trans. Soc. Rheol.*, 21, 429 (1977).

results obtained by m$_e$ans of these methods are sometimes at least controversial.

In considering the kinematics of deformation of polymer sample attention
should also be focused on non-uniform extension.* This case is realized when
one end of the cylindrical specimen is secured to the rotating roller and the
other is extended by the applied constant force. The main task here is to
establish the relation between the kinematics of uniform and non-uniform
deformation with the same initial tensile stress. The aim is to calculate the
specimen profile along the length on inhomogeneous strain from the known
mode of change of strain with time on uniform deformation.

Let the z axis be directed along the axis of the sample stretched and V
be the extensional velocity which depends on z. The velocity profile is expressed in the following manner:

$$V(z) = V_0 + \int_0^z \dot{\varepsilon}(\xi) \, d\xi$$

where $\dot{\varepsilon}$ is the rate of deformation and V_0 is the value of V at the initial point
at $z = 0$. If the volume rate that provides the formation of a polymer fibre
is Q, then $V = Q/\pi R^2$, where R is the radius of the extrudate, which depends
on z. Here

$$V = V_0 (R_0/R)^2 = V_0 (l/l_0) = V_0 \varkappa$$

where R_0 and l_0 are the initial values of the radius and length of the extrudate
and $\varkappa$ is the degree of elongation.

The relation between the time of deformation t in uniform straining and
the coordinate z in non-uniform extension is given by

$$z = \int_0^t V(t_1) \, dt_1 = V_0 \int_0^t \varkappa(t_1) \, dt_1$$

The profile of the specimen, i.e., the dependence $R(z)$, is calculated from the
last formula with account taken of $R = R_0 \varkappa^{-1/2}$ if the dependence $\varkappa(t)$
in uniform extension is known (experimentally determined).

The above-written relations determine the kinematics of non-uniform deformation of the polymer specimen being stretched.

7.3. Rheological Relations for Uniaxial Extension (Theory)

For a purely viscous liquid whose viscosity depends on the second invariant of the rate-of-strain tensor and whose apparent shear viscosity decreases with growth of the shear rate, the viscosity in extension (estimated as
$\sigma_{11}/\dot{\varepsilon}$) must also diminish with increasing elongational velocity gradient.
This conclusion contradicts what is known about the extension of polymeric
systems, whose viscosity may increase on extension. Therefore, the basic
regularities of the extension of polymers are due to their viscoelastic prop-

---

* The discussion of this problem is based on the results reported in: Prokunin, A. N.
and N. G. Proskurnina, *Inzhenerno-Fizich. Zhurn.*, 34, 629 (1 978); *Khim. Volokna*, 5, 22
(1978).

erties, i.e., to the fact that on extension the irrecoverable and rubbery deformations are superimposed. Of greatest importance is also the orientation effect which is intensified with increasing elongational velocity gradient.

7.3.1. Extension of Linear Viscoelastic Fluids

At sufficiently low stresses and rates of deformation the behaviour of polymeric systems is described by the relations of the theory of linear viscoelasticity, and all the peculiarities of the behaviour of the material under any deformation regime can be determined if its relaxation spectrum is known. The concept of linear viscoelasticity is an asymptotic concept of the real properties of the material at infinitesimally low stresses. Experimentally within the measurement error, the "linear region" covers a more or less wide range of deformation conditions. The boundary of the "linear" behaviour depends on the nature of the material: it may be in the range of very low stresses or it may be shifted to the side of large stresses, covering practically the entire region of the fluid state of the polymer. The latter is especially typical of narrow-distribution polymers with a molecular mass considerably exceeding M_c. It is important that for every polymer the "linear" behaviour in uniaxial extension is invariably preserved over a wider range of stresses than in simple shear.

Whether the behaviour of the material conforms to the theory of linear viscoelasticity may be judged from its integral characteristics, for example, the viscosity or the rubbery modulus. The constancy of such parameters is, however, a necessary but insufficient criterion of "linearity" since various nonlinear effects may reveal themselves in transient deformation regimes. Therefore, in order to judge about whether the behaviour of the material is "linear" it is necessary, in a general case, to confirm that such characteristics of the system as, for example, the relaxation or creep functions, are independent of the deformation regime.

Suppose that the rheological properties of a material are described by relations of the theory of linear viscoelasticity and are characterized by the creep function $\psi(t)$ or the relaxation function $\varphi(t)$. Then, under the deformation conditions $\dot{\varepsilon} = \dot{\varepsilon}_0 = \text{const}$ the variation of stresses with time is described by the formula:

$$\sigma(t) = 3\dot{\varepsilon}_0 \int_0^t \varphi(t')\, dt'$$

The rate of accumulation of irrecoverable deformation, $\dot{\varepsilon}_f$, is expressed in the following way:

$$\dot{\varepsilon}_f = \frac{\sigma(t)}{3\eta} = \frac{\dot{\varepsilon}_0}{\eta} \int_0^t \varphi(t')\, dt'$$

and the variation of the recoverable deformation with time, $\varepsilon_e(t)$, occurs (the instantaneous component being neglected) in accordance with the formula

$$\varepsilon_e(t) = \frac{1}{3} \int_0^t \frac{d\sigma(t')}{dt'}\, \varphi(t-t')\, dt'$$

At $t \to \infty$ there are obtained a number of obvious relations:

$$\dot{\varepsilon}_f = \dot{\varepsilon}_0$$

$$\lambda = \sigma(\infty)/\dot{\varepsilon}_0 = 3\eta$$

$$\varepsilon_e = \dot{\varepsilon}_0(\eta/G) = \sigma(\infty)/?G = \sigma(\infty)/E$$

where λ is the extensional viscosity defined as the ratio of the stress and rate of accumulation of irrecoverable elongational strain; E is the rubbery modulus in uniaxial extension; η and G are the values of viscosity and rubbery modulus (the integral characteristics of the relaxation spectrum) measured at low stresses (in the linear region) under shear deformation conditions.

The theory of linear viscoelasticity presupposes the constancy of viscosity. This means that the initial values of viscosity are considered, $\lambda_0 = 3\eta_0$, where λ_0 is the initial (Trouton) value of viscosity. Thus, within the framework of the theory of linear viscoelasticity, for a viscoelastic liquid the extensional viscosity is equal to the tripled viscosity measured in shear ($\lambda_0 = 3\eta_0$), and the rubbery modulus in extension is equal to the tripled shear modulus ($E = 3G$). In the prestationary deformation regime the viscosity remains constant and equal to λ_0. The theory of linear viscoelasticity does not predict any new results (as compared with the theory of the viscous Newtonian liquid and the Hookean elastic body) with respect to steady-state deformation regimes.

In the transient (prestationary) stage of deformation with the assignment of the $\dot{\varepsilon} = $ const conditions the variation of the stress with time is described by the formula

$$\sigma(t) = 3\dot{\varepsilon} \int_0^\infty F(\theta)(1 - e^{-t/\theta})\,d\theta$$

or

$$\frac{\sigma(t)}{\dot{\varepsilon}\lambda_0} = \frac{1}{\eta} \int_0^\infty F(\theta)(1 - e^{-t/\theta})\,d\theta$$

where $F(\theta)$ is the relaxation spectrum determined in shear deformation; η is the viscosity under steady-state shearing flow.

From the last formula it follows that if the polymer behaves as a linear viscoelastic body, then both under steady-state and transient deformation regimes (for example, under the prestationary regimes when the polymer passes from the state of rest to steady-state flow) the variation of $\sigma/\dot{\varepsilon}\lambda_0$ with time must be described invariantly to temperature, the rate of deformation, and the amount of deformation ε (time). The temperature invariancy is determined by the fact that the relaxation time spectra at different temperatures are similar and can be made to superpose by shifting them along the time axis. This shift is determined by the dependence of the initial viscosity on temperature, which corresponds to the normalization of time by the initial viscosity: $t_r = t/\eta$.

What has just been said is illustrated by Fig. 7.2 for 1,2-polybutadiene which was subjected to constant rates of deformation. The solid line in Fig. 7.2 was calculated from data on the relaxation spectrum at 25°C, which had been obtained on the basis of measurements of the components of the complex modulus under the regimes of small-amplitude shear deformation. The

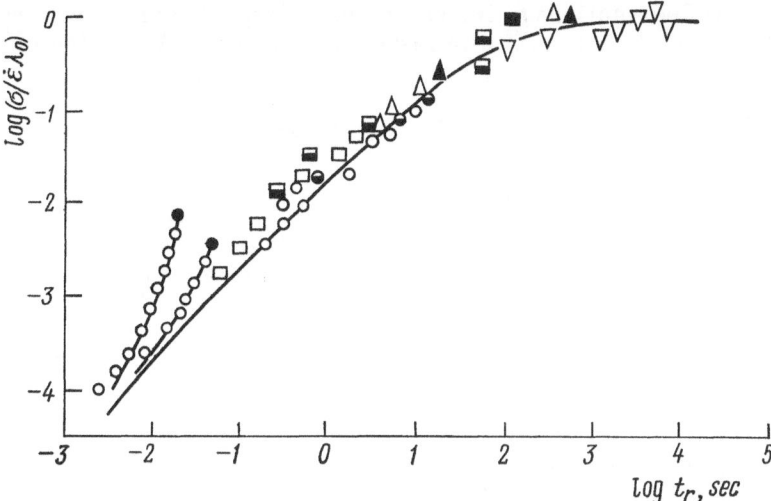

Fig. 7.2. Dependence of reduced viscosity on time for the extension of 1,2-polybutadiene (molecular mass 1.7×10^5, $\overline{M}_w/\overline{M}_n = 1.70$; the contents of 1,2-*cis* groups, 84 per cent; 1,4-*trans* groups, 8 per cent; T_g, 255 K) at various rates of deformation and temperatures (reduced to 25°C). The unfilled symbols correspond to the pre-fracture regimes or steady-state flow; the filled symbols correspond to the fracture of samples; the half-filled symbols correspond to the start-up of necking:

(○ = 0°C; □ = +10°C; △ = +25°C; ▽ = +50°C); left curve $\dot{\varepsilon} = 4 \times 10^{-1}$ sec^{-1} (0°C); next curve $\dot{\varepsilon} = 2 \times 10^{-1}$ sec^{-1} (0°C); right curve $\dot{\varepsilon} = 4 \times 10^{-4}$ to 4×10^{-1} sec^{-1} (0-50°C).

polybutadiene sample investigated had a high glass-transition temperature. Upon approach to T_g and with increasing rate of deformation there is observed an increasing departure from linear behaviour, which may be accompanied by the fracture of the samples. On the other hand, as the temperature mounts and the rate of deformation falls the attainment of steady-state flow is facilitated. To this there corresponds the condition $\lim\limits_{t \to \infty} (\sigma/\dot{\varepsilon}\lambda_0) = 1$.

Evidently, the theory of linear viscoelasticity makes it possible to consider the various cases of deformation. Thus, on deformation under the $V = = V_0 = $ const regime the variation of stresses and rates of accumulation of irrecoverable deformation are described by the formulas*:

$$\sigma\,(t) = 3\dot{\varepsilon}_0 \int\limits_0^t \varphi(t-t') \frac{dt'}{1+\dot{\varepsilon}_0 t'}$$

$$\dot{\varepsilon}_f\,(t) = \frac{\dot{\varepsilon}_0}{\eta} \int\limits_0^t \varphi\,(t-t') \frac{dt'}{1+\dot{\varepsilon}_0 t'}$$

where $\dot{\varepsilon}_0 = V_0/l_0$. In this case, the extensional viscosity remains to be equal to 3 η and does not vary from the start-up of deformation up to the attainment of steady-state flow conditions. The stresses first increase because of the increase of deformation and then, at $t \to \infty$, fall off down to zero since at $t \to \infty$ the rate of deformation decreases to zero and, accordingly, $\dot{\varepsilon}_f$ also decreases. Rubbery deformations stored in the material vary in the same way

---

* Vinogradov, G. V., A. I. Leonov, and A. N. Prokunin, *Rheol. Acta*, 8, 482 (1968).

since at low deformations ε_e increases and at large deformations, because of the decrease of the stress, it diminishes and at $t \to \infty$ the value of $\varepsilon_e \to 0$.

7.3.2. Extension of Viscoelastic Nonlinear Liquids

In order to quantitatively describe the dependence of extensional viscosity on the elongational velocity gradient, it is necessary to make use of a model of a viscoelastic body. A typical example is the behaviour of a viscoelastic liquid with one relaxation time θ (the Maxwell model) in uniaxial extension in which the possibility of large deformations is taken into account. Just as has been done in the preceding chapters in considering the effect of finite deformations on the stresses arising in steady-state shearing flow in this case the partial time derivative is replaced by the various differential operators describing the movement of the point and the associated coordinate system on deformations in space.

Thus, let the constitutive equation of a viscoelastic liquid be written in the form of an operator equation:

$$(1+\theta D)\sigma'_{ij} = 2\eta\dot{\gamma}_{ij} \tag{7.9}$$

where D is a certain differential operator; σ'_{ij} are the components of the stress tensor deviator; $\dot{\gamma}_{ij}$ are the components of the rate-of-strain tensor (the tensor $\{\dot{\gamma}\}$ is a deviator since its first invariant is equal to zero). In steady-state flow on extension with a constant elongational velocity gradient $\dot{\varepsilon}_0$ the diagonal components of the tensor $\{\dot{\gamma}\}$ are equal to $\dot{\gamma}_{11} = \dot{\varepsilon}_0$, $\dot{\gamma}_{22} = \dot{\gamma}_{33} = -\dot{\varepsilon}_0/2$, and all the non-diagonal components are equal to zero.

Let D be the Oldroyd linear operator. Then, for steady-state flow conditions at $\partial\sigma_{11}/\partial t = 0$, the constitutive equation (7.9) decomposes into the following three equalities:

$$\begin{cases} \sigma'_{11} - 2\theta\dot{\varepsilon}_0\sigma'_{11} = 2\eta\dot{\varepsilon}_0 \\ \sigma'_{22} + \theta\dot{\varepsilon}_0\sigma'_{22} = -\eta\dot{\varepsilon}_0 \\ \sigma'_{22} = \sigma'_{33} \end{cases} \tag{7.10}$$

In order to obtain the value of the stress σ_{11} from these equations, one must make use of the equality $\sigma_{ii} = -p + \sigma'_i$ and, neglecting the surface-tension forces, write

$$\sigma_{22} = \sigma_{33} = -p + \sigma'_{22} = 0$$

Then, from the second equation of the system (7.10) it follows that the hydrostatic pressure (which is far from being equal to the external pressure) is expressed in terms of the velocity gradient:

$$p = -\frac{\eta\dot{\varepsilon}_0}{1+\theta\dot{\varepsilon}_0}$$

Using the first equation of the system (7.10), one can obtain

$$\sigma_{11} = -p + \sigma'_{11} = \frac{\eta\dot{\varepsilon}_0}{1+\theta\dot{\varepsilon}_0} + \frac{2\eta\dot{\varepsilon}_0}{1-2\theta\dot{\varepsilon}_0} = \frac{3\eta\dot{\varepsilon}_0}{(1+\theta\dot{\varepsilon}_0)(1-2\theta\dot{\varepsilon}_0)}$$

It thus follows that the extensional viscosity on extension is expressed as a function of the velocity gradient:

$$\lambda = \frac{\sigma_{11}}{\dot{\varepsilon}_0} = \frac{3\eta}{(1+\theta\dot{\varepsilon}_0)\,(1-2\theta\dot{\varepsilon}_0)} \tag{7.11}$$

Theory predicts that at low elongational velocity gradients of the rates of extension (at $\dot{\varepsilon}_0 \ll \theta^{-1}$) the value of $\lambda = 3\eta = \lambda_0$, but as the velocity gradient increases the extensional viscosity increases monotonically and at $\dot{\varepsilon}_0 \to (2\theta)^{-1}$ the extensional viscosity increases infinitely: $\lambda \to \infty$. At velocity gradients higher than $(2\theta)^{-1}$ the steady-state extensional flow is found in general to be impossible.

Let us examine the case of uniaxial compression which is the reverse of uniaxial extension with respect to kinematics. For an ordinary viscous liquid, when the extension is replaced by compression, all the rheological characteristics of the material (with an accuracy of up to a digit) remain unchanged. But for a viscoelastic material, compression is not the reverse of extension. This is seen from the relations given below. To the compression there corresponds the following rate-of-strain tensor:

$$\{\dot{\gamma}\} = \begin{vmatrix} -1 & 0 & 0 \\ 0 & 0.5 & 0 \\ 0 & 0 & 0.5 \end{vmatrix} \dot{\varepsilon}_0$$

and therefore Eq. (7.10) is replaced by the following system of equations:

$$\begin{cases} \sigma'_{11} + 2\theta\dot{\varepsilon}_0\sigma'_{11} = 2\eta\dot{\varepsilon}_0 \\ \sigma'_{22} - \theta\dot{\varepsilon}_0\sigma'_{22} = -\eta\dot{\varepsilon}_0 \\ \sigma'_{22} = \sigma'_{33} \end{cases} \tag{7.12}$$

Repeating all the calculations carried out for uniaxial extension and determining the viscosity on compression, $\bar{\lambda}$, just in the same way as for any other deformation regime by means of the ratio $(\sigma_{11}/\dot{\varepsilon}_0)$, one can find that

$$\bar{\lambda} = \frac{3\eta}{(1-\theta\dot{\varepsilon}_0)\,(1+2\theta\dot{\varepsilon}_0)} \tag{7.13}$$

Thus, for the model (7.9) generalized to finite deformations according to Oldroyd, the viscosity on extension λ appears to be unequal to the viscosity on compression $\bar{\lambda}$. This result shows that, in principle, for a viscoelastic liquid with arbitrary rheological properties, in spite of the kinematic reversibility of extension and compression, the following inequality may be fulfilled: $\lambda \neq \bar{\lambda}$.

If we now recall (see Chapter 2) that a viscoelastic liquid, whose rheological properties are described by Eq. (7.9) with the Oldroyd derivative, does not display non-Newtonian behaviour in shear, the general picture of variation of the "viscosities" of this liquid, depending on the velocity gradient in the three deformation schemes under steady-state flow conditions, is shown in Fig. 7.3.

Thus, a liquid that does not exhibit non-Newtonian behaviour in shearing flow can show the effect of viscosity increase on extension as a result of the developing rubbery deformations.

The effect of non-Newtonian viscosity in shearing flow is described in a natural way when use is made of the constitutive equation (7.9) generalized to the case of finite deformations with the aid of the Jaumann derivative (see Chapter 1). But for uniaxial extension this model does not predict the appearance of any new effects different from those which are known to arise for a purely viscous liquid, i.e., for such a viscoelastic material

$$\lambda = \bar{\lambda} = 3\eta$$

This conclusion is physically substantiated by the fact that the effect of non-Newtonian viscosity in the Jaumann model arises because of the rotation of the coordinate system associated with a given point during the deformation of the body. In uniform uniaxial extension the elements of the body

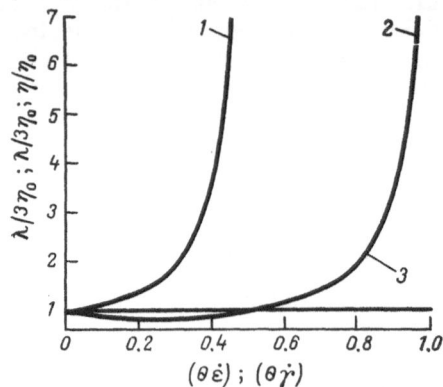

Fig. 7.3. Variation of the viscosity of a viscoelastic "Oldroyd" liquid with one relaxation time as a function of the rate of deformation:

1—extensional viscosity in extension, $\lambda/3\eta_0$; 2—extensional viscosity in compression, $\bar{\lambda}/3\eta_0$; 3—viscosity in shearing flow, η/η_0.

do not rotate and therefore the viscoelastic body behaves as a Newtonian liquid.

Thus, the use of the Jaumann derivative does not allow one to put forth any plausible arguments about the character of the rheological behaviour of a viscoelastic body on extension. Since the Oldroyd derivative does not permit one to do so with respect to shearing flow, it is obvious that none of these models can provide a correct description of the rheological properties of a viscoelastic liquid either in extension or in shear, and therefore, in a general case, they appear to be inadequate for the real properties of viscoelastic material.

Of the more complicated models of viscoelastic materials it is expedient to consider the Spriggs model, which is the model of a viscoelastic liquid with a known relaxation spectrum and is generalized to the case of finite deformations with the aid of a differential operator of a rather complex structure (see Chapter 1).

For the kinematics of a motion corresponding to uniaxial extension the extensional viscosity λ is found to be equal to:

$$\lambda = \frac{\sigma_{11}}{\dot{\varepsilon}_0} = 3\eta \, \frac{1}{\zeta(\alpha)} \sum_{p=1}^{\infty} \frac{1}{p^{\alpha} - (1+\varepsilon)(\theta_m \dot{\varepsilon}_0)} \tag{7.14}$$

where α is an exponent which characterizes the frequency of the relaxation-time distribution in the spectrum; θ_m is the maximum relaxation time; $\zeta(\alpha)$ is the Riemann zeta function.

Some special cases of interest directly follow from this model. If $\varepsilon = -1$, which conforms to the De Witt's model, then $\lambda = 3\eta$, as has been obtained above. If $\varepsilon = 0$, which corresponds to the Oldroyd generalized (nonlinear) model, then formula (7.14) predicts the growth of the extensional viscosity with increasing velocity gradient, this growth being the same as in the case of the use of the Oldroyd linear operator. In this model, however, the increase of the extensional viscosity is accompanied by a decrease of the apparent viscosity in shearing flow (see Chapter 2). This shows that there exist such methods of generalization of the constitutive equations of linear viscoelastic materials, which correctly describe the behaviour of a liquid both in extension and in shear.

Other nonlinear models of the rheological behaviour of viscoelastic liquids have also been discussed in the literature; these models provide a correct description of the behaviour of viscoelastic liquids both in shear deformations and in extension* (many other investigations in this line are also known).

The above-considered results of the use of rheological models of viscoelastic materials for the analysis of extensional flow refer to systems whose relaxation spectrum and, hence, their viscoelastic properties do not depend on the intensity of deformation. Incidentally, as is well known for the case of shear deformation, the increase of the stress intensity leads to the change of the relaxation properties of the system. The same effect must be observed in extension since the shear viscosity coefficient appearing in all formulas for extensional viscosity decreases with increasing intensity of mechanical action on the system.

A quantitative characteristic of the effect of the stress intensity on the coefficient η is its dependence on the second invariant T_2 of the rate-of-strain tensor. In extension (see Chapter 1)

$$T_2 = -\frac{3}{4}\dot{\varepsilon}_0^2$$

and in shear

$$T_2 = -\frac{1}{4}\dot{\gamma}_0^2$$

Therefore, the condition of the equivalency of the intensity of action on the material in shear and extension is fulfilled if

$$\dot{\varepsilon}_0 = \dot{\gamma}_0/\sqrt{3} \tag{7.15}$$

If one proceeds from the concept of the existence of an entanglement network in polymers, it may be said that the entanglement network density on deformation of polymers may decrease, which is accompanied by the decrease of the apparent viscosity and is usually treated as the destruction of the polymer structure. On the other hand, upon deformation of polymers the macromolecules may be oriented in the direction of strain, which is characteristic of the uniaxial extension and leads to the increase of the resistance to deformation and to the corresponding growth of the viscosity.

It is important to note that, though the orientation effect plays an essential part in uniaxial extension, in this case the polymer structure is likely to be

---

* Litvinov, V. G., *Mekhanika Polimerov*, 2, 326 (1967); White, J. L. and Y. Ide, *Appl. Polymer Symp.*, 27, 61 (1975); Agrawal, P. K., W. K. Lee, J. M. Lorntson, C. I. Richardson, K. F. Wissburn, and A. B. Metzner, *Trans. Soc. Rheol.*, 21, 355 (1977).

destroyed. Therefore, under various extension regimes and at various stages
of its development there may be observed an increase or decrease of the apparent viscosity; it may also remain constant.

7.3.3. Extensional Viscosity of Solutions (Molecular Models)

The consideration of the theory of extensional flow of dilute polymer solutions (though this flow regime is very difficult to realize in practice) allows
one to judge about the extent to which the viscoelastic properties of the macromolecule may be the original cause of the behaviour of polymeric systems
in uniaxial extension. This problem is analogous to the one discussed in the
analysis of the viscoelastic properties of individual polymeric chains in shear
(see Chapter 3) where the rheological properties of the system were explained
on the basis of the relaxation spectrum of separate macromolecules whose
motion is composed of independent displacements of segments. This approach
consists basically in constructing a physical model for a polymeric system and
in considering the behaviour of this model in shear and extension.

The simplest systems capable of simulating the properties and behaviour
of dilute polymer solutions are dilute suspensions.* In its turn, the simplest
shape of particles that can orient in the flow are ellipsoids. Therefore, the
behaviour of a suspension of rigid ellipsoids during flow in the field of velocities with the longitudinal and transverse gradients permits one to disclose
the effect of the orientation factor on the character of the dependences
$\eta\,(\dot{\gamma})$ and $\lambda\,(\dot{\varepsilon})$. Each particle in flow is acted by the viscous friction forces
exerted by the surroundings and the forces associated with the Brownian mo-
tion of the particle itself. Under the influence of the velocity gradient the
particles tend to be oriented in the stream in a strictly definite way. As a
result of this, in the steady-state flow there is set up a certain equilibrium
distribution of orientations of the axes of the particles, which depends both
on the inherent properties of the particles (their size, shape and the diffusion
coefficient) and on the velocity gradient. The sum of friction losses on deformation of such a suspension is determined by the distribution of the orientations of the axes of particles relative to the direction of the velocity gradient. The distribution of the orientations is possible only if the particles are
anisodiametric; in a suspension of spherical particles all the directions of
orientations are equally probable and the increase of the velocity gradient
does not alter the structure of the system.

It is essential that the equilibrium distribution of the orientations of ellipsoids in flow depends on the flow geometry.** Here the function $\eta\,(\dot{\varepsilon}_0)$
is decreasing, but the function $\lambda\,(\dot{\varepsilon}_0)$ is found to be increasing and its form
depends on the relations between the properties of the particles and the ve-
locity gradient. This theoretical result shows that a solution whose rheological properties are characterized in shear by non-Newtonian viscosity
(the apparent viscosity decreases with increasing shear rate) may behave in
extension so that the extensional viscosity increases with increasing velocity
gradient.

** Pokrovsky, V. N., *Statistical Mechanics of Dilute Suspensions,* Nauka Publishers,
Moscow, 1978 (in Russian).

* Takserman-Krozer, R., *J. Polymer Sci.,* A, 1, 2477 and 2487 (1963).

Thus, even the simplest model of a suspension of rigid ellipsoids enables one to predict qualitatively the fundamental difference in the behaviour of a polymer during shearing flow and extension and shows that the relationship between the viscosities in shear and extension may be sufficiently complicated and ambiguous.

In a more realistic model of a polymer the macromolecule is thought to be a viscoelastic thread of a permeable coil with a statistical distribution of segments relative to the centre of mass. The viscoelastic properties of such a model on shear have been considered in detail in Chapter 3. It has been shown that the apparent viscosity of the model is independent of the rate of shear within the framework of the theory of linear viscoelasticity. If we analyse the rheological properties of the molecular model in uniaxial extension, it will appear* that the extensional viscosity should be expected to increase with increasing velocity gradient. The exact form of the dependence $\lambda(\dot{\varepsilon})$ is determined by the numerical values of the parameters of the model.

The increase of the extensional viscosity with increase of the velocity gradient during the extension of a viscoelastic permeable coil is caused by two factors—the orientation mechanism analogous to the one described for a suspension of rigid ellipsoids (with the only difference that the molecular coil exhibits forced anisotropy created by the velocity gradient itself; anisotropy of this kind may be called "deformation anisotropy") and the relaxation mechanism associated with finite deformations of a viscoelastic material and analogous to the one that leads to the increase of the viscosity of a Maxwell liquid with a single relaxation time on large deformations. Quantitative predictions of the theory of the extensional flow of a suspension of viscoelastic random coils depend on the choice of the model of the coil itself and on the method used to take account of large elastic deformations. Therefore, theoretical results prove to be ambiguous, though, in principle, they permit one to explain and describe the observed form of the function $\lambda(\dot{\varepsilon})$ on the basis of the concept of the relaxation spectrum of the material.

Molecular models such as the Kargin-Slonimsky-Rouse and Kirkwood-Riseman-Zimm models (see Chapter 3) are models of viscoelastic materials with a discrete distribution of relaxation times θ_p. The variation of the extensional viscosity λ on extension for a liquid with a single relaxation time, with account taken of the Oldroyd finite strains, is predicted by formula (7.13). The superposition of various relaxation mechanisms leads to the summation of the contributions of each of them to the extensional viscosity. Therefore, for the model of the permeable coil with the relaxation time spectrum θ_p (with account taken of finite strains with the aid of the Oldroyd operator) the dependence $\lambda(\dot{\varepsilon})$ has the form (after Bird and coworkers):

$$\lambda(\dot{\varepsilon}) = 3N_0 kT \sum_p \frac{\theta_p}{(1+\theta_p\dot{\varepsilon})(1-2\theta_p\dot{\varepsilon})} \tag{7.16}$$

where N_0 is the number of chains in unit volume; k is Boltzmann's constant; T is the absolute temperature; η_0 is the initial (Newtonian) viscosity in shearing flow.

---

* A complete and detailed account of theoretical results on the motion of suspensions of viscoelastic particles having the shape of dumbbells under various deformation conditions is given in: Bird, R. B., H. R. Warner, Jr., and D. S. Evans, *Adv. Polymer Sci.*, **8**, 1 (1971).

The correspondence etween formulas (7.13) and (7.16) is obvious. This model, however, does not predict the effect of non-Newtonian viscosity in shearing flow, although in simple shear to this model there corresponds the appearance of normal stresses proportional to $\dot{\gamma}^2$.

Molecular models lead practically to the same quantitative results that are obtained by means of the phenomenological models* with the only difference that the constants contained in the final formulas are assigned a definite physical meaning. This outcome is natural since the molecular models manipulate with the same basic concepts and ideas used in the phenomenological models. The most important of them are: first, the concept of the relaxation spectrum of the system and the effect of strain intensity on the relaxation properties of the system and, second, the method of transformation from the convected to the space-fixed coordinate system. The first takes into account the specific response of a polymer to external effects as the viscoelastic relaxation; the second takes account of the geometric effects due to large elastic deformations of the material. The combination of these factors practically governs all the observed or theoretically considered specific features of the rheological properties of a polymer under any deformation regime. Depending on the strain geometry (for example, in extension or in shear) the mutual interference of these factors may be varied, which gives rise to differences in the manifestation of the rheological properties of the material, depending on the deformation regime.

7.4. Viscoelastic Properties of Fluid Polymers in Uniaxial Extension (Experiment)

7.4.1. The Prestationary Stage of Extension

In this section we shall be concerned with the variation of the extensional viscosity λ and other characteristics of the properties of fluid polymers in the prestationary stage of extension, when, depending on deformation, there occurs not only viscous flow but rubbery (recoverable) deformations are also developed. For an understanding of the general regularities of the behaviour of polymeric materials in transient regimes, the separation of the total deformation into the recoverable and irrecoverable components is of fundamental importance. The typical dependences of ε_e and ε_f on the total deformation at various velocity gradients (rates of straining) $\dot{\varepsilon}$, maintained constant during the extension, are given in Fig. 7.4. The maximum values of the total draw ratio $\varkappa$ in these experiments were approximately equal to the 9-fold elongation of the specimen.

To the attainment of steady-state flow conditions there corresponds the constancy of the quantity ε_e and, accordingly, the fulfillment of the equality $\dot{\varepsilon}_f = \dot{\varepsilon}$. Measurements of the irrecoverable and rubbery components of the total deformation enable one to calculate the variation of the extensional

---

* See, for example, the results of calculations of extensional viscosity for a model with one relaxation time: Everage, A. E., Jr., and R. L. Ballman, *J. Appl. Polymer Sci.*, **21**, 841 (1977).

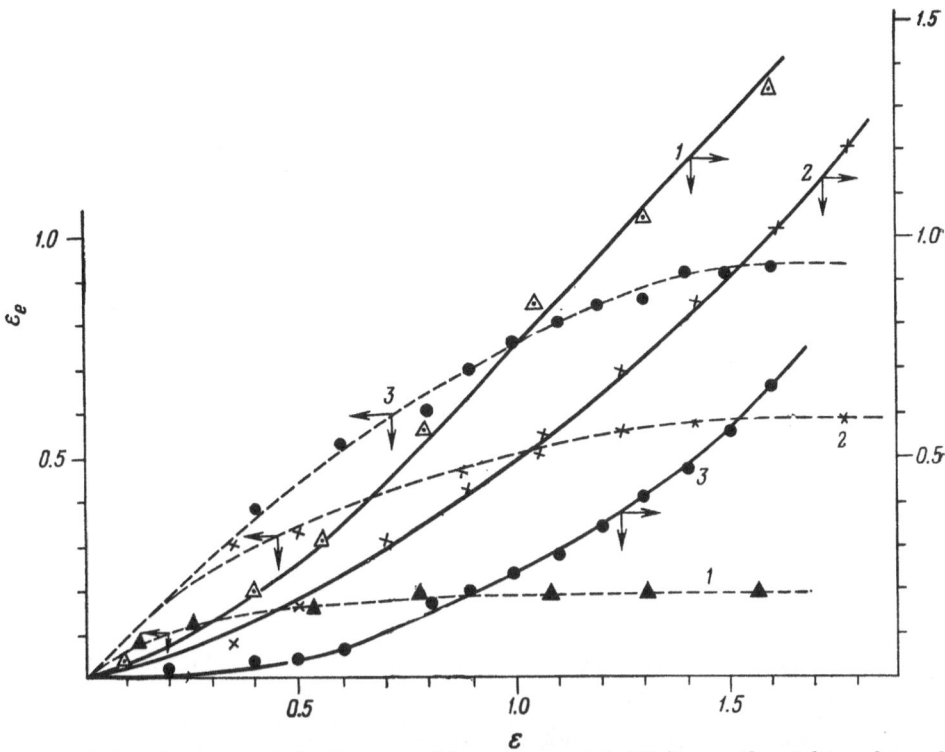

Fig. 7.4. Development of the irrecoverable component (solid lines—the right scale) and rubbery component (dashed lines—the left scale) of total deformation with a specified longitudinal velocity gradient for extension of polyisobutylene at 22°C. The values of $\dot{\varepsilon}_0 \cdot 10^3$ (in sec^{-1}):

1—1.1; *2*—2.9; *3*—19.6. [*Advances in Polymer Rheology*. Edited by G. V. Vinogradov, Moscow, Khimiya, 1970, pp. 9-23 (in Russian).]

viscosity, $\lambda = \sigma/\dot{\varepsilon}_f$, and of the rubbery modulus, $E = \sigma/\varepsilon_e$, at the various stages of deformation, including both the transient and the steady-state extensional flow. The mode of variation of the extensional viscosity as the deformation increases is shown in Fig. 7.5. The curves *1* and *2* in this figure correspond to such rates of deformation at which in steady-state flow conditions the viscosity proves to be independent of the stress and rate of deformation. The curves *3* and *4* describe the extension regimes under which steady-state flow is attained and the viscosity increases with increasing stress and rate of deformation. The curve *5* corresponds to the case where steady-state flow is not attained at all and the deformation of the polymer is accompanied by the continual increase stress and results in the fracture of the specimen.

The experimental data presented in Fig. 7.5 show that the stress-independent values of viscosity in steady-state flow may be preceded by the extremal character of the variation of λ (ε) at the prestationary stage of deformation. This result is an illustration of what has been said above, namely, that the constancy of the extensional viscosity in steady-state flow is not yet a proof of the linearity of the behaviour of the material under transient extension regimes.

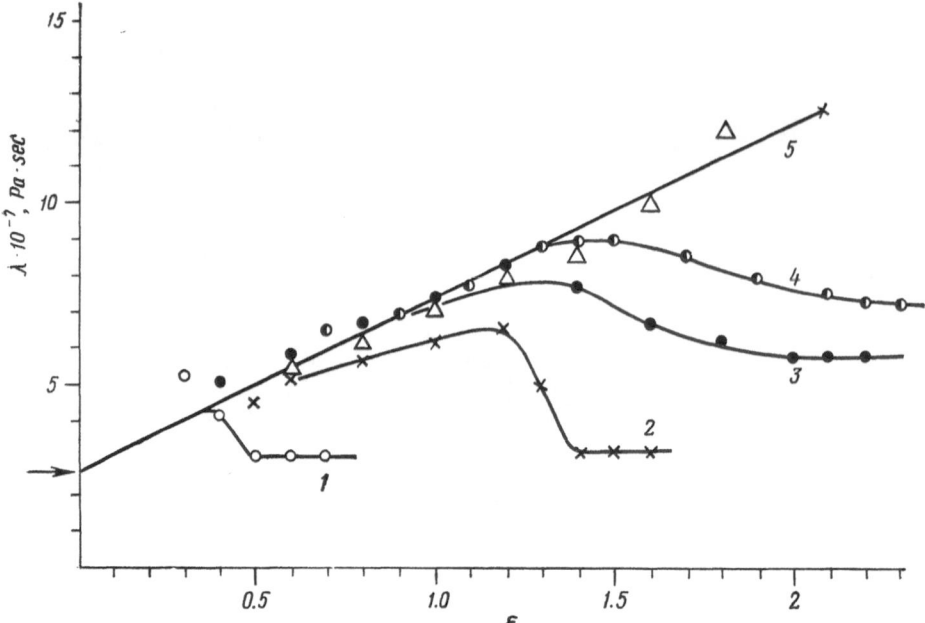

Fig. 7.5. Variation of the extensional viscosity of polystyrene at 130°C as a function of the amount of deformation at a constant velocity gradient. The values of $\dot{\varepsilon}_0 \cdot 10^3$ (in sec^{-1}): 1—0.137; 2—1.11; 3—2.98; 4—4.56; 5—5.7. [Vinogradov, G. V., V. D. Fikhman, B. V. Radushkevich and A. Ya. Malkin, J. Polymer Sci., A-2, 8, 657 (1970).]

As early as 1949 Kargin and Sogolova, while investigating the extension of polydisperse polymers, found that the extensional viscosity increases with increasing extension ratio, the variation of the viscosity being proportional to the deformation.* This effect is most spectacularly seen when the quantity ε_f is used as the argument (Fig. 7.6). It is characteristic that at $\varepsilon_f \to 0$ the extensional viscosity falls down to the initial value corresponding to the Trouton formula: $\lambda_0 = 3\eta_0$. The growth of extensional viscosity is typical for the initial stages of the prestationary region of deformation since, as seen from Fig. 7.5, at large deformations the extensional viscosity does not increase at all and may even decrease down to the initial value λ_0.

The increase of extensional viscosity in uniaxial extension is often treated as a necessary condition for large draw ratios of polymers to be realized. The growth of λ as the specimen is extended does really stabilize the process of uniaxial extrusion but large draw ratios can be achieved at high elongational velocity gradients again due to the development of rubbery deformations in linear polymers with the extensional viscosity being almost constant up to the fracture of the specimen. This is especially typical of polymers with narrow molecular-mass distributions, which do not display nonlinear effects at all because the strength of such polymers is lower than the stresses at which the viscosity should have been expected to vary in the course of deformation.** Therefore, for the possibility of stabilization of the proc ss to be explai-

* Kargin, V. A. and T. I. Sogolova, Zhurn. Fiz. Khim., 23, 530, 540 and 551 (1949).
** Malkin, A. Ya., V. V. Volosevich, and G. V. Vinogradov, A paper presented to the International Symposium on Chemical Fibres, Kalinin, Preprints, 1974, vol. 1, pages 196–200; J. Polymer Sci., Polymer Phys. Ed., 13, 1721 (1975).

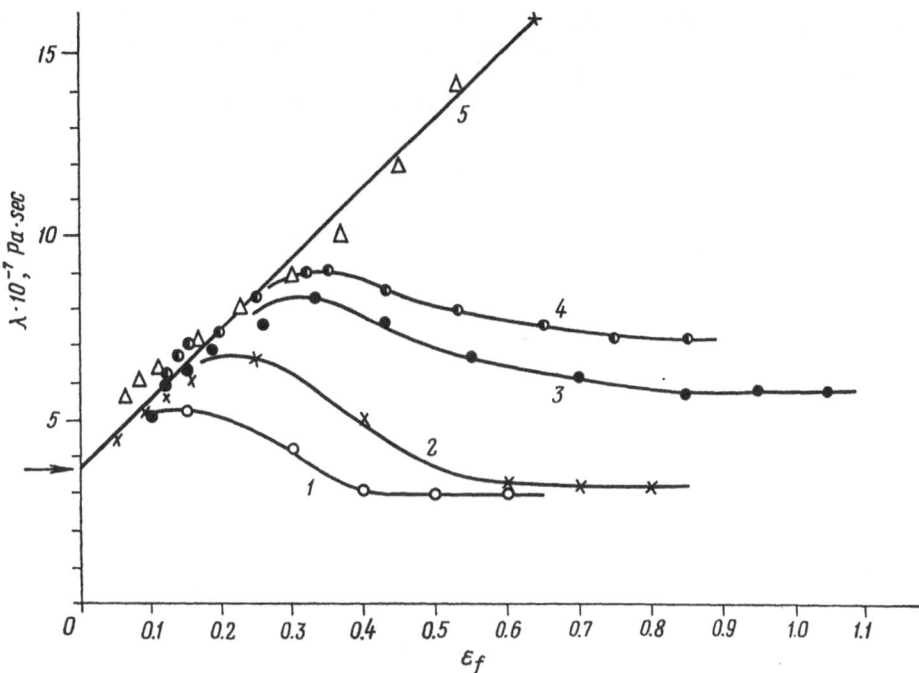

Fig. 7.6. Extensional viscosity at 130°C versus the deformation of viscous flow (for the designations and the reference, see Fig. 7.5).

ned it is sufficient to make the assumption of the viscoelastic properties of the liquids being extended (without increase of the extensional viscosity).* It should be stressed that here and afterwards we shall be concerned only with homogeneous extension which ends either with steady-state flow or with the fracture of the sample. When linear polymers are extended, there may develop other, more complex phenomena, such as a local reduction in cross-sectional area (a phenomenon referred to as "necking") and the appearance of self-oscillations with a periodic variation of the cross-sectional area of the thread. The problems associated with such extension regimes (and, in particular, the determination of the limits of the region of stability) will not be considered here.**

The character of stress development at constant $\dot{\varepsilon}$ as the polymer with a narrow molecular-mass distribution is extended is shown in Fig. 7.2. This refers to the case where the extension ends with the attainment of steady-state flow. As seen from Fig. 7.2, narrow-distribution polymers subjected to uniaxial extension behave as linear viscoelastic materials. This means, in particular, that the extensional viscosity of linear polymers remains constant

* Chang, R. and A. S. Lodge, *Rheol. Acta*, 10, 448 (1971). The possibility of development of large deformations on extension of polymer solutions and melts by way of producing predominantly rubbery deformations has also been discussed theoretically for a model of a viscoelastic liquid in: Leonov, A. I. and A. N. Prokunin, *Izv. Akad. Nauk SSSR, Ser.: Mekhanika Zhidkosti i Gaza*, 5, 25 (1973).
** For a theoretical discussion of the various types and causes of instability in extension of polymer threads on the basis of the data of earlier publications, see: White, J. L. and Y. Ide, *J. Appl. Polymer Sci.*, 22, 3057 (1978).

during the course of the entire process of extension, including steady-state flow, though here the rate of deformation varies over wide ranges and the draw ratio reaches many units.

In a number of works*, for the estimation of the prestationary extension regimes, use is made of the apparent viscosity $\sigma/\dot\varepsilon$, i.e., the ratio of the stresses varying with time to the constant rate of deformation. The dependence of this quantity on time is schematically shown in Fig. 7.7 for the case where the temperature of the experiment is far away from T_g. The arrow in the upper part of the figure indicates the direction in which the constant values of deformation rate $(\dot\varepsilon_0)$ specified in each experiment increase; the crosses indicate the moments when the specimens are fractured.

It should be emphasized that the quantity $\sigma/\dot\varepsilon$ is not viscosity since it does not characterize the dissipation of the work of external forces, except

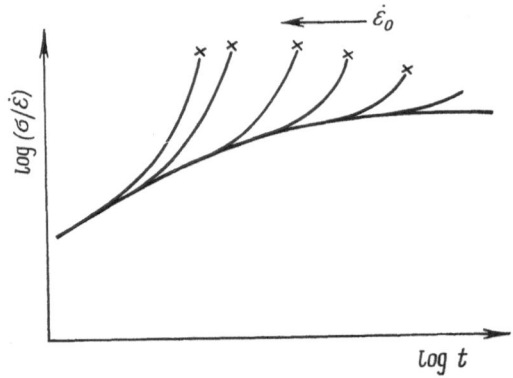

Fig. 7.7. Diagram illustrating the time dependence of apparent viscosity under the pre-stationary and fracture conditions of the deformation of polydisperse high-molecular-mass polymers at various rates of deformation.

for the steady-state flow conditions. For polydisperse polymers, this regime is usually attained at extremely low rates of deformation. In the case under discussion it corresponds to the right horizontal portion of the envelope.

From Fig. 7.7 it is seen that at the initial stage of extension as far as the amount of deformation is small the dependence of $\sigma/\dot\varepsilon$ on t is invariant to $\dot\varepsilon$ and the polymer being deformed behaves like a linear viscoelastic body. As the deformation progresses, if the rates of deformation are not extremely low (corresponding in the limit to the rate of deformation at which the right horizontal portion of the envelope is realized), there is observed a departure from the envelope, which becomes greater with increasing rate of deformation and, accordingly, begins to appear at the earlier stages of deformation of the specimens. Under these regimes the deformation of the polymer ends with the fracture of the specimens. An important conclusion follows from the fore-going: the departure from the linear viscoelastic behaviour is determined not only by the rate of deformation but also by the amount of deformation.

* Meissner, J., *Rheol. Acta*, 10, 230 (1971); Everage, A. E. and R. L. Ballman, *J. Appl. Polymer Sci.*, 20, 1173 (1976); Ide, Y. and J. L. White, *J. Appl. Polymer Sci.*, 22, 1061 (1978); Akutin, M. S., A. N. Prokunin, N. G. Proskurnina, and O. Yu. Sabsai, *Mekhanika Polimerov*, 2, 353 (1977).

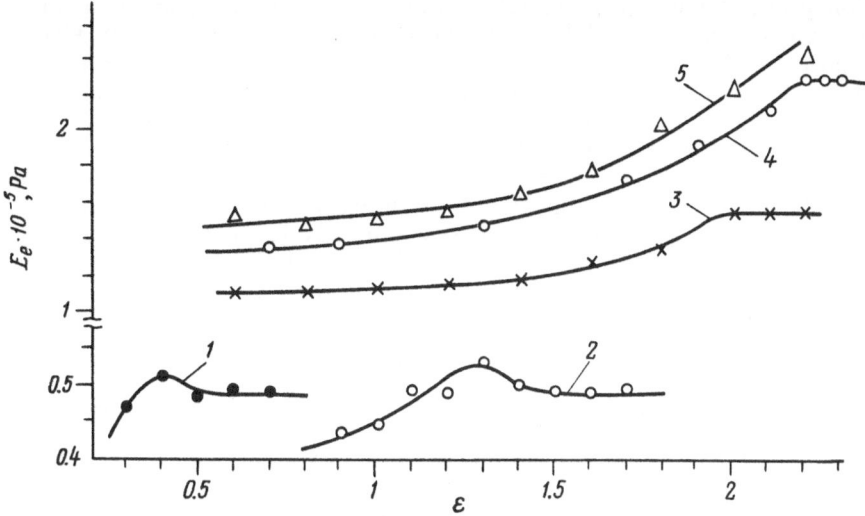

Fig. 7.8. The rubbery modulus of polystyrene at 130°C versus the amount of deformation at various longitudinal velocity gradients (for the designations and the reference, see Fig. 7.5).

The variation of the rubbery modulus at various stages of the deformation process, which takes place at different rates of strain, is shown in Fig. 7.8 for a polydisperse polymer. The curves *1* and *2* correspond to such deformation regimes under which steady-state flow with the viscosity equal to λ_0 is attained; the curves *3* and *4* correspond to deformation regimes under which the viscosity in steady-state flow increases with increasing rate of deformation, and the curve *5* shows the growth of the modulus for the case where steady-state flow is not attained because of the sample being fractured. The general pattern of variation of the rubbery modulus as a function of deformation qualitatively resembles the course of variation of extensional viscosity, though with increasing extensional viscosity (the curves *3* and *4*) the modulus may not reach a maximum. Special mention should be made of the continuous growth of the rubbery modulus when steady-state flow is not attained (the curve *5*).

The variation of the viscoelastic (relaxation) properties of linear polymers during uniaxial extension can be traced out in measurements of the dynamic characteristics of the sample under the conditions in which small-amplitude harmonic oscillations are superposed on extensional flow.* The data obtained in such experiments reflect the character of the above-described variations of the elastic and flow properties of the material during extension. Here the region of values of tan $\delta > 1$ may be conditionally defined as the fluid state of the polymer and the transition to the values of tan $\delta < 1$ may be treated as the attainment of the rubbery state at large deformations and/or rates of deformation.

The general character of the variation of the properties of the polymer in transient straining conditions shows that at the prestationary stage of the deformation process the relaxation spectrum of the polymer undergoes change. The character of the dependence of the viscosity, rubbery modulus, and rela-

* Briedis, I. P. and V. V. Leitland, *Mekhanika Polimerov*, 5, 901 (1977); 3, 507 (1978).

xation properties of the material on the amount of deformation in uniaxial extension shows that until steady flow is attained there occur structural changes. At the initial stage of deformation the rate of accumulation of rubber-like deformations exceeds the flow rate. Rubbery deformations are caused by the straightening of macromolecular chains in the direction of the applied stress. This results in a decrease of the conformational set of macromolecules and an increase in their rigidity. The orientation of chains in the direction of extension also contributes to the enhancement of inter-molecular interactions. All this leads eventually to an increase of viscosity and the rubbery modulus and to the retardation of relaxation processes since in the relaxation spectrum there appear new regions corresponding to slow relaxation processes.

The region beyond the viscosity maxima in the transient deformation region corresponds to the predominant effect of structural relaxation on the properties of the material, in spite of the continuing (but abruptly slowing-down) increase of the rubbery deformation. This process may be tied up with breaks in the network of intermolecular entanglements and with the short-ening of the lifetime of intermolecular contacts.

Upon completion of the development of rubbery deformations and struc-tural relaxation the polymer reaches an equilibrium state corresponding to steady-state flow in the field of the extensional velocity gradient. This state corresponds to the dynamic equilibrium of orientation and deorientation processes when the rates of formation and breakdown of intermolecular bonds are equal. Accordingly, the values of viscosity and relaxation characteristics of the material become constant and are no longer dependent on the deforma-tion of the material.

7.4.2. Extensional Viscosity and the Modulus of Rubber Elasticity in Steady-State Flow

The steady-state flow conditions are characterized by the fact that the corresponding parameters of the material do not depend on the way in which steady-state flow is attained (i.e., at $\dot{\varepsilon} = $ const or $\sigma = $ const) or the polymer is deformed. That the values of extensional viscosity are independent of the deformation mode in steady-state flow has been repeatedly confirmed by exper-iment.* An example of the variation of extensional viscosity as a function of the rate of deformation in steady-state flow is shown in Fig. 7.9 for the polydisperse polymer. The same figure shows the dependences of the viscosity η measured in shearing flow on the shear rate $\dot{\gamma}$. At low rates of deformation the Trouton law is always fulfilled: $\lambda = 3\eta_0$. The validity of this relation between λ and η_0 is seen, for example, from Fig. 7.9 for the region of low rates of deformation and has been confirmed many times for various polymers (see, for example, ref.*). Here the quantity λ does not depend on $\dot{\varepsilon}$, and η is independent of $\dot{\gamma}$. The corresponding range of deformation rates may be called linear since in this range the stresses are linearly related to the defor-mation rate. This refers, however, only to steady-state flow conditions since, as has been pointed out above, to linear steady-state deformation regimes

---

* Vinogradov, G. V., V. D. Fikhman, and B. V. Radushkevich, *Rheol. Acta*, 11, 286 (1972); Laun, H. M. and H. Münstedt, *Rheol. Acta*, 15, 517 (1976); *ibid*, 17, 415 (1978).

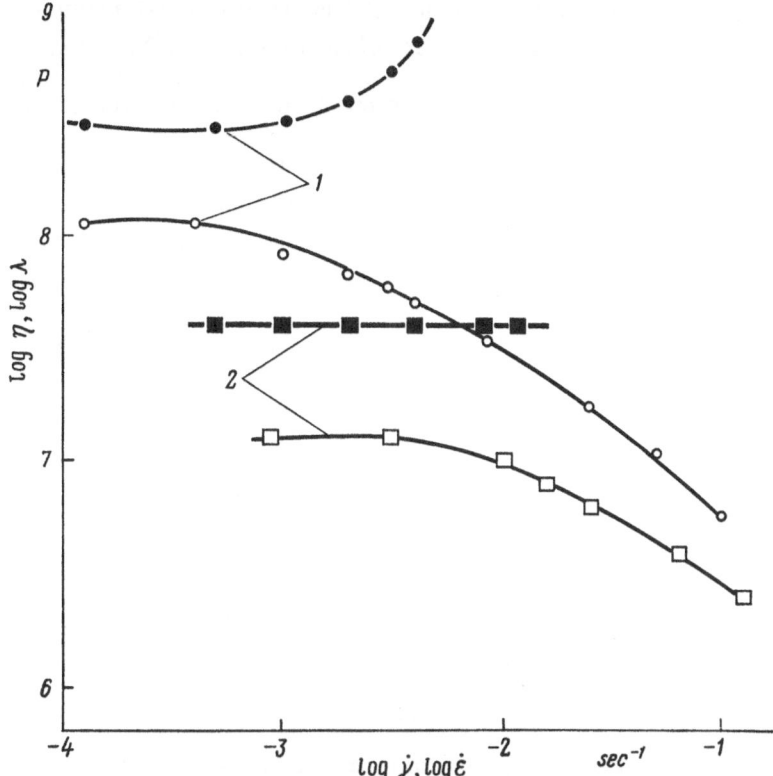

Fig. 7.9. Extensional viscosity (solid lines) and shear viscosity (dashed lines) of polystyrene versus velocity gradient in steady-state flow at 130°C (*1*) and 150°C (*2*) (see the reference to Fig. 7.5).

there may correspond the transient regions of the nonlinear behaviour of the material.

The transition to nonlinear steady-state flow occurs in substantially different ways for shear and extension: in the first case, the apparent viscosity decreases and in the second, it increases. Here, usually, the linear region corresponding to flow with constant viscosity in the field of the elongational velocity gradient continues up to much higher values of the rate of deformation than in shear.* In the nonlinear region where the viscosity η decreases and λ increases the ratio (λ/η) substantially increases with increasing rate of deformation.

The experimentally observed dependence λ ($\dot{\varepsilon}$) results from the superposition of two dependences: the decreasing φ ($\dot{\varepsilon}$) and increasing f ($\dot{\varepsilon}$) functions. The role of the orientation factor in measurements of extensional viscosity is found to be predominant as compared with the effect of the structural breakdown of the system and structural stress relaxation. Therefore, the function f ($\dot{\varepsilon}$) must be strongly increasing in order to cover the effect of the

---

* Ballman, R. L., *Rheol. Acta*, 4, 137 (1965); Cogswell, F. N., *Rheol. Acta*, 8, 187 (1968); Vinogradov, G. V., B. V. Radushkevitch, V. D. Fikhman, *J. Polymer Sci.*, A-2, 8, 1 (1970); Stevenson, J. E., *A.I.Ch.E. Journal*, 18, 540 (1972).

decrease of $\varphi(\eta)$. It may be assumed that the growth of extensional viscosity and, accordingly, the increasing nature of the function $\lambda(\dot{\varepsilon})$ are due to rubbery deformations that develop in extensional flow; these deformations characterize the degree of orientation of macromolecules during the stretching of polymers.

In a general case, as a result of the superposition of the effects of two factors—the increasing function $f(\dot{\varepsilon})$ and the decreasing function $\varphi(\dot{\varepsilon})$, the dependence $\lambda(\dot{\varepsilon})$ may be extremal. Indeed, in carefully conducted experiments*, on the $\lambda(\dot{\varepsilon})$ curve there was observed a distinctly pronounced maximum; the ratio λ/η_0 often reaches 7 at the point of the maximum (if the comparison is carried out on condition that shear and normal stresses are equal). Data of this kind have been repeatedly reported in the literature. It should,

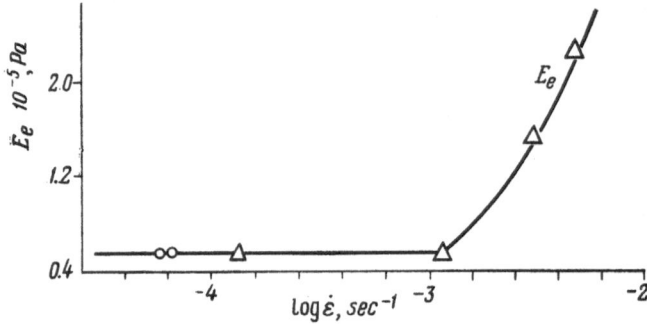

Fig. 7.10. Increase of the rubbery modulus corresponding to steady-state flow conditions in the extension of polystyrene at 130°C (see the reference to Fig. 7.5).

however, be borne in mind that they are often obtained through the use of different methods of measurement of extensional viscosity. Therefore, the dependence $\lambda(\dot{\varepsilon})$ must not be taken for granted if the procedure of determination of extensional viscosity has not been verified by direct measurements· This is especially important since the determination of λ is, strictly speaking, a rather subtle and time-consuming procedure.

The observed growth of extensional viscosity, which is associated with the orientation of macromolecules, correlates well with the increase of the rubbery modulus E (Fig. 7.10) since the orientation contributes to the increase of the rigidity of the material. In the region of low rates of deformation there is fulfilled the relation $E = 3G$ which follows from the theory of linear viscoelasticity.

Of interest in connection with the discussion of the problem of development of rubbery deformations are data* on the correlation between the dependences of the accumulation of rubbery deformations in uniaxial extension and in simple shear on the corresponding values of rates of deformation and stresses.

What has been said above refers mainly to polydisperse polymers. Monodisperse polymers of high molecular mass do not usually show nonlinear effects either in shear deformations or in uniaxial extension. The extensional viscosity of narrow-distribution polymers, like the shear viscosity of mono

---

* Laun, H. M. and H. Münstedt, *Rheol. Acta*, **17**, 415 (1978); *ibid*, **18**, 492 (1978).

disperse samples (see Chapter 2), does not depend on the rate of deformation. Therefore, in steady-state flow ($t \to \infty$) the ratio ($\sigma/\dot{\varepsilon}$) reaches the Trouton (initial) extensional viscosity λ_0 equal to $3\eta_0$ at all rates of deformation at which the extension ends up with the attainment of steady-state flow conditions.

7.4.3. The Relation Between Rubbery Deformations of Molten Polymers and Their Strength in the Glassy State

Orientation processes which can be judged by the increase of the rubbery modulus are the consequences of the rubber-like strains stored in the polymer. The orientational effect must be accompanied by the increase of the intermolecular interaction and, hence, of the strength of the polymer. In this connection, a series of investigations* have been carried out, in which the extended polymer samples were rapidly vitrified and their strength was determined in the tensile test machine. It has been found (Fig. 7.11) that the strength of the vitrified specimens is unambiguously determined by the large recoverable deformation accumulated in them before their vitrification (but not the total deformation), regardless of the way by which the given value

---

* Vinogradov, G. V., V. D. Fikhman, B. V. Radushkevich, and A. Ya. Malkin, *J. Polymer Sci.*, A-2, 8, 657 (1970); Vinogradov, G. V., V. D. Fikhman, and V. M. Alekseeva, *Polymer Eng. Sci.*, 12, 317 (1972); Vinogradov, G. V. *et al.*, *Rheol. Acta*, 17, 231 (1978).

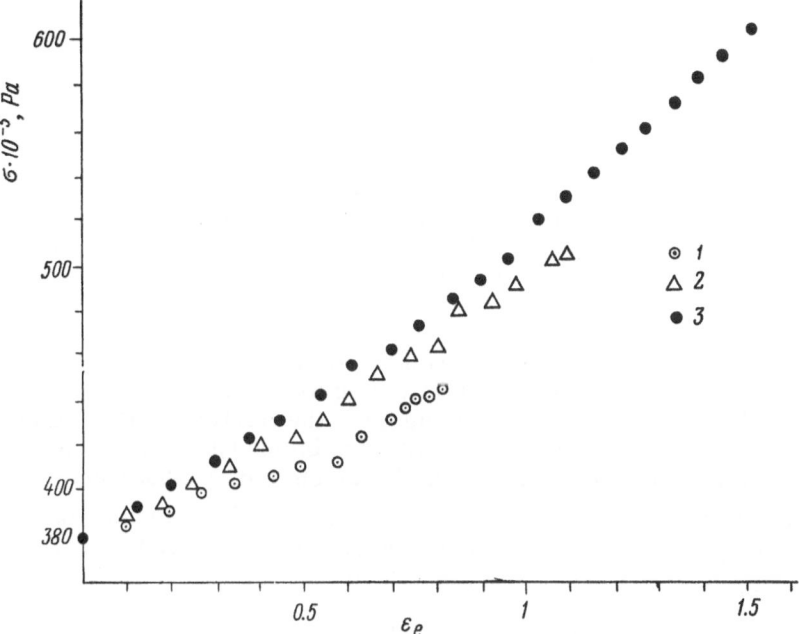

Fig. 7.11. The strength of vitrified samples of polystyrenes versus the rubbery deformations retained in the sample. The rate of deformation in flow preceding the "freezing" of deformations, $\dot{\varepsilon} \cdot 10^3$ (in sec^{-1}): *1*—1.11; *2*—2.98; *3*—5.7 (see the reference to Fig. 7.6).

of ε_e was attained. The strength is therefore determined not by the total draw ratio but only by the degree of extension corresponding to the rubbery deformation of the material.

The effect of the rate of deformation on the strength properties of a polymer is associated with the fact that with the same extension ratio the stored rubbery deformations depend on the rate of extension. At each specified rate of extension (after completion of the development of the rubbery deformation and attainment of steady-state flow) the subsequent accumulation of irrecoverable deformation does not affect the strength of vitrified samples. This is in agreement with the conception that the viscous flow proper under steady-state conditions cannot alter the state and structure of the polymer, which are in this case uniquely determined by the accumulation of rubbery deformation. But as the rate of deformation increases the strength corre-

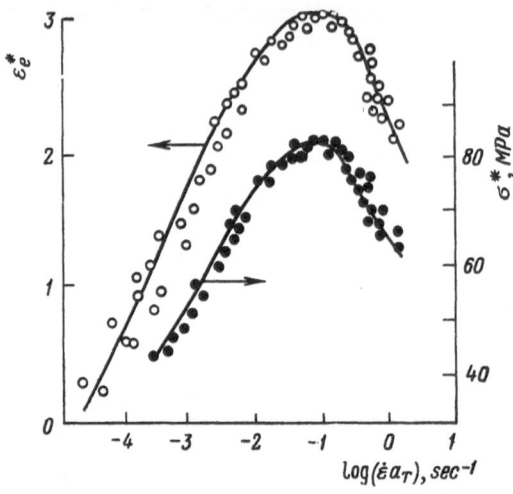

Fig. 7.12. Relation between the strength σ^* and the rubbery deformations stored at various rates of extension for polyvinyl chloride. [Vinogradov, G. V., V. D. Fikhman, and V. M. Alekseeva, *Polymer Eng. Sci.*, **12**, 317 (1972).]

sponding to the state of the material in steady-state flow increases since to this there corresponds the growth of the equilibrium value of rubbery deformation. But if the extension ends up with the fracture of the sample before steady-state flow is attained, then the maximum values of σ^* reached at a given rate of extension depend on the amount of rubbery deformation ε_e corresponding to the moment of fracture. Here, as the value of $\dot{\varepsilon}$ increases the values of ε_e may continue either to increase or decrease. In the latter case, the dependence $\sigma^* (\dot{\varepsilon})$ is extremal (Fig. 7.12); the function $\sigma^* (\dot{\varepsilon})$ varies in the same direction as does the dependence $\varepsilon_e (\dot{\varepsilon})$ and the point of the maximum corresponds to the transition from low rates of extension, at which steady-state flow is possible, to high rates of extension at which the extension ends up with the fracture of the sample.

7.5. The Ultimate Properties of Molten Linear Polymers and Uncured Elastomers

7.5.1. Relaxation Transitions of Polymers from the Fluid to the Rubbery and Glassy States on Deformation

The conception of the limiting states of polymers on deformation caused by their relaxation transitions from the fluid to the forced rubbery and glassy states has been systematically developed since the late sixties.[*] These transitions are induced by the increase of the rate of deformation. They are manifested most distinctly and simply in the case of high-molecular-mass flexible-chain linear polymers of narrow molecular-mass distribution. Therefore, it is exactly such polymers that form the basis of our treatment of ultimate properties.

The transitions in question are especially important in the case of extension. In simple shear (see Chapter 2, Sec. 2.6.6) the transition of a polymer to the forced rubbery state sets a limit on the possibility of flow since it upsets the conditions of sticking of the polymer to the walls relative to which shear flow takes place. Under the conditions of uniaxial extension the deformation of polymers can be effected at very high rates of deformation (though in such cases the deformation of polymers ends sooner or later with the fracture of the specimens).

It should be noted that the transition from the fluid to the rubbery state is somewhat influenced not only by the rate of deformation but also by the amount of deformation.[**] The effect of rubber-like deformation on such a transition has been also noted by Hudson and Ferguson.[***] But this kind of transition is, however, more appreciably affected by the change of the rate of deformation.

The generalized deformation characteristic of the polymers considered here is given in Fig. 7.13. The upper part of this figure shows the development of the recoverable, irrecoverable, and total deformations over a very wide range of deformation rates. This graph is invariant to temperature at $T \gg T_g$.

At deformation rates and temperatures limited from the right by a vertical dashed line the polymer is in the fluid state. Under these conditions it behaves as a linear viscoelastic body for which the elongational viscosity is constant $\lambda = \lambda_0 = \text{const}$. In this state, the attainment of steady-state flow is possible. In principle, it can continue for an infinitely long period of time. This deformation regime is, however, always limited since after the specimen is considerably thinned the extension becomes unstable. The specimen is fractured when the amplitude of surface waves that arise due to surface tension increases. The total length of the specimen before fracture occurs depends on the ratio of the viscous drag to the surface tension. The rubbery elasticity of

---

[*] Vinogradov, G. V., *Pure and Appl. Chem.*, **26**, 423 (1971); **42**, 527 and 942 (1975); in: *Progress in Heat and Mass Transfer*, ed. by W. R. Schowalter, vol. 5, Pergamon Press, Oxford, 1972; *Rheol. Acta*, **12**, 357 (1973); *Polymer*, **18**, 1275 (1977); *Polymer Eng. Sci.*, in press; Vinogradov, G. V., A. Ya. Malkin *et al.*, *J. Polymer Sci.*, A-2, **10**, 1061 (1972); *Appl. Polymer Symposia*, **27**, 47 (1975).

[**] Vinogradov, G. V., A. I. Isayev, and E. V. Katsyutsevitch, *J. Appl. Polymer Sci.*, **22**, 727 (1978).

[***] Hudson, N. F and J. Ferguson, *Trans. Soc. Rheol.*, **20**, 253 (1976).

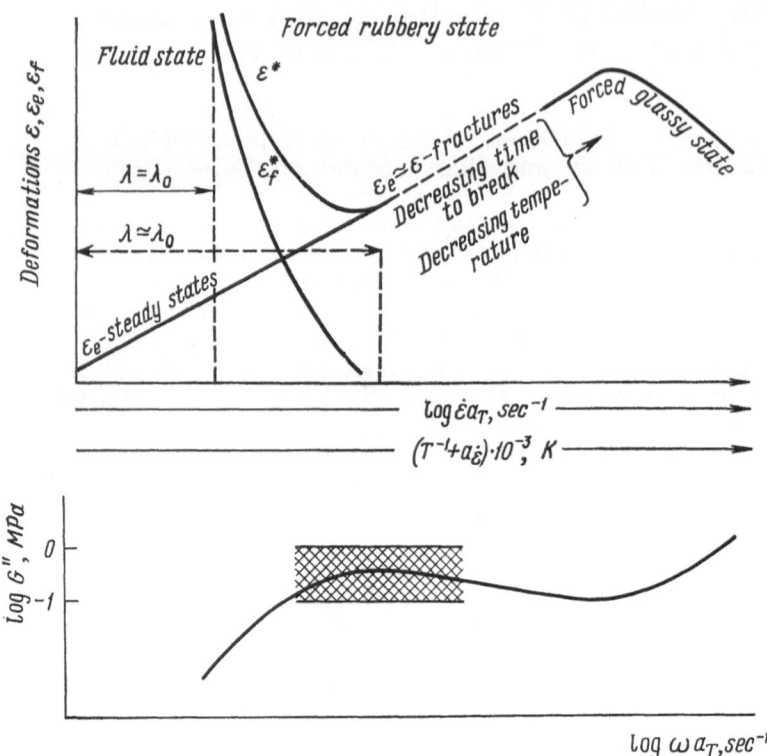

Fig. 7.13. The most important characteristics of the deformation behaviour of linear flexible-chain high-molecular-mass polymers of narrow distribution in a very wide range of deformation rates in the case of uniaxial extension. Their behaviour is shown under the conditions of low-amplitude oscillating shear; the hatched area corresponds to various homologous series of polymers of the type indicated.

polymers and their solutions at the same values of viscosity and surface tension as in the case of a Newtonian liquid has a bearing on the amount of ultimate elongation. But the course of the dependence of the ultimate elongation in this mode of fracture on the properties of the polymer is unknown in a general case.

When the critical rate of deformation ($\dot{\varepsilon}_{cr}$) is attained, which is depicted in Fig. 7.13 by the vertical dashed line, there begins a forced transition of the polymer to the rubbery state, in which the polymer behaves like cured elastomers. This transition is accompanied by a rapid decrease of the ability of the polymer to accumulate irrecoverable deformation, this being represented by the steeply falling values of ε_f. Since the recoverable deformation continues to increase, the total ultimate deformation passes through a distinctly pronounced maximum.

At $\dot{\varepsilon} > \dot{\varepsilon}_{cr}$, in the forced rubbery state the deformability of the polymer is limited, i.e., the specimen is always fractured on extension.

Of greatest importance are the critical rates and stresses at which the polymer passes to the rubbery state. In experiments carried out on various polymer-homologous series it has been found that $\dot{\varepsilon}_{cr}$ is proportional to λ_0^{-1}, or $\dot{\varepsilon}_{cr}$ is proportional to $M^{-3.5}$, or $\dot{\varepsilon}_{cr}$ is proportional to η_0^{-1}. From this it follows that $\dot{\varepsilon}_{cr} \cdot \lambda_0 = \text{const} = \sigma_{cr}$. This means that for high-molecular-mass polymers there exists one value of critical stress. When this stress is exceeded, the sample is invariably fractured. This critical stress is simply related to the critical shear stress τ_s at which the polymer stream undergoes a spurt during flow in the duct:

$$\sigma_{cr} = 3\tau_s$$

Special emphasis should be placed on the fact that with polymers of very narrow MMD the extensional viscosity exhibits constancy even in the rubbery state until the accumulation of irrecoverable deformation becomes negligibly small.

At deformation rates and temperatures, which are respectively higher and lower than their values corresponding to the minimum of the ε ($\dot{\varepsilon}$) curve, a neck is formed in the specimen. These regimes are represented by the dashed portion of the curve of ε vs. $\dot{\varepsilon}$. In what follows, special attention is focused on the fact that the parameters that correspond to the onset of necking obey all the regularities that describe the fracture of polymers in uniaxial extension.

Inspection of Fig. 7.13 will show that the portion of the ε vs. $\dot{\varepsilon}$ curve that corresponds to the conditions of neck formation has a limited extent. As the rate of deformation increases further, which corresponds to the transition of the polymer to the forced glassy (or more exactly, leathery) state, the polymer specimen is elongated homogeneously.

The maximum on the ε vs. $\dot{\varepsilon}$ curve is related to the uncoiling of macromolecules. Therefore the corresponding values of ε_m and $\dot{\varepsilon}_m$ are the intrinsic characteristics of the polymer.

Of great interest is the correlation between the conditions of uniaxial extension and oscillating small-amplitude shear. That such a correlation is basically possible is determined by the fact that for high-molecular-mass polymers the fracture characteristics are simply related to their initial viscosity, i.e., the polymers considered behave as linear viscoelastic materials over a wide range of deformation rates and values of deformation.

A typical curve of G'' versus ω is shown in the lower part of Fig. 7.13. The maximum on this curve corresponds to the completion of the transition of the polymer to the forced rubbery state. The position of the maximum along the frequency axis corresponds to deformation rates and temperatures at which the irrecoverable component of the total deformation becomes negligibly small and the amount of total deformation passes through a minimum.

For the homologous series of polymers investigated the value of G'' corresponding to the maximum on the G'' versus ω curve is within narrow ranges of variation (shown by the hatched band in Fig. 7.13); it varies by a factor of 10-20. In absolute value it usually does not exceed 1 MPa. It means that at stresses higher than 1 MPa the polymers must be not only rubberlike but also incapable of storing appreciable irrecoverable deformations, i.e., are known in advance to be incapable of flowing. Therefore, at stresses greater than 1 MPa the polymers must necessarily be fractured (depending on the nature of the polymer this may happen at lower stresses). The limiting stresses that the polymer can withstand without being fractured are related with the quantity M_e, the molecular mass of the dynamic segment (see Chapter 3, Sec. 3.3). Accordingly, fracture stresses appear to be greatest for polymers of

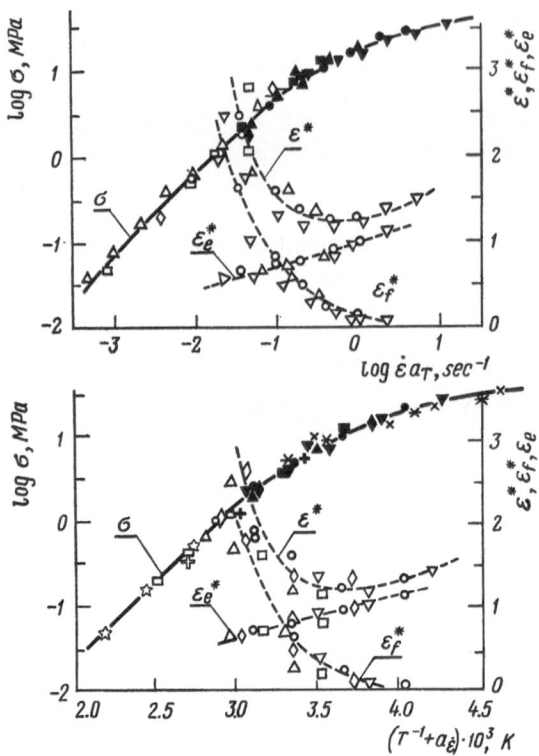

Fig. 7.14. Dependences of stresses, and total, rubbery, and irrecoverable deformations on the rate of deformation (reduction to 25°C) and on the inverse absolute temperature (reduction to the rate of deformation 0.01 sec^{-1}) for polyisoprene (molecular mass 5.75×10^5; M_w/M_n not greater than 1.1).

The upper graph: $\triangledown = -25$; $\bigcirc = 0$; $\triangle = +25$; $\square = +50$; $\diamondsuit = +75$°C; the lower graph: $\bigcirc = 0.002$; $\triangledown = 0.005$; $\triangle = 0.01$; $\square = 0.02$; $\diamondsuit = 0.05$; $\star = 0.1$; $+ = 0.2$; $\ast = 0.05$; $\times = 1.0$ sec^{-1}; the black symbols correspond to fractures of polymers.

the type of linear polyethylene and 1,4-polybutadiene and much lower for polystyrene or methyl methacrylate.

The fact that polymer melts cannot be deformed at stresses greater than 1 MPa has been noted* in the literature. The general explanation of this fact has been given above.

After the minimum on the G'' versus ω curve is attained the polymers pass first to the leathery and then to the glassy state.

An illustration of the master curve of the deformation behaviour of polymers in the region of their transition from the fluid to the rubbery state is given in Fig. 7.14. The left portions of the σ vs. $\dot{\varepsilon}$ and σ vs. $1/T$ curves, which are represented by the unfilled symbols, correspond to steady-state flow conditions. The filled symbols refer to the fracture of the specimens. It is interesting to note that if the ultimate stresses (σ^*) are referred not to the rate of total deformation but to the rate of accumulation of irrecoverable deformation, then the black circles will be shifted to the left, so that they will appear to be a continuation of the straight line describing steady-state Trou-

_____

* Cogswell, F. N., *Appl. Polymer Symposia,* 27, 1 (1975).

ton flow conditions. This substantiates the above statements that the viscosity remains constant until there is observed an appreciable accumulation of irrecoverable deformation.

The experimental data illustrating the correlation between the minima on the ε vs. $\dot{\varepsilon}$ and G'' vs. ω curves for two narrow-distribution polyisoprenes are shown in Fig. 7.15. The correlation becomes even better if with the use of formula (7.15) the frequency scale is shifted to the side of higher values by an amount $\sqrt{3}$.

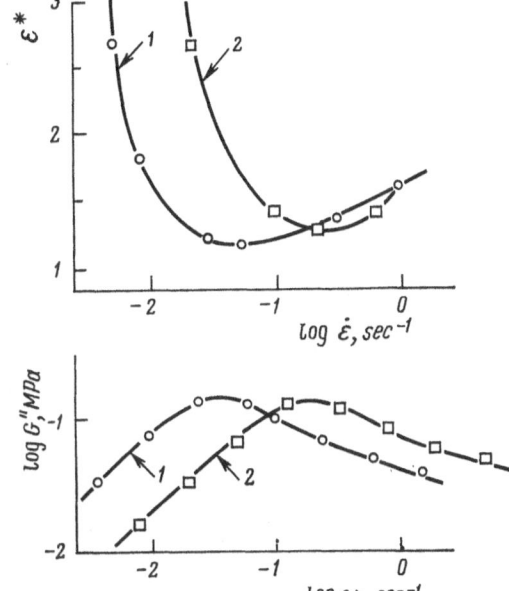

Fig. 7.15. Dependence of the total fracture deformation on the rate of deformation and of the loss modulus on frequency under low-amplitude oscillating shear for polyisoprenes (molecular masses: 1—5.75; 2—3.80×10^{5}) of narrow distribution (M_{w}/M_{n} not greater than 1.1).

Surprisingly, it appears that there exists, at least, a qualitative correlation between the forced transition of polymers of narrow MMD to the leathery state and the characteristics of small-amplitude oscillating shear. This is shown in Fig. 7.16. At stresses exceeding 1 MPa the Young modulus greatly increases. Hence, the polymer is no longer a linear viscoelastic body and the correlation between the deformation characteristics obtained at high values of σ or under small-amplitude oscillating shear must be regarded as purely empirical. The above-indicated dependence of the Young modulus is presented in Fig. 7.17.

The temperature superposition has been repeatedly used in this book in many special cases; the temperature reduction frequently conformed to the reduction by the initial viscosity. We shall now show that this method of superposition is of general importance and may be used within a given polymer-homologous series with respect to both temperature and molecular mass. For this purpose, reference must be made to the data presented in Fig. 7.18. This figure shows the dependences of stress on rate of deformation for samples of polyisoprenes ($\overline{M}_{w}/\overline{M}_{n}$ not greater than 1.1) at temperatures from $-25°C$ to $+75°C$. As before, the unfilled symbols designate steady-state flow and the filled symbols correspond to fracture conditions.

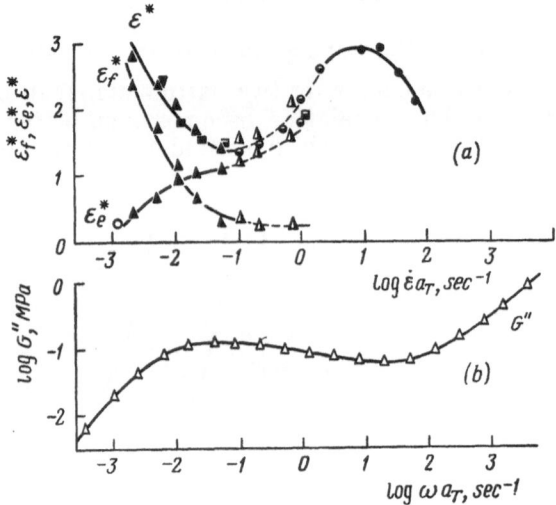

Fig. 7.16. Dependence of the total, rubbery and irrecoverable deformations on the rate of deformation (*a*) and of the loss modulus on the frequency (*b*) for 1,2-polybutadiene (for its characteristics, see the caption to Fig. 7.2); $\bigcirc$ = 0°C; $\blacksquare$ = 10; $\blacktriangle$ = 25; $\blacktriangledown$ = 50°C; the half-filled symbols refer to the start-up of necking; all the data are reduced to 25°C; $\triangle$ = 25°C.

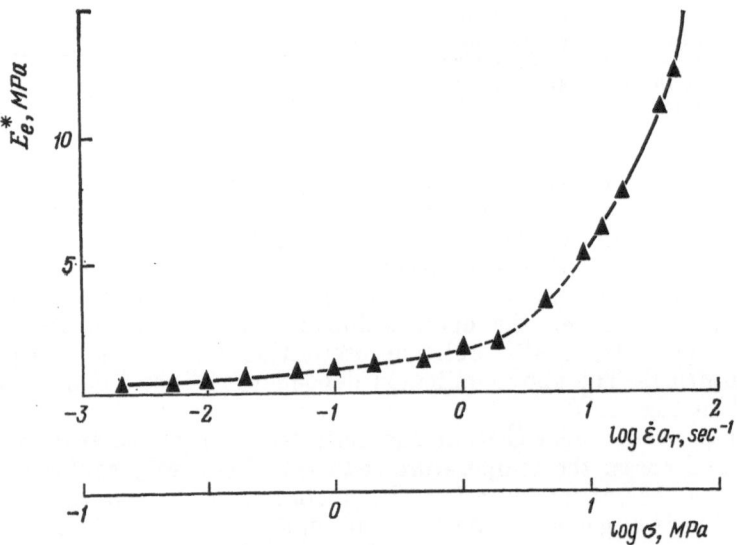

Fig. 7.17. Dependence of the rubbery modulus corresponding to the fracture of polymers before necking takes place (the dashed portion of the curve) on the reduced rate of deformation (reduction to 25°C) and the fracture stress for 1,2-polybutadiene.

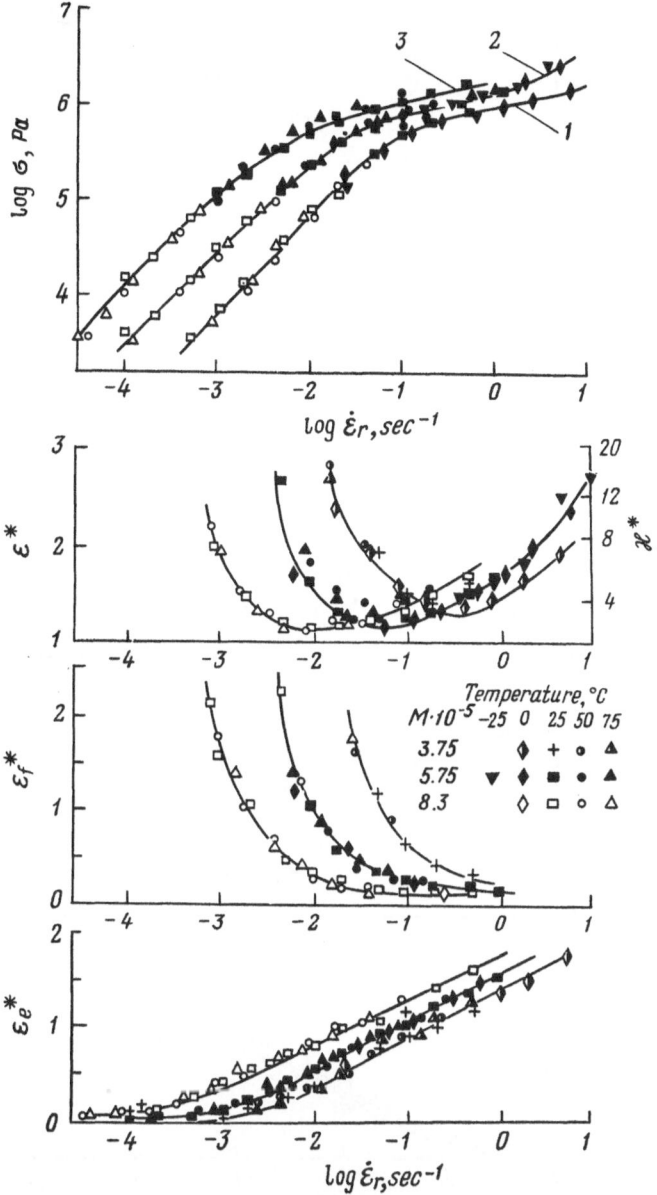

Fig. 7.18. Deformation characteristics of three high-molecular-mass polyisoprenes of narrow distribution over a wide temperature range (reduced to 25°C). The open symbols on the curve of stress vs. rate of deformation (the upper graph) correspond to the steady-state regime and the black symbols to fracture regime of deformation.

Figure 7.19 shows the results of reduction with respect to molecular mass and temperature. The success of the reduction of the ultimate characteristics on the basis of the use of the initial viscosity indicates that the fracture of linear polymers in the region of the transition from the fluid to the rubbery state is associated with the mechanism of viscoelastic relaxation caused by

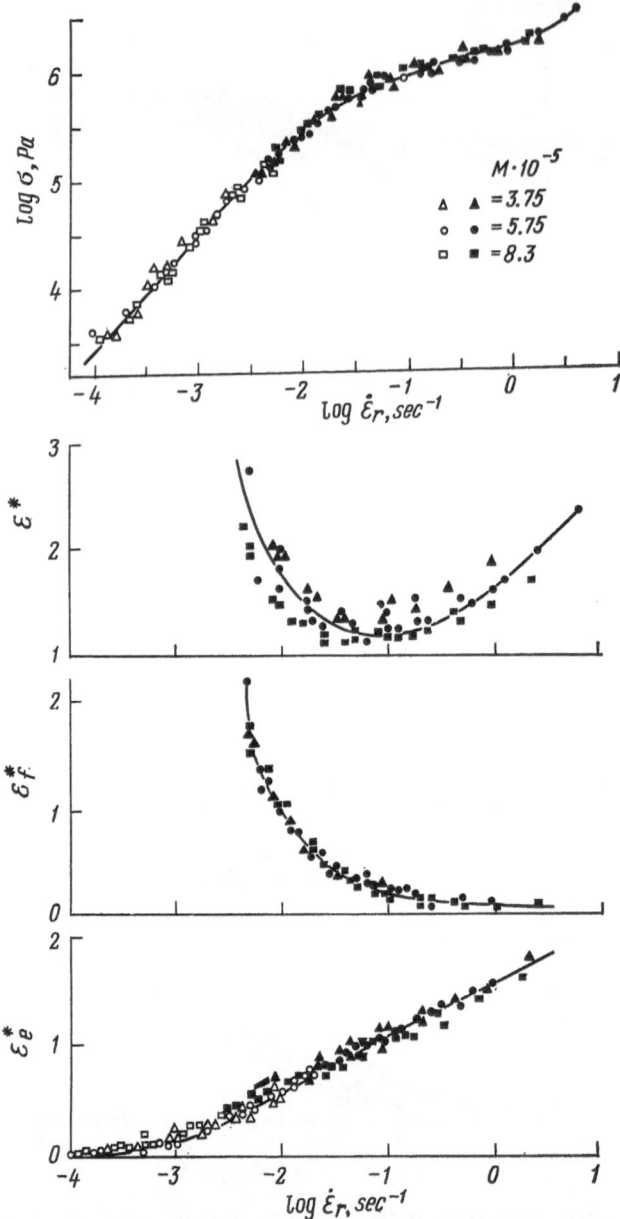

Fig. 7.19. Reduction of the deformation characteristics of polyisoprenes by temperatures (to 25°C) and molecular masses (to 3.8×10^5). The open symbols correspond to steady-state deformation conditions and the black symbols, to fracture deformation regimes.

the slipping of macromolecules past each other rather than with the mechanical destruction of macromolecules (in contrast to the well-known case of the fracture of rigid highly oriented plastics).

7.5.2. Generalized Strength Characteristics

The strength properties of polymers are determined by the ultimate strength (the dependence of stress on recoverable and total deformations at rupture) and by the long-term durability (the stress dependence of the time from the onset of loading to the fracture of specimens).

One of the most important characteristics of polymers is the failure envelope proposed by Smith*, which gives the dependence of ultimate stress on rubber-like deformation for cured elastomers. It has been lately shown** that it is possible to extend this approach to polymer melts and uncured elastomers in such a manner that it would cover data for the total and rubbery deformations. This is shown in Fig. 7.20a. Here the unfilled symbols represent the points obtained at constant rate of deformation. The filled symbols show the ultimate deformations recorded at constant true stresses. The half-filled symbols refer to the onset of necking. The region situated to the right of the failure envelope is forbidden for the deformation process. Figure 7.20b presents (apart from the dependence of σ vs ε) the dependence of

---

* Smith, T., in: *Fracture* (edited by H. Leibowitz), vol. 7, part IIIB, Academic Press, New York, 1972; in: *Rheology* (edited by F. Eirich), vol. 5, Academic Press, New York, 1969.

** Vinogradov, G. V., A. Ya. Malkin *et al.*, *J. Polymer Sci.*, A-2, **10**, 1061 (1972); Vinogradov, G. V., *Polymer*, **18**, 1275 (1977); Vinogradov, G. V., *Polymer Eng. Sci.*, in press.

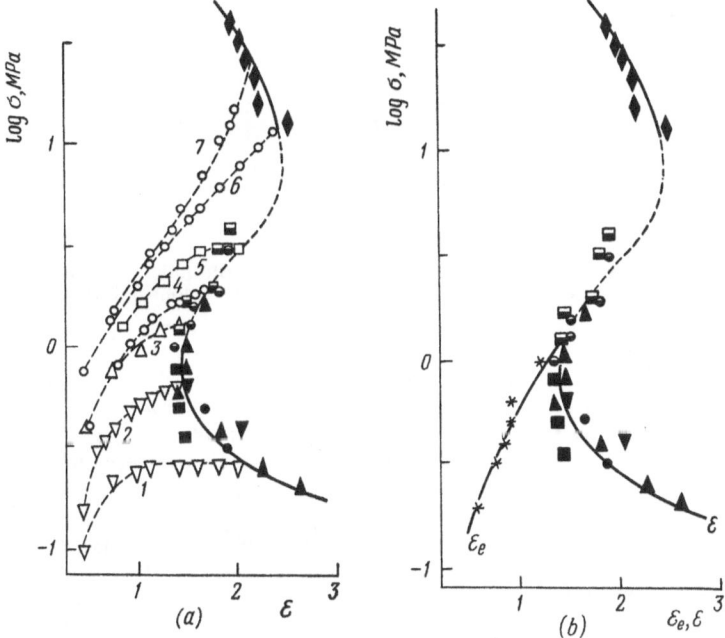

Fig. 7.20. The stress-strain relation for 1,2-polybutadiene (its characteristics are given in the caption to Fig. 7.2) under the pre-fracture regimes with assignment of constant rate of deformation (the open symbols). The fracture envelope obtained with specified constant stresses (the black symbols; half-filled symbols correspond to the start-up of necking). The temperatures and rates of deformation:

1 and *2*—50°C, 5×10^{-2} and 2×10^{-1} sec^{-1}; *3*—25°C, 1×10^{-1} sec^{-1}; *4, 6,* and *7*—0°C, 1×10^{-2}, 2×10^{-1}, and 4×10^{-1} sec^{-1}; *5*—0°C, 5×10^{-2} sec^{-1}; ◆ = 10°C. (a) the fracture envelope for the total and recoverable deformations; the symbols are the same as in (b).

ultimate stress on recoverable deformation. The curves of σ vs. ε and σ vs.
ε_e merge into the region corresponding to necking. This proves that not only
the Considère condition, which predicts the possibility of necking at a deformation equal to 1, is fulfilled but also that for a neck to appear the polymer
must undergo purely elastic deformation. The dependence σ (ε_e) at values
of the variables lower than shown in Fig. 7.20b changes into the envelope
which corresponds to the attainment of steady-state flow.

From what has been said above it follows that the ultimate strength is
independent of the loading conditions and temperature. For this statement
to be proved, experiments have been carried out on 1,2-polybutadiene and

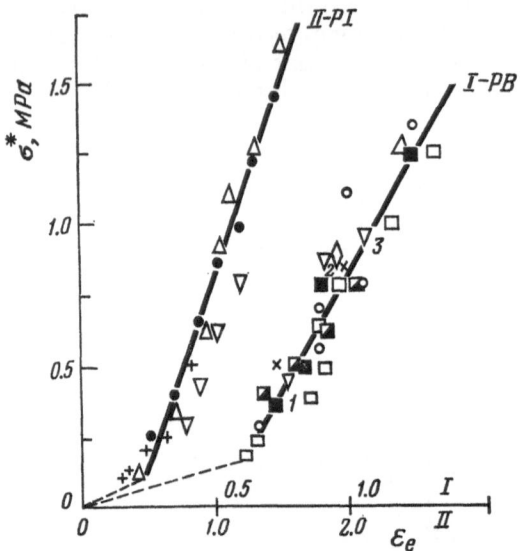

Fig. 7.21. Dependence of fracture stress on the rubbery component of the fracture deformation for polybutadienes (I) and polyisoprenes (II) under various deformation conditions:

I—1,2-PB, deformation conditions: σ = const (■ = + 10°C, □ = +25°C, ▨ = +50°C); $\dot{\varepsilon}$ = const
(◇ = +10°C, ○ = +25°C); constant rate of extension (× = + 25°C); constant force (△ = +25°C);
stepwise loading conditions at 25°C (∇_1 = deformation at $\dot{\varepsilon}_1$ = 2 × 10⁻³ sec⁻¹ for 30 sec, relaxation
at constant deformation for 60 sec; deformation at $\dot{\varepsilon}_2$ = 5 × 10⁻³ sec⁻¹ before fracture; ∇_2 = deformation at $\dot{\varepsilon}_1$ = 5 × 10⁻³ sec⁻¹ for 180 sec, relaxation at constant deformation for 60 sec; deformation
at $\dot{\varepsilon}_2$ = 2 × 10⁻² sec⁻¹ before fracture; ∇_3 = deformation at $\dot{\varepsilon}_1$ = 2 × 10⁻² sec⁻¹ for 30 sec, relaxation
at constant deformation for 30 sec; deformation at $\dot{\varepsilon}_2$ = 2 × 10⁻² sec⁻¹ before fracture). II—PI at
25°C, the $\dot{\varepsilon}$ = const regime (molecular mass, MM: ∇ = 3.75 × 10⁵, ● = 5.75 × 10⁵, △ = 8.3 × 10⁵); the
σ = const regime, + = 5.75 × 10⁵. [For the characteristics of PI, see Vinogradov, G. V., A. Ya. Malkin, and V. V. Volosevitch, J. Polymer Sci., Polym. Phys. Ed., 13, 1721 (1975); the characteristics of
PB are given in the caption to Fig. 7.2.]

polyisoprene. The results of the experiments are shown in Fig. 7.21, where
the ultimate stress σ* is plotted along the ordinate and the ultimate recoverable deformation ε_e^* (according to Hencky) along the abscissa. It can be
seen from Fig. 7.21 that uncured high polymers become capable of being fractured only after the "critical" recoverable deformation ε_e^{**} is reached. In the
case under consideration ε_e^{**} = 0.44. The conditions under which high

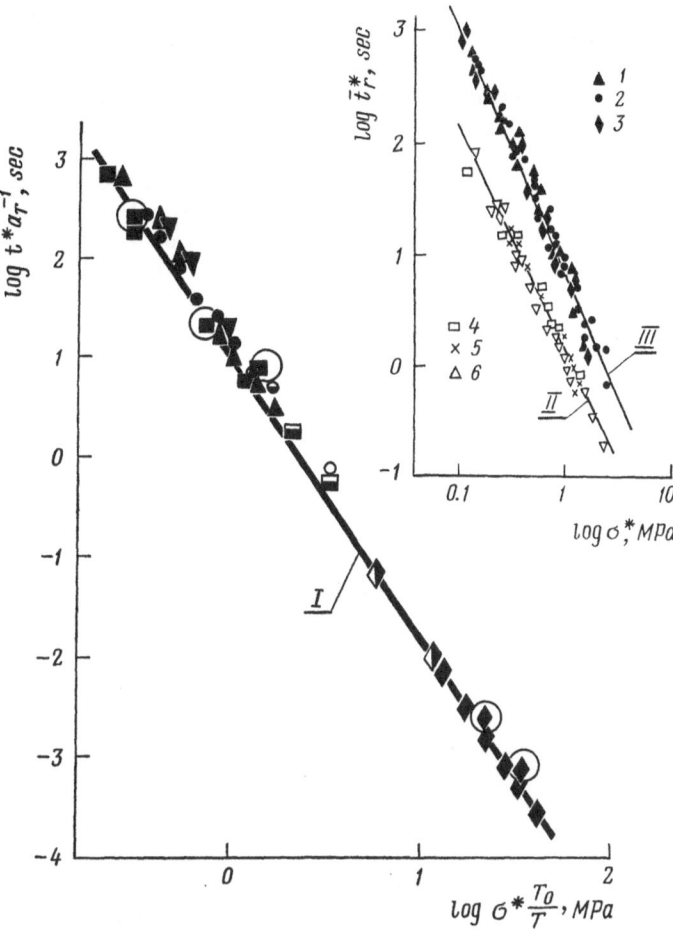

Fig. 7.22. Dependence of durability on fracture stresses for 1,2-polybutadiene (curve *I*; the characteristics of the polymer are given in the caption to Fig. 7.2), polyisoprenes, and polybutadienes (curves *II* and *III*; their characteristics are given in: Vinogradov, G. V. et al., *J. Polymer Sci., Polym. Phys. Ed.*, 13, 1721 (1975); Vinogradov, G. V., A. Ya. Malkin, and V. V. Volosevitch, *J. Appl. Polymer Sympos.*, 27, 47 (1975).
I. Temperatures, °C; ◇ = 10; ○ = 0; □ = 10; △ = 25; ▽ = 50; *II* and *III.* Molecular masses, $M \cdot 10^{-4}$: 3.75 (*1*); 5.75 (*2*); 8.30 (*3*); 1.20 (*4*); 1.95 (*5*); 2.90 (*6*).

polymers are fractured are determined in the following manner:

$$\frac{\sigma^*}{\varepsilon_e^* - \varepsilon_e^{**}} = E^* = \text{const} > 0$$

Here E^* is the slope of the straight line (Fig. 7.21) which expresses the dependence of stress on ultimate elastic deformation. Although σ^* and ε_e^* are determined by the relaxation characteristics of the material, from the data presented in Fig. 7.21 it explicitly follows that the Deborah number alone cannot predict the critical deformation regimes corresponding to the fracture of polymers.

That the ultimate stresses are independent of the route of deformation is a proof that polymer melts and uncured elastomers do not accumulate defects during the course of their deformation. This is a consequence of the relaxation mode of their fracture.

The relaxation nature of the fracture of polymer melts and uncured elastomers is most distinctly pronounced when one is dealing with the second most important characteristic of the strength properties of polymers—the long-term durability.

Figure 7.22 shows the results of measurements of the long-term durability of a number of polymers. The curve *I* presents the temperature-invariant dependence of the durability of 1,2-polybutadiene on ultimate stress over a very wide range of times; the curves *II* and *III* are the same dependences for

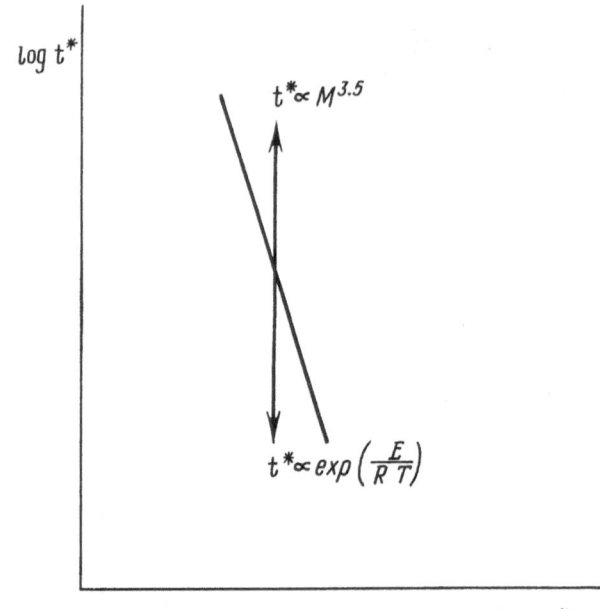

Fig. 7.23. Diagram showing the dependence of the durability of linear **high**-molecular-mass flexible-chain polymers on molecular mass and temperature.

several polyisoprenes and polybutadienes (reduction by molecular mass and temperature is made). The slope of the curves is determined by the mode of deformation. The data given in Fig. 7.22 have been obtained by specifying constant stresses. In this case, the slopes are close to 3. The slopes of these dependences, obtained under the conditions of constant rates of deformation are smaller. This is quite understandable since the constant deformation rate regime is milder than the σ = const regime.

The reductions of durability curves were made by means of the dependence of viscosity on temperature and molecular mass. The reduction procedure is schematically shown in Fig. 7.23. This may be regarded as a proof that the process of fracture of linear high-molecular-mass polymers as estimated by their long-term durability can be described within the framework of theory of linear viscoelasticity.

Author Index

A

Ablazova, T. I., 406, 407, 408
Adkins, J. E., 102
Agarwal, P. K., 398, 425
Aharoni, S. M., 155
Akers, L. C., 317
Aklonis, J. J., 270, 271
Akovali, G., 179
Akutin, M. S., 432
Aleksandrov, A. P., 239
Alekseenko, A. I., 391, 400
Alekseeva, V. M., 437
Altzitzer, V. S., 409
Andrade, E. N. da Costa, 106
Andrianova, G. P., 154, 194, 195
Arman, J., 173
Armstrong, R. C., 216, 302
Arrhenius, S., 106
Astarita, G., 102
Attanasio, A., 400
Auer, P. L., 231

B

Bagley, E. B., 354, 398
Baid, M., 415
Bakeev, N. F., 194, 391
Ballman, R. L., 176, 406, 428, 432, 435
Barancheeva, V. V., 410
Barlow, A. J., 249
Bartenev, G. M., 132, 214
Batchinski, A. J., 106
Belcher, H. V., 122
Belenky, B. G., 177
Belousova, T. A., 197
Belov, M. A., 85
Benoit, H., 155, 160
Berens, A. R., 154
Berezhnaya, G. V., 211, 323, 330, 337, 338, 362, 365, 368
Bernini, U., 400
Bernstein, B., 92
Berry, G. C., 157, 159, 160, 180, 188, 202, 205, 360
Bestul, A. B., 122
Bianchi, U., 183
Bird, R. B., 91, 216, 302, 322, 427
Bishop, E. T., 408
Blinova, N. K., 171. 203, 253
Blizard, R. B., 274
Bloch, R., 358
Bogdanovich A. E., 85

Boger, D. V., 379
Bogue, D. C., 95
Bohdanecký, M., 116, 191
Boltzmann, L., 62
Booij, H. C., 320
Borisenkova, E. K., 410
Bostandjiyan, S. A., 330
Botvinnik, G. O., 185, 188, 189
Bowles, W. A., 181, 212
Boyer, R. F., 111, 117, 125
Bradna, P., 116
Brauer, E., 165
Brenner, H., 381, 386
Breuer, H., 125
Briedis, I. P., 165, 279, 299, 433
Brizitsky, V. I., 324, 347, 379
Brodkey, R. S., 135
Brodnyan, J. G., 129, 206, 344
Budtov, V. P., 165, 190
Bueche, F., 141, 161, 162, 163, 200, 201, 224, 258, 259, 274
Bukhgalter, V. I., 165
Busse, W. F., 166
Butcher, A. F., 362

C

Cabbot, I. M., 354
Cancio, L. V., 417
Cantow, H.-J., 165
Casson, N., 401
Cauchy, A. L., 329
Cerf, R., 232
Chalykh, A. E., 127
Chan, F. S., 177
Chan, Y., 400
Chang, H., 431
Chang, W. V., 358
Charlesby, A. J., 179
Chee, K. K., 153
Chelko, A. J., 361
Childers, C. W., 408
Chistov, S. F., 166
Chitrangad, B., 186
Chompff, A. J., 259, 260, 261
Chong, T. S., 385
Christensen, R. M., 102
Chung, C. I., 412
Ciferri, A., 197
Clark, E. S., 325, 370
Cogswell, F. N., 417, 435, 442
Cohen, M. H., 112
Cohen, R. E., 410
Coleman, B. D., 102

Subject Index

A

Absolute rate theory, 105, 106, 130
Activation energy, of flow, 105, 106, 111, 113, 116-23, 130, 131, 184, 190, 191, 281
  and branching, 119
  and molecular mass, 118
  and solvent nature, 121
  apparent, 114, 115
  true, 115
Activation theory, 112, 113
Active fillers, 390, 391, 392, 395
Addition, of tensors, 11, 12
Adestructive method, 292
AFE formula, 106, 113
Amorphous polymers, 110, 111, 117
Amplitude, 56, 57
Amplitude ratio, 57
Anisotropy, 341, 342
  of viscosity, 197
Annular flow, 35, 36
Anomalously viscous liquids, 52
  (see also Non-Newtonian liquids)
Apparent viscosity, 104, 122, 127, 128, 149
  and deformation conditions, 132
  and rate of shear, 128 et seq., 134, 135, 136, 146, 147, 332
  and shear stress, 128, 134, 135
  (see also Non-Newtonian viscosity)
Aroclor, 106, 234
Arrhenius-Frenkel-Eyring formula, 106, 113
Average statistical molecular mass, 252

B

Barus effect, 375, 376
Batchinski formula, 106
Bead-spring models, 218-24
  retardation time, 221, 222
  variants, 218
  with inner viscosity, 218
Bestul-Belcher formula, 122
Bianchi equation, 183
Biaxial stress, 4
Binary relaxation function, 90, 91
Bingham viscoplastic body, 54
Bingham viscosity, 54
Birefringence, 340 et seq.
Birefringence ellipsoid, 341
Birefringence tensor, 341

BKZ theory, 92
Blizard-Marvin-Oser model, 275, 276, 277
Blizard model, 274, 275
Block copolymers, 117, 408-12
  thixotropic behaviour, 411
Body forces, 2
Boltzmann's superposition principle, 74 et seq., 82, 311
Boltzmann-Volterra equations, 63, 82
Boltzmann-Volterra principle, 73
Box-like relaxation spectrum, 272
Branched polymers
  viscosity, 178-82
Branching, chain, 153, 165, 179 et seq.
Brownian motion, 197, 382
Budtov formula, 190
Bueche formula, 141
Bueche model, 224-7, 274, 275
Bueche theory, 141, 161-4, 208, 226, 227
Bulk compression, 10, 19
Bulk modulus, 40, 41
Burgers-Frenkel model, 217, 218
Butyl rubber
  dynamic properties, 289, 290
  normal and shear stresses, 316, 324
  normal stress coefficient, 336
  plasticized, 289, 290
  relaxation times, 316
  rubbery modulus, 359
  rubbery properties, 361
  viscosity and temperature, 114
  viscosity coefficient, 336

C

Capillary flow, 172, 184, 374, 375, 376
Carbon black, 395, 396
Casson formula, 401
Cauchy's equality, 5
Cauchy-Green tensor, 25, 26, 27, 43, 92
  inverted, 26, 47
CDP model, 259
Cellulose acetate, 189, 193
Cellulose nitrate, 206
  flow curves, 206
  WLF constant, 124
Chompff-Duiser-Prins model, 259
Circular frequency, 56

ABOUT THE AUTHORS

G. V. VINOGRADOV was born in 1910 in Moscow. He graduated from the Moscow Higher Technical School (Department of Chemical Engineering). He received his Candidate's degree in 1941 from the Karpov Institute of Physical Chemistry for research on the application of nomographic methods of physico-chemical calculations. He has been engaged in research work in the field of rheology of lubricants since 1945. He obtained the degree of Doctor of Chemical Sciences in 1951 for his research work on the rheology of greases. In 1952 Dr. Vinogradov was appointed Professor of Chemistry and Materials Science in the Military Academy of Armoured Troops in Moscow. He has been working at the Institute of Petrochemical Synthesis,
Academy of Sciences, USSR, since 1960. His primary scientific interests were in the areas of the friction of metals and polymers. In the early sixties he discovered the determining role of the presence of oxygen in lubricants in preventing the scuffing and seizure of metals. In 1963 he was appointed Head of the Laboratory of Polymer Rheology at the Institute of Petrochemical Synthesis. Since 1948 sixty Candidate's and fifteen Doctor's degrees have been gained under his supervision. Professor Vinogradov is a member of the editorial boards of several International journals. He has participated in a large number of International Conferences and Symposia on Polymers and Rheology. He has published about 500 papers and review articles on friction, rheology of lubricants, polymeric systems and related topics.

A. Ya. MALKIN was born in 1937 in Novomoskovsk near Moscow. He graduated from the Moscow Institute of Chemical Engineering with an Honours Diploma in Plastics Processing in 1959. After graduating he was engaged in research on the extrusion of plastics. He later worked as a senior research worker in the Laboratory of Polymer Rheology at the Institute of Petrochemical Synthesis, where he received his Candidate's degree in 1965. He obtained the degree of Doctor of Physico-Mathematical Sciences in 1971 for research on the rubbery elasticity and viscoelasticity of polymers in the fluid state. He is Head of a laboratory at the Institute of Plastics, Moscow. Dr. Malkin has
contributed over 200 papers to various International and Soviet journals devoted to the rheological properties of polymers and polymer-based compositions, experimental methods and devices and the technology of plastics and elastomers. He has co-authored 4 monographs: *Polystyrene. Physico-chemical Foundations of Production and Processing* (1975), *Pheology of Polymers* (1977), *Methods of Measurement of the Mechanical Properties of Polymers* (1977), and *Diffusion and Viscosity of Polymers. Methods of Measurement* (1979). He has published a large number of review articles. The author has participated in a large number of International and USSR conferences, including the 6th World Congress on Rheology (Lyons, 1972) and IUPAC Conferences on Polymers (since 1965). Sixteen Candidate's degrees have been obtained under his supervision. Dr. Malkin was appointed Professor of Polymer Technology and Processing at the Institute of Plastics in 1979. Currently, Professor Malkin's research interests are centered around the rheology of polymerizable systems tied up with the kinetics of chemical reactions and problems of polymer technology.